Signaling and Communication in Plants

Series Editor

František Baluška, IZMB, Department of Plant Cell Biology, University of Bonn, Bonn, Nordrhein-Westfalen, Germany

More information about this series at http://www.springer.com/series/8094

Mirza Hasanuzzaman · Mohsin Tanveer
Editors

Salt and Drought Stress Tolerance in Plants

Signaling Networks and Adaptive Mechanisms

Springer

Editors
Mirza Hasanuzzaman
Department of Agronomy
Sher-e-Bangla Agricultural University
Dhaka, Bangladesh

Mohsin Tanveer
School of Land and Food
University of Tasmania
Hobart, TAS, Australia

ISSN 1867-9048 ISSN 1867-9056 (electronic)
Signaling and Communication in Plants
ISBN 978-3-030-40276-1 ISBN 978-3-030-40277-8 (eBook)
https://doi.org/10.1007/978-3-030-40277-8

This Springer imprint is published by the registered company Springer Nature Switzerland AG
The registered company address is: Gewerbestrasse 11, 6330 Cham, Switzerland

Preface

Salinity, drought, extreme temperature, toxic metals/metalloids, waterlogging etc. are common stress conditions that adversely affect plant growth and crop production. The cellular and molecular responses of plants to environmental stress have been studied intensively. Understanding the mechanisms by which plants perceive environmental signals and transmit the signals to cellular machinery to activate adaptive responses is of fundamental importance to biology. Knowledge about stress signal transduction is also vital for the continued development of rational breeding and transgenic strategies to improve stress tolerance in crops. Drought and soil salinity are major environmental setbacks, which are reducing agricultural productivity worldwide. According to recent reports, salinity stress is costing agricultural sector over US $27.3 billion per annum in lost revenues, thus is further aggravating the food security problems. Worldwide more than 0.8 billion hectares of arable land are now saline (Munns and Tester 2008). In parallel to this, the world population is increasing uncontrollable and is expected to cross 9.3 billion by 2050 (FAO 2013). Based on these facts and figures, the development of salinity and drought tolerant crop plants is an urgent task to deal with the predicted food demand. Salinity also brings water deficit conditions in soil, thus aggravating further. Compared to salt stress, the problem of drought is even more pervasive and economically damaging. In this regard, drought stress signaling certainly merits separate treatment. Nevertheless, most studies on water stress signaling have focused on salt stress primarily because plant responses to salt and drought are closely related and the mechanisms overlap. Because cell signaling controls plant responses and adaptation, it is probably not an exaggeration to state that stress signaling has in large part shaped the flora on earth. Because of the complex and overlapping nature of salinity and drought stress tolerance mechanism in plants, it is more practical option to target signaling pathways to improve salinity and drought stress tolerance. Hundreds of researchers are working worldwide to improve salinity and drought stress in plants but still there is a need to look for more options or to examine overlooked traits in context of salinity and drought tolerance in plants. There was a time when the green revolution saved millions of lives by improving crop yield production however this came along with the loss of numerous key physiological traits or genes in modern cultivars, which may play an important role in sustaining crop yield under saline conditions. Rather than investing billions of dollars in genetic engineering or developing mutants to grow crop plant better under saline conditions, which is not much cost and time effective approach, finding out key players which govern salinity and drought tolerance in tolerant plants can show us better direction in order to salinity tolerance in salt sensitive crop plants. Different signaling networks such as hormonal signaling, ROS signaling, or up-regulation of stress specific transcription factors work together in order to shape stress specific responses and to regulate growth and development under salinity and drought conditions. Other than these signaling networks, QTL mapping of different stress tolerant traits also showed significant improvement in stress tolerance. In this research topic, we have collected 16 chapters related to salinity and drought stress signaling in plants, including reviews of the role of transcription factors, phytohormones, and ROS in regulating salinity and drought stress tolerance in plants. Moreover, overviews of stress adaptive mechanisms have also been provided.

We, the editors, would like to give special thanks to the authors for their outstanding and timely work in producing such fine chapters. Our profound thanks also to Dr. Md. Mahabub Alam, Khursheda Parvin, and Sayed Mohammad Mohsin for his critical review and valuable support in formatting and incorporating all editorial changes in the manuscripts. We are highly thankful to Prof. Dr. František Baluška, IZMB, Department of Plant Cell Biology, University of Bonn, Germany for inviting us in editing this volume. We are also thankful to Ms. Banu Dhayalan, Project Coordinator and all of the technical editors and book publishing staff of Springer Nature for their continuous support and timely advice during the course of the preparation of this volume.

Dhaka, Bangladesh — Mirza Hasanuzzaman
Hobart, Australia — Mohsin Tanveer

References

Munns R, Tester M (2008) Mechanisms of salinity tolerance. Annu Rev Plant Biol 59:651–81

FAO (2013) FAO statistical yearbook (2013). Food and Agriculture Organization of the United Nations, Rome

Contents

About the Editors

Mirza Hasanuzzaman is a Professor of Agronomy at Sher-e-Bangla Agricultural University, Dhaka, Bangladesh. He received his Ph.D. on 'Plant Stress Physiology and Antioxidant Metabolism' from the United Graduate School of Agricultural Sciences, Ehime University, Japan with Japanese Government (MEXT) Scholarship. Later, he completed his postdoctoral research in Center of Molecular Biosciences (COMB), University of the Ryukyus, Okinawa, Japan with 'Japan Society for the Promotion of Science (JSPS)' postdoctoral fellowship. Subsequently, he joined as an Adjunct Senior Researcher at the University of Tasmania with Australian Government's Endeavour Research Fellowship. Professor Hasanuzzaman has been devoting himself in research in the field of Crop Science, especially focused on Environmental Stress Physiology since 2004. Professor Hasanuzzaman published over 100 articles in peer-reviewed journals and books. He has edited 12 books and written 35 book chapters on important aspects of plant physiology, plant stress responses, and environmental problems in relation to plant species. These books were published by the internationally renowned publishers. Professor Hasanuzzaman is a research supervisor of undergraduate and graduate students and supervised 20 M.S. students so far. He is Editor and Reviewer of more than 50 peer-reviewed international journals and recipient of 'Publons Global Peer Review Award 2017, 2018, and 2019'. Professor Hasanuzzaman is an active member of about 40 professional societies and acting as Publication Secretary of Bangladesh Society of Agronomy. He has been honored by different authorities due to his outstanding performance in different fields like research and education. He received the World Academy of Science (TWAS) Young Scientist Award 2014. He attended and presented 25 papers and posters in national and international conferences in different countries (USA, UK, Germany, Australia, Japan, Austria, Sweden, Russia, etc.).

Mohsin Tanveer is a researcher at Stress Physiology Research Group, School of Land and Food, University of Tasmania, Australia. His research is focused on 'tissue specificity of ROS production and ROS signaling in context of salinity stress tolerance in halophyte versus glycophyte'. He obtained his B.Sc. (Hons.) degree in Agronomy from the University of Agriculture, Faisalabad, Pakistan in 2012. He also received some advanced training and courses on plant molecular biology, soil and plant nutrient analysis from China. He published over 30 articles in peer-reviewed journals. He has written 5 book chapters on important aspects of plant physiology, plant stress responses, and environmental problems in relation to plant species.

An Overview of Salinity Tolerance Mechanism in Plants

Waqas-ud-Din Khan, Mohsin Tanveer, Rabia Shaukat, Mohsin Ali and Fiza Pirdad

Abstract Salinity stress is a worldwide dilemma and salinity affected hectares of arable land is increasing inevitable particularly due to climate change. Developing salinity stress tolerance in plants is very complex because of the complex nature of salinity stress in plants. Previously, major focus regarding improving salinity stress tolerance in plants has been given to Na^+ exclusion or Na^+ compartmentalization, or enhanced antioxidant defense system and redox regulation. Moreover, ameliorative effects of different substance such as hormones, amino acids, nutrients, and organic osmolytes have also been extensively reported however still we are at our infancy stage in understanding salinity stress tolerance in plants. Exploring traits in wild genotype or particular in halophytes may help in finding better solutions. Therefore, in this book chapter, we have tried to discuss the role of few overlooked physiological mechanisms, which might be potential indicators of salinity stress tolerance in plants. Particularly, the role of potassium retention in leaf mesophyll, xylem ion loading, and potassium efflux as potential physiological mechanisms in regulating plant growth during salinity stress conditions has been discussed. Moreover, the role of cell wall lignification and potential role of jasmonic acid and ethylene has also been narrated.

1 Introduction

Soil salinity, being one of most detrimental abiotic stresses, critically damages crop plants and causes significant reduction in yield (Munns et al. 2006). More or less, 7% of total land area (1,000 million ha) and 20% of the irrigated arable land has been distressed by salt stress (Li et al. 2014) and imposes major constraints to the sustainability of crop yield. Salt stress subjugates crop performance when salts concentration exceeds normal limits. Sodium Chloride (NaCl) is common salt and can potentially

W.-D. Khan · M. Tanveer (✉)
Tasmanian Institute of Agriculture, University of Tasmania, Hobart, Australia
e-mail: mohsin.tanveer@utas.edu.au

W.-D. Khan · R. Shaukat · M. Ali · F. Pirdad
Sustainable Development Study Centre, Government College University Lahore, Lahore 54000, Pakistan

M. Hasanuzzaman and M. Tanveer (eds.), *Salt and Drought Stress Tolerance in Plants*, Signaling and Communication in Plants, https://doi.org/10.1007/978-3-030-40277-8_1

lead to disruption in metabolism, growth retardation, and cessation of photosynthesis in plants. Salinity stress is characterized with some deleterious effects like lowering osmotic potential of soil solution, nutritional imbalance in plant roots, specific ion effect on plant root–soil solution interface; ultimately leads to stunted plant growth. In addition, salt stress hampered plant cell membrane permeability and its integrity leads to leakage of electro and non-electrolytes. Under severe salinity levels, death of plant is directly associated with higher amount of total soluble and exchangeable salts and high osmotic potential of soil solution in the immediate vicinity of plant.

It is well documented that soil salinity not only directly affects growth but also alters different factors that play substantial role in growth improvement. Kulkarn et al. (2000) reported salinity as one of the major constrains that affects rhizobia symbiotic relationships with plants via: interfering survival of rhizobia in soil, root colonization, infection, and development of threads, thus retarding nodulation growth. Later on, reduction in nodulation was ascribed to heavy accumulation of toxic ions in nodule, decrease in leg hemoglobin, and salt accumulation in leaves (Rao et al. 2002), thus salinity is considered as a major growth limiting factor in legumes. Salt stress leads to a significant reduction in growth and yield related attributes in terms of plant height, root density, grain yield, and biological yield. Salinity also inhibited cell expansion in aerial and below parts of plants (Tanveer and Shah 2017). Cell expansion depends on influx of water, which is associated with hydraulic conductivity of plant, uptake of ions or solutes and surrounding cell volume, thus decrease in cell expansion is a consequence of ionic toxicity and nutrient deficiency due to salinity. Detrimental effects of salt stress include loss of intracellular water (Fatnassi et al. 2011), thus its effects when accompanied with water loss are more harmful than other abiotic stresses. In addition, water deficit is coupled with high osmotic stress. The high amount of toxic ions interferes with the uptake of many essential nutrients via competition or affecting membrane permeability of a particular nutrient. It is clear from the negative effect of high amount of Na^+ on uptake of potassium K^+ ion, that Na^+ attenuated the influx of K^+ (Alleva et al. 2006). These toxic ions also reduce transport of nutrients across plasmalemma of root cells. With the increase in Na^+, a significant reduction in accumulation of K^+ in radical and plumule of maize is reported (Khan et al. 2015), thus affects its seed germination. Sever salinity stress may lead to oxidative stress, associated with production of reactive oxidative species (ROS) during photosynthesis (Kawano et al. 2001), thus reduces photosynthetic efficiency (Tanveer et al. 2018). Under saline conditions, formation of reactive oxygen species (ROS) is a prominent response of oxidative effect induced by high concentration of salts (Tanveer and Shabala 2018). These oxidative species deteriorate cellular macromolecules, such as proteins, amino acids, lipids, and nucleic acids. Salinity changes photosynthetic parameters, including osmotic and leaf water potential, transpiration rate, leaf temperature, and relative leaf water content (RWC). Previous studies were primarily focused on Na exclusion or vacuolar Na sequestration; therefore, some of key and novel physiological processes were forgotten. In this draft, we highlighted some key overlooked mechanisms such as K^+-efflux and K^+ retention in leaf mesophyll, which should be focused in future breeding programs to improve salinity tolerance in plants.

2 ROS-Ca^{2+} Hub

Upon salinity stress, increase in cytosolic Ca^{2+} is an important mechanism to activate stress signaling. The elevation of cytosolic Ca^{2+} is regulated by three ion channels; (i) depolarization of activated Ca^{2+} permeable channels, (ii) hyperpolarization of activated Ca^{2+} permeable channels, and (iii) voltage independent activated Ca^{2+} permeable channels (also known as non-selective cation channels). All these channels are regulated by several genes such as cyclic nucleotide-gated channels (CNGC), annexin, glutamate receptors, and mechano-sensitive channels (Demidchik and Shabala 2018). Nonetheless, an increase in cytosolic Ca^{2+} is not always beneficial for cell metabolism, and prolonged cytosolic Ca^{2+} is detrimental to normal cell metabolism. Such a prolonged increase in cytosolic Ca^{2+} results in the production of ROS, thus reduction in plant growth. In order to reduce cytosolic Ca^{2+} toxicity and high ROS production, plant developed two mechanisms (i) Ca^{2+} efflux system and a complex self-amplification mechanism in order to maintain cytosolic Ca^{2+} at basal level and ROS production at a minimum level. Moreover, signal transduction from cell to cell is also an important mechanism in this regard. In this section, we focused on self-amplification mechanism termed as 'ROS-Ca^{2+} hub' in literature. According to this mechanism, excessive ROS production in response to soil salinity leads to the influx of Ca^{2+} (for signaling purpose) by activating ROS activated Ca^{2+} permeable channels and elevation of cytosolic Ca^{2+} in return activates NADPH oxidase to trigger ROS production (Demidchik 2015; Choi et al. 2016). This process develops a self-amplification in plants, which further amplifies weak signals into fast ROS-Ca^{2+} mediated signals. This mechanism is up-regulated by self-activation of HACC, glutamate receptors, and phosphorylation of NADPH at plasma membrane (Shabala et al. 2015).

Several key roles of ROS-Ca^{2+} hub that have been reported in the literature include hormonal signal transduction (HST), osmotic adjustment, and osmoregulation, mineral nutrition, cell elongation, and programmed cell death. HST is an important mechanism in plants involved in early stress signaling cascades. This mechanism is also controlled by ROS-Ca^{2+} hub, which play very crucial role in the regulation of plant growth and its stress tolerance. Numerous hormones and regulators (abscisic acid, ethylene, brassinolide, auxins, methyl jasmonate, polyamines, and salicylic acid) involved in ROS-Ca^{2+} hub have been reported by different scientists (Wang et al. 2013; Peer et al. 2013; Manzoor et al. 2013; Pottosin et al. 2014; Xia et al. 2015; Zhu et al. 2016; Deng et al. 2016; Lu et al. 2016). It has been suggested that signaling by hormones and ROS-Ca^{2+} hub can overlap and simultaneously initiate the stress signaling cascades in plants during early stress signaling (Sewelam et al. 2016). Another classical example and first identified role of ROS-Ca^{2+} hub was Ca^{2+} loading at root elongation zone (Foreman et al. 2003) and later this was also identified in pollen tubes as well (Potocký et al. 2007). This process is regulated by ROS production in polar and is governed by both NADPH and ROS activated Ca^{2+} permeable channels (Foreman et al. 2003). Rounds and Bezanilla (2013) suggested that concentration of cytosolic free Ca^{2+} in elongation zone is regulated by ROS-Ca^{2+}

hub at high levels, which concomitantly stimulates the development of longitudinal cytoskeleton bundles and exocytosis. According to other researchers, activation of CNGC gene and respiratory burst oxidase (RbohC) in Arabidopsis also forms ROS-Ca^{2+} hub (Gobert et al. 2006; Wilkins et al. 2016). Moreover, Arabidopsis CNGC18 (Gao et al. 2016), GLR1.2, and GLR3.7 (Michard et al. 2011) combined with RbohH and RbohJ (Kaya et al. 2014) function as a 'ROS-Ca^{2+} hub' in elongating pollen tube. Future studies are required to find out the molecular basis of ROS-Ca^{2+} hub mediated physiological processes in plants.

3 Role of Potassium Efflux in Stress Signaling

Cytosolic potassium (K^+) homeostasis is very critical in terms of salinity stress tolerance in plants and various plant tissues show various K^+ retention abilities. Leakage of K^+ from cell under saline conditions induces the activation of caspase like proteases and endonucleases which further lead to programmed cell death (Demidchik et al. 2010). For decades, it has been extensively reported that salinity stress induces tremendous K^+ efflux from plant tissues and breeders and plant physiologist tried to reduce K^+ loss either by genetic engineering or foliar applications of hormones or chemical. Nonetheless, recent evidence suggested that K^+ efflux may be equally important as maintaining cytosolic K^+ homeostasis in order to control plant growth and development under stress conditions. In this section, we have suggested (i) the unique signaling role of K^+ efflux for cytosolic K^+ changes and (ii) the concept of energy distribution based on metabolic switch mechanism.

Soil salinity induces K^+ leakage from cells and some electrophysiological studies showed that K^+ selective depolarization activated outward rectifying channels (KORK or GORK in *Arabidopsis*) is one of major salinity induced K^+ efflux pathways from plant cells (Pottosin and Dobrovynskaya 2014). GORK channels that belong to Shaker family of transporter are highly voltage sensitive and acted upon membrane depolarization under salinity stress or ROS stress (Demidchik et al. 2010; Wang et al. 2016). The essential nature of K^+ retention in cells is important under saline conditions and the ability of plant tissues to retain K^+ seems to be a common feature of salt tolerant plant species (Wu et al. 2015a, b). One may raise questions here that (I) If K^+ retention is so important then why nature has given plant the GORK channels? (II) Can we just simply knock down GORK channels and retain more K^+ under saline conditions? (III) Why do plants invest in K^+ efflux pathway?

K^+ plays as a charge balancing ion in plant cells and under saline conditions inevitable influx of Na^+ induces an abrupt membrane depolarization, which affects the transport of major essential elements across plasma membrane. To restore membrane potential, plants have to pump out H^+ (operated by H^+-ATPase) or to use K^+ efflux to balance charge in cytosol. The first option is energy expensive and is up-regulated for long term but the second option is less expensive and also regulated for short term, thus the second option can be preferred under saline conditions. In an experiment, Chakraborty et al. (2016) reported that GORK transcript levels were increased under salinity stress in brassica roots along with HAK5 (high affinity potassium transporter) transcript levels, suggested a compensation mechanism by which brassica first showed K^+ leakage through GORK channel and then compensated K^+ loss by increasing K^+ uptake by HAK5 channels. This process could be highly tissue specific, thus should be examined in future research.

Salinity tolerance in salt tolerant plant species was related to higher H^+-ATPase activity in order to restore its membrane potential. This is an expensive process and some plant species (e.g. *Chenopodium quinoa*) relied more on K^+-efflux to balance charges in cytosol than investing in Hcytosol than investing pumping (Tanveer et al. unpublished). K^+ efflux may also represent a 'metabolic-switch mechanism' inhibiting energy consuming anabolic reaction and saving energy for adaptation and repair (Demidchik 2014). According to this mechanism, plant redirects more ATPs toward plant defense system under salinity stress, even though ATP production reduces dramatically. However under control conditions, plants use cytosolic K^+ to activate numerous enzymes to produce ATP and to perform other cellular metabolic processes. There may be a competition between the energy supply for metabolic activity and for defense system under salinity stress but only option plants have is to shut cell metabolism and supply all energy in defense to protect and survive under salinity stress. However this process is of short term and is confirming the signaling role of K^+ efflux in regulating plant growth.

4 Potassium Retention in Leaf Mesophyll

Potassium retention in mesophyll cells is a major strategy for the existence of salt tolerant plants. Osmotic adjustment of sodium and potassium 'tissue tolerance' is mediated through Na^+/H^+ NHX exchangers that require H-P Pasetonoplast, H-pump to energize (Shabala 2013). Potassium concentration in cytosol greatly affects the activity of H-PPase. High apoplastic salinity causes K^+ efflux and results in huge decline of cytosolic potassium from 150 to 59 mM (Percey et al. 2014). Photosynthetic abilities of chloroplast get affected through K^+ imbalance in the cytosol of mesophyll cells. Outward rectifying K^+ KOR channels and permeable non-selective cation channel (NSCC) facilitate the efflux of potassium ions due to high salinity stress. ROS stimulates NSCC and KOR in mesophyll cells.

Sodium sequestration into vacuoles is beneficial for plants to the removal of apoplastic sodium. Sodium not only maintains turgor pressure inside the vacuole but also causes the relocation of K^+ into cytosol. Sodium sequestration in vacuole maintains plasma membrane potential and removes positive charge from the cytosol and helps in promoting K^+ retention. NHX and H-PPase regulation maintains K^+ retention in mesophyll cells through the regulation of salinity stress, thus maintained in Arabidopsis. Better retention of K^+ ions in mesophyll cells results in less ROS production thus prevents further leak of K^+ ions. Different light responses of plant species disturbed by salt stress are also related to the amount of energy available in cells. K^+ ions retention without H^+ ions fluxes is also an important plant trait of energy conservation under salinity tolerance among plants (Percey et al. 2016).

Cytosolic K^+ plays its vital role in plant homeostasis under salinity stress by maintaining tolerance along with its switch between defense mechanism and metabolic pathways. Total energy pool of a cell is 100% under normal conditions. When there is high cytosolic K^+, 70% of bulk starts directing on metabolism. As stress increases, the energy pool left for defense mechanism reduced to zero level and K^+ efflux occupies as initial energy and allocates 30% to the metabolism. This channeling helps in delaying cell death under severe stress attacks (Shabala 2017).

5 Cell Wall Lignification and Subernization

Glycerol, fatty acids, and fluvic acid derivatives combine to produce suberin monomers which are intermediates of polypropanoid pathway. They are transported to apoplast through ATP binding cassette ABC and then suberin form by the polymerization of peroxidases. Lignin content in plant enhances through Si and provides protection to leaves plant against rice blast by accelerating the activities of polyphenol oxidase and peroxidase. During anaerobic and reducing conditions, low oxygen surrounding is created by aerenchyma cells. Dense waxy substance, suberin deposit, and lignin formulate on the outer walls as a barrier against radial oxygen loss (ROL) in the basal root cells. Silicon application not only improved maize growth, its chlorophyll fluorescence, and photosynthetic parameters under salinity stress (Khan et al. 2018) but also enhance the binding of Na^+ with the cell wall and lignifies the rhizosphere that prevents the salts to transfer into wheat shoot (Saqib et al. 2008).

Suberization and lignification of sclerenchyma cells enhances due to silicon treatment that develops Casparian strips in rice plants that increase oxidation power of mature roots and reduce ROL (Fleck et al. 2010; Suzuki et al. 2012). Transcription of lignin and suberin genes also gets triggered through silicon that provides a barrier in apoplastic Na^+ transport in roots. Sodium followed transcriptional bypass flow to move into shoot and deposits in endodermal root cells by the suberin deposition. For the underdeveloped endodermal barriers during salinity stress, Si provides bypass flow mechanism and translocates minimum Na^+ and Cl^- into shoots.

S-Adenosyl-L-methionine synthase uses ATP and L-methionine to synthesize AdoMet. Various transmethylation phases in eukaryotes and prokaryotes occur through major methyl group, AdoMet which is utilized in the biosynthesis of numerous cell products. In the presence of salinity stress, SAM1 and SAM3 isogenes formulate during transcriptional phase. SAM enzymes can affect the production of phenyl propanoid compounds, amino acids, and polyamines but lignin production is suggested as major sing for the consumption of AdoMet in the vascular bundles, since the constituents of cell wall could be facilitated through lignification. Tomato plant undergo extensive network of lignin development to meet up the regulatory requirement for the AdoMetin response to salinity and water stress conditions.

6 Role of Jasmonate and Ethylene in Salinity Tolerance

6.1 JA Biosynthesis and Signaling

Brief mechanisms of Biosynthesis and signaling of JA in plants that are following Production of Jas is a series reaction done by different types of plastids, cytoplasmic enzymes, and peroxisomes (Wasternack 2014; Kazan and Manners 2013). In stress conditions, Coronatine Insensitive1-Jasmonate Zim Domain (COI1-JAZ) receptor complex starts degradation of JAZ proteins following the production of jasmonyl-isoleucine (Ja-Ile). JAZ proteins have repressor effects on transcription factors such as MYC2 known as a master regulator. These repressor effects initiate the binding of MYC2 and corresponding proteins with G-Box promoters to regulate the JA response. In recent studies, many bHLH proteins identified as repressors of JA responses (Fonseca et al. 2014; Song et al. 2013). It is suggested that activators and repressors work side by side to minimize the damaging effects of stress in plants. PFT1/MED25, encoding a subunit of the plant mediator complex, also plays an important role in the regulation of JA responses by interacting with several JA-responsive TFs (Kidd et al. 2009; Evik et al. 2012).

6.2 Role of JAs in Salt Tolerance

Salinity is abiotic stress that has an inauspicious consequence on plant growth and development. In salinity stress, plant faces ionic, osmotic, and oxidative stress (Kumar et al. 2013; Golldack et al. 2014). According to recent researchers, JAs act as a positive regulator for plant development in salinity stress (Qiu et al. 2014; Zhao et al. 2014). TaAOC1 gene in wheat (*Triticum aestivum*) encoding on AOC (Allene Oxide cyclase) enzyme in Arabidopsis promotes JA level to tolerate high salinity level. This expression shows that Jas has positive effect on salt tolerance. Transgenic expression of TAOC1 shows the same salt tolerance phenotype in the ABA deficient mutant as observed in ample ABA type, this expression suggests that ABA presence does not have not any influence on TAOC1 gene expression; on the other hand, MYC2 mutant background influences the TAOC1 expression against salinity tolerance. So, in JA biosynthesis pathway, MYC2 influences the AOC-catalyzed branch but ABA does not have any effect on JAs biosynthesis pathway against salinity stress (Zhao et al. 2014) (Fig. 1). 12-oxo-phytodienoate reductase 1 uses to mediate the salinity tolerance upon overexpression of TaOPR1 in Arabidopsis. TaOPR1 is ABA dependent on salinity tolerance and it might be possible that the enhanced capability of TaOPR1 in detoxification of ROS is its dependence on ABA and overexpression in plant to mediate salt tolerance (Dong et al. 2013). It have been proved, JAs also augment the activities of antioxidant enzymes such as SOD, CAT, APX, and POD in wheat plant (Qiu et al. 2014). TFs like ERF1 and MYC2 use to regulate JAs signaling through interactions with mediator subunit PFT1/MED25 (Evik et al.

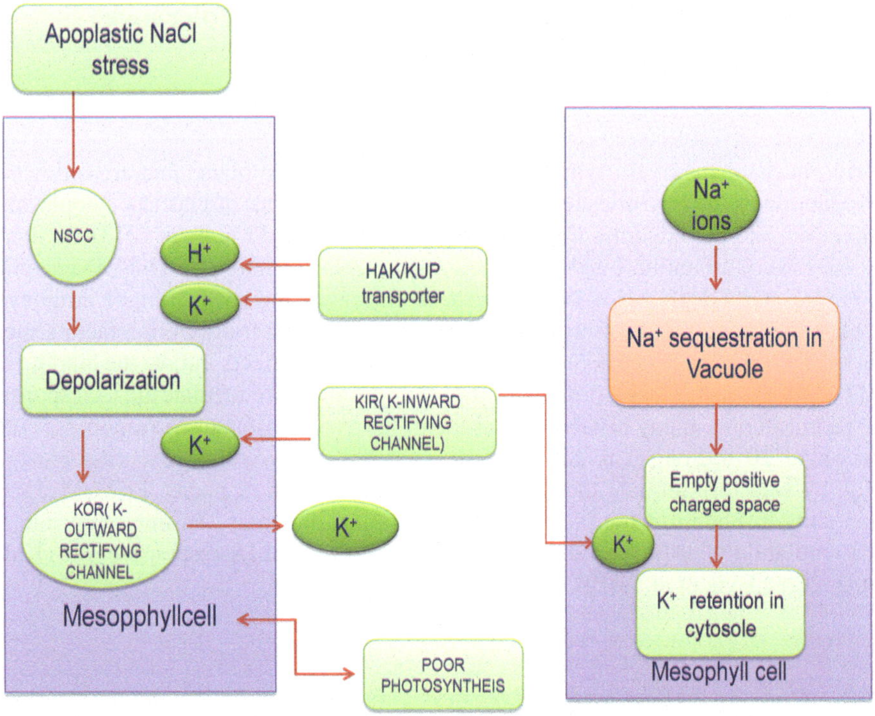

Fig. 1 Potassium ions leakage and its retention from mesophyll cells under salinity stress (Percey et al. 2016, 2014)

2012; Chen et al. 2012). These mediator subunits also regulate salinity tolerance in self-reliant ABA manner by interaction with AP2/ERF/TF/ DRE-Binding Protein 2A (DREB2A). Observations suggests that PFT1/MED25 acts as positive regulator against salinity stress in Arabidopsisplant and enhanced signaling of JAs due to high sensitivity for salts (Elfving et al. 2011). In *Oryza sativa*, the nuclear protein RICE SALT SENSITIVE3 (RSS3) promotes cell elongation in the root tips for better growth by suppressing the root JA responses against salinity. Inhibited root growth under salinity stress has been observed due to lost function of RSS3 which allows in expressing JA-responsive genes in root tips. In *Oryza sativa*, JA signaling components are OsbHLH089, OsbHLH094, OsJAZ9, and OsJAZ11 with them RSS3 interacts to form a stable complex which is required by plant to repress the transcriptional activity of OsbHLH094 (Toda et al. 2013). It has been observed that OsJAZ9 acts as negative regulator of salinity tolerance, this negative regulator suppresses OsbHLH062 which in result activates the arch genes like OsSKC1, OsHAK21, and OsHKA27 expressions to regulate ion homoeostasis in plants (Ye et al. 2009; Wu et al. 2015a, b).

6.3 ET Biosynthesis and Signaling

Ethylene is the gaseous hormone. It plays an essential role in the regulation and development of plant in stress conditions (Groen and Whiteman 2014). The whole process of ethylene biosynthesis consists of three steps: Firstly, methionine is converted into S-adenosyl methionine (SAM). Secondly, SAM is further converted into ACC by ACC synthases (ACS) which act as the direct precursor of ethylene (Fig. 2). Production of ethylene is an exothermic process that undergoes in the presence of ethylene. Biosynthesis of ethylene is regulated at the level of ACS enzymes, which are also under post-translational control: they can be phosphorylated before ubiquitin-mediated protein degradation by, for instance, ETO1 and CUL3 (Thomann et al. 2009; Yoon 2015). Environmental factors trigger and stop the ethylene accumulation by controlling ACS induction and activation (Cao et al. 2007; Dubois et al. 2017). MKK9 and MPK3/6 are involved in activation of ACS2 and ACS6 by phosphorylation of MAPK using post-translational method (Xu et al. 2016). Conjugates

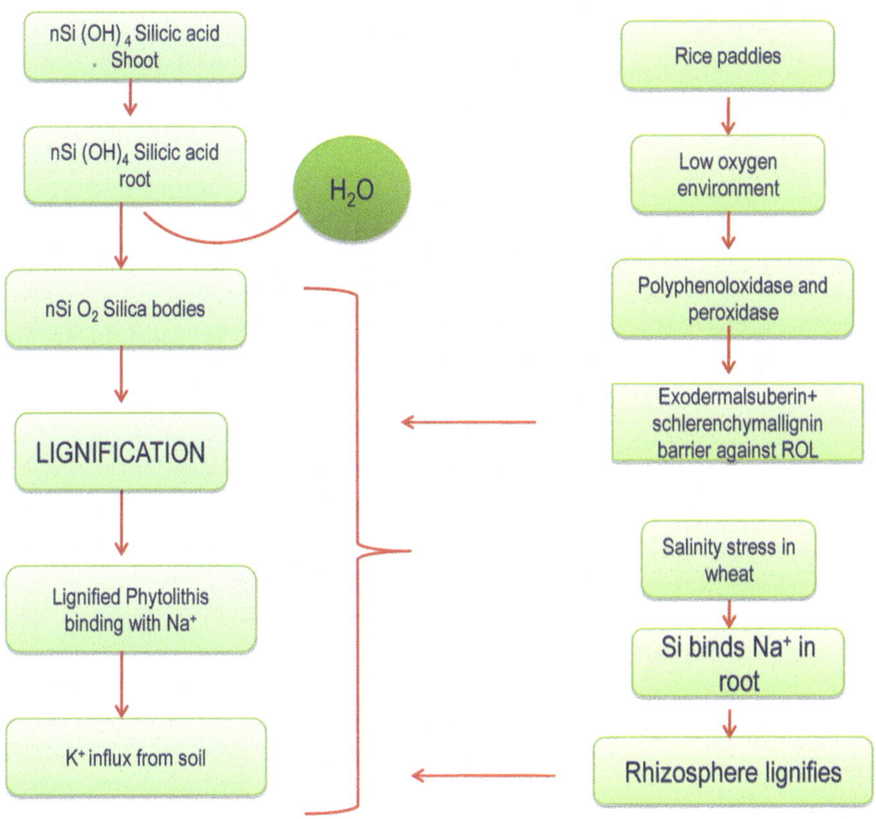

Fig. 2 Lignification and subernization in rice and wheat under stress (Khan et al. 2017; Fleck et al. 2010)

of malonyl or jasmonyl ACC are used to regulate the ACC levels by controlling the conjugates (Poel and Straeten 2014). The soluble ethylene precursor ACC can be taken up by the amino acid transporter LHT1 and further transported through the plant via the xylem (Fig. 1) (Shin et al. 2015). In the destination organ, ethylene triggers a signaling cascade initiated by ethylene receptors in the ER and Golgi membrane: ERS1 (ETHYLENE RESPONSE SENSOR 1), ERS2, ETR1 (ETHYLENE RESISTANCE 1), ETR2, and EIN4 (ETHYLENE INSENSITIVE 4). ETHYLENE RESPONSE 1 (ETR1), ETR2, ETHYLENE RESPONSE SENSOR 1(ERS1), ERS2, and ETHYLENE INSENSITIVE 4 (EIN4) receptors are used to perceive ET by Arabidopsis. EIL1 and EIN3 are TFs, play a role in the regulation of different gene families TFs (like WRKY, AP2/ERF, NAC TF) whereas EIN2 is a protein associated with membrane. In the absence of ET, proteasomic-mediated degradation of arch signaling components EIN2, EIN3, and EIL1 is caused by activating CTR1 and F-Box proteins EIN3, EBF2, EBF1 in response to negative signaling pathway. Ethylene after binding with receptors initiates MAPKKK (Mitogen Activated Protein Kinase Kinase Kinase) to preclude the proteasome-mediated degradation of EIN2, EIN3, and EIL1 (Resnick et al. 2006; Shi et al. 2016) (Fig. 3).

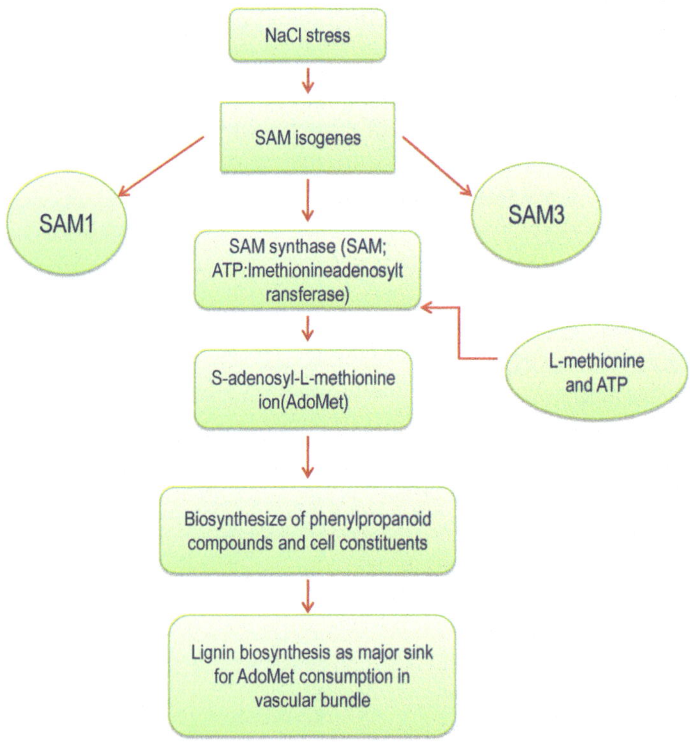

Fig. 3 Genetic stimulation of lignin under salinity stress in tomato (*Solanum lycopersicum*) plant (Liu et al. 2018)

6.4 Roles of ET in Salt Tolerance

Ethylene has the potential to affect plants in both negative and positive ways against salt tolerance. Ethylene have positive effects on leaf growth. In very low concentrations of ethylene, increased rates of leaf elongations are observed in *Poa alpina*, *Poa compressa*, and in sunflower (*Helianthus annuus*) (Lee and Reid 1997). In contrast to leaf elongation, promoter ET has a negative correlation with ethylene sensitivity

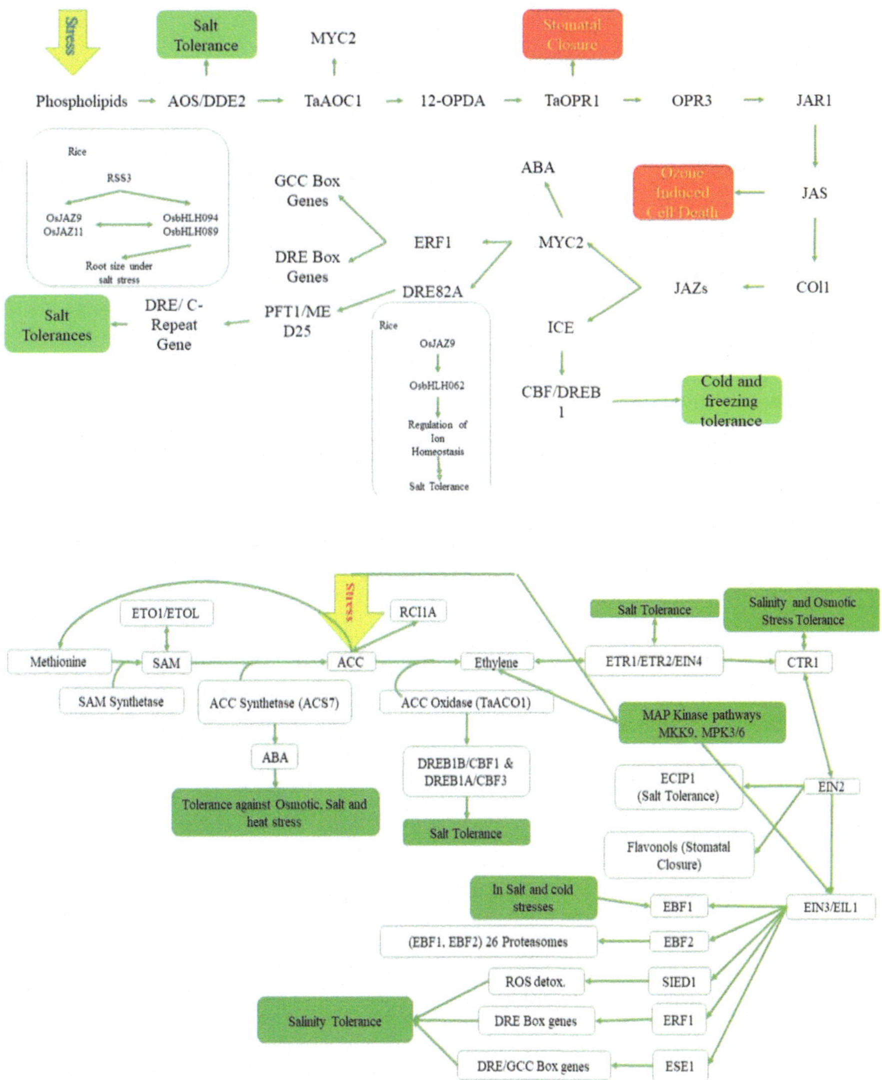

Fig. 4 Role of ethylene and ABA under salinity stress

and leaf growth and it acts as Growth repressing hormone in high concentrations above than its low growth promoting optimum concentrations. Effects of ethylene are very organ specific and it affects the cell divisions in different levels like in early developmental stages of apical hook, ethylene stimulates the cell divisions (Raz and Koornneef 2001). Moreover, in Arabidopsis stems, during the development of vascular structures, ethylene and its downstream transcription factors like (TFs) ERF018 and ERF019 have promotive effect on cell division (Etchells et al. 2012). During salinity stress, a great number of ET-responsive genes show alterations (Shen et al. 2014). In some researches, ET has negative effects on plants in salinity conditions. A negative correlation between low salt tolerance and high ACC levels was found in Arabidopsis (Xu et al. 2008; Dong et al. 2011) (Fig. 4).

7 Conclusion

Salinity stress tolerance is a very complex process which involves the activation of numerous physiological mechanisms together to induce salinity stress adaptation. Analysis of literature showed that K retention in leaf mesophyll and role of K-efflux as metabolic switch player are two most important traits which have been overlooked in the past. Their regulation may be highly tissue specific; therefore, future studies are required to further identify the molecular players behind those physiological mechanisms. Moreover, the role of JA and ET also addressed at molecular level to further underpin their role in salinity stress tolerance.

References

Alleva K, Niemitz CM, Maurl C, Parisi M, Tyeman SD, Amodeo G (2006) Plasma membrane of *Beta vulgaris* L. storage root shows high water channel activity regulated by cytoplasmic pH and a dual range of calcium concentrations. J Exp Bot 7:609–621

Cao WH, Liu J, He XJ, Mu RL, Zhou HL, Chen SY, Zhang JS (2007) Modulation of ethylene responses affects plant salt-stress responses. Plant Physiol 143:707–719

Chakraborty K, Bose J, Shabala L, Shabala S (2016) Difference in root K^+ retention ability and reduced sensitivity of K^+-permeable channels to reactive oxygen species confer differential salt tolerance in three Brassica species. J Exp Bot 67:4611–4625

Chen R, Jiang H, Li L, Zhai Q, Qi L, Zhou W, Liu X, Li H, Zheng W, Sun J, Li C (2012) The *Arabidopsis* mediator subunit MED25 differentially regulates jasmonate and abscisic acid signaling through interacting with the MYC2 and ABI5 transcription factors. Plant Cell 24:2898–2916

Choi WG, Hilleary R, Swanson SJ, Kim SH, Gilroy S (2016) Rapid, long-distance electrical and calcium signaling in plants. Annu Rev Plant Biol 67:287–307

Demidchik V (2014) Mechanisms and physiological roles of K^+ efflux from root cells. J plant Physiol 171(9):696–707

Demidchik V (2015) Mechanisms of oxidative stress in plants: from classical chemistry to cell biology. Environ Exp Bot 109:212–228

Demidchik V, Shabala S (2018) Mechanisms of cytosolic calcium elevation in plants: the role of ion channels, calcium extrusion systems and NADPH oxidase-mediated 'ROS-Ca^{2+} Hub'. Funct Plant Biol 45(2):9–27

Demidchik V, Cuin TA, Svistunenko D, Smith SJ, Miller AJ, Shabala S, Sokolik A, Yurin V (2010) Arabidopsis root K^+-efflux conductance activated by hydroxyl radicals: single-channel properties, genetic basis and involvement in stress-induced cell death. J Cell Sci 123(9):1468–1479

Deng XG, Zhu T, Zou LJ, Han XY, Zhou X, Xi DH, Zhang DW, Lin HH (2016) Orchestration of hydrogen peroxide and nitric oxide in brassinosteroid-mediated systemic virus resistance in *Nicotiana benthamiana*. Plant J 85(4):478–493

Dong H, Zhen Z, Peng J, Chang L, Gong Q, Wang NN (2011) Loss of ACS7 confers abiotic stress tolerance by modulating ABA sensitivity and accumulation in Arabidopsis. J Exp Bot 62:4875–4887

Dong W, Wang M, Xu F, Quan T, Peng K, Xiao L, Xia G (2013) Wheat oxophytodienoate reductase gene *TaOPR1* confers salinity tolerance via enhancement of abscisic acid signaling and reactive oxygen species scavenging. Plant Physiol 161:1217–1228

Dubois M, Claeys H, Broeck VDL, Inzé D (2017) Time of day determines Arabidopsis transcriptome and growth dynamics under mild drought. Plant Cell Environ 40(2):180–189

Elfving N, Davoine C, Benlloch R, Blomberg J, Brännström K, Müller D, Nilsson A, Ulfstedt M, Ronne H, Wingsle G, Nilsson O (2011) The *Arabidopsis thaliana* Med25 mediator subunit integrates environmental cues to control plant development. Proc Natl Acad Sci USA 108:8245–8250

Etchells JP, Provost CM, Turner SR (2012) Plant vascular cell division is maintained by an interaction between PXY and ethylene signalling. PLoS Genet 8:e1002997

Evik V, Kidd BN, Zhang P, Hill C, Kiddle S, Denby KJ, Holub EB, Cahill DM, Manners JM, Schenk PM, Beynon J (2012) MEDIATOR25 acts as an integrative hub for the regulation of jasmonate-responsive gene expression in Arabidopsis. Plant Physiol 160:541–555

Fatnassi IC, Jebara SH, Jebara M (2011) Selection of symbiotic efficient and high salt-tolerant rhizobia strains by gamma irradiation. Ann Microbiol 61:291–297

Fleck AT, Nye T, Repenning C, Stahl F, Zahn M, Schenk MK (2010) Silicon enhances suberization and lignification in roots of rice (*Oryza sativa*). J Exp Bot 62(6):2001–2011

Fonseca S, Calvo FP, Fernandez GM, Díaz DM, Ibanez GS, Vidriero LI, Godoy M, Barbero FG, Leene VJ, Jaeger DG, Zorrilla FJM (2014) bHLH003, bHLH013 and bHLH017 are new targets of JAZ repressors negatively regulating JA responses. PLoS ONE 9:e86182

Foreman J, Demidchik V, Bothwell JH, Mylona P, Miedema H, Torres MA, Linstead P, Costa S, Brownlee C, Jones JD, Davies JM (2003) Reactive oxygen species produced by NADPH oxidase regulate plant cell growth. Nature 422(6930):442

Gao QF, Gu LL, Wang HQ, Fei CF, Fang X, Hussain HJ, Sun SJ, Dong JY, Liu H, Wang YF (2016) Cyclic nucleotide-gated channel 18 is an essential Ca^{2+} channel in pollen tube tips for pollen tube guidance to ovules in *Arabidopsis*. Proc Natl Acad Sci USA 113(11):3096–3101

Gobert A, Park G, Amtmann A, Sanders D, Maathuis FJ (2006) Arabidopsis thaliana cyclic nucleotide gated channel 3 forms a non-selective ion transporter involved in germination and cation transport. J Exp Bot 57(4):791–800

Golldack D, Li C, Mohan H, Probst N (2014) Tolerance to drought and salt stress in plants: unraveling the signaling networks. Front Plant Sci 5:151

Groen SC, Whiteman NK (2014) The evolution of ethylene signaling in plant chemical ecology. J Chem Ecol 40:700–716

Kawano T, Kawano N, Muto S, Lapeyrie F (2001) Cation-induced superoxide generation in tobacco cell suspension culture is dependent on ion valence. Plant Cell Environ 24:1235–1241

Kaya H, Nakajima R, Iwano M, Kanaoka MM, Kimura S, Takeda S, Kawarazaki T, Senzaki E, Hamamura Y, Higashiyama T, Takayama S (2014) Ca^{2+}-activated reactive oxygen species production by *Arabidopsis* RbohH and RbohJ is essential for proper pollen tube tip growth. Plant Cell 26(3):1069–1080

Kazan K, Manners JM (2013) MYC2: the master in action. Mol Plant 6:686–703

Khan WUD, Aziz T, Warriach EA, Khalid M (2015) Silicon application improves germination and vegetative growth in maize grown under salt stress. Pak J Agric Sci 52(4):937–944
Khan WUD, Aziz T, Hussain I, Ramzani PMA, Reichenauer TG (2017) Silicon: a beneficial nutrient for maize crop to enhance photochemical efficiency of photosystem II under salt stress. Arc Agron Soil Sci 63(5):599–611
Khan WUD, Aziz T, Maqsood MA, Farooq M, Ramzani PMA, Bilal HM (2018) Silicon nutrition improved salt resistance in maize by modulating ionic accumulation in plant organs and improving antioxidant enzymes and gas exchange parameters. Photosynthetica. https://doi.org/10.1007/s11099-018-0812-x
Kidd BN, Edgar CI, Kumar KK, Aitken EA, Schenk PM, Manners JM, Kazan K (2009) The mediator complex subunit PFT1 is a key regulator of jasmonate-dependent defense in Arabidopsis. Plant Cell 21:2237–2252
Kulkarn S, Surange S, Nautiyal CS (2000) Crossing the limits of Rhizobium existence in extreme conditions. Curr Microbiol 41:402–409
Kumar K, Kumar M, Kim SR, Ryu H, Cho YG (2013) Insights into genomics of salt stress response in rice. Rice 6:27
Lee SH, Reid DM (1997) The role of endogenous ethylene in the expansion of Helianthus annuus leaves. Can J Bot 75:501–508
Li J, Pu L, Han M, Zhu M, Zhang R, Xiang Y (2014) Soil salinization research in China: advances and prospects. J Geogr Sci 24(5):943–960
Michard E, Lima PT, Borges F, Silva AC, Portes MT, Carvalho JE, Gilliham M, Liu LH, Obermeyer G, Feijó JA (2011) Glutamate receptor-like genes form Ca^{2+} channels in pollen tubes and are regulated by pistil D-serine. Science 332(6028):434–437
Liu Q, Luo L, Zheng L (2018) Lignins: biosynthesis and biological functions in plants. Int J Mol Sci 19(2):335
Lu M, Zhang Y, Tang S, Pan J, Yu Y, Han J, Li Y, Du X, Nan Z, Sun Q (2016) AtCNGC2 is involved in jasmonic acid-induced calcium mobilization. J Exp Bot 67:809–819
Manzoor H, Kelloniemi J, Chiltz A, Wendehenne D, Pugin A, Poinssot B, Garcia-Brugger A (2013) Involvement of the glutamate receptor AtGLR3.3 in plant defense signaling and resistance to *Hyaloperonospora arabidopsidis*. Plant J 76:466–480
Munns R, James RA, Lauchli A (2006) Approaches to increasing the salt tolerance of wheat and other cereals. J Exp Bot 57:1025–1043
Peer WA, Cheng Y, Murphy AS (2013) Evidence of oxidative attenuation of auxin signalling. J Exp Bot 64(9):2629–2639
Percey WJ, Shabala L, Breadmore MC, Guijt RM, Bose J, Shabala S (2014) Ion transport in broad bean leaf mesophyll under saline conditions. Planta 240(4):729–743
Percey WJ, Shabala L, Wu Q, Su N, Breadmore MC, Guijt RM, Bose J, Shabala S (2016) Potassium retention in leaf mesophyll as an element of salinity tissue tolerance in halophytes. Plant Physiol Biochem 109:346–354
Poel BV, Straeten DVD (2014) 1-Aminocyclopropane-1-carboxylic acid (ACC) in plants: more than just the precursor of ethylene! Front Plant Sci 5:640
Potocký M, Jones MA, Bezvoda R, Smirnoff N, Zárský V (2007) Reactive oxygen species produced by NADPH oxidase are involved in pollen tube growth. New Phytol 174:742–751
Pottosin I, Dobrovinskaya O (2014) Non-selective cation channels in plasma and vacuolar membranes and their contribution to K^+ transport. J Plant Physiol 171(9):732–742
Pottosin I, Buendía VAM, Bose J, Jazo ZI, Shabala S, Dobrovinskaya O (2014) Cross-talk between reactive oxygen species and polyamines in regulation of ion transport across the plasma membrane: implications for plant adaptive responses. J Exp Bot 65(5):1271–1283
Qiu Z, Guo J, Zhu A, Zhang L, Zhang M (2014) Exogenous jasmonic acid can enhance tolerance of wheat seedlings to salt stress. Ecotoxicol Environ Saf 104:202–208
Rao DLN, Giller KE, Yeo AR, Flowers TJ (2002) The effects of salinity and sodicity upon nodulation and nitrogen fixation in Chickpea (*Cicer arietinum* L.). Ann Bot 89:563–570

Raz V, Koornneef M (2001) Cell division activity during apical hook development. Plant Physiol 125:219–226

Resnick JS, Wen CK, Shockey JA, Chang C (2006) *REVERSION-TO-ETHYLENE SENSITIVITY1*, a conserved gene that regulates ethylene receptor function in Arabidopsis. Proc Natl Acad Sci USA 103:7917–7922

Rounds CM, Bezanilla M (2013) Growth mechanisms in tip-growing plant cells. Annu Rev Plant Biol 64:243–265

Saqib M, Zörb C, Schubert S (2008) Silicon-mediated improvement in the salt resistance of wheat (*Triticum aestivum*) results from increased sodium exclusion and resistance to oxidative stress. Funct Plant Biol 35(7):633–639

Sewelam N, Kazan K, Schenk PM (2016) Global plant stress signaling: reactive oxygen species at the cross-road. Front Plant Sci 7:187

Shabala S (2013) Learning from halophytes: physiological basis and strategies to improve abiotic stress tolerance in crops. Ann Bot 112(7):1209–1221

Shabala S (ed) (2017) Plant stress physiology. CABI

Shabala S, Wu H, Bose J (2015) Salt stress sensing and early signaling events in plant roots: current knowledge and hypothesis. Plant Sci 241:109–119

Shen X, Wang Z, Song X, Xu J, Jiang C, Zhao Y, Ma C, Zhang H (2014) Transcriptomic profiling revealed an important role of cell wall remodeling and ethylene signaling pathway during salt acclimation in Arabidopsis. Plant Mol Biol 86:303–317

Shi J, Drummond BJ, Wang H, Archibald RL, Habben JE (2016) Maize and Arabidopsis ARGOS proteins interact with ethylene receptor signaling complex, supporting a regulatory role for ARGOS in ethylene signal transduction. Plant Physiol 171:2783–2797

Shin K, Lee S, Song WY, Lee RA, Lee I, Ha K, Koo JC, Park SK, Nam HG, Lee Y, Soh MS (2015) Genetic identification of ACC-RESISTANT2 reveals involvement of LYSINE HISTIDINE TRANSPORTER1 in the uptake of 1-aminocyclopropane-1-carboxylic acid in Arabidopsis thaliana. Plant Cell Physiol 56:572–582

Song S, Qi T, Fan M, Zhang X, Gao H, Huang H, Wu D, Guo H, Xie D (2013) The bHLH subgroup IIId factors negatively regulate jasmonate-mediated plant defense and development. PLoS Genet 9:e1003653

Suzuki S, Ma JF, Yamamoto N, Hattori T, Sakamoto M, Umezawa T (2012) Silicon deficiency promotes lignin accumulation in rice. Plant Biotech 29(4):391–394

Tanveer M, Shabala S (2018) Targeting redox regulatory mechanisms for salinity stress tolerance in crops. In: Salinity responses and tolerance in plants, vol 1. Springer, Cham, pp 213–234

Tanveer M, Shah AN (2017) An insight into salt stress tolerance mechanisms of Chenopodium album. Environ Sci Pollut R 24(19):16531–16535

Tanveer M, Shahzad B, Sharma A, Biju S, Bhardwaj R (2018) 24-Epibrassinolide; an active brassinolide and its role in salt stress tolerance in plants: a review. Plant Physiol Biochem 130:69–79

Thomann A, Lechner E, Hansen M, Dumbliauskas E, Parmentier Y, Kieber J, Scheres B, Genschik P (2009) Arabidopsis CULLIN3 genes regulate primary root growth and patterning by ethylene-dependent and-independent mechanisms. PLoS Genet 5:e1000328

Toda Y, Tanaka M, Ogawa D, Kurata K, Kurotani KI, Habu Y, Ando T, Sugimoto K, Mitsuda N, Katoh E, Abe K (2013) RICE SALT SENSITIVE3 forms a ternary complex with JAZ and class-C bHLH factors and regulates jasmonate-induced gene expression and root cell elongation. Plant Cell 25(5):1709–1725

Wang YF, Munemasa S, Nishimura N, Ren HM, Robert N, Han M, Puzõrjova I, Kollist H, Lee S, Mori I, Schroeder JI (2013) Identification of cyclic GMP-activated nonselective Ca^{2+}-permeable cation channels and associated CNGC5 and CNGC6 genes in Arabidopsis guard cells. Plant Physiol 163(2):578–590

Wang F, Chen ZH, Liu X, Colmer TD, Shabala L, Salih A, Zhou M, Shabala S (2016) Revealing the roles of GORK channels and NADPH oxidase in acclimation to hypoxia in *Arabidopsis*. J Exp Bot 68(12):3191–3204

Wasternack C (2014) Action of jasmonates in plant stress responses and development—applied aspects. Biotechnol Adv 32:31–39

Wilkins KA, Matthus E, Swarbreck SM, Davies JM (2016) Calcium-mediated abiotic stress signaling in roots. Front Plant Sci 7:1296

Wu H, Ye H, Yao R, Zhang T, Xiong L (2015a) OsJAZ9 acts as a transcriptional regulator in jasmonate signaling and modulates salt stress tolerance in rice. Plant Sci 232:1–12

Wu H, Zhu M, Shabala L, Zhou M, Shabala S (2015b) K^+ retention in leaf mesophyll, an overlooked component of salinity tolerance mechanism: a case study for barley. J Integr Plant Biol 57(2):171–185

Xia XJ, Zhou YH, Shi K, Zhou J, Foyer CH, Yu JQ (2015) Interplay between reactive oxygen species and hormones in the control of plant development and stress tolerance. J Exp Bot 66(10):2839–2856

Xu L, Tang Y, Gao S, Su S, Hong L, Wang W, Fang Z, Li X, Ma J, Quan W, Sun H, Li X, Wang Y, Liao X, Gao J, Zhang F, Li L, Zhao C (2016) Comprehensive analyses of the annexin gene family in wheat. BMC Genomics 17:415

Xu J, Li Y, Wang Y, Liu H, Lei L, Yang H, Liu G, Ren D (2008) Activation of MAPK kinase 9 induces ethylene and camalex in biosynthesis and enhances sensitivity to salt stress in Arabidopsis. J Biol Chem 283:26996–27006

Ye H, Du H, Tang N, Li X, Xiong L (2009) Identification and expression profiling analysis of TIFY family genes involved in stress and phytohormone responses in rice. Plant Mol Biol 71:291–305

Yoon GM (2015) New insights into the protein turnover regulation in ethylene biosynthesis. Mol Cells 38:597–603

Zhao Y, Dong W, Zhang N, Ai X, Wang M, Huang Z, Xiao L, Xia G (2014) A wheat allene oxide cyclase gene enhances salinity tolerance via jasmonate signaling. Plant Physiol 164:1068–1076

Zhu C, Yang N, Guo Z, Qian M, Gan L (2016) An ethylene and ROS-dependent pathway is involved in low ammonium-induced root hair elongation in *Arabidopsis* seedlings. Plant Physiol Bioch 105:37–44

Plant Responses and Tolerance to Combined Salt and Drought Stress

Waleed Fouad Abobatta

Abstract Agricultural production is exposed to different environmental challenges such as salinity, drought, and global warming; plants have used different physiological and biochemical responses to adapt and survive under abiotic stress conditions. Drought and salinity are the most important abiotic constraints to plant survival and to crop productivity; they furthermore have main harmful effects on the plant tissue such as the negative effects on the cellular energy supply and redox homeostasis that are stable by re-programming of plant primary metabolism and modification of cellular composition. The agricultural sector is considered the main user of freshwater resources in numerous regions of the world, with growing population and increasing water scarcity in various regions of the world, thus reducing the amount of crop production, which affects the availability of global food quantity in the near future. Drought can affect plants in various ways such as decrease in photosynthesis and growth inhibition; accumulation of abscisic acid (ABA), proline, mannitol, sorbitol; formation of radical scavenging compounds (ascorbate, glutathione, α-tocopherol etc.); and synthesis of new proteins and mRJNAs; also, water stress results in stomatal closure and reduced transpiration rates, a reduce in the water potential of plant tissues, and therefore, plants under drought stress use two processing to control the relation between photosynthetic potential and relative water content of leaves; first one through decreased stomatal conductance and reduced photosynthesis and decreased CO_2 concentration inside the leaf, while the other one by elevated CO_2 which decreases progressively as relative water content (RWC) declines, and reduced gas exchange; with both type there is reducing in metabolic processing in leaf tissue. Salinity stress causes changes in numerous physiological and metabolic processes in plant tissue, unfortunately, the majority of economic crop species are glycophytes, therefore, salinity inhibits crop productivity worldwide. Salinity stress causes changes in various physiological and metabolic processes, depending on severity and duration of the stress, and eventually hamper crop production; glycophytes plant have different physiological mechanisms such as ion homeostasis, compatible solute, antioxidant regulation, and polyamines production.

W. F. Abobatta (✉)
Citrus Department, Horticulture Research Institute, Agriculture Research Center, Giza 12112, Egypt
e-mail: wabobatta@yahoo.com

M. Hasanuzzaman and M. Tanveer (eds.), *Salt and Drought Stress Tolerance in Plants*, Signaling and Communication in Plants, https://doi.org/10.1007/978-3-030-40277-8_2

Generally, plants could combine a range of response to avoiding drought and salinity stresses injuries by different mechanisms to be able to complete their life cycle. By using different strategies such as using maximum available resources, they store reserves in plant organs and use them for fruit production; also, plants can tolerate stress conditions by avoiding tissue dehydration.

Abbreviations

ROS	Reactive oxygen species
UV	Ultraviolet
ASAL	Arid and semi-arid regions
NO	Nitric oxide
CAM	Crassulacean acid metabolism
PEP	Phosphoenol pyruvate
DSP	Dimethyl sulfonium propionate
WF	Water fraction
ABA	Abscisic acid
V-ATPase	Vacuolar type H^+-ATPase
V-PPase	Vacuolar pyrophosphatase
SOS	Salt Overly Sensitive
CMO	Choline monooxygenase
ETC	Electron transport chains
BADH	Betaine aldehyde dehydrogenase
$^1O_2$	Singlet oxygen
OH^-	Hydroxyl radical
O_2^-	The superoxide radical
H_2O_2	Hydrogen peroxide
SOD	Superoxide dismutase
CAT	Catalase
GPX	Glutathione peroxidase
APX	Ascorbate peroxidase
GR	Glutathione reductase
Si	Silicon
PA	Polyamines
PUT	Diamine Putrescine
SPD	Triamine spermidine
SPM	Tetra-amine spermine
ODC	Ornithine decarboxylase
ADC	Arginine decarboxylase
SNP	Sodium nitroprusside
MDA	Malondialdehyde
DNA	Deoxyribonucleic acid
Rubisco	Ribulose 1,5-bisphosphate carboxylase/oxygenase

RO	Alkoxy radicals
SA	Salicylic acid
RWC	Relative water content

1 Introduction

Climate change has a remarkable effect on the growth and productivity of various plants, there are different environmental factors that affect annual and perennial plants, which includes salinity, drought, high temperature, heat waves, cool temperature and frost, and rising carbon dioxide (CO_2) levels (Li et al. 2009). Climate change represents a dangerous challenge for human civilization; under climate change conditions, the world faces major challenges to produce adequate food crops to more than 50% by 2050, also, the world faces major challenges to producing enough food quantity for feeding the growing world population, which will reach to 9 billion in 2050 and 12 billion in 2100 (FAO 2017). Under climate change conditions plants face various abiotic and biotic stress; previous research on plants subjected to abiotic stress have provided important information over the last decades. Recently, significant progress has been made in understanding the physiological, metabolic, and molecular responses of several plant species to a combination of different abiotic stress conditions (Zandalinas et al. 2016); also, plants under a combination of two or more stress conditions use specific mechanisms to adapt and complete the life cycle.

Plant injuries varied from one plant species to another, depending on plant habitat, duration, and severity of stress conditions, and there are various morphological and physiological conditions that take places in the plant to adapt to stress conditions (Flowers 2004). Abiotic stress disrupts plant metabolism and causes physiological disorders which leads to decrease plant growth and loss of crop productivity (Rahnama et al. 2010). Plants use various mechanisms to avoid injuries of stress conditions including physiological and biochemical mechanisms to survive under different stress conditions like high salt concentration in water, soils, and drought conditions. Salinity is one of the main biotic stresses that affect crop productivity; more than 20% of arable land is affected by salt stress and this area is increasing yearly all over the world particularly in arid and semi-arid regions (ASAL) (FAO 2009). Salinity stress causes various morphological, physiological, and metabolic process changes in plant in the form of osmotic stress and ion toxicity (Rozema and Flowers 2008; Abobatta 2018), the hazards of salinity depending on the growth stage, severity, and duration of stress; generally, salinity stress reduces roots water absorption capacity and mostly inhibits plant growth and crop productivity.

Plants classified based on adaptive to salinity conditions:

I. Halophytes which grow and reproduce under high salinity conditions (200 mM NaCl) (Flowers and Colmer 2008).
II. Glycophytes plants that represent the majority of economic crop species, glycophytes cannot survive under high salinity conditions and finally die.

There are different reasons for increased saline soil which includes drought high temperature, hydrological and pedological processes, and also water management (Wanjogu et al. 2001), deforestation (Bui et al. 1996), accumulation of water-borne salts in soils (Bond 1998; Bouwer 2002) excessive fertilizing could also increase soil salinization (Pessarakli and Huber 1991), and overgrazing (Szaboles 1994).

Under salinity conditions, glycophytes use various mechanisms to decrease adverse effects on plant growth and productivity including physiological and biochemical mechanisms like ion homeostasis and salt tolerance, ion transport and uptake, biosynthesis of osmo-protectants and compatible solutes, induce synthesis of antioxidant compounds, synthesis of polyamines, and compatible solute accumulation. Synthesis of organic osmolytes in plants varying from one plant species to another, for example, proline (amino acid) accumulated in various plant cells under stress conditions (Hanson et al. 1994). The amount and concentration of compatible solutes remain balanced in the cell and the remaining is balanced through synthesis and depletion or by the irreversible synthesis of the solutes (Hasegawa et al. 2000; Matysik et al. 2002).

Drought is considered to be a limiting factor for plant growth; it has affected negatively on plant growth and crop production; drought stress is compounded by depletion of water in the rooting zone and increased atmospheric vapor pressure deficit (Ahanger et al. 2014). Drought conditions induced loss in crop yield probably exceeds losses from all other abiotic stress, therefore, severity and longevity of drought stress and interaction between drought and other environmental factors are very important to crop productivity; the reduction in productivity varied depending on plant species, growth stage, longevity, and severity of the drought. The negative impact of drought predominates at all developmental stages (Valliyodan and Nguyen 2006) while the duration, severity, and growth stage certainly had critical roles in determining how the plant responded to water stress, drought-affected cell membrane to disturb ion exchange, and also causing an imbalance in the composition of nutrients in the plant cell, also, transpiration is inhibited by drought, but this may not necessarily affect nutrient uptake in a similar manner (Peuke et al. 2002).

Under drought stress, plants use various adaptive mechanisms to reduce the adverse effects of drought on plant growth and reproduction like morphological mechanism (Krasensky and Jonak 2012) which includes decrease in leaf size, reduce stem elongation root proliferation, physiological mechanism like increase water-use efficiency, control of gas exchange, closing stomatal, reduce turgor pressure, and molecular mechanism including synthesis of specific proteins, the increased production of osmolytes within the plant system protect the bio-chemicals against their deformation.

2 Objectives of This Chapter

The present chapter describes the plant responses and tolerance to salinity and drought stress.

Also, we talk about halophytes and various strategies and mechanisms which is used under salinity conditions, talking also about the glycophytes growth under salt stress and the potential tolerance strategies that is used to adapt to saline conditions and improve plant growth and crop productivity. We discuss the role of plant hormones under salinity stresses and the negative effects of salinity on the growth of glycophytes.

In this chapter, we discussed the effect of drought stress on plant growth and crop productivity, and also we discussed various adapted morphological, physiological, and molecular mechanisms used by plants to reduce the harmful effects of drought on plant growth and reproduction like decrease leaf size, reduced stem elongation root proliferation, increase in water-use efficiency, control of gas exchange, closing stomatal, reduce turgor pressure, increased ABA biosynthesis, and increased proline content. In this chapter, we compiled the last advances in understanding plant responses to combined salinity and drought, focusing on recent molecular, physiological, and metabolic aspects. Also, we further highlight the importance of using this information to develop tolerant crops to the effects of the future impact of predicted climate changes through discussion on 150 references in this chapter approximately.

3 Plant Stress Definitions

Stress is defined as any external abiotic (heat, water, salinity) or biotic (herbivore, pests, pathogens) constraint that limits a plant's ability to convert energy to biomass (Mahajan and Tuteja 2005). Previous studies have reported that stress interactions, including drought and heat, salinity and heat, ozone and salinity, ozone and heat, nutrient stress and drought, nutrient stress and salinity, ultraviolet (UV) and heat, UV and drought, and high light intensity combined with heat, drought or chilling have a significantly higher negative impact on crop productivity than each of the different stress components applied individually (Mittler and Blumwald 2010; Suzuki et al. 2014). Abiotic stress in plants defined as any change in growth conditions, in the plant's natural habitat that change or disrupts metabolic homeostasis. Like changes in growth, the condition requires a modification of metabolic pathways, to achieving new homeostasis, in a process that is usually referred to as acclimation (Ahmad and Prasad 2012), it is one of the major important factors that affect plant growth and it is responsible for greatest losses in the crop production, where the growth reductions may achieve to 50% in different crops (Munns 2005); also, abiotic stress is considered an extra challenge inducing a strong effect on plants and adding to the damage through biotic stress like pests or pathogen attack (Rozema and Flowers 2008).

Plant hazards depend on duration and severity of stress conditions, and there are various physiological and metabolic processes that occur in plant which decrease crop production (Flowers 2004). Plants use various mechanisms to avoid injuries of stress conditions including physiological and biochemical mechanisms to survive under different stress conditions like high salt concentration in water, soils, and drought conditions. Due to the recent climate change conditions, there is more probability that plants will encounter, new or more severe combinations of abiotic stresses in the near future than past anticipated (Rizhsky et al. 2004). So, previous and recent researches on plants subjected to a single abiotic stress condition such as drought, salinity or heat, have provided important information over the last decades; many of these papers cannot be used to infer the effects of a combination of two or more different stresses on plants (Suzuki et al. 2014). Recently, significant progress has been made in understanding the physiological, metabolic, and molecular responses of several plant species to a combination of different abiotic stress conditions (Zandalinas et al. 2016). The previous works have indicated that every combination of two or more stress conditions imposes a specific set of requirements on the plant. Therefore, various stress combinations required the tailoring of unique metabolic and signaling responses including photosynthesis, antioxidant mechanisms, hormone signaling, and osmolyte synthesis (Rasmussen et al. 2013; Pandey et al. 2015).

Plants in their native habitats adapted with stress conditions like drought, salinity, and heat by using various mechanisms beginning with transient responses to minimal soil humidity to main survival mechanisms of escape by early flowering in absence of seasonal precipitation. However, crop plants designated by humans to yield product like grain, vegetable, or fruit in favorable environments with high inputs of water and fertilizer are expected to yield an economic product in response to inputs, these plants designated for survival under drought stress through mechanisms that maintain crop yield. Exposure of plants to abiotic stress induces a disruption in plant metabolism implying physiological disorder which lead to reduced plant growth and eventually a reduction in crop productivity (Rahnama et al. 2010). Abiotic stress is one of the main important features that has an effect on plant growth and productivity; it has a massive impact on growth and is responsible for highest losses in the crop production; the growth reductions may reach up to 50% in most plant species (Munns 2005). Abiotic stress is considered as an additional challenge inducing a strong effect on plants and adding to the damage through biotic stress like pests or pathogen attack (Ahmad and Prasad 2012). So, the plant's response to simultaneous stresses leads to a more complicated scenario. From the perception of the stress to the ultimate response in cells, plants use various signaling pathways after perception of the stimulus and depending on the challenge severity it seems that plants respond in a specific manner when they have to face combined stress simultaneously. Therefore, cross-tolerance between environmental and organic phenomenon stress could induce a positive result and increased resistance in plants and have important agricultural implications.

4 Salinity Stress

Salinity is a major stress limiting the increase in the food crops productivity; more than 20% of cultivated land worldwide is affected by salt stress and this area is increasing yearly; therefore salinity stress reduce crop production and causes different physiological and metabolic process changes in plant tissue; this change depends on the severity and longevity of stress; ultimately, salinity stress may inhibit crop productivity; also, initially soil salinity is known to represses plant growth in the form of osmotic stress which is then followed by ion toxicity (Rozema and Flowers 2008; Abobatta 2018).

Plants based on adaptive evolution can be classified generally into two major types:

Halophytes are plants that can survive and reproduce in environments with high salinity conditions (200 mM NaCl) (Rozema and Flowers 2008), unfortunately, halophytes represent about 1% of the world's flora.

Glycophytes: Glycophytes (that cannot withstand salinity and eventually die), unfortunately, the majority of economic crop species belong to this second category; therefore, salinity is one of the most brutal environmental stresses that hamper crop productivity worldwide (Flowers and Colmer 2008).

Soil salinity is one of the main elements for soil degradations, it can arise from natural factors or results from human activities like excessive use of chemical fertilizers; therefore, salinity stress decreases agricultural production in arid and semi-arid regions (ASAL) particularly.

5 Salinity Causes

5.1 Primary Salinization

There are various natural reasons for increased saline soil, like hydrological and pedological processes, climatic factors, and water management which could increase soil salinization particularly in ASAL; also, evapotranspiration has a very vital role in the pedogenesis of saline particularly in ASAL (Wanjogu et al. 2001).

5.2 Secondary Salinization

Secondary salt-affected soils are those that have been salinized by human-caused factors, mainly as a consequence of improper methods of irrigation, usage of low quality water for irrigation, and also, it includes, but are not limited to, the following reasons:

(a) Deforestation (Bui et al. 1996).
(b) Accumulation of air-borne or water-borne salts in soils (Bond 1998; Bouwer 2002).
(c) Salinization caused by contamination with chemicals (Pessarakli and Huber 1991).
(d) Overgrazing (Szaboles 1994).

6 Salinity Effects on Plants

Salinity is considered one of the most critical biotic stresses that inhibit productivity of glycophytes plants all over the world (FAO 2009). Salinity stress causes both ionic and osmotic effects on plants and most of the known responses of plants to salinity are linked to these effects (Yeo et al. 1999). Salt accumulation in leaves accelerate premature senescence (Fig. 1), decreasing the delivery of assimilates to the growing regions and thus reducing plant growth (Munns et al. 1995). Growth inhibition is the main injury that leads to other symptoms; it is considered as an immediate effect on plant growth through inhibition of cell elongation; also, salt stress induces the synthesis of abscisic acid (ABA) which is transported to guard cells and closed stomata, then photosynthesis declines, and photo inhibition and oxidative stress occurs, although under severe salinity shock cells die directly. Sodium and chloride are the most regular ions in saline soils or water (Levitt 1980), accumulation of excessive sodium ions at the soil solution around root surface reduces plant potassium nutrition, due to strong inhibitory effects of sodium ions on potassium uptake by roots.

Glycophytes plants have low-affinity system which has low potassium/sodium selectivity due to similarity nature of sodium and potassium ions, the strong negative effects of sodium occur under the low-affinity system, therefore, under sodium stress plants use both low- and high-affinity systems for potassium uptake to preserve proper potassium nutrition, and reduce the inhibitory effects of sodium ions on plant growth. When sodium gets into cytoplasm, it deactivates various enzymes, the high ratio of potassium in the cytoplasm is considered the determined factor for activating of many enzymes; however, a high sodium/potassium ratio is the most damaging, even in the halophytes that accumulate large quantities of sodium inside the cell, their cytosolic enzymes have sodium sensitive like enzymes of glycophytes (Noble and Rogers 1992).

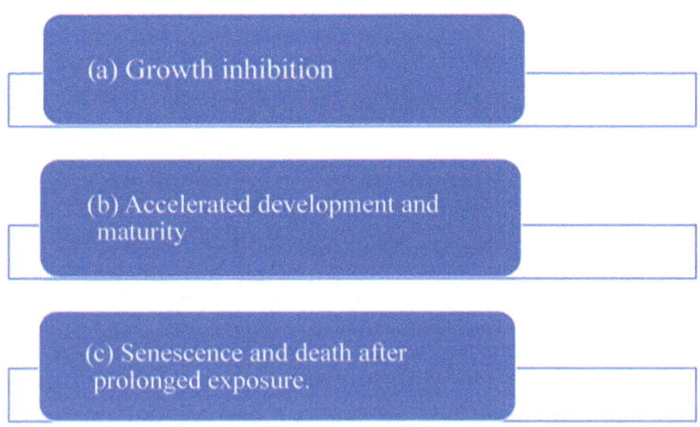

Fig. 1 General symptoms of damage by salt stress

7 Plant Responses to Salinity

Plant use different mechanisms to reduce negative effects of salinity on plant growth and productivity including physiological and biochemical mechanisms.

7.1 *Changes in Metabolism Under Salt Stress*

During salinity stress some of the enzymatic machinery for Crassulacean acid metabolism (CAM), like phosphoenol pyruvate (PEP) carboxylase, is induced after a short period of salt stress, the major benefit of the CAM metabolism has improved water use efficiency through control on time of opening stomata at night only when evaporative water loss is at a minimum. During salt stress conditions photosynthesis mode change in succulent plants from C3 to CAM mode; it occurs in ice plant (Mesembryanthemum crystallinum) and after a short period under salt stress conditions mode of photosynthesis changes from C3 to CAM; this may be due to the induced enzymatic machinery of CAM metabolism, like PEP carboxylase after salt stress. The major advantage of the CAM metabolism is improved water use efficiency by controlling the timing of stomata to open only at night when evaporative water loss is at minimum.

Under salt stress, there are various low molecular weight organic solutes accumulated in the plant cell, these solutes include.

1. Linear polyols (glycerol, mannitol or sorbitol).
2. Cyclic polyols (inositol or pinitol and other mono- and dimethylated inositol derivatives)
3. Amino acids (glutamate or proline).
4. Betaines (glycine betaine or alanine betaine).

Also, other plants that often suffer nitrogen limitation accumulates sulfonium compounds such as dimethyl sulfonium propionate (DSP), which are equivalent to the nitrogen-containing betaines.

Generally, all organic solutes at any concentrations are not injuries to cellular structure or plant enzyme; under high concentrations of these solutes it plays an important role in osmotic adjustment, therefore, these organic solutes are often referred to as compatible osmolytes; however, in halophytes organic solutes often accumulate in the cell with high concentrations from 0.5 to 4.0 mol/L of compatible osmolytes.

7.2 *Role of Organic Solutes*

The organic solutes play an important role in cell protection from salt stress as follows:

1. The high concentration of organic solutes in the cell balances the high concentration of salt outside the cell.
2. Organic solutes with high concentration counter the high concentrations of sodium and chloride ions in the vacuole.
3. It has a protective achieve against injuries by toxic ions or dehydration.

Also, the low amount of compatible osmolytes could protect plants by scavenging oxygen-free radicals produced by salt stress, and organic solutes may also have a protective effect against damage by toxic ions or dehydration, therefore, these solutes (organic compounds) are mostly called as compatible osmolytes.

From another side, the protective effect cannot be completely understood by the osmotic adjustment theory because in many cases the transgenic plants only produce additional several millimoles per liter of the engineered osmolytes, and concentration is very low to affect significantly to osmotic adjustment.

7.3 Salinity Stress and Their Effects on Water Relations

Their many literatures described the effects of salinity stress on a plant's water relations (Noble and Rogers 1992), there are various components of a plant's water relation as follows:

1. Water potential
2. Hydraulic conductivity
3. The water fraction (WF)

Water potential refers to the potential energy of water relative to pure water, and it determines the direction of water movement, where water moves from a location with a higher water potential to a location with a lower water potential; water potential in the leaves could be measured by using a pressure chamber (Scholander et al. 1965). **Hydraulic conductivity** refers to the ease with which water can flow from one location to another and therefore affects the rate of water movement, under high salinity conditions, the plant's survival depends on the control in these two components; hydraulic conductance in the roots could be determined by a high-pressure flow meter according to the method of Tyree et al. (1995). Salt-tolerant plants could decrease the hydraulic conductance of their roots, then reducing the supply salty water to the shoots, so decreases the leaves water potential (Gama et al. 2009). **The water fraction (WF)**: Water fraction of a tissue is the water content of the shoot (under controlled conditions) as a fraction of the fresh mass of the shoot. The WF under stress conditions relative to control conditions are better able to maintain its water content in the shoot upon salt stress, Hsiao and Xu (2000) reported that, plants under salinity stress often lose some water from their tissues, which can have fast effects on cell expansion, cell division, stomatal opening, ABA accumulation, etc. Most of these effects become evident without any change in turgor pressure, also, water potential turn to be more negative as a result to osmotic potential becoming more negative. It has suggested that the plant ability to maintain regular rate of transpiration under salinity is an essential factor of salt tolerance, particularly because transpiration is correlated with normal rates of CO_2 uptake for photosynthesis (Harris et al. 2010).

8 The Main Salinity Tolerance Mechanisms

Plants use various physiological and biochemical mechanisms to avoid salinity hazards and survive under salinity stress in water and soils. Salinity is considered one of the most critical biotic stresses that decreased glycophytes productivity worldwide (Hanson et al. 1994). The changes in physiological and metabolic processes depend on severity and duration of the salinity stress, and ultimately inhibit crop production (Flowers 2004). Plants used different mechanisms to survive under high salinity conditions like.

(1) Ion homeostasis and salt tolerance.
(2) Compatible solute accumulation and osmotic protection.
(3) Ion transport and uptake.
(4) Biosynthesis of osmoprotectants and compatible solutes.
(5) Activation of antioxidant enzyme and synthesis of antioxidant compounds.
(6) Synthesis of polyamines.
(7) Generation of NO.
(8) Hormone modulation.

8.1 Ion Homeostasis and Salt Tolerance

Ion uptake and compartmentalization is considered crucial for maintaining ion homeostasis and to reserve regular plant growth, and also it is an essential process for plant growth within salt stress conditions (Hasegawa 2013). Irrespective of their nature, both glycophytes and halophytes cannot tolerate high salt concentration in their cytoplasm, therefore, the excess salt is either transported to the vacuole or sequestered in older leaves which finally are dropping to shield the plant from salinity conditions (Zhu 2003). In different regions affected by salinity the main form of salt present in the soil is Na Cl, so the transport mechanism of Na^+ ion and its compartmentalization is considered the main point for salinity researches in arid and semi-arid regions, therefore, sodium ion after entering the cytoplasm directly moves to vacuole through Na^+/H^+ antiporter (Wang et al. 2001).

There are two types of H^+ pumps that are available in the vacuolar membrane:

(a) Vacuolar-type H^+-ATPase (V-ATPase)
(b) Vacuolar pyrophosphatase (V-PPase)

However, the type V-ATPase is the most dominant H^+ pump in the plant cell.

Under regular conditions, H^+ pumps play an essential role in maintaining solute homeostasis, energizing secondary transport and facilitating vesicle fusion; however, under the stressed condition the survivability of the plant depends upon the activity of V-ATPase (Dietz et al. 2001). De Lourdes et al. (2001) noticed that the activity of V-ATPase pump increased when exposed to seedlings of *Vigna unguiculata* to salinity stress conditions; however, the activity of V-PPase was inhibited under the same conditions. Whereas Suaeda salsa (halophyte) V-ATPase activity was upregulated and V-PPase played a minor role (Wang et al. 2001).

8.1.1 Salt Overly Sensitive (SOS)

There are increasing evidence that demonstrates the roles of a Salt Overly Sensitive (SOS) stress signaling pathway in ion homeostasis and salt tolerance (Sanders 2000). The SOS signaling pathway consists of three major proteins as follows:

1. SOS1: which encodes a plasma membrane Na^+/H^+ antiporter.
2. SOS2: which encodes a serine/threonine kinase.
3. SOS3: is a myristoylated Ca^{2+} binding protein and contains a myristoylation site at its N-terminus.

SOS1 protein is necessary for adaptable Na^+ efflux at the cellular level; it also facilitates long-distance transport of Na^+ from root to shoot, therefore increasing the quantity of SOS1 protein essential for salt tolerance in plants (Shi et al. 2002). SOS1 protein consists of long cytosolic C-terminal tail, about 700 amino acids long, which consist of a putative nucleotide binding motif and an autoinhibitory domain. SOS2 is activated by salt stress elicited Ca^{2+} signals; this protein includes a well-developed N-terminal catalytic domain and a C-terminal regulatory domain (Liu et al. 2000). SOS3 is the third type of protein involved in the SOS stress signaling pathway; it is a myristoylated Ca^{2+} binding protein and contains a myristoylation site at its N-terminus. This site plays a vital role in conferring salt tolerance (Ishitani et al. 2000). There is an interaction between SOS2 and SOS3 for the activation of the kinase; the activated kinase and phosphorylates SOS1 protein increase its transport activity which was primarily identified in yeast (Quintero et al. 2002). C-terminal regulatory domain of SOS2 protein contains a FISL motif (also known as NAF domain), which is about 21 amino acid long sequence and serves as a site of interaction for Ca^{2+} binding SOS3 protein, this interaction between SOS2 and SOS3 protein results in the activation of the kinase. The autoinhibitory domain for SOS1 is the target site for SOS2 phosphorylation.

The SOS3 protein interacts and activates SOS2 protein by releasing its self-inhibition. The SOS3–SOS2 complex is then loaded onto the plasma membrane where it phosphorylates SOS1, the phosphorylated SOS1 results in increased Na^+ efflux, reducing Na^+ toxicity (Martınez-Atienza et al. 2007). Various plants have used efficient processes to keep a low level of the ion concentration in their cytoplasm, also, membranes play an important role with their associated components in maintaining ion concentration within the cytosol under stress conditions by adaptable ion uptake and transport.

8.2 *Compatible Solute Accumulation and Osmotic Protection*

Compatible solutes, or compatible osmolytes, are a group of chemically diverse organic compounds that are uncharged, polar, and soluble in nature and do not interfere with the cellular metabolism even at high concentration, the main contents are proline (Tahir et al. 2012), glycine betaine (Wang and Nii 2000), sugar (Bohnert et al. 1995), and polyols (Saxena et al. 2013). Synthesis and accumulated amount of organic osmolytes in plant varying from one plant species to another, for example, proline (amino acid) accumulated in a various plant cell under stress conditions; however, accumulation of quaternary ammonium compound beta alanine betaine's is restricted only in some of the Plumbaginaceae plants (Hanson et al. 1994). Compatible solutes concentration in the cell is remaining balanced through synthesis and degradation or by irreversible synthesis of the solutes. Accumulation of biochemical osmolytes in the cell under salt stress must be proportional with external osmolarity to preserve cell structure and conserve balanced osmotic pressure into the cell via continuous water supply (Hasegawa et al. 2000), in the same time some free amino acid like methionine, arginine, and cysteine are decreased under salinity stress, meanwhile, proline concentration is increased when exposed to salinity conditions, therefore, proline accumulation is a regular measure adopted for the alleviation of salinity stress (Matysik et al. 2002).

8.3 *Intracellular Proline*

Accumulated intracellular proline plays a vital role as a reservoir of organic nitrogen under salt stress and provides stress tolerance; proline functions as an oxygen quencher thereby revealing its antioxidant capability; Ben Ahmed et al. (2009) reported that, proline application increased salt tolerance in olive (*Olea europaea)* by amelioration of some antioxidative enzyme activities, photosynthetic activity, and plant growth and the preservation of a suitable plant water status under stress conditions (Brito et al. 2019). Plant cell synthesize proline from glutamine or ornithine; however, under osmotical stress glutamate is used as the primary precursor for proline synthesis, the biosynthetic pathway comprises two major enzymes, pyrroline carboxylic acid synthetase and pyrroline carboxylic acid reductase. It has been indicated that proline enhance salt tolerance in Nicotiana tabacum by improving the activity of enzymes involved in antioxidant defense system. Deivanai et al. (2011) reported that rice seedlings from seeds treated with proline exhibited enhanced growth during salt stress.

8.4 Glycine Betaine

Glycine betaine is an amphoteric quaternary ammonium compound ubiquitously, it is highly soluble in water and due to its exceptional structural features it interacts both with hydrophobic and hydrophilic domains of the macromolecules, such as enzymes and protein complexes. Glycine betaine is a nontoxic cellular osmolyte that raises the osmolarity of the cell under stress conditions, therefore, it plays an essential function in stress mitigation. Glycine betaine also protects the cell by osmotic adjustment, stabilizes proteins, and protects the photosynthetic apparatus from stress damages, and reduction of ROS (Gupta and Huang 2014). Accumulation of glycine betaine is found in a wide variety of plants belonging to different taxonomical background. Glycine betaine is synthesized within the cell from either choline or glycine. Synthesis of glycine betaine from choline is a 2-step reaction involving two or more enzymes. In the first step choline is oxidized to betaine aldehyde which is then again oxidized in the next step to form glycine betaine. In higher plants the first conversion is carried out by the enzyme choline monooxygenase (CMO), whereas the next step is catalysed by betaine aldehyde dehydrogenase (BADH) (Ahmad et al. 2013). Another pathway which is observed in some plants, mainly halophytic, demonstrated the synthesis of glycine betaine from glycine. Here glycine betaine is synthesized by three successive N-methylation and the reactions are catalysed by two S-adenosyl methionine-dependent methyl transferases, glycine sarcosine N-methyl transferase (GSMT), and sarcosine dimethylglycine N-methyl transferase (SDMT). These two enzymes have overlapping functions as GSMT catalyses the first and the second step while SDMT catalyses the second and third step. Rahman et al. (2002) reported the positive effect of glycine betaine on the ultrastructure of *Oryza sativa* seedlings when exposed to salt stress.

Under stressed condition, (150 mM NaCl) the ultrastructure of the seedling shows several damages such as swelling of thylakoids, disintegration of grana and intergranal lamellae, and disruption of mitochondria. However, these damages were largely prevented when seedlings were pretreated with glycine betaine. When glycine betaine is applied as a foliar spray in a plant subjected to stress, it led to pigment stabilization and increase in photosynthetic rate and growth (Cha-Um and Kirdmanee 2010). Polyols are compounds with multiple hydroxyl functional groups available for organic reactions. Sugar alcohols are a class of polyols functioning as compatible solutes, as low molecular weight chaperones and as ROS scavenging compounds, they can be classified into two major types, cyclic (e.g., pinitol) and acyclic (e.g., mannitol). Mannitol synthesis is induced in plants during stressed period via action of NADPH dependent mannose-6-phosphate reductase. These compatible solutes function as a protector or stabilizer of enzymes or membrane structures that are sensitive to dehydration or ionically induced damage. It was found that the transformation with bacterial mltD gene that encodes for mannitol-1-phosphate dehydrogenase in both Arabidopsis and tobacco (*N. tabacum*) plants confer salt tolerance, thereby maintaining normal growth and development when subjected to high level of salt stress (Thomas et al. 1995). Pinitol is accumulated within the plant cell when the plant is subjected to salinity stress. The biosynthetic pathway consists of two major steps, methylation of myo-inositol which results in formation of an intermediate compound, ononitol, which undergoes epimerization to form pinitol. Inositol methyl transferase enzyme encoded by IMT gene plays a major role in the synthesis of pinitol. Transformation of IMT gene in plants shows a result similar to that observed in the case of mltD gene. Thus it can be said that pinitol also plays a significant role in stress alleviation. Accumulation of polyols, both straight-chain metabolites such as mannitol and sorbitol or cyclic polyols such as myo-inositol and its methylated derivatives, is correlated with tolerance to drought and/or salinity, based on polyol distribution in many species, including microbes, plants, and animals (Bohnert et al. 1995). Accumulations of carbohydrates such as sugars (e.g., glucose, fructose, fructans, and trehalose) and starch occur under salt stress (Parida et al. 2004).

The major role played by these carbohydrates in stress mitigation involves osmoprotection, carbon storage, and scavenging of reactive oxygen species. It was observed that salt stress increases the level of reducing sugars (sucrose and fructose) within the cell in a number of plants belonging to different species, besides being a carbohydrate reserve, trehalose accumulation protects organisms against several physical and chemical stresses including salinity stress. They play an osmoprotective role in physiological responses (Saxena et al. 2013). Sucrose content was found to increase in tomato (*Solanum lycopersicum*) under salinity due to increased activity of sucrose phosphate synthase. Sugar content (Gao et al. 1998), during salinity stress, has been reported to both increase and decrease in various rice genotypes (Alamgir and Yousuf-Ali 1999). In rice roots it has been observed that starch content decreased in response to salinity while it remained fairly unchanged in the shoot. Decrease in starch content and increase in reducing and nonreducing sugar content were noted in leaves of *Bruguiera parviflora* (Parida et al. 2004).

8.5 Antioxidant Regulation of Salinity Tolerance

Abiotic stress in plants can cause overflow, deregulation, or even disruption of electron transport chains (ETC) in chloroplasts and mitochondria. Under these conditions molecular oxygen (O_2) acts as an electron acceptor, giving rise to the accumulation of ROS. Singlet oxygen ($^1O_2$), the hydroxyl radical (OH_-), the superoxide radical (O_2^-), and hydrogen peroxide (H_2O_2) are all strongly oxidizing compounds and therefore are potentially harmful for cell integrity (Groß et al. 2013). Antioxidant metabolism, including antioxidant enzymes and nonenzymatic compounds, plays critical parts in detoxifying ROS induced by salinity stress. Salinity tolerance is positively correlated with the activity of antioxidant enzymes, such as superoxide dismutase (SOD), catalase (CAT), glutathione peroxidise (GPX), ascorbate peroxidase (APX), and glutathione reductase (GR) and with the accumulation of nonenzymatic antioxidant compounds (Gupta et al. 2005). Gill et al. (2013) reported a couple of helicase proteins (e.g., DESD-box helicase and OsSUV3 dual helicase) functioning in plant salinity tolerance by improving/maintaining photosynthesis and antioxidant machinery; also, Kim et al. (2014) showed that Si application to rice root zone influenced the hormonal and antioxidant responses under salinity stress.

8.6 Roles of Polyamines in Salinity Tolerance

Polyamines (PA) are small, low molecular weight, ubiquitous, polycationic aliphatic molecules widely distributed throughout the plant kingdom. Polyamines play a variety of roles in normal growth and development such as regulation of cell proliferation, somatic embryogenesis, differentiation and morphogenesis, dormancy breaking of tubers and seed germination, development of flowers and fruit, and senescence (Knott et al. 2007); it also plays a crucial role in abiotic stress tolerance including salinity and increases in the level of polyamines are correlated with stress tolerance in plants (Kovacs et al. 2010). The most common polyamines that are found within the plant system are diamine putrescine (PUT), triamine spermidine (SPD), and tetraamine spermine (SPM) (Shu et al. 2012). The PA biosynthetic pathway has been thoroughly investigated in many organisms including plants and has been reviewed in detail (Rambla et al. 2010). PUT is the smallest polyamine and is synthesized from either ornithine or arginine by the action of enzyme ornithine decarboxylase (ODC) and arginine decarboxylase (ADC), respectively (Hasanuzzaman et al. 2019).

8.7 Roles of Nitric Oxide in Salinity Tolerance

Nitric oxide is a small volatile gaseous molecule, which is involved in the regulation of various plant growth and developmental processes, such as root growth, respiration, stomata closure, flowering, cell death, seed germination and stress responses, as well as a stress signaling molecule (Zhao et al. 2009). NO directly or indirectly triggers expression of many redox-regulated genes. NO reacts with lipid radicals thus preventing lipid oxidation, exerting a protective effect by scavenging superoxide radical and formation of peroxynitrite that can be neutralized by other cellular processes. It also helps in the activation of antioxidant enzymes (SOD, CAT, GPX, APX, and GR) (Bajgu 2014). Exogenous NO application has been found to play roles in stress mitigation (Xiong et al. 2010); Kopyra and Gwozdz (2003) indicated that the effects of NO depend on concentration, exogenous application of sodium nitroprusside (SNP); a NO donor, on *Lupinus luteus* seedlings subjected to salt stress enhanced seed germination and root growth, also, there is promoting effect of NO on seed germination that persisted in the presence of heavy metals (Pb and Cd) and NaCl. Pretreatment of maize seedlings with 100 μM SNP increases dry matter of roots and shoots under salinity stress; however, when the concentration of SNP was increased to 1000 μM shoot and root dry weight decreased (Zhang et al. 2006). Thus, this paper highlighted both the protective effects of low NO concentration and the toxic effect of high NO concentration on plants. The positive effects of NO on salinity tolerance or stress mitigation have been attributed to antioxidant activities and modulation of ROS detoxification system. Improved plant growth under salinity stress by exogenous application of NO was associated with increase in antioxidant enzymes such as SOD, CAT, GPX, APX, and GR (Zhao et al. 2004), and suppression of malondialdehyde (MDA) production or lipid peroxidation (Nalousi et al. 2012).

Effects of NO on salinity tolerance are also related to its regulation of plasma membrane H^+-ATPase and Na^+/H^+ ratio. NO stimulates H^+-ATPase (H^+-PPase), thereby producing a H^+ gradient and offering the force for Na^+/H^+ exchange. Such an increase of Na^+/H^+ exchange may contribute to K^+ and Na^+ homeostasis, although NO acts as a signal molecule under salt stress and induces salt resistance by increasing PM H^+-ATPase activity (Crawford 2006), research results from Zhang et al. (2007) with calluses from *Populus euphratica* also indicated that NO cannot activate purified PM H^+-ATPase activity, at least in vitro. They initially hypothesized ABA or H_2O_2 might be downstream signal molecules to regulate the activity of PM H^+-ATPase. Further results indicated that H_2O_2 content increased greatly under salt stress. Since H_2O_2 might be the candidate downstream signal molecule, Zhang et al. (2007) tested PM H^+-ATPase activity and K^+ to Na^+ ratio in calluses by adding H_2O_2. The results suggested that H_2O_2 inducing an increased PM H^+-ATPase activity resulted in an increased K^+ to Na^+ ratio leading to NaCl stress adaptation.

8.8 Hormone Regulation of Salinity Tolerance

Abscisic acid is an phytohormones whose application to plant ameliorates the effect of stress condition(s). It has long been recognized as a hormone which is upregulated due to soil water deficit around the root. Salinity stress causes osmotic stress and water deficit, increasing the production of ABA in shoots and roots, the accumulation of ABA can mitigate the inhibitory effect of salinity on photosynthesis, growth, and translocation of assimilates (Cabot et al. 2009). The positive relationship between ABA accumulation and salinity tolerance has been at least partially attributed to the accumulation of K^+, Ca^{2+} and compatible solutes, such as proline and sugars, in vacuoles of roots, which counteract with the uptake of Na^+ and Cl^- (Gurmani et al. 2011). ABA is a vital cellular signal that modulates the expression of a number of salt and water deficit-responsive genes. Fukuda and Tanaka (2006) demonstrated the effects of ABA on the expression of two genes, *HVP1* and *HVP10*, for vacuolar H^+-inorganic pyrophosphatase, and of HvVHA-A, for the catalytic subunit (subunit A) of vacuolar H^+-ATPase in Hordeum vulgare under salinity stress. ABA treatment in wheat induced the expression of MAPK4-like, *TIP 1*, and *GLP 1* genes under salinity stress (Keskin et al. 2010).

8.8.1 Drought Stress

Drought or water deficit defined as a situation in which plant water potential and turgor are reduced enough to interface with normal functions. The impact of water stress is compounded by depletion of water in the rooting zone and increased atmospheric vapor pressure deficit (Ahanger et al. 2014). Therefore, water deficit is considered to be a limiting factor for crop productivity, it is affected negatively on plant growth and productivity, drought is considered the main factor for sustainable yield under climate change conditions; it will be challenging to provide the food demands of growing global population, the world faces major challenges to increase its food production to more than 50% by 2050 (Li et al. 2009). Drought conditions induced loss in crop yield probably exceeds losses from all other abiotic stress; therefore, severity and longevity of drought stress and interaction between drought and other environmental factors are very important to crop productivity. The reduction in productivity varied depending on plant species, growth stage, longevity and severity of drought.

The negative impact of drought predominates at all developmental stages, starting from germination to seed maturation (Valliyodan and Nguyen 2006), although the timing, duration, severity and growth stage undoubtedly had crucial roles in determining how the plant responded to water deficit; water stress affected cell membrane to disturb ion exchange and also causing an imbalance in the composition of nutrients in the tissues, and also transpiration is inhibited by drought, as shown for beech (Peuke et al. 2002), but this may not necessarily affect nutrient uptake in a similar manner. Comparing results from different studies is complex due to interspecific differences in the response of stomatal conductance and photosynthesis to leaf water potential and/or relative water content; the parameters mostly used to evaluate the drought severity (Cornic and Massacci 1996); there is a positive correlation between stomata closing and drought progress, followed by decreasing photosynthesis. It is clear that water availability is not only the controlled factor for stomatal conductance, but there are a complicated interaction of intrinsic and extrinsic factors. Also, drought stress disturbed the ionic balance of the cell and thus impacting the physiological processes of the plant (Khan et al. 2015).

8.8.2 Drought Tolerance

Drought tolerance is defined as the ability to reserve leaf area and growth under prolonged vegetative stage stress; however, the major in drought is structure of root system that provide enough water for plant growth (Nguyen et al. 1997). Plants use different mechanisms on a morphological, physiological, and molecular basis to survive under drought stress (Fig. 2), which include drought avoidance, drought tolerance, drought escape, and drought recovery; these mechanisms facilitate the plants to overcome stress (Reddy et al. 2004; Cruz De Carvalho 2008). There are various mechanisms and strategies to survive under drought stress and the anatomical adaptations include.

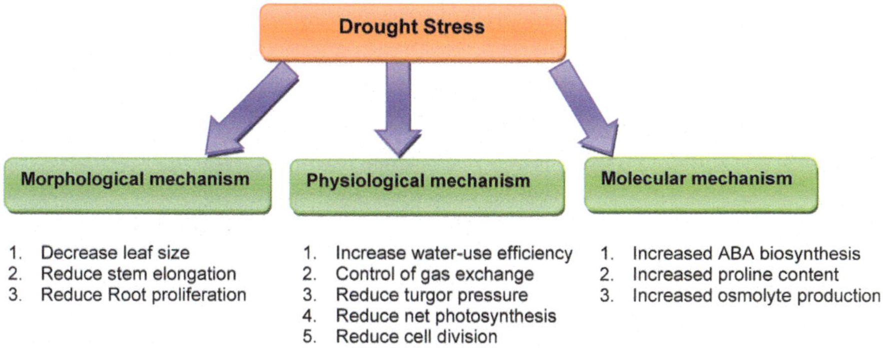

Fig. 2 Different plant mechanisms for adapting with drought stress (Reddy et al. 2004)

1. **Morphological mechanism**: includes decreased leaf size, reduced stem elongation root proliferation (Valliyodan and Nguyen 2006).
2. **Physiological mechanism**: like increased water-use efficiency, control of gas exchange, closing stomatal, reduce turgor pressure, also, net photosynthesis and metabolic processing for cell division reduced under drought stress (Peuke et al. 2002).
3. **Molecular mechanism**: including synthesis of specific proteins increased ABA biosynthesis and increased proline content; the increased production of osmolytes within the plant system protects the bio-chemicals against their deformation (Krasensky and Jonak 2012).

The plants strive to mitigate the effects of water stress by over-production of scavenging phyto-hormones to scavenging free radicals to avoid damaging effects of ROS and deoxyribonucleic acid (DNA) (Gill and Tutej 2010), activation of antioxidants system (Ahmadi and Baker 2001) upregulating osmotic adjustment through accumulation of excessive quantities of organic solute for adjustment of osmotic pressure (Farooq et al. 2010). Moreover, Hussain et al. (2015) reported that compatible solutes were efficacious in protecting cell organelles from the generation of free radicals under water stress.

9 Effect of Drought on Plant Growth and Productivity

Water scarcity affect ion balanced in the cell; it inhibits the uptake of K^+ ion and increases absorption of Na^+ ion in various plant like wheat, maize, and barely (Farooq et al. 2015). However, Chandra et al. (2007) found that drought stress caused reduction in soluble protein contents in sorghum and tomato crops due to the breakdown of protein into amino acids under water stress condition.

9.1 Effects of Drought Stress on Plant Growth and Yield

There are various effects of drought on plant varied from morphological or physiological and even at molecular levels, where drought can affect at any growth stage depending on timing, longevity, and severity. Plant growth is one of the most drought-sensitive physiological processes due to the reduction of turgor pressure, cell expansion can only occur when turgor pressure is greater than the cell wall threshold, therefore, water stress greatly suppresses cell expansion and cell growth due to the low turgor pressure. Drought stress has a severe effect on plant growth; the plant shoots and roots are the major components that are adapted during water stress, therefore, at a morphological base plants reduce the leaves area and leaf number to reduce water loss and improve crop yield tolerant drought species bearing small and needle leaves; these plants survive and withstand drought very well; however, their growth rate and biomass are moderately low (Schuppler et al. 1998), roots play an important role in tolerating drought conditions, therefore, the root growth, density, proliferation, and size are the determining factor for tolerant drought stress (Ball et al. 1994). Drought-induced yield reduction has been reported in many crop species, which depends upon the severity duration of the stress period and growth stage (Table 1); crop productivity could be inhibited under drought stress conditions from 50 to 73% (Berry et al. 2013).

9.2 Effect of Drought on Plant Water Relations

All components of plants water relations, leaf temperature and canopy temperature are important characteristics that influence plant water relations. Under drought stress, plant–water relation are affected by reduced availability of water, reduced leaf water potential, relative water content, transpiration rate and increased leaf temperature, and also, stomatal opening and closing is more strongly affected by drought stress (Kavar et al. 2007). Comparing results from previous studies is complex due to interspecific differences in the response of stomatal conductance and photosynthesis to leaf water potential and/or relative water content, the parameters mostly used to evaluate the drought severity (Nguyen et al. 1997); there is a positive correlation between stomata closing and drought progress, followed by decreasing photosynthesis. It is clear that water availability is not only the controlled factor for stomatal conductance, but there are complicated interactions of intrinsic and extrinsic factors; and also, drought stress disturbed the ionic balance of the cell and thus impacting the physiological processes of the plant (Farooq et al. 2009).

Table 1 Effect of drought stress on yield reduction in some field crops

Crop	Scientific name	Growth stage	Yield reduction (%)	References
Barley	*Hordeum vulgare*	Seed filling	49–57	Samarah (2005)
Maize	*Zea mays*	Grain filling	79–81	Monneveux et al. (2008)
Maize	*Zea mays*	Reproductive	63–87	Kamara et al. (2003)
Maize	*Zea mays*	Reproductive	70–47	Chapman and Edmeades (1999)
Maize	*Zea mays*	Vegetative	25–60	
Maize	*Zea mays*	Reproductive		
Rice	*Oryza Sativa*	Reproductive (mild stress)	53–92	Lafitte et al. (2007)
Rice	*Oryza Sativa*	Reproductive (severe stress)	48–94	Lafitte et al. (2007)
Rice	*Oryza Sativa*	Grain filling (mild stress)	30–55	Basnayake et al. (2006)
Rice	*Oryza Sativa*	Grain filling (severe stress)	60	Basnayake et al. (2006)
Rice	*Oryza Sativa*	Reproductive	24–84	Venuprasad et al. (2007)
Chickpea	*Cicer arietinum*	Reproductive	45–69	Nayyar et al. (2006)
Pigeon pea	*Cajanus cajan*	Reproductive	40–55	Nam et al. (2001)
Common beans	*Phaseolus vulgaris*	Reproductive	58–87	Martínez et al. (2007)
Soybean	*Glycine max*	Reproductive	46–71	Samarah et al. (2006)
Cowpea	*Vigna unguiculata*	Reproductive	11–60	Ogbonnaya et al. (2003)
Sunflower	*Helianthus*	Reproductive	60	Mazahery-Laghab et al. (2003)
Canola	*Brassica napus*	Reproductive	30	Sinaki et al. (2007)
Potato	*Solanum tuberosum*	Flowering	13	Kawakami et al. (2006)
Cotton	*Gossypium hirsutum*	Flowering		McWilliams (2003)

*Modified from Farooq et al. (2009)

9.3 Effect of Drought on Plant Nutrient Relations

Drought stress inhibited absorption of different nutrients by reducing available energy for assimilation mineral nutrient forms to organic form that can be used in metabolic processing; this processes require energy to convert mineral nutrients to proper form that can be used in plant growth. McWilliams (2003) reported that cotton plants could not absorb N and K under drought conditions, and also, P content is depleted in plant tissue under water shortage, due to low moisture availability which reduce PO3-4 mobility (Peuke and Rennenberg 2004; Grossman and Takahashi 2001); therefore, proper nutrition could reduce adverse effect of water shortage and enhance crop productivity (Garg 2003).

9.4 Influence of Drought on Photosynthesis

The main influence of drought is decreasing photosynthesis by stomatal closure which is considered the major element for reduced photosynthesis, and also, drought stress affects pigments components, reduces activities of Calvin cycle enzymes, which are essential factors of decreased crop productivity; and also, under mild to moderate drought stress leaf area decreases and occurs early senescence of leaves (Yokota et al. 2002). Drought stress disturbs the balance between the production of reactive oxygen species and the antioxidant defense, causing accumulation of reactive oxygen species which induces oxidative stress in cellular components; the reduction in photosynthesis consequently reduces other processes like CO_2 uptake and nutrients, limitation of CO_2 directly decreases the carboxylation and increase electrons which forms reactive oxygen species (Fu and Huang 2001; Reddy et al. 2004). There is a correlation between drought severity and decreasing photosynthesis due to an increase in the activity of RuBisCO binding inhibitors which reduce ribulose 1,5-bisphosphate carboxylase/oxygenase (RuBisCO) activity, furthermore, non-cyclic electron transport is down-regulated to match the reduced requirements of NADPH production and thus reduces the ATP synthesis. During water stress, activities of the various enzymes like phosphoenol pyruvate carboxylase, RuBisCO, fructose-1, 6-bisphosphatase, and pyruvate orthophosphate dikinase decreased linearly with lowered leaf water potential (Farooq et al. 2010).

Negative effects of drought on photosynthesis

1. Decrease in leaf turgor.
2. Decrease water potential (Ludlow and Muchow 1990).
3. Closure of their stomata to prevent transpirational water loss (Mansfield and Atkinson 1990).
4. Increased leaf heat as the rate of transpiration decreases, the amount of heat that can be dissipated increases (Fahad et al. 2017).

9.5 Oxidative Damage

Plants under certain environmental stresses start generation of reactive oxygen species, including superoxide anion radicals (O_{-2}), hydroxyl radicals (OH), hydrogen peroxide (H_2O_2), and alkoxy radicals (RO). ROS may react with proteins, lipids and deoxyribonucleic acid, causing oxidative damage and impairing the normal functions of cells (Munne-Bosch and Penuelas 2003).

All above negative effects decreased plant growth and reduced crop productivity.

10 Drought Resistance Mechanisms in Plants

Drought stress tolerant as mentioned above depends on root system, the roots are the main organs that adapts with drought stress, the distribution and structure, and not a number of roots determines the most efficient strategy for extracting water during the crop-growing season. Selection for a deep and extensive root system has been advocated to increase the productivity of food legumes under moisture-deficit conditions as it can optimize the capacity to acquire water (Subbarao et al. 1995). Under drought stress conditions plants adapted to survive by using a different morphological, physiological, and biochemical responses. Tolerant species have different changes in plant tissue, and at physiological and molecular basis to tolerant drought conditions, to cope with the drought, tolerant plants initiate defense mechanisms against water deficit (Chaves and Oliveira 2004).

10.1 Morphological Mechanisms

Under drought stress, plants use various morphological mechanisms, the manifestation of a single or a combination of inherent changes determines the ability of the plant to sustain itself under limited moisture supply; an account of various morphological mechanisms operative under drought conditions is given below.

Drought escape is attained through a shortened life cycle or growing season, allowing plants to reproduce before the environment becomes harshly dry; flowering earlier is one of drought adaptation techniques which lead to drought escape; however, there are various factors such as genotype and environmental conditions that determine crop ability to escape from drought stress (Araus et al. 2002). Drought escape occurs when phenological development is successfully matched with periods of soil moisture availability, where the growing season is shorter and terminal drought stress predominates. In robusta coffee under low moisture start dropping their leaves from older to youngest as response to drought stress; however, matching growth duration of plants to soil moisture availability is critical to realize high seed yield

(DaMatta 2004). Plants may escape drought stress by cutting short their growth duration, and avoid the stress with the maintenance of high tissue water potential either by reducing water loss from plants or improved water uptake, or both. **Drought avoidance** means reduce water loss from plants through stomatal control of transpiration, also reserve water uptake through an extensive and prolific root system (Berry et al. 2013; Cheserek and Gichimu 2012). The root system's characters such as biomass, length, density, and depth are the key drought avoidance traits that determine the final yield under drought conditions; therefore, a deep and thick root system is helpful for extracting water from considerable depths. Glaucousness or waxy bloom on leaves helps with maintenance of high tissue water potential and is therefore considered as a desirable trait for drought tolerance; different levels of glaucousness led to increased water-use efficiency but did not affect total water use or crop index (Lazaridou and Koutroubas 2004).

10.2 Physiological Mechanisms

Osmotic adjustment, osmoprotection, antioxidation, and a scavenging defense system have been the most important bases responsible for drought tolerance; there are different physiological mechanisms to explain drought tolerance on physiological bases, and some of these mechanisms are described below.

10.2.1 Cell and Tissue Water Conservation

It has been identified that among various mechanisms, osmotic adjustment, abscisic acid, and induction of dehydrins may confer tolerance against drought injuries by maintaining high tissue water potential. Riccardi et al. (2001) suggested that water potential was not the defining feature of the tolerance; osmotic adjustment allows the cell to decrease osmotic potential and, as a consequence, increases the gradient for water influx and maintenance of turgor. Improved tissue water status may be achieved through osmotic adjustment and/or changes in cell wall elasticity; this is essential for maintaining physiological activity for extended periods of drought with the accumulation of solutes, the osmotic potential of the cell is lowered, which attracts water into the cell and helps with turgor maintenance; this is consistent with other studies of species with elastic cell walls. Osmotic adjustment helps to maintain the cell water balance with the active accumulation of solutes in the cytoplasm, thereby minimizing the harmful effects of drought (Kramer and Boyer 1995). Osmotic adjustment has vital role in delaying dehydrative damage under water stress (Fig. 3) by continued maintenance of the cell water balance and to maintain cell turgor and physiological processes (Taiz and Zeiger 2006). The osmotic adjustment, also, facilitates a better translocation of pre-anthesis carbohydrate partitioning during grain filling, while high turgor maintenance leads to higher photosynthetic rate and growth (Moinuddin and Khannu-Chopra 2004).

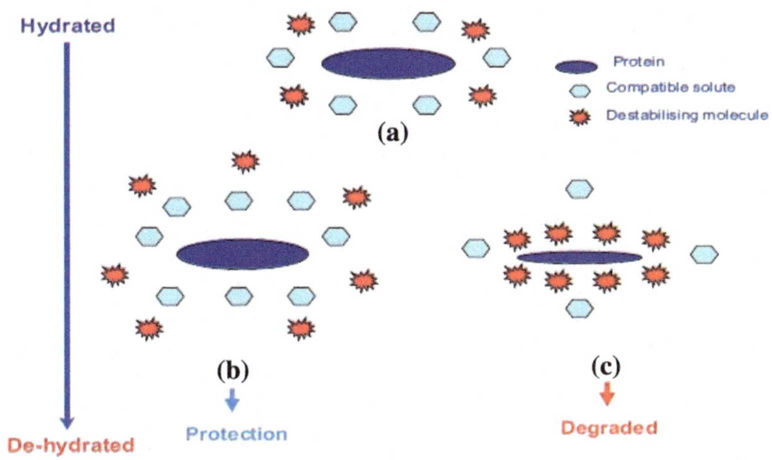

Fig. 3 Role of compatible solutes in drought tolerance (adapted from Taiz and Zeiger 2006)

10.2.2 Role of Compatible Solutes in Drought Tolerance

In the hydrated state, (Fig. 3) the presence of water reduces the interaction of destabilizing molecules (a); in tolerant cells the synthesis of compatible solutes preferentially excludes the binding of destabilizing molecules and stabilizes native protein conformation (b) and in sensitive cells the lack of compatible solutes results in the preferential binding of destabilizing molecules to the protein surface, leading to degradation (c) (Taiz and Zeiger 2006). Variation in osmotic adjustment among chickpea cultivars in response to soil drought has been observed, and seed yield of chickpea was correlated with the degree of osmotic adjustment when grown under a line-source irrigation system in the field (Moinuddin and Khannu-Chopra 2004). Contrarily, Serraj and Sinclair (2002) found no yield advantage from the osmotic adjustment in any crop. Nevertheless, further investigations are imperative to establish this controversy.

10.2.3 Role of Proline in Plant Tissue

Proline is one among the most vital cytosolutes and its free accumulation is a widespread response of higher plants, algae, animals, and bacteria to low water potential and other stress conditions (Yamada et al. 2005). Under water stress leaves synthesis proline by a combination of increased biosynthesis and slow oxidation in mitochondria. There are many physiological roles that have been assigned to free proline including, stabilization of macromolecules, a sink for excess reductant and a store of carbon and nitrogen for use after relief of water deficit (Morgan 1990). Drought-tolerant petunia (Petunia hybrida) varieties were reported to accumulate free proline under a drought that acted as an osmoprotectant and induced drought tolerance (Yamada et al. 2005), and also, Alexieva et al. (2001) found that increasing in proline contents in pea cultivars under drought stress.

10.2.4 Role of Salicylic Acid in Drought Resistance

Salicylic acid (SA), a plant growth hormone and important signaling molecule, has a great agronomic potential to improve the drought tolerance of plants. SA modulates the plant responses to environmental stresses by regulating plant growth, development, ripening, and defense responses. The role of SA in regulation of drought was also supported by the induction of SA inducible genes *PR1* and *PR2* by drought stress (Kang et al. 2013). Plants under drought conditions increased synthesis of salicylic acid to alleviate stress; in *Phillyrea angustifolia* plants increased endogenous levels of SA up to fivefold (Munne-Bosch and Penuelas 2003), and also SA level reach to twofold in barley roots under water deficit conditions (Bandurska and Stroiński 2005). Various researchers have reported that oxidative stress caused by generation of ROS under drought stress could be quenched by increasing the quantum of salicylic acid in the plant system; and also, Khan et al. (2015) reported that salicylic acid improve antioxidative system and reduce undesirable effects of abiotic stress, which also improved drought tolerance in *Arabidopsis* (Miura et al. 2013). SA treatment increased the membrane stability and levels of proline and ABA in water stressed barley conferring plants with stress tolerance (Bandurska and Stroiński 2005), and also SA positively influenced the ascorbate glutathione cycle in pretreated wheat leading to enhancement in tolerance to stress and alleviating substantial water loss (Singh and Usha 2003). Exogenous SA treatment increase antioxidantive metabolism and alleviated negative effects on photosynthesis under salinity stress (Kang et al. 2013); in addition, SA treatment improve defense system of *zea mays* plants under water stress (Saruhan et al. 2012).

11 Molecular Mechanisms

Drought stress reduced plant cellular water, and under these conditions, changes in gene expression (up- and down-regulation) take place.

Various genes are induced in response to drought at the transcriptional level, and these gene products are thought to function in tolerance to drought (Kavar et al. 2007). Gene expression may be triggered directly by the stress conditions or result from secondary stresses and/or injury responses.

11.1 Aquaporins

Aquaporins is one of a min-intrinsic membrane protein; in plants, aquaporins are present abundantly in the plasma membrane and in the vacuolar membrane. The structural analysis of aquaporins has revealed the general mechanism of protein-mediated membrane water transport (Tyerman et al. 2002). Although the discovery of aquaporins in plants has resulted in a prototype shift in the understanding of plant water relations (Maurel et al. 2002), the relation between aquaporins and plant drought resistance is still elusive (Aharon et al. 2003). Nevertheless, it is believed that they can regulate the hydraulic conductivity of membranes and potentiate a ten- to 20-fold increase in water permeability. There was evident from a number of reports on mercury which may be a potential inhibitor of aquaporins, mercury-induced decrease root hydraulic conductivity, which substantiated that aquaporins play a major role in overall root water uptake, and play a role in cellular osmoregulation of highly compartmented root cells (Javot and Maurel 2002; Chaumont and Tyerman 2014). Recently, efforts have been concentrated on investigating the function and regulation of plasma membrane's intrinsic protein aquaporins. The aquaporins play an important role in controlling transcellular water transport. For example, they are abundantly expressed in roots where they mediate soil water uptake, and transgenic plants down regulating one or more prolactin-inducible protein genes that had lesser root water absorption (Javot et al. 2003).

11.2 Stress Proteins

Synthesis of stress proteins is a ubiquitous response to cope with prevailing stressful conditions including water deficit. Most of the stress proteins are soluble in water and therefore contribute toward the stress tolerance phenomena by hydration of cellular structures (Hoekstra et al. 2001), Synthesis of a variety of transcription factors and stress proteins is exclusively implicated in drought tolerance (Zhu 2002). The dehydration-responsive element-binding genes are involved in the abiotic stress signaling pathway. It was possible to engineer stress tolerance in transgenic plants by manipulating the expression of dehydration-responsive element-binding genes (Agarwal et al. 2006). Membrane-stabilizing proteins and late embryogenic abundant proteins are another important protein group responsible for conferring drought tolerance; this enhances the water holding capacity by providing protective conditions for other protein structures, and it also plays a major role in the sequestration of ions that are concentrated during cellular dehydration (Kovtun et al. 2000). These proteins help to protect the partner protein from degradation and proteinases that function to remove denatured and damaged proteins.

11.3 Signaling and Drought Stress Tolerance

General responses to stress involve signaling stress detection via the redox system, checkpoints arresting the cell cycle, and deoxyribonucleic acid repair processes stimulated in response to deoxyribonucleic acid damage. The complication of signaling proceedings associated with the sensing of stress and the launch of defense and acclimation pathway is believed to include ROS species, calcium, calcium-regulated proteins, mitogen-activated protein kinase cascades, and cross-talk between different transcription factors (Gorantla et al. 2006; Chen et al. 2002). Different chemical signals and plant hormones play an important role in raising stress tolerance by using transduction cascades and induce genomic programming (Joyce et al. 2003).

12 Physiological Responses to a Combination of Drought and Salinity Stress

There are different responses that occurred in plant under stress conditions at various levels to alleviated negative effects, thus including reduced CO_2 assimilation through stomatal closure, destroy cell membrane and disruption of enzyme activity particularly those of adenosine triphosphate (ATP) synthesis and CO_2 fixation (Wahid et al. 2007). The water stress induced either by drought or salinity is the most damaging factor influencing productivity to a greater extent across the globe (Davenport et al. 2003). There are estimates that about 30–60% of total water applied to soil are lost through evaporation in the arid and semi-arid regions (Ashraf 2010). Globally, the agricultural production areas are facing a continuous decline in irrigation water supplies and also a concurrent increase in salinization of soils due to the application of underground brackish water (Cai et al. 2013). The frequent occurrence of drought condition is inevitable, but its ill-effects could be alleviated by the adoption of certain short- and long-term strategies. The plants being immobile in nature and thereby adapt to certain diverse physiological, morphological, biochemical, and molecular measures to cope with environmental stresses.

13 Concluding Remarks

Agricultural production exposed to different environmental challenge such as salinity, drought, global warming, plants have used different physiological and biochemical responses to adapt and survive under abiotic stress conditions. Drought and salinity are the most important abiotic constraints to plant survival and to crop productivity; they furthermore have main harmful effects on the plant tissue such as the negative effects on the cellular energy supply and redox homeostasis that are stable by re-programming of plant primary metabolism and modification of cellular composition. The agricultural sector is considered as the main user of freshwater resources in numerous regions of the world, with growing population and increasing water scarcity in various regions of the world, thus reducing the amount of crop production, which affects the availability of global food quantity in the near future. Drought can affect plants in various ways such as decrease in photosynthesis and growth inhibition, accumulation of ABA, proline, mannitol, sorbitol, formation of radical scavenging compounds (ascorbate, glutathione, α-tocopherol etc.), and synthesis of new proteins and mRJNAs, and also, water stress results in stomatal closure and reduced transpiration rates, reduces the water potential of plant tissues, therefore, plants under drought stress use two processing to control relation between photosynthetic potential and relative water content of leaves; first one through decreased stomatal conductance and reduced photosynthesis and decreased CO_2 concentration inside the leaf, while the other one by elevated CO_2 decreases progressively as relative water content (RWC) declines, and reduced gas exchange, with both type there is reducing in metabolic processing in leaf tissue. Soil salinity considered a major stress limiting the increase in food crops production; there are more than 20% of arable land worldwide that is affected by salt stress and this ratio increased yearly all over the world due to environmental factors and human activity. Salinity stress causes changes in numerous physiological and metabolic processes in plant tissue, unfortunately, the majority of economic crop species are glycophytes, and therefore, salinity inhibits crop productivity worldwide. Salinity stress causes changes in various physiological and metabolic processes, depending on severity and duration of the stress, and eventually hamper crop production, glycophytes plants have different physiological mechanisms such as ion homeostasis, compatible solute, antioxidant regulation, and polyamines production. Generally, plants could combine a range of response to avoiding drought and salinity stresses and injuries by different mechanisms to be able to complete their life cycle. By using different strategies such as using maximum available resources, they store reserves in plant organs and use them for fruit production, and also, plants can tolerate stress conditions by avoiding tissue dehydration.

References

Abobatta WF (2018) Some physiological mechanisms of salt tolerance in the glycophytes plant: overview. Acta Sci Agric 2(12):154–156

Agarwal PK, Agarwal P, Reddy MK, Sopory SK (2006) Role of DREB transcription factors in abiotic and biotic stress tolerance in plants. Plant Cell Rep 25:1263–1274

Ahanger MA, Tyagi SR, Wani, MR, Ahmad P (2014) Drought tolerance: role of organic osmolytes, growth regulators, and mineral nutrients. In: Physiological mechanism and adaptative strategies in plants under changing environment. Springer, pp 25–55

Aharon R, Shahak Y, Wininger S, Bendov R, Kapulnik Y, Galili G (2003) Overexpression of a plasma membrane aquaporins in transgenic tobacco improves plant vigour under favourable growth conditions but not under drought or salt stress. Plant Cell 15:439–447

Ahmad P, Prasad MNV (2012) Abiotic stress responses in plants: metabolism, productivity and sustainability. Springer, New York, NY

Ahmad R, Lim CJ, Kwon SY (2013) Glycine betaine: a versatile compound with great potential for gene pyramiding to improve crop plant performance against environmental stresses. Plant Biotechnol Rep 7:49–57
Ahmadi A, Baker DA (2001) The effect of water stress on the activities of key regulatory enzymes of the sucrose to starch pathway in wheat. Plant Growth Regul 35:81–91
Alamgir ANM, Yousuf-Ali M (1999) Effect of salinity on leaf pigments, sugar and protein concentrations and chloroplast ATPase activity of rice (*Oryza sativa* L.). Bangladesh J Bot 28 (2):145–149
Alexieva V, Sergiev I, Mapelli S, Karanov E (2001) The effect of drought and ultraviolet radiation on growth and stress markers in pea and wheat. Plant Cell Environ 24:1337–1344
Araus JL, Slafer GA, Reynolds MP, Royo C (2002) Plant breeding and drought in C3 cereals: what should we breed for? Ann Bot 89:925–940
Ashraf M (2010) Inducing drought tolerance in plants: recent advances. Biotchnol Adv 28:169–183. https://www.mdpi.com/2223-7747/8/7/232/htm
Bajgu A (2014) Nitric oxide: role in plants under abiotic stress. In: Physiological mechanisms and adaptation strategies in plants under changing environment, pp 137–159
Ball RA, Oosterhuis DM, Mauromoustakos A (1994) Growth dynamics of the cotton plant during water-deficit stress. Agron J 86:788–795
Bandurska H, Stroiński A (2005) The effect of salicylic acid on barley response to water deficit. Acta Physiol Plant 27:379–386
Basnayake J, Fukai S, Ouk M (2006) Contribution of potential yield, drought tolerance and escape to adaptation of 15 rice varieties in rainfed lowlands in Cambodia. In: Proceedings of the Australian Agronomy Conference, Australian Society of Agronomy, Birsbane, Australia
Ben Ahmed C, Ben Rouina B, Sensoy S, Boukhriss M, Ben Abdullah F (2009) Exogenous proline effects on photosynthetic performance and antioxidant defense system of young olive tree. J Agric Food Chem 58(7):4216–4222
Berry P, Ramirez-villegas J, Branseley H (2013) Regional impacts of climate change on agriculture and the role of adaptation. Plant Genet Res Clim Chang 4:78
Bohnert HJ, Nelson DE, Jensen RG (1995) Adaptations to environmental stresses. Plant Cell 7(7):1099–1111
Bond WJ (1998) Effluent irrigation—an environmental challenge for soil science. Aust J Soil Res 36:543–555
Bouwer H (2002) Integrated water management for the 21st century: problems and solutions. J Irrig Drain Eng 28:193–202
Brito C, Dinis LT, Moutinho-Pereora J, Correia CM (2019) Drought stress effects and olive tree acclimation under a changing climate. Plants 8(7):232. https://doi.org/10.3390/plants8070232
Bui EN, Smettem KRJ, Moran CJ, Williams J (1996) Use of soil survey information to assess regional salinization risk using geographical information systems. Environ Qual 25:433–439
Cabot C, Sibole JV, Barcelo J, Poschenrieder C (2009) Abscisic acid decreases leaf Na^+ exclusion in salt-treated *Phaseolus vulgaris* L. J Plant Growth Regul 28(2):187–192
Cai X, Molden D, Mainuddin M, Sharma B, Mohyddin A, Karini P (2013) Producing more food with less water in the changing world: assessment of water productivity in 10 major river basins. Water Intl 36:421–462
Chandra A, Anand A, Dubey A (2007) Effect of salicylic acid on morphological and biochemical attributes in cowpea. J Environ Biol 28:193–196
Chapman SC, Edmeades GO (1999) Selection improves drought tolerance in tropical maize populations. II. Direct and correlated responses among secondary traits. Crop Sci 39:1315–1324
Cha-Um S, Kirdmanee C (2010) Effect of glycinebetaine on proline, water use, and photosynthetic efficiencies, and growth of rice seedlings under salt stress. Turkish J Agric Forest 34(6):517–527
Chaumont F, Tyerman SD (2014) Aquaporins: highly regulated channels controlling plant water relations. Plant Physiol 164:1600–1618
Chaves MM, Oliveira MM (2004) Mechanisms underlying plant resilience to water deficits: prospects for water-saving agriculture. J Exp Bot 55:2365–2384

Chen W, Provart NJ, Glazebrook J, Katagiri F, Chang HS, Eulgem T, Mauch F, Luan S, Zou G, Whitham SA, Budworth PR, Tao Y, Xie Z, Chen X, Lam S, Kreps JA, Harper JF, Si-Ammour A, Mauch-Mani B, Heinlein M, Kobayashi K, Hohn T, Dang JL, Wang X, Zhu T (2002) Expression profile matrix of Arabidopsis transcription factor genes suggests their putative functions in response to environmental stresses. Plant Cell 14:559–574

Cheserek JJ, Gichimu BM (2012) Drought and heat tolerance in coffee: a review. Int Res J Agric Sci Soil Sci 2(12):498–501

Cornic G, Massacci A (1996) Leaf photosynthesis under drought stress. In: Baker NR (ed) Photosynthesis and the environment. Kluwer Academic Publishers, The Netherlands

Crawford NM (2006) Mechanisms for nitric oxide synthesis in plants. J Exp Bot 57(3):471–478

Cruz De Carvalho MH (2008) Drought stress and reactive oxygen species: production, scavenging and signaling. Plant Signal Behav 3:156–165

DaMatta FM (2004) Exploring drought tolerance in coffee: a physiological approach with some insights for plant breeding. Braz J Plant Physiol 16:1–6

Davenport SB, Gallego SM, Benavides MP, Tomaro ML (2003) Behaviour of anti-oxidant defence system in the adaptive response to salt stress in *Helianthus annuus* L. Cell Plant Growth Regul 40:81–88

De Lourdes OOM, Menezes Sobreira AC, De Aragao FME, Orellano EG, Silva Lima D-GM, De Melo FD (2001) Saltmodulation of vacuolarH^+-ATPase and H^+-Pyrophosphatase activities in *Vigna unguiculata*. J Plant Physiol 158(5):545–551

Deivanai S, Xavier R, Vinod V, Timalata K, Lim OF (2011) Role of exogenous proline in ameliorating salt stress at early stage in two rice cultivars. J Stress Physiol Biochem 7:157–174

Dietz KJ, Tavakoli N, Kluge C (2001) Significance of the Vtype ATPase for the adaptation to stressful growth conditions and its regulation on the molecular and biochemical level. J Exp Bot 52(363):1969–1980

Fahad S, Bajwa AA, Nazir U, Anjum SA, Farooq A, Zohaib A, Sadia S, Nasim W, Adkins S, Saud S, Ihsan MZ, Alharby H, Wu C, Wang D, Huang J (2017) Crop production under drought and heat stress: plant responses and management options. Front Plant Sci 8:1147

FAO (2009) High level expert forum—how to feed the world in 2050. Economic and Social Development, Food and Agricultural Organization of the United Nations, Rome, Italy

FAO (2017) Faostat [online]. https://www.faostate.fao.org/

Farooq M, Hussain M, Wakeel A, Siddique KHM (2015) Salt stress in maize: effects, resistance mechanisms, and management. A review. Agron Sustain Dev 35:461–481

Farooq M, Wahid A, Kobayashi N, Fujita D, Basra SMA (2009) Plant drought stress: effects, mechanisms and management. Agron Sustain Dev 29:185–212

Farooq MA, Lee DJ, Wahid A, Cheema S, Aziz T (2010) Drought stress: comparative time course action of the foliar applied glycine betaine, salicylic acid, nitrous oxide, brassinosteroids and spermine in improving drought resistance of rice. J Agron Crop Sci 196:336–345

Flowers TJ (2004) Improving crop salt tolerance. J Exp Bot 55(396):307–319

Flowers TJ, Colmer TD (2008) Salinity tolerance in halophytes. New Phytol 179:945–963

Fu J, Huang B (2001) Involvement of antioxidants and lipid peroxidation in the adaptation of two cool-season grasses to localized drought stress. Environ Exp Bot 45:105–114

Fukuda A, Tanaka Y (2006) Effects of ABA, auxin, and gibberellin on the expression of genes for vacuolar H^+-inorganic pyrophosphatase, H^+-ATPase subunitA, and Na^+/H^+ antiporter in barley. Plant Physiol Biochem 44(5–6):351–358

Gama PBS, Tanaka K, Eneji AE, Eltayeb AE, El Siddig K (2009) Salt-induced stress effects on biomass, photosynthetic rate, and reactive oxygen species-scavenging enzyme accumulation in common bean. J Plant Nutr 32:837–854

Gao Z, Sagi M, Lips SH (1998) Carbohydrate metabolism in leaves and assimilate partitioning in fruits of tomato (*Lycopersicon esculentumL.*) as affected by salinity. Plant Sci 135(2):149–159

Garg BK (2003) Nutrient uptake and management under drought: nutrient-moisture interaction. Curr Agric 27:1–8

Gill SS, Tutej N (2010) Reactive oxygen species and antioxidant machinery in abiotic stress tolerance in crop plants. Plant Physiol Biochem 48:909–930

Gill SS, Tajrishi M, Madan M, Tuteja N (2013) A DESD box helicase functions in salinity stress tolerance by improving photosynthesis and antioxidant machinery in rice (*Oryza sativa* L. cv. PB1). Plant Mol Biol 82(1–2):1–22

Gorantla M, Babu R, Lachagari VBR, Reddy AMM, Wusirika R, Bennetzen JL, Reddy AR (2006) Identification of tressresponsive genes in an indica rice (*Oryza sativa* L.) using ESTs generated from drought-stressed seedlings. J Exp Bot 58:253–265

Groß F, Durner J, Gaupels F (2013) Nitric oxide, antioxidants and prooxidants in plant defence responses. Front Plant Sci 4:419

Grossman A, Takahashi H (2001) Macronutrient utilization by photosynthetic eukaryotes and the fabric of interactions. Annu Rev Plant Phys 52:163–210

Gupta B, Huang B (2014) Mechanism of salinity tolerance in plants: physiological, biochemical, and molecular characterization. Int J Genom 701596:18

Gupta KJ, Stoimenova M, Kaiser WM (2005) In higher plants, only root mitochondria, but not leaf mitochondria reduce nitrite to NO, in vitro and in situ. J Exp Bot 56(420):2601–2609

Gurmani AR, Bano A, Khan SU, Din J, Zhang JL (2011) Alleviation of salt stress by seed treatment with abscisic acid (ABA), 6-benzylaminopurine (BA) and chlormequat chloride (CCC) optimizes ion and organic matter accumulation and increases yield of rice (*Oryza sativa* L.). Aust J Crop Sci 5 (10): 1278–1285

Hanson AD, Rathinasabapathi B, Rivoal J, Burnet M, Dillon MO, Gage DA (1994) Osmoprotective compounds in the Plumbaginaceae: a natural experiment in metabolic engineering of stress tolerance. Proc Natl Acad Sci USA 91(1):306–310

Harris BN, Sadras VO, Tester M (2010) A water-centred framework to assess the effects of salinity on the growth and yield of wheat and barley. Plant Soil 336:377–389

Hasanuzzaman M, Alhaithloul HAS, Parvin K, Bhuyan MHM, Tanveer M, Mohsin SM, Fujita M (2019) Polyamine action under metal/metalloid stress: regulation of biosynthesis, metabolism, and molecular interactions. Int J Mol Sci 20:3215

Hasegawa PM (2013) Sodium (Na^+) homeostasis and salt tolerance of plants. Environ Exp Bot 92:19–31

Hasegawa PM, Bressan RA, Zhu J-K, Bohnert HJ (2000) Plant cellular and molecular responses to high salinity. Ann Rev Plant Biol 51:463–499

Hoekstra FA, Golovina EA, Buitink J (2001) Mechanisms of plant desiccation tolerance. Trends Plant Sci 6:431–438

Hsiao TC, Xu LK (2000) Sensitivity of growth of roots versus leaves to water stress: biophysical analysis and relation to water transport. J Exp Bot 51:1595–1616

Hussain S, Saleem MF, Iqbal J, Ibrahim M, Ahmad M, Nadeem SM, Ali A, Atta S (2015) Abscic acid mediated biochemical changes in sunflower (*Helianthus annuus* L.) grown under drought and well-watered field conditions. J Anim Plant Sci 25(2):406–416

Ishitani M, Liu J, Halfter U, Kim CS, Shi W, Zhu JK (2000) SOS3 function in plant salt tolerance requires Nmyristoylation and calcium binding. Plant Cell 12(9):1667–1677

Javot H, Maurel C (2002) The role of aquaporins in root water uptake. Ann Bot 90:301–313

Javot H, Lauvergeat V, Santoni V, Martin-Laurent F, Guclu J, Vinh J, Heyes J, Franck KI, Schaffner AR, Bouchez D, Maurel C (2003) Role of a single aquaporin isoform in root water uptake. Plant Cell 15:509–522

Joyce SM, Cassells AC, Mohan JS (2003) Stress and aberrant phenotypes in vitro culture. Plant Cell Tiss Organ Cult 74:103–121

Kamara AY, Menkir A, Badu-Apraku B, Ibikunle O (2003) The influence of drought stress on growth, yield and yield components of selected maize genotypes. J Agric Sci 141:43–50

Kang GZ, Li GZ, Liu GQ, Xu W, Peng XQ, Wang CY (2013) Exogenous salicylic acid enhances wheat drought tolerance by influence on the expression of genes related to ascorbate-glutathione cycle. Biol Plant (Prague) 57:718–724

Kavar T, Maras M, Kidric M, Sustar-Vozlic J, Meglic V (2007) Identification of genes involved in the response of leaves of *Phaseolus vulgaris* to drought stress. Mol Breed 21:159–172
Kawakami J, Iwama K, Jitsuyama Y (2006) Soil water stress and the growth and yield of potato plants grown from microtubers and conventional seed tubers. Field Crops Res 95:89–96
Keskin BC, Sarikaya AT, Yuksel B, Memon AR (2010) Abscisic acid regulated gene expression in bread wheat (*Triticum aestivum* L.). Aust J Crop Sci 4(8):617–625
Kim YH, Khan AL, Waqas M, Shim JK, Kim DH, Lee KY, Lee IJ (2014) Silicon application to rice root zone influenced the phytohormonal and antioxidant responses under salinity stress. J Plant Growth Regul 33:137–149
Knott JM, Romer P, Sumper M (2007) Putative spermine synthases from *Thalassiosira pseudonana* and *Arabidopsis thaliana* synthesize thermospermine rather than spermine. FEBS Lett 581(16):3081–3086
Kopyra M, Gwozdz EA (2003) Nitric oxide stimulates seed germination and counteracts the inhibitory effect of heavy metals and salinity on root growth of *Lupinus luteus*. Plant Physiol Biochem 41(11–12):1011–1017
Kovacs Z, Simon-Sarkadi L, Szucs A, Kocsy G (2010) Differential effects of cold, osmotic stress and abscisic acid on polyamine accumulation in wheat. Amino Acids 38(2):623–631
Kovtun Y, Chiu WL, Tena G, Sheen J (2000) Functional analysis of oxidative stress-activated mitogen-activated protein kinase cascade in plants. Proc Natl Acad Sci USA 97:2940–2945
Kramer PJ, Boyer JS (1995) Water relations of plants and soils. Academic Press, San Diego
Krasensky J, Jonak C (2012) Drought, salt, and temperature stress-induced metabolic rearrangements and regulatory networks. J Exp Bot 1–16
Lafitte HR, Yongsheng G, Yan S, Li ZK (2007) Whole plant responses, key processes, and adaptation to drought stress: the case of rice. J Exp Bot 58:169–175
Lazaridou M, Koutroubas SD (2004) Drought effect on water use efficiency of berseem clover at various growth stages. New directions for a diverse planet. In: Proceedings of the 4th international crop science congress, Brisbane, Australia, 26 Sept–1 Oct 2004
Levitt J (1980) Responses of plants to environmental stresses: water, radiation, salt, and other stresses. Academic Press, New York, pp 365–488
Li YP, Ye W, Wang M, Yan XD (2009) Climate change and drought: a risk assessment of crop-yield impacts. Clim Res 39:31–46
Liu J, Ishitani M, Halfter U, Kim C-S, Zhu J-K (2000) The Arabidopsis thaliana SOS2 gene encodes a protein kinase that is required for salt tolerance. Proc Natl Acad Sci USA 97(7):3730–3734
Ludlow MM, Muchow RC (1990) A critical evaluation of traits for improving crop yields in water-limited environments. Adv Agron 43:107–153
Mahajan S, Tuteja N (2005) Cold, salinity and drought stresses: an overview. Arch Biochem Biophys 444(2):139–158
Mansfield TJ, Atkinson CJ (1990) Stomatal behaviour in water stressed plants. In: Alscher RG, Cumming JR (eds) Stress responses in plants: adaptation and acclimation mechanisms. Wiley-Liss, New York, pp 241–264
Martınez-Atienza J, Jiang X, Garciadeblas B (2007) Conservation of the salt overly sensitive pathway in rice. Plant Physiol 143(2):1001–1012
Martínez JP, Silva H, Ledent JF, Pinto M (2007) Effect of drought stress on the osmotic adjustment, cell wall elasticity and cell volume of six cultivars of common beans (*Phaseolus vulgaris* L.). Europ J Agron 26:30–38. https://doi.org/10.1016/j.eja.2006.08.003
Matysik J, Alia A, Bhalu B, Mohanty P (2002) Molecular mechanisms of quenching of reactive oxygen species by proline under stress in plants. Curr Sci 82(5):525–532
Maurel C, Javot H, Lauvergeat V, Gerbeau P, Tournaire C, Santoni V, Heyes J (2002) Molecular physiology of aquaporins in plants. Int Rev Cytol 215:105–148
Mazahery-Laghab H, Nouri F, Abianeh HZ (2003) Effects of the reduction of drought stress using supplementary irrigation for sunflower (*Helianthus annuus* L.) in dry farming conditions. Pajouhesh va Sazandegi Agron Hortic 59:81–86

McWilliams D (2003) Drought strategies for cotton, cooperative extension service circular, vol 582. College of Agriculture and Home Economics, New Mexico State University, USA
Mittler R, Blumwald E (2010) Genetic engineering for modern agriculture: challenges and perspectives. Annu Rev Plant Biol 61:443–462
Miura K, Okamoto H, Okuma E, Shiba H, Kamada H, Hasegawa PM (2013) SIZ1 deficiency causes reduced stomatal aperture and enhanced drought tolerance via controlling salicylic acid-induced accumulation of reactive oxygen species in Arabidopsis. Plant J 49:79–90
Moinuddin KHM, Khannu-Chopra R (2004) Osmotic adjustment in chickpea in relation to seed yield and yield parameters. Crop Sci 44:449–455
Monneveux P, Sanchez C, Tiessen A (2008) Future progress in drought tolerance in maize needs new secondary traits and cross combinations. J Agric Sci 146:287–300
Morgan PW (1990) Effects of abiotic stresses on plant hormone systems. In: Stress responses in plants: adaptation and acclimation mechanisms. Wiley-Liss, Inc., pp 113–146
Munne-Bosch S, Penuelas J (2003) Photo- and antioxidative protection, and a role for salicylic acid during drought and recovery in field grown *Phillyrea angustifolia* plants. Planta 217:758–766
Munns R (2005) Genes and salt tolerance: bringing them together. New Phytol 167:645–663
Munns R, Schachtman DP, Condon AG (1995) The significance of a two-phase growth response to salinity in wheat and barley. Aust J Plant Physiol 22:561–569
Nalousi AM, Ahmadiyan S, Hatamzadeh A, Ghasemnezhad M (2012) Protective role of exogenous nitric oxide against oxidative stress induced by salt stress in bell-pepper (*Capsicum annum* L.). Am-Eurasian J Agric Environ Sci 12(8):1085–1090
Nam NH, Chauhan YS, Johansen C (2001) Effect of timing of drought stress on growth and grain yield of extra-short-duration pigeonpea lines. J Agri Sci 136:179–189. https://doi.org/10.1017/S0021859601008607
Nayyar H, Kaur S, Singh S, Upadhyaya HD (2006) Differential sensitivity of Desi (small-seeded) and Kabuli (large-seeded) chickpea genotypes to water stress during seed filling: effects on accumulation of seed reserves and yield. J Sci Food Agric 86:2076–2082. https://doi.org/10.1002/jsfa.2574
Nguyen HT, Babu RC, Blum A (1997) Breeding for drought resistance in rice: physilogy and molecular genetics considerations. Crop Sci 37:1426–1434
Noble CL, Rogers ME (1992) Arguments for the used of physiological criteria for improving the salt tolerance in crops. Plant Physiol 146:99–107
Ogbonnaya CI, Sarr B, Brou C, Diouf O, Diop NN, Roy-Macauley H (2003) Selection of cowpea genotypes in hydroponics, pots, and field for drought tolerance. Crop Sci 43:1114–1120
Pandey P, Ramegowda V, Senthil-Kumar M (2015) Shared and unique responses of plants to multiple individual stresses and stress combinations: physiological and molecular mechanisms. Front Plant Sci 6:1–14
Parida AK, Das AB, Mohanty P (2004) Investigations on the antioxidative defence responses to NaCl stress in a mangrove, *Bruguiera parviflora*: differential regulations of isoforms of some antioxidative enzymes. Plant Growth Regul 42(3):213–226
Pessarakli M, Huber J (1991) Biomass production and protein synthesis by alfalfa under salt stress. J Plant Nutrit 14:283–293
Peuke AD, Rennenberg H (2004) Carbon, nitrogen, phosphorus, and sulphur concentration and partitioning in beech ecotypes (*Fagus sylvatica* L.): phosphorus most affected by drought. Trees 18:639–648
Peuke AD, Hartung W, Schraml C, Rennenberg H (2002) Identification of drought sensitive beech ecotypes by physiological parameters. New Phytol 154:373–388
Quintero FJ, Ohta M, Shi H, Zhu J-K, Pardo JM (2002) Reconstitution in yeast of the Arabidopsis SOS signaling pathway for Na^+ homeostasis. Proc Natl Acad USA 99:9061–9066
Rahman S, Miyake H, Takeoka Y (2002) Effects of exogenous glycinebetaine on growth and ultrastructure of salt-stressed rice seedlings (*Oryza sativa* L.). Plant Prod Sci 5(1): 33–44
Rahnama A, James RA, Poustini K, Munns R (2010) Stomatal conductance as a screen for osmotic stress tolerance in durum wheat growing in saline soil. Funct Plant Biol 37(3):255–263

Rambla JL, Vera-Sirera F, Blazquez MA, Carbonell J, Granell A (2010) Quantitation of biogenic tetraamines in *Arabidopsis thaliana*. Anal Biochem 397(2):208–211

Rasmussen S, Barah P, Suarez-Rodriguez MC, Bressendorff S, Friis P, Costantino P, Bones AM, Nielsen HB, Mundy J (2013) Transcriptome responses to combinations of stresses in *Arabidopsis*. Plant Physiol 161:1783–1794

Reddy AR, Chaitanya KV, Vivekanandan M (2004) Drought-induced responses of photosynthesis and antioxidant metabolism in higher plants. J Plant Physiol 161:1189–1202

Riccardi L, Polignano GB, de Giovanni C (2001) Genotypic response of faba bean to water stress. Euphytica 118:39–46

Rizhsky L, Liang H, Shuman J, Shulaev V, Davletova S, Mittler R (2004) When defense pathways collide. The response of Arabidopsis to a combination of drought and heat stress. Plant Physiol 134:1683–1696

Rozema J, Flowers T (2008) Ecology: crops for a salinized world. Science 322(5907):1478–1480

Samarah NH (2005) Effects of drought stress on growth and yield of barley. Agron Sustain Dev 25(1):145–149

Samarah NH, Mullen RE, Cianzio SR, Scott P (2006) Dehydrin-like proteins in soybean seeds in response to drought stress during seed filling. Crop Sci 46:2141–2150. https://doi.org/10.2135/cropsci2006.02.0066

Sanders D (2000) Plant biology: the salty tale of *Arabidopsis*. Curr Biol 10(13):R486–R488

Saruhan N, Saglam A, Kadioglu A (2012) Salicylic acid pretreatment induces drought tolerance and delays leaf rolling by inducing antioxidant systems in maize genotypes. Acta Physiol Plant 34:97–106

Saxena SC, Kaur H, Verma P (2013) Osmoprotectants: potential for crop improvement under adverse conditions. Plant acclimation to environmental stress. Springer, New York, NY, USA, pp 197–232

Scholander PF, Hammel HT, Bradstreet ED, Hemmingsen EA (1965) Sap presuure in vascuar plants. Proc Natl Acad Sci USA 52:119–125

Schuppler U, He PH, John PCL, Munns R (1998) Effects of water stress on cell division and cell-division-cycle-2-like cell-cycle kinase activity in wheat leaves. Plant Physiol 117:667–678

Serraj R, Sinclair TR (2002) Osmolyte accumulation: can it really help increase crop yield under drought conditions? Plant Cell Environ 25:333–341

Shi H, Quintero FJ, Pardo JM, Zhu JK (2002) The putative plasma membrane Na^+/H^+ antiporter SOS1 controls longdistance Na^+ transport in plants. Plant Cell 14(2):465–477

Shu S, Guo SR, Yuan LY (2012) A review: polyamines and photosynthesis. In: Najafpour MM (ed) Advances in photosynthesis—fundamental aspects. EdInTech, Rijeka, Croatia, pp 439–464

Sinaki JM, Heravan EM, Rad AHS, Noormohammadi G, Zarei G (2007) The effects of water deficit during growth stages of canola (*Brassica napus* L.). Am Eurasian J Agric Environ Sci 2:417–422

Singh B, Usha K (2003) Salicylic acid induced physiological and biochemical changes in wheat seedlings under water stress. Plant Growth Regul 39:137–141

Subbarao GV, Johansen C, Slinkard AE, Rao RCN, Saxena NP, Chauhan YS (1995) Strategies and scope for improving drought resistance in grain legumes. Crit Rev Plant Sci 14:469–523

Suzuki N, Rivero RM, Shulaev V, Blumwald E, Mittler R (2014) Abiotic and biotic stress combinations. New Phytol 203:32–43

Szaboles I (1994) Soils and salinization. In: Pessarakli M (ed) Handbook of plant and crop stress. Marcel Dekker, New York, pp 3–11

Tahir MA, Aziz T, Farooq M, Sarwar G (2012) Silicon induced changes in growth, ionic composition, water relations, chlorophyll contents and membrane permeability in two salt stressed wheat genotypes. Arch Agron Soil Sci 58(3):247–256

Taiz L, Zeiger E (2006) Plant Physiol, 4th edn. Sinauer Associates Inc., Publishers, Massachusetts

Thomas JC, Sepahi M, Arendall B, Bohnert HJ (1995) Enhancement of seed germination in high salinity by engineering mannitol expression in Arabidopsis thaliana. Plant Cell Environ 18:801–806

Tyerman SD, Niemietz CM, Brameley H (2002) Plant aquaporins: multifunctional water and solute channels with expanding roles. Plant Cell Environ 25:173–194

Tyree MT, Patino S, Bennink J, Alexander J (1995) Dynamic measurements of root hydraulic conductance using a high-pressure flowmeter in the laboratory and field. J Exp Bot 46:83–94
Valliyodan B, Nguyen HT (2006) Understanding regulatory networks and engineering for enhanced drought tolerance in plants. Curr Opin Plant Biol 9:189–195
Venuprasad R, Lafitte HR, Atlin GN (2007) Response to direct selection for grain yield under drought stress in rice. Crop Sci 47:285–293
Wahid A, Gelani S, Ashraf M, Foolad MR (2007) Heat tolerance in plants: an overview. Environ Exp Bot 61:199–223
Wang Y, Nii N (2000) Changes in chlorophyll, ribulose bisphosphate carboxylase-oxygenase, glycine betaine content, photosynthesis and transpiration in *Amaranthus tricolor* leaves during salt stress. J Hort Sci Biotechnol 75(6):623–627
Wang B, Luttge U, Ratajczak R (2001) Effects of salt treatment and osmotic stress on V-ATPase and V-PPase in leaves of the halophyte *Suaeda salsa*. J Exp Bot 52:2355–2365
Wanjogu SN, Muya EM, Gicheru PT, Waruru BK (2001) Soil degradation: management and rehabilitation in Kenya. In: Proceedings of the FAO/ISCW expert consultation on management of degraded soil in Southern and Eastern Africa (MADS-SEA) 2nd networking meeting, Pretoria, South Africa PR, pp 102–113
Xiong J, Fu G, Tao L, Zhu C (2010) Roles of nitric oxide in alleviating heavy metal toxicity in plants. Archiv Biochem Biophys 497:13–20
Yamada M, Morishita H, Urano K, Shiozaki N, Yamaguchi-Shinozaki K, Shinozaki K, Yoshiba Y (2005) Effects of free proline accumulation in petunias under drought stress. J Exp Bot 56:1975–1981
Yeo AR, Flowers SA, Rao G, Welfre K, Senanayake N, Flowers TJ (1999) Silicon reduces sodium uptake in rice (*Oryza sativa* L.) in saline conditions and this is accounted for by a reduction in the transpirational bypass flow. Plant Cell Environ 22:559–565
Yokota A, Kawasaki S, Iwano M, Nakamura C, Miyake C, Akashi K (2002) Citrulline and DRIP-1 protein (ArgE homologue) in drought tolerance of wild watermelon. Ann Bot 89:825–832
Zandalinas SI, Balfagon D, Arbona V, Gomez-Cadenas A, Inupakutika MA, Mittler R (2016) ABA is required for the accumulation of APX1 and MBF1c during a combination of water deficit and heat stress. J Exp Bot 67:5381–5390
Zhang Y, Wang L, Liu Y, Zhang Q, Wei Q, Zhang W (2006) Nitric oxide enhances salt tolerance in maize seedlings through increasing activities of proton-pump and Na^+/H^+ antiport in the tonoplast. Planta 224(3):545–555
Zhang F, Wang Y, Yang Y, Wu H, Wang D, Liu J (2007) Involvement of hydrogen peroxide and nitric oxide in salt resistance in the calluses from *Populus euphratica*. Plant Cell Environ 30(7):775–785
Zhao L, Zhang F, Guo J, Yang Y, Li B, Zhang L (2004) Nitric oxide functions as a signal in salt resistance in the calluses from two ecotypes of reed. Plant Physiol 134(2):849–857
Zhao MG, Chen L, Zhang LL, Zhang WH (2009) Nitric reductase-dependent nitric oxide production is involved in cold acclimation and freezing tolerance in *Arabidopsis*. Plant Physiol 151(2):755–767
Zhu JK (2002) Salt and drought stress signal transduction in plants. Ann Rev Plant Biol 53:247–273
Zhu JK (2003) Regulation of ion homeostasis under salt stress. Curr Opin Plant Biol 6:441–445

Special Adaptive Features of Plant Species in Response to Salinity

Parinita Agarwal, Mitali Dabi, Kasturi Kinhekar, Doddabhimappa R. Gangapur and Pradeep K. Agarwal

Abstract Plants are the primary producers of any organic material for food, via their pigment-light harvesting process, utilizing carbon dioxide and water. Salinity has negative influence on plant's growth, development, and productivity as it limits the plant from giving its full yield potential. The occurrence of salinity is one of the most substantial abiotic stresses in agriculture. Halophytes are plants that exhibit high salt tolerance, allowing them to survive and complete their life cycle under extremely saline conditions; the family Chenopodiaceae has the highest number of halophytic population. Studies have elucidated the role and adaptive features of various halophytic species required for their survival in high salinity conditions, including secretion of salt through the salt glands and bladders, succulent nature, regulation of cellular ion homeostasis and osmotic pressure, detoxification of reactive oxygen species, and changes in membrane composition. Also, several stress-responsive genes/transcription factors have been isolated and characterized in vitro as well as *in planta* via advanced technologies. In this chapter, we discuss the different adaptive mechanisms employed by halophytes to attain normal growth and metabolism under salt stress, with emphasis on two important halophytes of the Gujarat coast, a salt secreting grass *Aeluropus lagopoides* and a salt accumulating succulent *Salicornia brachiata*.

Abbreviations

MDA	Malondialdehyde
NaCl	Sodium chloride

P. Agarwal (✉) · M. Dabi · K. Kinhekar · D. R. Gangapur · P. K. Agarwal
Division of Biotechnology and Phycology, CSIR-Central Salt and Marine Chemicals Research Institute (CSIR-CSMCRI), Council of Scientific & Industrial Research (CSIR), Gijubhai Badheka Marg, Bhavnagar 364002, Gujarat, India
e-mail: parinitaa@csmcri.res.in

M. Dabi · D. R. Gangapur · P. K. Agarwal
Academy of Scientific and Innovative Research, CSIR-CSMCRI, Gijubhai Badheka Marg, Bhavnagar 364002, Gujarat, India

M. Hasanuzzaman and M. Tanveer (eds.), *Salt and Drought Stress Tolerance in Plants*, Signaling and Communication in Plants, https://doi.org/10.1007/978-3-030-40277-8_3

KCl	Potassium chloride
H_2O_2	Hydrogen peroxide
MAPK	Mitogen activated protein kinase
DREB	Dehydration responsive element binding proteins
EST	Expressed sequence tag
CAT	Catalase

1 Introduction

Plants have intricate mechanisms to integrate a wide range of tissue, developmental, and environmental signals to regulate complex patterns of gene expression established over a long period of evolution as sessile organisms (Wu et al. 2007). Plants with autotrophic lifestyle play a major role in keeping a stable environment on the globe. They regulate global climate and surroundings in many ways at different levels such as molecular, cellular, organ, individual, community, regional, ecosystem, and global ecosystem levels.

The problem of salinity for agriculture imposes a huge challenge worldwide. High salinity slows down overall plant development and hampers the agriculture productivity significantly (Munns and Tester 2008). About one billion hectares of land is affected by salinity worldwide with an annual increase of 10% (Fageria et al. 2012). It is estimated that from a 5.2 billion ha of dry land used for agriculture, about 3.6 billion ha of the land is deteriorated by erosion, soil degradation, infiltration, and accumulation of NaCl in the soil (Riadh et al. 2010). The United Nations Environment Programme (UNEP) reported that about 20% of agricultural land under irrigation has been degraded because of excessive salt accumulation (Nellemann and Corcoran 2009). The salinization leads to an increase in soil Na^+ concentrations above 40 mM (Munns 2005).

Salinity is an important ecological factor, greatly influencing the life forms on earth. Salinity indicates the presence of a high concentration (>4 dS/m) of soluble salts in the soil. The saline soils are classified as (i) sodic (or alkali), (ii) saline, and (iii) saline/sodic soils. The sodic soils are found in arid and semi-arid regions, and are characterized by the presence of a crust of insoluble sodium carbonate and bicarbonate. Sodic soils with low soluble salt content have an EC < 4 dS/m, an exchangeable sodium percentage (ESP) > 15%, and a pH > 8.5. The saline soils found in arid regions, estuaries, and coastal fringes, have a high concentration of water-soluble salts with an electrical conductivity (EC) > 4 dS/m, an (ESP) < 15% and a pH < 8.5. The saline-sodic soils are found in arid and semi-arid regions and possess an EC > 4 dS/m, ESP > 15%, and pH < 8.5 (Rengasamy 2002). The soil salinization process can be primary or secondary depending on the source of soil salinization. Primary or natural salinization occurs via natural processes such as weathering of minerals, soil from saline parent rocks, seawater incursions, and varied climatic changes. The secondary salinization results from anthropogenic activities including irrigation, intensive cropping, and deforestation or overgrazing (Rengasamy 2006).

Saline soils with high concentrations of sodium (Na^+) and chloride (Cl^-) ions result in toxicity to the plants and cause degraded soil structure. The plants' response to salinity comprises two phases: the early shoot ion-independent osmotic response effects water relations causing stomatal closure and the inhibition of leaf expansion (Munns and Termaat 1986). The second, the ion-dependent delayed response involves accumulation of ions to toxic concentrations leading to early senescence of leaves, reduced yield and cell death (Munns and Tester 2008).

Thus, the increased salt concentrations generate both hyperionic and hyperosmotic conditions, which impair the ability of the plant to take up the water and nutrients leading to physiological drought conditions.

2 Halophytes and Their Adaptation

Plants have evolved various adaptive features and mechanisms to survive in saline environments. Halophytes are plants specifically favorable for saline environments, they can survive and complete their life cycles with 200 mM NaCl (Flowers et al. 1986; Flowers and Colmer 2008) (Figs. 1, 2 and 3).

The growth response and/or physiological proficiency of halophytes towards salinity categorize them as either "facultative" or "obligate halophytes" (Mitsch and Gosselink 2000). Facultative halophytes do not require salinity for their survival but tolerate high salinity, show optimum growth during non-saline or low salinity environments and can survive in freshwater also. The obligate halophytes require salinity, varying from low to highly saline environment (Flowers et al. 2010). The halophytes are also classified based on the characteristics of naturally saline habitats (Waisel 1972; Le Houérou 1993) or the chemical composition of the shoots ("physiotypes"; Albert and Kinzel 1973; Albert et al. 2000) or the ability to secrete ions (recreto-halophytes, Breckle 1983). Breckle (1995) classified halophytes on the basis of their salt resistance mechanisms: pseudo-halophytes (salt-excluders), euhalophytes (salt-accumulators or halosucculants), and recretohalophytes or salt-includers (exorecretohalophytes and endorecretohalophytes). Approximately 0.25% of angiosperms, representing more than 600 taxa are halophytic (Flowers et al. 2010). The Chenopodiaceae family has the largest number of halophytes, other families including Poaceae, Fabeaceae, and Asteraceae also have a prominent number of halophytes (Aronson 1989). An online database eHALOPH (http://www.sussex.ac.uk/affiliates/halophytes) is enlisting halophytic plant species worldwide. These halophytes being rich in proteins, minerals, amino acids, antioxidants etc. are used by local people as raw vegetable and also cooked or pickled. Their potential to tolerate salinity (a multigene trait) is an important character which is difficult to develop via both conventional breeding as well as genetic engineering, thus halophytes can be domesticated and transformed to useful crops for the saline regions (Tanveer and Shah 2017). In India, Gujarat state has large coastline of 1600 km long, which

(a)

(b)

Fig. 1 Halophytes growing in their natural habitat **a** *Aeluropus lagopoides* near Hathab village in Bhavnagar district. **b** *Salicornia brachiata* near Sartanpur village in Bhavnagar district

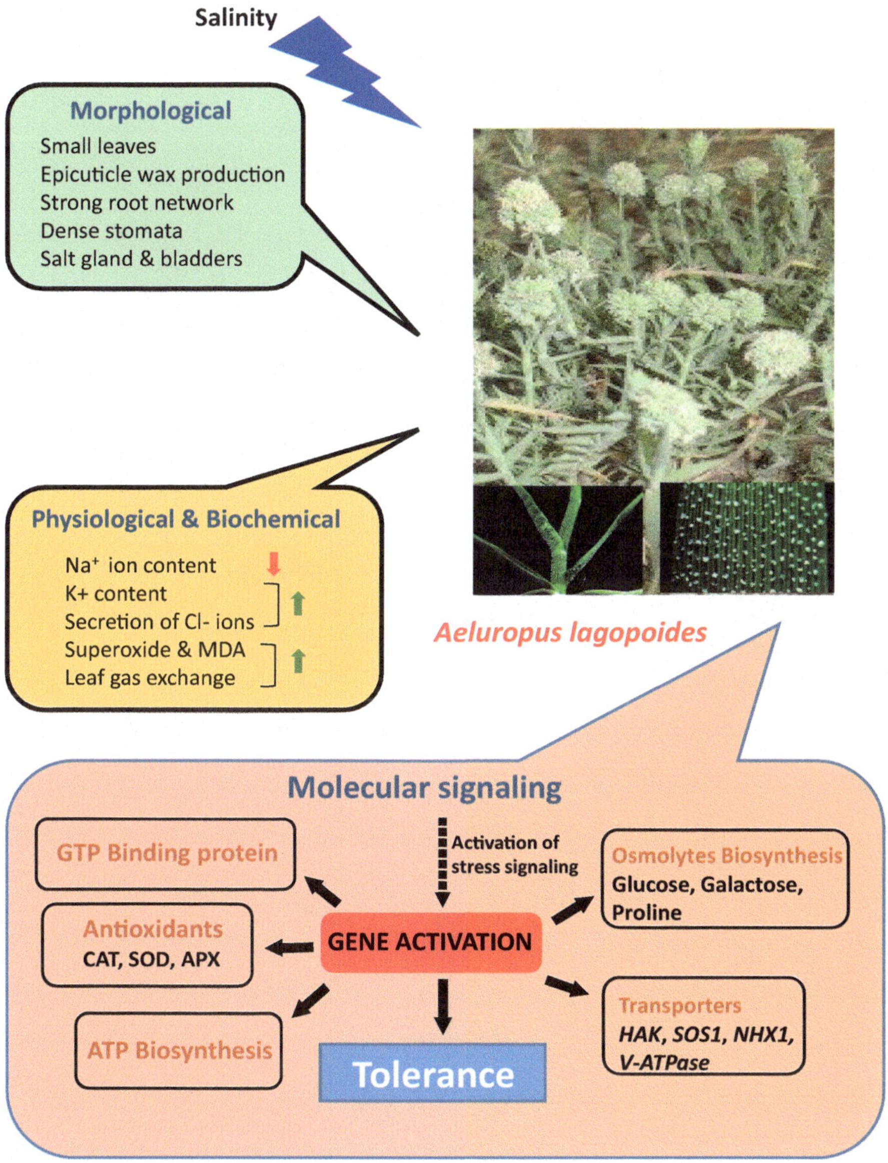

Fig. 2 A generalized schematic representation explaining salinity stress tolerance mechanisms in non-succulent halophyte, *Aeluropus lagopoides*. The box in the picture shows the magnified image of Aeluropus stem excreting salt crystals

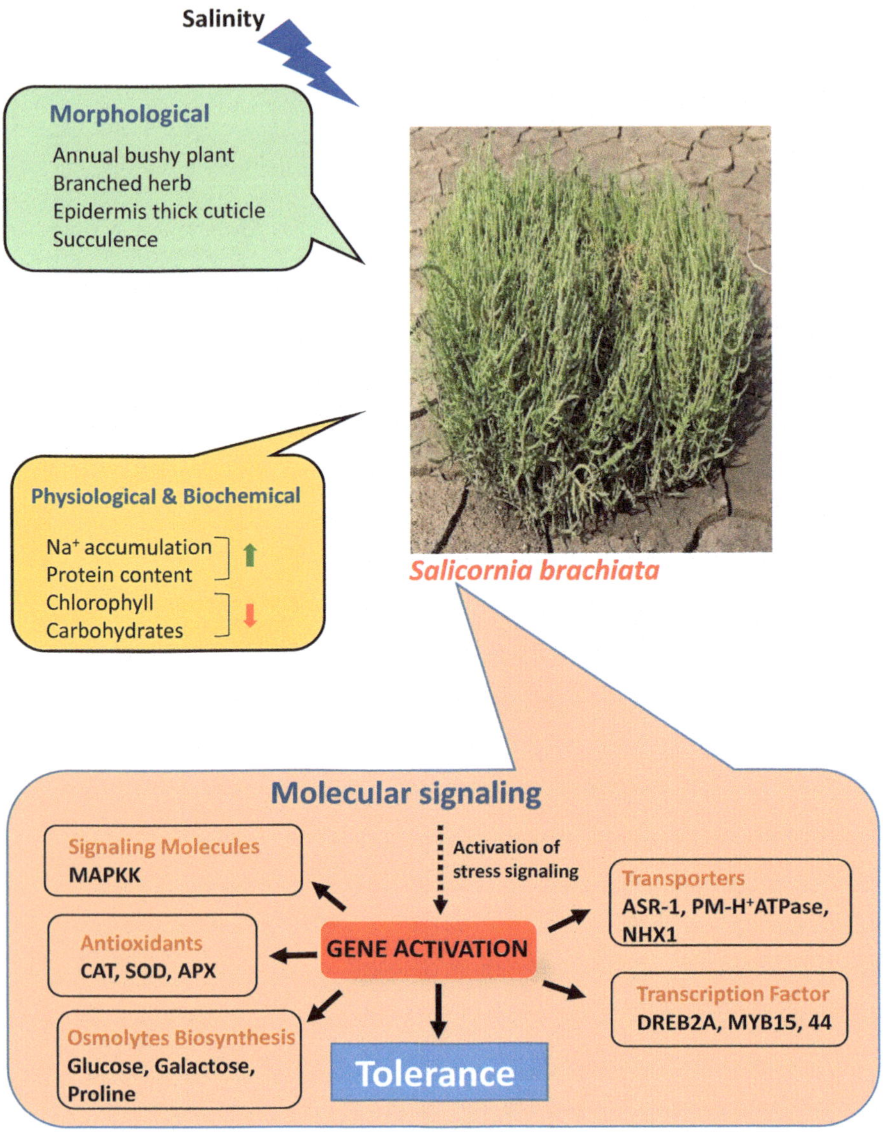

Fig. 3 A generalized schematic representation explaining salinity stress tolerance mechanisms in succulent halophyte, *Salicornia brachiata*

eventually makes it highly saline and adversely impact on agriculture productivity. In this area, due to presence of several salt farms, less annual rainfall, and heavy demand for groundwater by ever-spreading industries in the coastal area, the soil is becoming increasingly saline and salt ingress has become a regular feature. Different halophyilic plants species are able to adapt and survive under these conditions, we have in this chapter discussed in detail on the morpho-physio-biochem-molecular adaptations of a salt secreting grass *Aeluropus lagopoides* and a salt accumulating succulent *Salicornia brachiata*.

2.1 Physiological and Biochemical Responses

Plants develop various physiological and biochemical mechanisms in order to survive in soils with high salt concentration.

2.1.1 Regulation of Transporter Genes in Halophytes

Ionic imbalance resulting from salinity stress impairs plant metabolism in both glycophytes and halophytes; however, halophytes have varied and better mechanisms for ion homeostasis at morphological, anatomical, physiological, and biochemical levels during salinity (Aslam et al. 2011; Shabala 2013). *Thellungiella salsuginea*, a halophyte, considered to be a close relative of *Arabidopsis* (glycophyte) showed high tolerance under salinity stress compared to the latter one as revealed by microarray studies (Bartels and Dinakar 2013; Taji et al. 2004). Taji et al. (2010) studied abiotic-stress-related genes, transcription factor genes, and protein phosphatase 2C genes in *Thellungiella* and *Arabidopsis*. Both plants had similar 5′-UTR but, the motifs were somewhat different. Furthermore, a short splicing variant of *T. halophila* salt overly sensitive 1 (ThSOS1), designated ThSOS1S was identified. ThSOS1S contains the transmembrane domain similar to the ThSOS1; however, the C-terminal hydrophilic region is absent. Moreover, several other genes that participate in Na^+ excretion, compartmentation, and diffusion (SOS1, SOS2, NHX1, and HKT1) showed higher expression in *Thellungiella* as compared to *Arabidopsis*. The early response during salinity stress was studied in *Cakila maritima* (halophyte) and *Arabidopsis thaliana* (glycophyte), the two closely related species of Brassicaceae. An improved physiological and antioxidant status was induced in *C. maritima* as compared to *A. thaliana* under salinity stress (Ellouzi et al. 2011). Since sodium chloride is the most abundant soluble salt in nature, plants have evolved mechanisms to regulate its accumulation and differentially uptake other ions present at low concentrations, such as K^+. In most plants, Na^+ and Cl^- are effectively removed from roots (Munns 2005); however, leaf blade shows higher Na^+ toxicity for most plants (Munns 2002). The Na^+ ion that enters the cytoplasm is then transported to the vacuole via Na^+/H^+ antiporter. Cellular ion homeostasis involving net intracellular sodium and chloride uptake and subsequent vacuole compartmentalization without toxic ion accumulation in the cytosol is a key factor regulating salinity tolerance (Munns 2005). These vacuolar antiporters were studied in various halophytes. Overexpression of *Aeluropus littoralis*, *AlNHX* gene in tobacco permits the compartmentalization of more Na^+ in roots and high K^+/Na^+ ratio in leaves (Zhang et al. 2008a). Rice transgenic overexpressing *AgNHX1* gene from halophyte *Atriplex gmelini* exhibits higher activity (eightfold) of the vacuolar-type Na^+/H^+ antiporter as compared to wild type plants under 300 mM NaCl (Ohta et al. 2002). Similarly, expression of *Salsola soda*, *SsNHX1* improved the salinity tolerance in transgenic alfalfa for 50 days, which is the highest level of salt tolerance reported in transgenic. The mechanism demonstrates the transfer of Na^+ ions from the cytoplasm of leaves to the vacuoles by the exogenous Na^+/H^+ antiporter (Li et al. 2011). Hu et al. (2012), characterized vacuolar H^+-pyrophosphatase and a B subunit of H^+-ATPase from succulent halophyte *Halostachys caspica* and studied role against salinity. The report showed the increased Na^+ content in leaves of Arabidopsis transgenic indicating increased accumulation in vacuoles. Overexpression of *Thellungiella halophile ThNHX1*, enhances salinity tolerance in transgenic *Arabidopsis*. Moreover, silencing of the gene in parent plant results in salt-sensitive transgenic plants indicating that the gene is playing as a tonoplast Na^+/H^+ antiporter (Wu et al. 2007). A comparative transcript expression analysis revealed a higher expression of antiporter SOS1 gene in *Thellungiella* compared to *Arabidopsis* (Oh et al. 2010). Similarly, a betaine aldehyde dehydrogenase gene from the two different species of halophytic plant *Atriplex* (AhBADH and AcBADH) was induced by salinity stress along with the increase in glycine betaine level (Yin et al. 2002; Jia et al. 2002).

2.1.2 Regulation by Antioxidative Enzymes

Salinity results in reduced photosynthetic activity in plants, thereby increasing the reactive oxygen species, ultimately increasing the vigorous antioxidative enzymes activity (Tanveer et al. 2018). Antioxidants can be further classified into three categories: water-soluble (ascorbates); lipid-soluble (tocopherols and carotenes); and antioxidative enzymes (superoxide dismutase SOD, ascorbate peroxidase APX, catalase CAT, peroxidases POD). The synchronized activity of all achieves a balance between the rate of formation and diminishing of ROS, and thus maintaining the internal ROS at the optimum level for cell signaling (Gill and Tuteja 2010; Tanveer and Shabala 2018). Various halophyte showed higher expression of genes involved in antioxidative pathways to combat salinity stress. Among them, *Avicennia marina*, AmMDHAR involved in ascorbate regeneration and ROS scavenging in transgenic tobacco (Kavitha et al. 2010). Ss.sAPX and SsCHLAPX, isolated from halophytic plant *Suaeda salsa*, were involved in salt stress tolerance in transgenic *Arabidopsis* (Li et al. 2012; Pang et al. 2011). Similarly, *Tamarix and rossowii*, *TaMnSOD* gene was also participated in scavenging of ROS in transgenic Populus plant (Wang et al. 2010). Tuteja et al. (2014) have described a couple of helicase proteins (DESD-box helicase and OsSUV3 dual helicase) functioning in plant salinity tolerance by improving/maintaining photosynthesis and antioxidant machinery. Glutathione is also an important antioxidant playing a crucial role in stress mitigation. It can react with toxic radicals generated during stress conditions thereby acting as a free radical scavenger. It can also play a role in ascorbate-glutathione cycle during regeneration of ascorbate (Foyer et al. 1997).

2.2 Transcriptional Regulation in Halophytes

Plants are severely affected by high salt concentrations in soils. Many reports are there to unravel the key components of the plant salt-tolerance network. Recent studies have shown that stress sensing and signaling components can play important roles in regulating the plant salinity stress response. Many stress-associated transcription factors (TFs) are of great importance as they are involved in different complex signaling pathways (Hennig 2012). More than 30 families of transcription factors have been predicted for *Arabidopsis*. Members of DREB or CBF, MYB, bZIP, and zinc-finger families have been well characterized with roles in the regulation of plant defence and stress responses (Agarwal et al. 2017; Thomashow 1999). Salt-tolerant genes from different halophytes such as *Suaeda maritima, Atriplex spp, T. halophile, S. brachiata, Cakile maritima,* and *Aeluropus* spp have studied for regulatory mechanism (Wang et al. 2004; Sahu and Shaw 2009). According to studies on promoter from halophytes, it has been proved that this class of plant can serve as good candidate for revising salinity tolerance mechanism in crops. Interestingly, there exists no sequence similarity among promoters isolated from halophytes with their close relatives as glycophytes. The *SOS1* promoters from *T. parvula* (*TpSOS1*) and *T. salsuginea* (*TsSOS1*) halophytes showed conserved motifs; however, no homology was observed between *AtSOS1* from Arabidopsis and *TpSOS1* or *TsSOS1* promoters (Oh et al. 2010). Overexpression of *Avicennia marina* AmMYB1 (a salt responsive MYB protein), in transgenic tobacco confers better salt tolerance and further regulates the expression of other genes in salt stress (Ganesan et al. 2012). The promoter SIBADH isolated from *S. liaotungensis* showed higher expression in salt stress (Zhang et al. 2008b). Likewise, TsVP1 gene promoter isolated from *T. halophile* showed maximum GUS gene expression in *Arabidopsis* under salinity stress (Sun et al. 2010). Similar results were found in SlPEAMT gene extracted from halophyte *S. liaotungensis* under NaCl stress (Li et al. 2016). Constitutive expression of AhAL1 from halophyte *Atriplex hortensis* improves tolerance to salinity stress in transgenic *Arabidopsis* (Tao et al. 2018). The overexpression of AlSAP gene from halophyte *Aeluropus littoralis* conferred tolerance to ionic and osmotic stress in transgenic tobacco (Saad et al. 2011). A DREB2A from S. brachiata showed better seed germination and growth characteristics in both hyperionic and hyperosmotic stresses in the tobacco overexpression lines. The transgenic lines revealed higher expression of downstream heat shock genes (Hsp18, Hsp26 and Hsp70), TFs (AP2 domain containing TF, HSF2 and ZFP), signalling components (PLC3 and Ca2+/calmodulin) and dehydrins (ERD10B, ERD10D and LEA5) under different abiotic stress treatments (Gupta et al. 2014). Overexpression of SbSI-1 showed higher expression of antioxidative genes *NtSOD*, *NtCAT,* and *NtAPX*, under salinity condition (Kumari et al. 2017). The *SbUSP* gene from halophyte Salicornia brachiata showed higher expression of *NtSOD*, *NtCAT,* and *NtAPX* under salinity stress (Udawat et al. 2016). SbSDR1 gene is involved in the transcriptional regulation of host stress-responsive genes and transcription factors, such as *NtAP2*, *NtPLC*, *NtLEA*, *NtP5CS,* and *NtERF8* under salt stress. Zhu et al. (2018), showed that *SlbZIP1*, via ABA-mediated pathway, plays an essential role in salt and drought stress tolerance. SbNHX1 and SbGST genes in transgenic tobacco leads to enhanced seed germination and growth under salt stress (Jha et al. 2011a, b). SbNHX1 transgenic plants showed higher photosynthetic pigments and electrolyte leakage, proline, and malondialdehyde (MDA) content under 200 mM NaCl stress (Pandey et al. 2016). Thus, engineering salt-tolerant crop might be a better option to study and establish salinity tolerance in agriculture.

2.3 Proteomic and Metabolic Under Salt Stress

Recently, considerable progress in elucidation of salt tolerance phenomenon, especially salt ion signaling and transport, has been attained owing to the exploitation of modern genetic approaches (Yamaguchi-Shinozaki and Shinozaki 2006) and due to highly gained methods of functional genomics (i.e., transcriptomics, proteomics, metabolomics). In C_3 plants such as *S. aegyptiaca*, an increased net photosynthetic rate under salt stress can be accomplished by other adaptive mechanisms such as an enhanced accumulation of D2 protein and glycine betaine which can significantly contribute to stabilization of the photosynthetic machinery under salt stress (Askari et al. 2006). Krishnamurthy et al. (2017) showed that *S. maritima* can withstand up to 200 mM NaCl by upregulating proteins that are mainly involved in protein transport, vesicle trafficking, iron binding, protein folding and assembly, chromosome segregation and cell maintenance. Wang et al. (2014a, b) demonstrated upregulation of proteins related to photosynthesis, carbohydrate and energy metabolism, and stress, in Chenopodiaceae halophyte *Halogeton glomeratus* under 100 mM NaCl condition. Thus, proteomic and metabolomics technologies are effective approaches to conduct large-scale analyses under stress conditions.

3 *Aeluropus lagopoides*

3.1 Origin and Taxonomy

Aeluropus lagopoides belongs to Poaceae family, which is a perennial grass with C_4 type photosynthesis mainly distributed in the regions with high salinity and semi-desert climate (Bor 1970; Breckle 1983) and is a wild relative of wheat (Razavi 2005). *Aeluropus lagopoides* plants are naturally occurring in Mediterranean region, Red Sea coasts through southwest Asia to Central Asia, India, and Sri Lanka. Several reports on ecological and physiological distribution on this plant have been found (Waghmode and Joshi 1982; Waghmode and Hegde 1984; Sher et al. 1994; Bhaskaran and Selvaraj 1997; Abarsaji 2000). *Aeluropus lagopoides*, are commonly found in Gujarat, Maharashtra, Goa, Karnataka, Tamilnadu, and other states of Indian coastlines.

3.2 Morphological and Anatomical Adaptations

Aeluropus lagopoides is a perennial species that produce flowers throughout the year (Watson and Dallwaitz 1992). *Aeluropus* is having the most appropriate and adaptive characteristics to withstand the saline environment which includes vital seed germination, slow vegetative propagation, vegetative propagation by stolon, epicuticle wax production, salt secreting structures, small waxy leaves, and strong root network which aids in survival of the plant in stressful condition (Mohsenzadeh et al. 2006). Halophytes grown in saline conditions usually accumulate inorganic ions in the vacuoles to decrease the cell water potential because the energy consumption from absorbing inorganic ions is far less than that from synthesizing organic compounds (Munns 2002). Naz et al. (2013) studied the anatomical adaptations in root, stem, and leaves of the salt stress plants. The plants growing at high salinity showed specific anatomical adaptations, which enable the plants to prevent the water loss from leaf surface. Stem cross section studied reveals that the plants growing in high salinity develops an endodermis-like layer, which have a protective function. Leaf anatomical alterations involve thickening of the epidermis layer and cuticle in response to the high salt concentrations. Arranged bulliform cells, highly dense micro-hairs, and trichomes were observed in leaf cross sections under salinity. Aeluropus successfully adapts the property of xeric plants in the means to grow under stress conditions via balancing the proper dense stomata and reducing the area (Martinez et al. 2007). Another aspect of tolerance was an increase in leaf fluffiness which is non-specific for the leaf, which can be a prompt feature of Aeluropus. This strategy also helps in avoiding water loss from the leaf (Abernethy et al. 1998; Wahid 2003). *Aeluropus* being a C_4 grass have a photosynthetic pathway linked to specialized Kranz leaf anatomy, consisting of two photosynthetic cells type, a bundle sheath cell, and mesophyll cells, which helps it to with stand in stress conditions such as drought, high temperature, high irradiance, and particularly low CO_2 concentrations (Wang et al. 2007).

3.3 Economic Importance

Aeluropus lagopoides has its excellent potential value as a feeder plant for animals and serve as a food in some part of India, due to its low sodium content and proved to be good candidate for saline agricultural (Torbatinejad et al. 2000) and in prevention of soil erosion (Tewari 1970). Gulzar et al. (2003) also suggested the role of *Aeluropus* as a lawn grass especially for the area having water scarcity to balance the sandy particles. It has proven that *Aeluropus* is an ideal plant system from the grass family, to study, understand, and develop salinity tolerance in cereals (Flowers and Colmer 2008). On the other side, somatic hybridization technique was employed for enhancing salt tolerance in wheat crops (Yue et al. 2001). Further, Joshi and Bhoite

(1988), suggested the part of *Aeluropus* plantation in salt-affected area used as a fodder purpose due to its high protein content and salinity tolerance.

3.4 Physiological and Biochemical Strategies

Aeluropus lagopoides plants have attracted numerous researchers to investigate its salt-tolerant ability in last few years. *Aeluropus* seed germination was observed 30% under 500 mM NaCl and can survive up to 1 M NaCl as reported by (Khan and Gul 1999; Gulzar et al. 2003). *Aeluropus* accumulates a large amount of Na^+ and Cl^- ions in shoots and roots and lower amount of K^+ and Ca^{++} ions. Naaz et al. (2009) showed that *Aeluropus*, under high saline level, increases the K^+ content and secretion of Cl^- ions in constant concentration in shoots. Binzel et al. (1988), relate it with the capacity of plants to accumulate the Cl^- ion in root/shoot or ions compartmented in vacuoles in order to maintain osmotic potential in the cells. The strategies employed by most of the halophytes to maintain the internal ion level is salt elimination. The feature is also found in those species having salt gland and bladders (Ramadan 2001). Likewise, the plants balanced their growth pattern and ions accumulation by storing excess of ions in vacuoles (Li et al. 2008) or secreting them through their specialized salt secretory structures (Barhoumi et al. 2007). But, *Aeluropus* enhanced the secretion of ions via roots and transport them to aerial parts (Flowers et al. 1986). Therefore, the NaCl ions were well efficiently eliminated from the plant leaves irrespective of its concentration (Gulzar et al. 2002). These species maintain to avoid the transpiration loss through these specialized stomata under stress conditions. Similarly, Naz et al. (2010) reported that the controlled transpiration rate and water use efficiency level, regulated by stomatal aperture further helps the *Aeluropus* to survive in adverse salinity conditions. Gorham et al. (1980) stated that osmoregulatory role of sugars in many halophyte species in salt stress is different. Likewise, concentration of glucose and galactose moiety in seedlings of *Aeluropus* were greater (Joshi et al. 2005). Ion content analysis of *AlHKT2*; 1 harboring yeast cells, when grown in high NaCl medium supplemented with KCl, showed that salt tolerance is correlated with accumulation of K^+ ions during salt stress conditions. At higher salinity levels, the inorganic and organic osmolytes, superoxide radicals, and MDA content increase to prevent the toxicity condition in *A. lagopoides* (Sanadhya et al. 2015). Superoxide dismutase has the main role in salinity stress tolerance in *Aeluropus*, and it could be concluded that, in *A. lagopoides*, the conversion of H_2O_2 by SOD is a very important strategy to detoxify. Also, *Aeluropus* thought to maintain internal osmotic adjustment through biosynthesis of amino acids. These amino acids were formed during glycolysis in the shoot cells which results in the formation of ATP molecule and intermediate sugars (Sobhanian et al. 2010). Naz et al. (2010) studied the stomatal distribution and functional characteristics under salinity in *Aeluropus* and concluded that plants tolerate the high salt concentrations by controlled transpiration rate and high water use efficiency. Reduction in leaf gas exchange was also evident in *A. lagopoides* plants grown in high salt concentration (Ahmed et al. 2013).

3.5 Molecular Mechanism

Aeluropus lagopoides showed maximum identity with mainly C_4 plants like pearl millet (*Pennisetum glaucum,* 90%), maize (*Zea mays,* 90%), and foxtail millet (*Setaria italic,* 91%). *Aeluropus* seedlings regulate the shoot sodium ion level by maintaining the secretion of salts up to 500 mM NaCl. According to proteomics data 2.1% of proteins were upregulated and these were mostly linked with ATP molecule synthesis, macromolecule biosynthesis, C_4 photosynthesis, and detoxification under salinity stress in *Aeluropus*. Ahmed et al. (2013) elucidated the expression of two important transport proteins (V/NHX and PM/NHX) which was compartmentalizing the Na^+ at salinity ranges higher than 200 mM. When the upregulation of glycolysis related proteins were examined in shoots of *A. lagopoides* under salt stress, alteration in the arrangements of photosynthesis activity was observed which demands osmotic adjustment (Dooki et al. 2006). The salt tolerance mechanism in in *Aeluropus*, can be attributed towards upregulation of ATP biosynthesis catalysing enzymes and also the chloroplast heat shock protein 70. Also, the same report suggested the important role of glyoxalase I and calmodulin as a stimulator of glyoxalase as a detoxifying enzyme in salt-stressed shoots. Further, the *A. lagopoides* transcripts showed differential regulation of the important transporter genes AlHKT2; 1, HAK, SOS1, NHX1, and V-ATPase under salinity stress (Sanadhya et al. 2015). The Rab7, a small GTP-binding protein reported at the vacuolar membrane regulates the vesicle fusion with the vacuole and facilitates recycling of the molecules was isolated from *A. lagopoides* (Rajan et al. 2015). The proteome and metabolome study demonstrated that major strategies used by *A. lagopoides* during salt stress are upregulation of ATP and amino acid biosynthesis pathway enzymes, C_4 photosynthesis-related proteins, glutathione S-transferase enzyme, and defence-related protein.

4 Salicornia brachaita

4.1 Origin and Taxonomy

Salicornia brachiata is a leafless annual small bushy plant with a height of 50 cm is a highly salt-tolerant plant. *Salicornia brachiata* is an obligatory halophyte with a distinctive genetic setup and can grow in wide range of salt concentration. *Salicornia brachiata* is recognized as pickle weed, crows foot greens, and sea asparagus and the name itself was originated from Latin word meaning “salt” *S. brachiata* being a hygrohalophyte is distributed among temperate and subtropical regions of the northern hemisphere. They are largely spread only in southern Africa of the southern hemisphere. *Salicornia* are known to be periodically found in wet inlands salt pans and in coastal salt marshes. In southern Africa, *Salicornia* have traditionally been known as *S. meyeriana* (Slenza et al. 2013). *Salicornia brachiata* is classified under class dicotyledons, in which flowering plants or angiosperms were formerly divided and have two embryonic leaves or cotyledons, sub class Apetalae. *Salicornia brachiata* belongs to family Chenopodiaceae (Amaranthaceace) Amaranthaceace and Genus *Salicorni*a and species brachiata. *Salicornia brachiata*, a succulent halophyte, are commonly found in surrounding area of Bhavnagar (Hathab, Sartanpur, Mahuva, Victor Port) and Amreli (Jaffrabad) districts of Gujarat and sea coast of India.

4.2 Morphological and Anatomical Adaptations

Salicornia brachiata is a succulent plant bearing tiny flowers and green jointed stem. *Salicornia brachiata* has branched herbs in which branches are slender opposite to phylloclade and fleshy; they usually have spikes of about 1–2 inches. S*alicornia brachiata* accumulates 40–50% of dry weight (Joshi 2011 monograph on Indian halophytes). Stem includes sub-parts which are enclosed by single-layered structure called epidermis. Epidermis is covered with thick layer of cuticle. Cortex which is present in the middle portion is differentiated into palisade parenchyma and spongy parenchyma; they are composed of two rows of cell and the palisade cells cylindrical shaped. They are arranged compactly. Their function is to transport water to peripheral layers and it serves for accumulation of air or water. Under endodermis is another layer of pericyclic which consists of six vascular bundles which are present in circular pattern. Above parenchyma, medulla is present in center; old and middle-aged parenchymatous tissues are loosely arranged. The vascular bundle are collateral and present in the inter space of endodermis. Xylem and phloem are arranged side by side. *Salicornia brachiata's* morphological and anatomical studies will help in further research to increase and develop coastal saline soil plant population (Rao and Murty 2013).

4.3 Economic Importance

Salt-tolerant plants such as *S. brachiata* act as a crop plant for future agricultural practices. *Salicornia brachiata* are widely studied for edible, nutraceutical, and pharmaceuticals purpose. The plant accumulates 30–40% NaCl of its dry weight. Due to the high content of salt, its biomass serves as a nutrient salt of plant origin (US patent no. 6,929,809). *Salicornia brachaita* are used as herbal plants; they are milled and dried and sold on large scale. *Salicornia* species capsules are sold with a promise of cure of inflammatory anti-hyperepidemic and anti-diabetic and anti-carcinogenic effects. Halophytes' extracts or compounds applications are occupied with different industries such as agriculture, animal production, and food industries. *Salicornia* seeds have a high content of poly saturated compounds, which share maximum properties with sunflower oil fatty acids (Buhmann and Papenbrock 2013). *Salicornia* have a wide range of benefits as it enhances the economical income of the stakeholders involved during cultivation process such as drying and packaging. *Salicornia* and other species halophytes can help in coastal development and protection. *Salicornia* cultivation can help in recovery of ecological losses; they can also help in the making of cheap biomass. *Salicornia brachiata* are also used in renewable energy, climate improvement, and CO_2 sequestration (Stanley 2008).

4.4 Physiological and Biochemical Strategies

Salicornia brachiata an extreme halophyte serves as a great tool for studying biochemical analysis. Plant growth and development have negative effects due to extreme salinity. High salinity disturbs intracellular ion homeostasis, which results in membrane differentiation, changes, which inhibits growth and induces cell death. Electrostatic forces that provide stability to protein structure disruption of electrostatic forces above 0.3 M NaCl reduces all enzymes activity. Other mechanisms such as photosynthesis under a combination of super oxide and hydrogen peroxide mediated oxidation. Plant adjusts to environmental stresses via an abundance of responses, including the activation of molecular mechanisms that regulates stress (Yadav et al. 2012). *Salicornia brachiata* was screened for different biochemical and physiological analysis to study the growth characteristics by soaking in wastewater and results concluded an increased fresh weight after multiple cycles, it was observed that fresh weight and dry weight of the plant decreased with an increase in salt concentration (Santhanakrishnan et al. 2013). The study reported that vacuole absorbs sodium and chloride and suggested the osmotic balance. Chlorophyll level is reported to be decreased due to reduction in pigment biosynthesis or enzymatic chlorophyll degradation analysis (Santhanakrishnan et al. 2013). During sugar content analysis *S. brachiata* observed a decrease in carbohydrate content. The total decrease of carbohydrates content was considered due to salinity condition and proline and glycine betaine synthesis (Santhanakrishnan et al. 2013). Protein content was high and inspected that over a period of high salinity it induces protein accumulation. Salinity helps in the conversion of inorganic nitrogen to protein and promoting protein synthesis. Another parameter was proline where proline being a primary defence response maintains osmotic pressure in cells (Santhanakrishnan et al. 2013).

4.5 Molecular Mechanisms

Tolerance caused by abiotic stress (such as temperature, extreme drought, and salinity) develop the adaptation to plants and elaborate their mechanism to make respective changes to external signals and express a response towards morphological, physiological biochemical alteration. The ESTs were identified in response to salinity in the extreme halophyte plant *S. brachiata* with 90 ESTSs showing unknown function (Jha et al. 2009). Some of the identified and unknown genes are being functionally validated in planta (Table 1). ESTs are important to understand and study the known genes which are responsible as well as unknown genes. PM- H^+ ATPase, Abscisic acid, and stress level ripening protein are salt-related stress genes which show a certain level of expression even in control condition (Jha et al. 2009). Some important transporter genes as well as transcription factors (TFs) are characterized. The *SbSOS1* gene demonstrates that in addition to the Na^+ efflux outside the plasma membrane, its transporter also helps to maintain different concentrations of Na^+ in various organs (Yadav et al. 2012). Furthermore, *S. brachiata* SOS1 promoter showed upregulation of GUS expression in stems and leaves by salt stress but not by ABA and cold stresses (Goyal et al. 2013). Mitogen–activated protein kinase cascade (MAPK) is studied for signaling toward developmental, hormonal, biotic, and abiotic stresses (Agarwal et al. 2010). Dehydration responsive element binding (DREB) is an ABA-independent TF which plays a vital role in plant stress signal transduction pathway. *SbDREB2A* was isolated from *S. brachiata* and it showed binding to DRE elements and improved salinity and dehydration tolerance to tobacco transgenics (Gupta et al. 2010, 2014). The important TF SbMYB15, R2R3-type MYB showed induction with different stresses, and conferred stress tolerance in transgenic tobacco by regulating antioxidative enzymes and also the expression of stress-responsive genes (Shukla et al. 2015b). Another TF SbMYB44 showed binding to dehydration-responsive cis elements (RD22 and MBS-1), and its overexpression enhanced the growth of yeast cells under both ionic and osmotic stresses (Shukla et al. 2015a). Some of these genes/TFs after validation in tobacco are being transformed in important crop plants like cumin, peanut, isabgol, tomato, etc.

5 Conclusion and Perspectives

With the increasing population, urbanization, soil degradation, and salinization, the need to exploit marginal lands for agriculture production is required. With this aim, it is of prime importance to develop crops with salinity tolerance. The traditional domesticated crops show limited genetic variability for salinity trait, therefore deciphering the salinity tolerance mechanisms from halophytes is of significance. Efforts have been made to identify certain halophytes for commercial utilization, like in Australia, 26 halophyte species have been identified as valuable to agriculture (Barrett-Lennard 2002). Quinoa, that can be used for bread, noodles, etc. is produced and

Table 1 The different genetic resource from *Aeluropus lagopoides* and *Salicornia brachiata* showing their transcript regulation and the potential of transgenics in response to stress

Sr. No.	Gene/ORF	Transcript response		Transgenic plant	Response of transgenics	References
		Salt	Drought			
	Aeluropus lagopoides					
1	*AlRab7*/621 bp	Yes	Yes	–	–	Rajan et al. (2015)
2	*AlHKT2*;1/1581 bp	Yes	Yes	–	–	Sanadhya et al. (2015)
3	*AlNAC4*/936 bp	Yes	Yes	Tobacco	Oxidative stress tolerance	Khedia et al. (2018)
	Salicornia brachiata					
1	*SbMAPKK*/1023 bp	Yes	Yes	–	–	Agarwal et al. (2009)
2	*SbDREB2A*/1062 bp	Yes	Yes	Tobacco	Salinity and dehydration tolerance	Gupta et al. (2010)
3	*SbNHX1*/1683 bp	Yes	–	Tobacco	Salinity tolerance	Jha et al. (2011a, b)
				Jatropha	Salinity tolerance	Jha et al. (2013)
				Castor	Salinity tolerance	Patel et al. (2015)
				Cumin	Salinity tolerance	Pandey et al. (2016)
4	*SbGST*/693 bp	Yes	Yes	Tobacco	Salinity tolerance	Jha et al. (2011a, b)
5	*SbSI-1*/971 bp	Yes	Yes	–	Salinity and dehydration tolerance	Yadav et al. (2012), Kumari et al. (2017)
6	*SbSOS1*/3480 bp	Yes	Yes	Tobacco	Salinity tolerance	Yadav et al. (2012)
7	*SbMT*-2/237 bp	–	–	Tobacco	Salt, drought and metal stress tolerance	Chaturvedi et al. (2014)
8	SbpAPX/1249 bp	Yes	Yes	Tobacco	Salinity and dehydration tolerance	Singh et al. (2014)
9	*SbSI-2*/537 bp	Yes	Yes	Tobacco	Salinity and osmotic tolerance	Yadav et al. (2014)

(continued)

Table 1 (continued)

Sr. No.	Gene/ORF	Transcript response		Transgenic plant	Response of transgenics	References
		Salt	Drought			
10	*SbASR1*/606 bp	Yes	Yes	Tobacco	Salinity and dehydration tolerance	Tiwari et al. (2015)
11	*SbMYB15*/1197 bp	Yes	Yes	Tobacco	Salinity and dehydration tolerance	Shukla et al. (2015a)
12	*SbMYB44*/810 bp	Yes	Yes	–	–	Shukla et al. (2015b)
13	*SbSLSP*/444 bp	Yes	Yes	Tobacco	Salinity and dehydration tolerance	Singh et al. (2016a)
14	*SbSDR1*/483 bp	Yes	Yes	Tobacco	Salinity and osmotic tolerance	Singh et al. (2016b)
15	*SbUSP*/486 bp	Yes	Yes	Tobacco	Salinity and osmotic tolerance	Udawat et al. (2016)
16	*SbRPC5L*/1202 bp	Yes	–	Tobacco	Salinity and osmotic tolerance	Kumari and Jha (2019)

grown under saline conditions and can be considered as an alternative crop for saline agriculture. Furthermore, with increasing Molecular biology research and "Omics" approach, the understanding of the underlying co-ordinated mechanism of salt tolerance in halophytes is being understood. The sequencing of certain halophytes from Chenopodiaceae family, the dominant halophyte family will open avenues for realizing the gene duplication, lineage-specific largely functionally uncharacterized genes and epigenomic modifications of halophytes. The highly orchestrated targeting external sequestration of ions in salt bladders, targeting internal Na^+ sequestration in vacuoles, controlling stomatal aperture, optimizing water use efficiency by targeting stomatal density, targeting xylem ion loading of halophytes needs to be studied and accordingly made functionally practical in other crops for their suitable existence under salinity conditions.

References

Abarsaji GA (2000) Identification and investigation on some of eco physiological characteristics of *Aeluropus* spp. in saline and alkaline rangelands in the north of Gorgan. Pajouhesh-Va-Sazandegi 46:21–25

Abernethy GA, Fountain DW, McManus MT (1998) Observations on the leaf anatomy of Festuca novae-zelandiae and biochemical responses to a water deficit. N Z J Bot 36:113–123

Agarwal PK, Gupta K, Jha B (2009) Molecular characterization of the Salicornia brachiata SbMAPKK gene and its expression by abiotic stress. Mol Biol Rep 37:981–986

Agarwal PK, Gupta Kapil, Jha B (2010) Molecular characterization of the Salicornia brachiata SbMAPKK gene and its expression by abiotic stress. Mol Biol Rep 2:981

Agarwal PK, Gupta K, Lopato S, Agarwal P (2017) Drought responsive element binding transcription factors and their applications for the engineering of stress tolerance. J Exp Bot 68(9):2135–2148

Ahmed MZ, Shimazaki T, Gulzar S, Kikuchi A, Gul B, Khan MA, Watanabe KN (2013) The influence of genes regulating transmembrane transport of Na^+ on the salt resistance of *Aeluropus lagopoides*. Funct Plant Biol 40:860–871

Albert R, Kinzel H (1973) Different physiotypes among halophytes from region east of Lake Neusiedlersee (Austria). Z Pflanzenphysiol 2:138–157

Albert R, Pfundner G, Hertenberger G, Kastenbauer T, Watzka M (2000) The physiotype approach to understanding halophytes and xerophytes. Ergebnisse weltweiter ökologischer Forschung, vol 7. Verlag Günter Heimbach, Stuttgart, Germany, pp 69–87

Aronson J (1989) Economic halophytes-a global review. In: Plants for arid lands, pp 177–188

Askari H, Edqvist J, Hajheidari M, Kafi M, Salekdeh GH (2006) Effects of salinity levels on proteome of Suaeda aegyptiaca leaves. Proteomics 6:2542–2554

Aslam R, Bostansup N, Mariasup M, Safdar W (2011) A critical review on halophytes: salt tolerant plants. J Med Plant Res 33:7108–7118

Barhoumi Z, Djebali W, Chaïbi W, Abdelly C, Smaoui A (2007) Salt impact on photosynthesis and leaf ultrastructure of *Aeluropus littoralis*. J Plant Res 120:529–537

Barrett-Lennard EG (2002) Restoration of saline land through revegetation. Agric Water Manag 53:213–226

Bartels D, Dinakar C (2013) Balancing salinity stress responses in halophytes and non-halophytes: a comparison between *Thellungiella* and *Arabidopsis thaliana*. Funct Plant Biol 40:819–831

Bhaskaran C, Selvaraj T (1997) Seasonal incidence and distribution of VA-mycorrhizal fungi in native saline soils. J Environ Biol 18:209–212

Binzel ML, Hess FD, Bressan RA, Hasegawa PM (1988) Intracellular compartmentation of ions in salt adapted tobacco cells. Plant Physiol 86:607–614

Bor NL (1970) Aeluropus. Flora Iranica 70:419–423

Breckle SW (1983) Temperate deserts and semi-deserts of Afghanistan and Iran. In: West NE (ed) Ecosystems of the world, temperate deserts and semi-deserts. Elsevier, pp 271–319

Breckle SW (1995) How do halophytes overcome salinity. Biol Salt Toler Plants 23:199–203

Buhmann A, Papenbrock J (2013) An economic point of view of secondary compounds in halophytes. Funct Plant Biol 9:952–967

Chaturvedi AK, Patel MK, Mishra A, Tiwari V, Jha B (2014) The *SbMT-2* gene from a halophyte confers abiotic stress tolerance and modulates ROS scavenging in transgenic tobacco. PLoS ONE 9:e111379

Dooki AD, Mayer-Posner FJ, Askari H, Zaiee AA, Salekdeh GH (2006) Proteomic responses of rice young panicles to salinity. Proteomics 6:6498–6507

Ellouzi H, Hamed KB, Cela J, Munné-Bosch S, Abdelly C (2011) Early effects of salt stress on the physiological and oxidative status of *Cakile maritima* (halophyte) and *Arabidopsis thaliana* (glycophyte). Physiol Plant 1422:128–143

Ephraim, Utah. Proc. RMRS-P-11, Ogden, UT: U.S. Department of Agriculture, Forest Service, Rocky Mountain Research Station, pp 227–230

Fageria NK, Gheyi HR, Moreira A (2012) Nutrient bioavailability in salt affected soils. J Plant Nutr 7:945–962

Flowers TJ, Colmer TD (2008) Salinity tolerance in halophytes. New Phytol 179:945–963

Flowers TJ, Hajibagheri MA, Clipson NJW (1986) Halophytes. Q Rev Biol 61:313–337

Flowers TJ, Galal HK, Bromham L (2010) Evolution of halophytes: multiple origins of salt tolerance in land plants. Funct Plant Biol 7:604–612

Foyer CH, Lopez-Delgado H, Dat JF, Scott IM (1997) Hydrogen peroxide- and glutathione-associated mechanisms of acclamatory stress tolerance and signalling. Plant Physiol 2:241–254

Ganesan G, Sankararamasubramanian HM, Harikrishnan M, Ashwin G, Parida A (2012) A MYB transcription factor from the grey mangrove is induced by stress and confers NaCl tolerance in tobacco. J Exp Bot 63:4549–4561

Gill SS, Tuteja N (2010) Reactive oxygen species and antioxidant machinery in abiotic stress tolerance in crop plants. Plant Physiol Biochem 12:909–930

Gorham J, Hughes LL, Jones RW (1980) Chemical composition of salt-marsh plants from Ynys Môn (Anglesey): the concept of physiotypes. Plant Cell Environ 3:309–318

Goyal E, Singh RS, Kanika K (2013) Isolation and functional characterization of salt overly sensitive 1 (SOS1) gene promoter from *Salicornia brachiata*. Biol Plant 57:465–473

Gulzar S, Khan MA, Ungar IA (2002) Salt tolerance of a coastal salt marsh grass. Commun Soil Sci Plan 34:2595–2605

Gulzar S, Khan MA, Ungar IA (2003) Effect of salinity on growth, ionic content, plant-water status in *Aeluropus lagopoides*. Commun Soil Sci Plan 34:1657–1668

Gupta K, Agarwal PK, Reddy MK, Jha B (2010) SbDREB2A, an A-2 type DREB transcription factor from extreme halophyte *Salicornia brachiata* confers abiotic stress tolerance in *Escherichia coli*. Plant Cell Rep 29:1131–1137

Gupta K, Jha B, Agarwal PK (2014) A dehydration-responsive element binding (DREB) transcription factor from the succulent halophyte *Salicornia brachiata* enhances abiotic stress tolerance in transgenic tobacco. Mar Biotechnol 6:657–673

Hennig L (2012) Plant gene regulation in response to abiotic stress. Biochim Biophys Acta 1819:85

Hu YZ, Zeng YL, Guan B, Zhang FC (2012) Overexpression of a vacuolar H^+-pyrophosphatase and a B subunit of H^+-ATPase cloned from the halophyte *Halostachys caspica* improves salt tolerance in *Arabidopsis thaliana*. Plant Cell Tiss Organ Cult 108:63–71

Jha B, Agarwal PK, Reddy PS, Lal S, Sopory SK, Reddy MK (2009) Identification of salt-induced genes from *Salicornia brachiata*, an extreme halophyte through expressed sequence tags analysis. Genes Genet Syst 2:111–120

Jha A, Joshi M, Yadav NS, Agarwal PK, Jha B (2011a) Cloning and characterization of the Salicornia brachiata Na^+/H^+ antiporter gene SbNHX1 and its expression by abiotic stress. Mol Biol Rep 3:1965–1973

Jha B, Sharma A, Mishra A (2011b) Expression of *SbGSTU* (tau class glutathione S-transferase) gene isolated from *Salicornia brachiata* in tobacco for salt tolerance. Mol Biol Rep 38:4823–4832

Jha B, Mishra A, Jha A, Joshi M (2013) Developing transgenic Jatropha using the *SbNHX1* gene from an extreme halophyte for cultivation in saline wasteland. PLoS ONE 8:e71136

Jia GX, Zhu ZQ, Chang FQ, Li YX (2002) Transformation of tomato with the BADH gene from Atriplex improves salt tolerance. Plant Cell Rep 21:141–146

Joshi AJ (2011) Monograph on Indian halophytes. Bhavnagar (India). Department of Life Sciences, Bhavnagar University, Bhavnagar, Gujarat

Joshi AJ, Bhoite AS (1988) Fluctuations of mineral ions in saline soils and halophytic grass *Aeluropus lagopoides* L. Ann Arid Zone 27:191–196

Joshi AJ, Mali BS, Harsha H (2005) Salt tolerance at germination and early growth of two forage grasses growing in marshy habitats. Environ Exp Bot 54:267–274

Kavitha K, George S, Venkataraman G, Parida A (2010) A salt-inducible chloroplastic monodehydroascorbate reductase from halophyte *Avicennia marina* confers salt stress tolerance on transgenic plants. Biochimie 92:1321–1329

Khan MA, Gul B (1999) Seed bank strategies of coastal populations at Arabian Sea coast. In: Proceedings: Shrubland Ecotones, pp 12–14

Khedia J, Agarwal P, Agarwal PK (2018) AlNAC4 transcription factor from halophyte Aeluropus lagopoides mitigates oxidative stress by maintaining ROS homeostasis in transgenic tobacco. Front Plant Sci 9

Krishnamurthy P, Mohanty B, Wijaya E, Lee DY, Lim TM, Lin Q, Xu J, Loh CS, Kumar PP (2017) Transcriptomics analysis of salt stress tolerance in the roots of the mangrove *Avicennia officinalis*. Sci Rep 7:10031

Kumari A, Jha B (2019) Engineering of a novel gene from a halophyte: potential for agriculture in degraded coastal saline soil. Land Degrad Dev 30:595–607

Kumari J, Udawat P, Dubey AK, Haque MI, Rathore MS, Jha B (2017) Overexpression of SbSI-1, a nuclear protein from Salicornia brachiata confers drought and salt stress tolerance and maintains photosynthetic efficiency in transgenic tobacco. Front Plant Sci 8:1215

Le Houérou HN (1993) Forage halophytes and salt-tolerant fodder crops in the Mediterranean Basin. In: Halophytes as a resource for livestock and for rehabilitation of degraded lands. Springer, Dordrecht, pp 123–137

Li N, Chen S, Zhou X, Li C, Shao J, Wang R, Fritz E, Hüttermann A, Polle A (2008) Effect of NaCl on photosynthesis, salt accumulation and ion compartmentation in two mangrove species, Kandelia candel and Bruguiera gymnorhiza. Aquat Bot 88:303–310

Li W, Wang D, Jin T, Chang Q, Yin D, Xu S (2011) The vacuolar Na^+/H^+ antiporter gene *SsNHX1* from the halophyte *Salsola soda* confers salt tolerance in transgenic alfalfa (*Medicago sativa* L.). Plant Mol Biol Rep 29:278–290

Li K, Pang CH, Ding F, Sui N, Feng ZT, Wang BS (2012) Over expression of *Suaeda salsa* stroma ascorbate peroxidase in *Arabidopsis* chloroplasts enhances salt tolerance of plants. S Afr J Bot 78:235–245

Li QL, Xie JH, Ma XQ, Li D (2016) Molecular cloning of Phosphoethanolamine N-methyltransferase (PEAMT) gene and its promoter from the halophyte *Suaeda liaotungensis* and their response to salt stress. Acta Physiol Plant 38:39

Martinez JP, Silva H, Ledent JF, Pinto M (2007) Effect of drought stress on the osmotic adjustment, cell wall elasticity and cell volume of six cultivars of common beans (*Phaseolus vulgaris* L.). Eur J Agron 26:30–38

Mitsch WJ, Gosselink JG (2000) The value of wetlands: importance of scale and landscape setting. Ecol Econ 1:25–33

Mohsenzadeh S, Malboobi KMA, Razavi Farrahi-Aschtiani S (2006) Physiological and molecular responses of *Aeluropus lagopoides* (Poaceae) to water deficit. Environ Exp Bot 56:314–322

Munns R (2002) Comparative physiology of salt and water stress. Plant Cell Environ 25:239–250

Munns R (2005) Genes and salt tolerance: bringing them together. New Phytol 167:645–663

Munns R, Termaat A (1986) Whole-plant responses to salinity. Funct Plant Biol 13:143–160

Munns R, Tester M (2008) Mechanisms of salinity tolerance. Annu Rev Plant Biol 59:651–681

Naaz N, Hameed M, Wahid A, Arshad M, Aqeel Ahmad MS (2009) Patterns of ion excretion and survival in two stoloniferous arid zone grasses. Physiol Plant 135:185–195

Naz N, Hameed M, Sajid Aqeel Ahmad M, Ashraf M, Arshad M (2010) Is soil salinity one of the major determinants of community structure under arid environments? Community Ecol 11:84–90

Naz N, Hameed M, Nawaz T, Batool R, Ashraf M, Ahmad F, Ruby T (2013) Structural adaptations in the desert halophyte *Aeluropus lagopoides* (Linn.) Trin. Ex Thw. under high salinity. Eur J Biol Res 19:150

Nellemann C, Corcoran E (2009) Blue carbon: the role of healthy oceans in binding carbon: a rapid response assessment. UNEP/Earthprint

Oh DH, Dassanayake M, Haas JS, Kropornika A, Wright C, d'Urzo MP, Hong H, Ali S, Hernandez A, Lambert GM, Inan G (2010) Genome structures and halophyte-specific gene expression of the extremophile *Thellungiella parvula* in comparison to *Thellungiella salsuginea* (*Thellungiella halophila*) and Arabidopsis. Plant Physiol 1:110

Ohta M, Hayashi Y, Nakashima A, Hamada A, Tanaka A, Nakamura T, Hayakawa T (2002) Introduction of a Na^+/H^+ antiporter gene from *Atriplex gmelini* confers salt tolerance to rice. FEBS Lett 532:279–282

Pandey S, Patel MK, Mishra A, Jha B (2016) In planta transformed cumin (*Cumin umcyminum* L.) plants, overexpressing the *SbNHX1* gene showed enhanced salt endurance. PloS One 7:e0159349

Pang CH, Li K, Wang B (2011) Overexpression of SsCHLAPXs confers protection against oxidative stress induced by high light in transgenic *Arabidopsis thaliana*. Physiol Plant 143:355–366

Patel MK, Joshi M, Mishra A, Jha B (2015) Ectopic expression of SbNHX1 gene in transgenic castor (Ricinus communis L.) enhances salt stress by modulating physiological process. Plant Cell Tiss Organ Cult 122:477–490

Rajan N, Agarwal P, Patel K, Sanadhya P, Khedia J, Agarwal PK (2015) Molecular characterization and identification of target protein of an important vesicle trafficking gene AlRab7 from a salt excreting halophyte *Aeluropus lagopoides*. DNA Cell Biol 34:83–91

Ramadan T (2001) Dynamics of salt secretion by *Sporobolus spicatus* (Vahl) Kunth from sites of differing salinity. Ann Bot 87:259–266

Rao GN, Murty PP (2013) Morphological and anatomical features of *Salicornia brachiata* Roxb. J Biol Chem 2:887–891

Razavi K (2005) Identification and characterization of mRNA transcripts differentially expressed in response to high salinity using DD-AFLP in *Aeluropus lagopoides*. In: 4th national biotechnology congress, 15–17 August, Kerman, Iran

Rengasamy P (2002) Transient salinity and subsoil constraints to dryland farming in Australian sodic soils: an overview. Aust J Exp Agr 42:351–361

Rengasamy P (2006) World salinization with emphasis on Australia. J Exp Bot 57:1017–1023

Riadh K, Wided M, Hans-Werner K, Chedly A (2010) Responses of halophytes to environmental stresses with special emphasis to salinity. Adv Bot Res 53:117–145

Saad RB, Romdhan WB, Zouari N, Azaza J, Mieulet D, Verdeil JL, Guiderdoni E, Hassairi A (2011) Promoter of the AlSAP gene from the halophyte grass *Aeluropus littoralis* directs developmental-regulated, stress-inducible, and organ-specific gene expression in transgenic tobacco. Transgenic Res 20:1003–1018

Sahu BB, Shaw BP (2009) Isolation, identification and expression analysis of salt-induced genes in *Suaeda maritima*, a natural halophyte, using PCR-based suppression subtractive hybridization. BMC Plant Biol 9:69

Sanadhya P, Agarwal P, Khedia J, Agarwal PK (2015) A low-affinity K^+ transporter AlHKT2; 1 from recretohalophyte *Aeluropus lagopoides* confers salt tolerance in yeast. Mol Biotech 57:489–498

Santhanakrishnan D, Perumal RK, Kanth SV, Jonnalagadda RR, Bangaru C (2013) Studies on the physiological and biochemical characteristics of *Salicornia brachiata*: influence of saline stress due to soaking wastewater of tannery. Desalin Water Treat 52:31–33

Shabala S (2013) Learning from halophytes physiological basis and strategies to improve abiotic stress tolerance in crops. Ann Bot 7:1209–1221

Sher M, Sen DA, Mohmmed S (1994) Seasonal variations in sugar and protein content of halophytes in Indian desert. Ann Arid Zone 33:249–251

Shukla PS, Agarwal P, Gupta K, Agarwal PK (2015a) Molecular characterization of an MYB transcription factor from a succulent halophyte involved in stress tolerance. AoB Plants 1:7

Shukla PS, Gupta K, Agarwal P, Jha B, Agarwal PK (2015b) Overexpression of a novel SbMYB15 from *Salicornia brachiata* confers salinity and dehydration tolerance by reduced oxidative damage and improved photosynthesis in transgenic tobacco. Planta 242:1291–1308

Singh N, Mishra A, Jha B (2014) Over-expression of the peroxisomal ascorbate peroxidase (SbpAPX) gene cloned from halophyte salicornia brachiata confers salt and drought stress tolerance in transgenic tobacco. Mar Biotechnol 16:321–332

Singh D, Yadav NS, Tiwari V, Agarwal PK, Jha B (2016a) A SNARE-like superfamily protein SbSLSP from the halophyte Salicornia brachiata confers salt and drought tolerance by maintaining membrane stability, K^+/Na^+ ratio, and antioxidant machinery. Front Plant Sci 7:737

Singh VK, Mishra A, Haque I, Jha B (2016b) A novel transcription factor-like gene *SbSDR1* acts as a molecular switch and confers salt and osmotic endurance to transgenic tobacco. Sci Rep 6:31686

Slenzka A, Ladislav M, Gudrun K (2013) *Salicornia* L. (Amaranthaceace) in South Africa and Namibia: rapid spread and ecological diversification of cryptic species. J Bot Linean Soc 172:175–186

Sobhanian H, Motamed N, RastgarJazii F, Nakamura T, Komatsu S (2010) Salt stress induced differential proteome and metabolome response in the shoots of *Aeluropus lagopoides* (Poaceae), a halophyte C_4 plant. J Prot Res 2:882–897
Stanley OD (2008) Bio prospecting marine halophyte *Salicornia brachiata* for medical importance and salt encrusted land development. J Coast Dev 2:62–69
Sun Q, Gao F, Zhao L, Li K, Zhang J (2010) Identification of anew 130 bp cis-acting element in the TsVP1 promoter involved in the salt stress response from *Thellungiella halophila*. BMC Plant Biol 10:90
Taji T, Seki M, Satou M, Sakurai T, Kobayashi M, Ishiyama K, Narusaka Y, Narusaka M, Zhu JK, Shinozaki K (2004) Comparative genomics in salt tolerance between Arabidopsis and Arabidopsis related halophyte salt stress using *Arabidopsis microarray*. Plant Physiol 135:1697–1709
Taji T, Komatsu K, Katori T, Kawasaki Y, Sakata Y, Tanaka S, Kobayashi M, Toyoda A, Seki M, Shinozaki K (2010) Comparative genomic analysis of 1047 completely sequenced cDNAs from an Arabidopsis-related model halophyte, *Thellungiella halophila*. BMC Plant Biol 10:261
Tanveer M, Shabala S (2018) Targeting redox regulatory mechanisms for salinity stress tolerance in crops. In: Salinity responses and tolerance in plants. Springer, Cham, pp 213–234
Tanveer M, Shah AN (2017) An insight into salt stress tolerance mechanisms of Chenopodium album. Environ Sci and Poll Res 24:16531–16535
Tanveer M, Shahzad B, Sharma A, Biju S, Bhardwaj R (2018) 24-Epibrassinolide; an active brassinolide and its role in salt stress tolerance in plants: a review. Plant Physiol Biochem 130:69–79
Tao JJ, Wei W, Pan WJ, Lu L, Li QT, Ma JB, Zhang WK, Ma B, Chen SY, Zhang JS (2018) An Alfin-like gene from Atriplex hortensis enhances salt and drought tolerance and abscisic acid response in transgenic Arabidopsis. Sci Rep 8:2707
Tewari A (1970) A note on the value of *Sporobolus coromandelianus* (Trin.) Kunth and *Aeluropus lagopoides* (Linn.) as feed and soil binder. Plant Sci 2:135–136
Thomashow MF (1999) Plant cold acclimation: freezing tolerance genes and regulatory mechanisms. Annu Rev Plant Biol 50:571–599
Tiwari V, Chaturvedi AK, Mishra A, Jha B (2015) Introgression of the SbASR-1 gene cloned from a halophyte Salicornia brachiata enhances salinity and drought endurance in transgenic groundnut (*Arachis hypogaea*) and acts as a transcription factor. PLoS ONE 10:e0131567
Torbatinejad N, Maghsoodlowrad H, Gharabash AM (2000) Nutritive value of Aeluropus littoralis and Aeluropus lagopoides in sheep, pp 31–46
Tuteja N, Banu MS, Huda KM, Gill SS, Jain P, Pham XH, Tuteja R (2014) Pea p68, a DEAD-box helicase, provides salinity stress tolerance in transgenic tobacco by reducing oxidative stress and improving photosynthesis machinery. PLoS ONE 9:e98287
Udawat P, Jha RK, Sinha D, Mishra A, Jha B (2016) Overexpression of a cytosolic abiotic stress responsive universal stress protein (SbUSP) mitigates salt and osmotic stress in transgenic tobacco plants. Front Plant Sci 7:518
Waghmode AP, Hegde BA (1984) Effect of sodium chloride on pyruvate orthophosphate dikinase of a saline grass *Aeluropus lagopoides* (Linn.). Trin Biovigyanam 10:209–210
Waghmode AP, Joshi GV (1982) Photosynthetic and photorespiratory enzymes and metabolism of 14 C-substrates in isolated leaf cells of the C4 species of *Aeluropus lagopoides* L. Photosynthetica 16:17–21
Wahid AB (2003) Physiological significance of morphoanatomical features of halophytes with particular reference to Cholistan flora. I J Agri Biol 5:207–212
Waisel Y (1972) Biology of halophytes. Academic Press, New York
Wang RZ (2007) Plant functional types and their ecological responses to salinization in saline grasslands, Northeastern China. Photosynthetica 42:511–519
Wang ZL, Li PH, Fredricksen M, Gong ZZ, Kim CS, Zhang C, Bohnert HJ, Zhu JK, Bressan RA, Hasegawa PM, Zhao YX (2004) Expressed sequence tags from *Thellungiella halophila*, a new model to study plant salt-tolerance. Plant Sci 166:609–616

Wang YC, Qu GZ, Li HY, Wu YJ, Wang C, Liu GF, Yang CP (2010) Enhanced salt tolerance of transgenic populus plants expressing a manganese superoxide dismutase from Tamarix androssowii. Mol Biol Rep 37:1119–1124
Wang J, Li B, Meng Y, Ma X, Lai Y, Si E, Yang K, Ren P, Shang X, Wang H (2014a) Transcriptomic profiling of the salt-stress response in the halophyte Halogeton glomeratus. BMC Genomic 16:169
Wang TT, Ren ZJ, Liu ZQ, Feng X, Guo RQ, Li BG, Li LG, Jing HC (2014b) SbHKT1; 4 a member of the high-affinity potassium transporter gene family from Sorghum bicolor, functions to maintain optimal Na^+/K^+ balance under Na^+ stress. J Integr Plant Biol 56:315–332
Watson L, Dallwitz MJ (1992) The grass genera of the world. CAB International, Wallingford
Wu G, Shao HB, Chu LY, Cai JW (2007) Insights into molecular mechanisms of mutual effect between plants and the environment. A review. Agron Sustain Dev 27:69–78
Yadav NS, Shukla PS, Jha A, Agarwal PK, Jha B (2012) The *SbSOS1* gene from the extreme halophyte *Salicornia brachiata* enhances Na^+ loading in xylem and confers salt tolerance in transgenic tobacco. BMC Plant Biol 1:188
Yadav NS, Singh VK, Singh D, Jha B (2014) A novel gene SbSI-2 encoding nuclear protein from a halophyte confers abiotic stress tolerance in *E. coli* and tobacco. PloS ONE (9):101926
Yamaguchi-Shinozaki K, Shinozaki K (2006) Transcriptional regulatory networks in cellular responses and tolerance to dehydration and cold stresses. Annu Rev Plant Biol 57:781–803
Yin X, Zhao Y, Luo D, Zhang H (2002) Isolating the promoter of a stress-induced gene encoding betaine aldehyde dehydrogenase from the halophyte *Atriplex centralasiatica*. Biochim Biophys Acta 1577:452–456
Yue W, Xia GM, Zhi DY, Chen HM (2001) Transfer of salt tolerance from *Aeluropus littoralissinensis* to wheat (*Triticum aestivum* L.) via asymmetric somatic hybridization. Plant Sci 161:259–266
Zhang GH, Su Q, An LJ, Wu S (2008a) Characterization and expression of a vacuolar Na^+/H^+ antiporter gene from the monocot halophyte *Aeluropus littoralis*. Plant Physiol Biochem 46:117–126
Zhang Y, Yin H, Li D, Zhu W, Li Q (2008b) Functional analysis of BADH gene promoter from *Suaeda liaotungensis* K. Plant Cell Rep 27(3):585
Zhu M, Meng X, Cai J, Li G, Dong T, Li Z (2018) Basic leucine zipper transcription factor SlbZIP1 mediates salt and drought stress tolerance in tomato. BMC Plant Biol 18:83

Special Adaptive Features of Plant Species in Response to Drought

Asif Iqbal, Shah Fahad, Mazhar Iqbal, Madeeha Alamzeb, Adeel Ahmad, Shazma Anwar, Asad Ali Khan, Amanullah, Muhammad Arif, Inamullah, Shaheenshah, Muhammad Saeed and Meizhen Song

Abstract Under drought stress conditions, plants need to adapt themselves by manipulating key morphological, physiological, biochemical, and molecular processes. By doing so plants enhance their water uptake and storage, limit water loss and prevent tissues from wilting. There are many mechanisms and every crop has its own way of adaptive mechanisms to overcome drought stress. Drought stress adversely affects many physiological aspects of the plants, especially photosynthetic rate and other gaseous exchange traits. Prolonged drought stress severely diminishes crop growth and productivity. The physiological and molecular mechanisms related to drought stress tolerance and adaptation are widely studied. Different adaptive mechanisms maintaining appropriate metabolomic and biochemical homeostasis to prevent excessive damage caused by drought stress are also discussed. The mechanisms that regulate plants for adaptation to drought stress through special adaptive features are the main subject of the current chapter. It was concluded that combinations of these different features enhance the plant's adaptation to drought condition. To understand how these mechanisms are regulated and how to overcome the adverse effect of drought on plant productivity, will give information to enhance adaptation of plants, which will ultimately improve quality and yield of the crops.

A. Iqbal · A. Ahmad · M. Song
State Key Laboratory of Cotton Biology, Institute of Cotton Research of CAAS, Anyang 455000, Henan, People's Republic of China
e-mail: songmeizhencaas@163.com

S. Fahad (✉)
Department of Agriculture, The University of Swabi, Swabi, Pakistan
e-mail: shah_fahad80@yahoo.com

M. Iqbal
Department of Botany, Shaheed Benazir Bhutto University Sheringal Dir (U), Sheringal, Pakistan

M. Alamzeb · S. Anwar · A. A. Khan · Amanullah · M. Arif · Inamullah · Shaheenshah
Department of Agronomy, The University of Agriculture Peshawar, Peshawar 25130, Pakistan

M. Saeed
Department of Weed Science, The University of Agriculture Peshawar, Peshawar 25130, Pakistan

M. Hasanuzzaman and M. Tanveer (eds.), *Salt and Drought Stress Tolerance in Plants*, Signaling and Communication in Plants, https://doi.org/10.1007/978-3-030-40277-8_4

Abbreviations

CK	Cytokinin
GA	Gibberellic acid
ABA	Abscisic acid
RWC	Relative water content
VPD	Vapor pressure deficit
CO_2	Carbon dioxide
ROS	Reactive oxygen species
RuBP	Ribulose 1,5-bisphosphate
RuBisCO	Ribulose-1,5-bisphosphate carboxylase/oxygenase
HYR	Higher Yield Rice
WUE	Water use efficiency
OA	Osmotic adjustment
BADH	Betaine aldehyde dehydrogenase
P5CR	Pyrroline-5-carboxylate reductase
OAT	Ornithine ornithine δ-aminotransferase
GC/MS	Gas Chromatography Mass Spectrometry
SOD	Superoxide dismutase
CAT	Catalase
GSH	Glutathione
TCA	Tricarboxylic acid cycle
OPP	Oxidative pentose phosphate
NADPH	Nicotinamide adenine dinucleotide phosphate
GABA	Gamma-aminobutyric Acid
G1P	Glucose-1-phosphate
LEAs	Late embryogenesis
HSPs	Heat shock proteins
IPT	Isopentenyl transferase
IAA	Indole-3-acetic acid
QTL	Quantitative Trait Loci
DRO1	DEEPER ROOTING 1
mRNA	Messenger RNA
JA	Jasmonic acid
SA	Salicylic acid
DREB	Dehydration responsive transcription factors
CDSP	Chloroplastic drought-induced stress protein
DRE/CRT	Hydration responsive element/C-repeat
CBF	C-repeat binding factor
ODC	Ornithine decarboxylase
ADC	Arginine decarboxylase
SAMDC	S-adenosylmethionine decarboxylase
SPDS	Spermidine synthase
SPMS	Spermine synthase

ASR	Abscisic acid, stress and ripening
REF	Rubber elongation factor
SRPP	Small rubber particle protein

1 Introduction

Among the abiotic stress, drought is the most important one for crop production as most of them are grown in the semi-arid and tropical regions (Tanveer et al. 2018). It affects all living organisms, including human being. Limitation of water in the soil during drought affects evaporation and evapotranspiration and ultimately precipitation. It occurs when the temperature is high and soil moisture and relative humidity is low. Plants faced drought when the water demand is higher than the supply, which is also determined by soil available water and its depth. Plants demand for water is also dependent on evaporation and evapotranspiration. Plants need evapotranspiration for photosynthesis but most of the water transpired without utilization (Blum 2011). All the terrestrial plants faced short or long-term water stress during their lifecycle that's why plants have evolved various adaptive strategies to cope with drought stress (Hussain et al. 2018). Among the plants some adapt very easily than others and compete in the changing environment. It may be moderate and for short duration or severe for long duration, which affect the plant life (Pereira and Chaves 1995). According to the UN report, one-third of the world population is livening in the areas having poor water resources (Watkins 2006), which restricted crop production periodically (Kramer 1980; Flowers 1989). The poor availability of water affects the physiology of the plants as well as their environment (Kramer 1980), directly or indirectly (Akinci 1997). These physiological responses of plants depend mostly on crop species, soil type, nutrients availability, and climate. Throughout the life cycle, plants need water from germination till maturity. The primary effect of drought on plants is retarded growth. This retards growth as a result of shoot growth inhibition and limits metabolic demands. Similarly, root development also changes with drought stress, to assess the plant in taking excess water from deep soil layers. Thus more assimilates are partitioned to the roots for further growth into deep soil. The inhibition of lateral roots is also an adaptive response of plants, which promotes primary root growth and enable the plants to extract more water from deep soil layers (Xiong et al. 2006). This growth inhibition is developed due to loss of turgidity by non-availability of enough water to cell because of poor hydraulic conductance. This reduction in hydraulic conductance also caused decrease in nutrient supply to the above parts and prevents embolism in xylem which is one of the adaptive responses of plants against drought stress.

Generally, to enhance yield under drought stress were mostly depends on secondary traits like root architecture, osmotic adjustment, leaf water potential, relative water content at vegetative stage, which may not positively correlate to yield (Jongdee et al. 2002). It has been recommended that the yield potential of crop plants must be evaluated under both normal and stress conditions, as the correlation between yield under normal and stressed condition is positive (Guan et al. 2010). Moreover, combining higher yield under normal condition with a good yield under stress is the ideal trait for comparison. Therefore, the main focus should be on the identification of mechanisms, traits, and respective genes regulating yield under drought stress and are free from yield limitation in normal condition. As three NAC family transcription factors has been noticed for yield regulation under both normal and stress conditions. Moreover, a tremendous progress in the understanding of drought adaptive strategies in plants like morphophysiological alterations were studied (Yordanov et al. 2000). The alterations in these adaptive mechanisms are regulated by the expressions of genes (Osakabe et al. 2014). In the plants, there is a large diversity within the species and genotypes against drought adaptation, as some genotypes are able to cope better than other during drought. Therefore, genotypes having contrasting adaptive mechanisms used to be an important source to know the variation in drought adaptation of plants. This variation must be exploited to improve drought and yields of crop varieties through understanding of these adaptive mechanisms, which would further used as selection traits for drought-tolerant genotypes (Reynolds and Tuberosa 2008).

In this chapter, we discuss different morphological, physiological, biochemical, and molecular mechanisms in the various organs of plants grown under drought conditions and regulating these mechanisms for improving growth, yield, and quality of plants.

2 Adaptation to Drought Stress

For adaptation to drought, plants use different mechanisms like morphological, physiological, biochemical, and molecular transitions in gene expression. It might be considered as a result of many traits evolved at various levels to give a specific function. Whereas one trait maybe related to many functions and multiple traits in combination can give a specific aim with the development of definite strategies. Due to these different strategies, the adaptation mechanism becomes complex moving from wet to arid areas. Focusing on the strategies to withstand water shortage, the adaptation mechanism becomes more and more developed and complex moving from wet areas toward arid areas, which leads to the development of diversified strategies (Monneveux and Belhassen 1996). To avoid dryness of the tissues, there is an adaptive link between leaf gas exchange and hydraulic efficiency in the plant body. This type of relation between photosynthetic capacity and hydraulic supply of water to plant leaves has been studied in conifers and angiosperm species (Brodribb and Feild 2000).

Generally, plants adapt to drought through different ways like drought escape, tolerance, and avoidance strategies, which has no clear cut boundaries (Levitt 1980a, b). Drought escape is the ability of a plant to complete their lifecycle before the onset of drought and therefore does not face water shortage. In this case, the plants do not experience water deficit because they are able to restrain their vegetative and reproductive phenology during the normal seasons (Aronne and Wilcock 1997), through the mechanisms of early phenological development and developmental plasticity (Jones 1981). Plants having rapid phenological development grow fast, produce some seeds before the water shortage without any special morphological, physiological, or biochemical adaptations. However, plants having developmental plasticity grow slowly in dry season with few flowers and seeds but grow indeterminately and produce more seeds when the season becomes normal. Drought avoidance is the ability of plants to maintain high water potential through many adaptive ways like reducing water loss and improve water uptake (Levitt 1980a, b). This is done through many adaptive traits comprising low transpiration rate in water-saving plants and osmatic adjustment in water spender plants to avoid desiccation (Levitt 1980a, b). Under drought stress, water spenders increased root growth, hydraulic conductance, etc. to maintain high tissue water status. However, water saver plants limit water loss by decreasing transpiration, transpiration area, and radiation absorption. Plants having drought tolerance strategies maintain low tissue water potential by osmotic adjustment and producing more compact and stiff tissue. Resurrection plants found in angiosperm species can tolerate extreme desiccation until fully recovered after rehydration. These traits can maintain cell turgor through cell osmotic adjustment and elasticity thus increasing protoplasm resistance (Morgan 1984). Plants of perennial nature rely on drought tolerance instead of drought escape (Kozlowski and Pallardy 2002). Drought tolerance can be established either through desiccation avoidance or tolerance. Desiccation is achieved through high water potential within the plant. The mechanisms work through morphological adaptations such as root, stem, and leaves and physiological adaptation through transpiration which increase water storage and reduce dehydration during water limiting period (Vinod 2012). Reduction in tissue hydration is performed through osmotic adjustment by solutes accumulation and stomatal closure with the production of abscisic acid which maintain normal physiological functions of the plants (Repellin et al. 1994). Generally, plant adaptation depends on the sensitivity of plant to drought stress, which varies among plants at species (rubber and cashew), genotype, phenological development or organ levels (leaves) (Chen et al. 2009).

Many studies on plants tolerance against drought showed that different morphological, physiological, and biochemical adaptations contribute to show the tolerance (Rajagopal et al. 1990). Therefore, to understand the tolerance level many physiological traits like pre-dawn and middy leaf water potential, photosynthetic rate, stomatal conductance, transpiration rate, intercellular carbon dioxide concentration have been used to know the tolerance of the plants. However, biochemical traits such as ascorbic acid, glutathione, chlorophylls content, tocopherols, amino acids, carotenoids, and soluble sugar have been used to check the tolerance level of the plants against drought stress. Moreover, morphological and physiological attributes like root depth and architecture, stomatal conductance and water use efficiency are generally used for genotypes selection under drought conditions.

3 Morphological Adaptations

Adaptations of plants to drought stress widely depend on either the evolutionary modification of specific organ like taproot system, storage organs, or specific habit. Deserts plants having good morphophysiological traits can withstand long dry periods of drought due to severe reduction in leaves and modification into thorns and development of succulent stem, which help in photosynthesis and store water in parenchyma. In semi-arid environment, most of the plants are evergreen sclerophyllous, drought deciduous, and seasonally dimorphic shrubs, which show the key adaptive approaches of perennial species to drought stress. The principal role of shrubs in semi-arid ecosystem exists in the fact that these plants can grow in environmental stress condition where trees cannot survive (Wilson 1995). It is interesting that, although it is known that water shortage is the main limiting factor in Mediterranean-type ecosystems, here maximum perennials assume extremely resource demanding processes during summer. As one-fifth of the perineal plants in Mediterranean region avoid aridity by reproducing in spring (Aronne and Wilcock 1997), other plants complete their lifecycle in summer including Sclerophylly with specialized xylem and root traits that enhance plant survival (De Micco and Aronne 2008).

3.1 Germination and Seedling Establishment

Adaptive strategies in many arid and semi-arid plants were studied against drought. However, less information is known about the defense strategies in the plants during early phases after germination (Fenner 1999). Plant populations are first formed by seed dispersal, which execute the habitat where the plants grow in, and then by the influence of environmental factors on seed survival, germination, seedling establishment and growth (Schupp 1995). Seedling establishment appears to be one of the utmost critical phases in the regeneration process by sexual reproduction (Liu et al. 2018). Once new gene combinations have been produced with the development of seeds capable to germinate, persistence, and growth at early stages of plant development are major bottlenecks to effectively complete the reproductive stage and to attain canopy occupancy as established in tropical and cool temperate forests (Sánchez-Gómez et al. 2006). If a species has a very specialized reproductive system that permits effective seed production and dispersal, other ecological factors may constrain seed germination and following seedling establishment, growth and survival (Traveset et al. 2001). Seedling survival relies on the capability to cope with many environmental factors such as water availability, temperature, radiation, pathogens, herbivory, and competitive interactions (Moles and Westoby 2004). However, the main reason for seedling mortality is drought, which limits the recruitment processes in time and space (Moles and Westoby 2004). Therefore, in arid areas recruitment occurs in restricted rainfall periods or in limited wet areas (Padilla et al. 2007). During dry periods combined with some rain events, two features are important in order to ensure fast seedling establishment, rapid anchoring of juvenile seedlings to the substrate and immediate water uptake (Aronne and De Micco 2004). High desiccation tolerance of the embryo in the dry seed is lost during germination, and survival extremely depends on the rapidity in surpassing the early life stages. Greater seedling survival is mostly connected to higher biomass distribution to root part. It increases the chances of better water and nutrient uptake from deep soil layers (Padilla et al. 2007). Therefore, survival chance of deep-rooted plants are more than shallow rooted, which varies from species to species (Padilla and Pugnaire 2007). Along with this larger seeds have also more survival chance due to more storage reserves which help in the development of deeper roots, establishment of seedling as compared to small seeds (Fenner 1999; Padilla et al. 2007). It was also reported that drought-tolerant plants and plants growing in dry area has higher capacity of root elongation showing that root plasticity is genetically controlled (Sharp et al. 2004). For example, seedlings of wild species showed deep rooting in response to drought, especially in those species which regenerate from seeds (Reader et al. 1993). Moreover, growth responses to water scarcity can be affected by other factors like irradiance. It was noted that the impact of light limitation is more negative in drought conditions and the low level of photosynthates produced conflicts between the concurrent demands of both above and below plant parts (Kubiske et al. 1996). Other experiments reported that the effect of drought is less under shadier conditions (Sánchez-Gómez et al. 2006) like nurse plants which enhance growth and development of other species under their canopy. Moreover, it creates a more favorable micro condition for seed germination and seedling establishment. Therefore, this has been used to reestablish vegetation in arid and semi-arid regions (Ren et al. 2008).

3.2 *Root Morphology*

In many important agricultural crops, drought stress is first perceived through the root system, which continues to grow under the soil even though shoot growth is hindered under these conditions (Anjum et al. 2017a). Although primary root growth is not affected by drought, the growth of lateral roots is significantly reduced, mainly by suppression of activation of lateral root meristems (Deak and Malamy 2005). The *Arabidopsis* R2R3-type MYB TF MYB96 model regulates the activation of a lateral root meristem through the abscisic acid (ABA) signaling cascade, with a demonstrated activation mutant showing improved drought with reduced lateral root formation (Seo and Park 2009). It has also been shown that the plant miR393 microRNA plays a role in root adaptation to drought-induced stress response through attenuation of auxin (Chen et al. 2012). In addition to lateral roots, the presence of small roots is also an adaptive strategy to increase water absorption by providing a more absorbent surface. The presence of specialized tissues such as rhizodermis, with a thick outer cell wall or suberized exodermis, or a reduction in the number of cortical layers is an adaptive advantage of survival from stress. Hydrotropism is another adaptive measure taken by plants to cope with stress. Studies have shown that degradation of amyloplasts in the columella cells of plant roots when exposed to drought stress increase hydrotropism (Takahashi et al. 2003). Hormonal interference by auxin, cytokinin (CK), gibberellic acid (GA), and abscisic acid (ABA) has been implicated as a potential chemical signal in response to water stress to modulate root system architecture (Blilou et al. 2005). The expression of enzymes associated with root morphology is induced by mild drought stress, while other structural proteins are down regulated, which are closely related to root growth and thus an increase in the surface area of water absorption. Changes in the expression of these proteins are positively related to lateral development, which in turn affects photosynthesis (Sengupta and Reddy 2011). More lateral root and root hair development was found in lines that possess QTL, qDTY12.1, only when they are under drought (Dixit et al. 2015). These traits, which are expressed only under drought stress, have greater potential to enhance grain yield under drought. Moreover, drought stress triggers a wide range of anatomical traits expressed for different levels and patterns in different species and even in different varieties within species (Henry et al. 2012). For example, suberization and compaction of sclerenchyma cells were reduced in rice, increasing water retention under drought stress (Henry et al. 2012). Apart from root/shoot biomass ratio and root length, many other morphological traits affect plant adaptation to drought. It has been shown that the development of specialized tissues such as hypocotyl hair, during seedling emergence has positive effects including facilitating water absorption and immediate physical support (Aronne and De Micco 2004).

4 Anatomical Adaptations

The vast majority of anatomical adaptations to drought are the modifications to decrease the loss of water from plants during drought conditions. This is achieved mainly by controlling the loss of transpiration by regulating stomata. Stomatal response may be one of the most complex behaviors in plants. External factors such as light, air humidity, soil water content, and nutrient status, and internal factors such as ABA concentration, and leaf water status are known to apply a direct effect on the behavior of stomatal cells. Stomatal behavior is a major indicator of drought tolerance assessment in crops. Amazing differences in stomatal function during drought have been reported in many plant species (Passos et al. 2006; Gomes et al. 2008). Such specific anatomical characteristics are created in response to drought in order to promote safer but slower water transport. Analyzing the variation in anatomical properties and the formation of isotopes along the IADFs, which is associated with the regulation of the stomata, has proved useful for reconstructing phenological events from wood, thus unrevealing how the plants differ in their role in absorbing carbon in the different environmental conditions resulting from drought (Battipaglia et al. 2010).

4.1 Xylem Cavitation

In drought conditions, the xylem cavitation occurs when sap pressure drops below the threshold leading to an irreversible breakdown of xylem pathways. In the cavitation resistance, the stomata closure occurs in advance in response to increased hydraulic resistance of the plant, which prevents water loss, thus maintaining the pressure of the sap above the threshold (Sperry and Tyree 1988). In the rubber trees, a relatively high vulnerability of xylem tubes was found in the cavitation. The associated responses between the stomatal behavior and appearance of the cavitation and the large genotypic variation were found to be susceptible to cavitation, suggesting that the whole plant level had drought adaptation mechanisms (Sangsing et al. 2004). Cavitation resistance is therefore a parameter that relates more to the survival of the drought. Xylem embolism analysis, especially in petioles, may provide a valid criterion for assessing genetic behavior under drought conditions. Hydraulic conductivity was found to be directly correlated with the rate of transpiration, which may help to regulate water loss through transpiration and maintain the sap pressure above the threshold to avoid the xylem cavitation (Tausend et al. 2000). The drought-tolerant genotype should therefore be able to maintain a high relative water content (RWC) under conditions of moisture stress. It has been reported that by effectively controlling transpiration along with the low-elasticity of the cell wall could elevate RWC of leaves under drought (Pinheiro et al. 2004). A slight shift in the turgor due to loss of water from the subsequent drought can signal leaves to retain a high RWC and high symplast volume. Moreover, maintaining a large symplast volume may be crucial to maintaining the exchange of gases under drought, so maintaining a high proportion of RWC is crucial in granting drought tolerance (DaMatta 2004).

4.2 Leaf Abscission

Abscission of leaves happens in many crops when approaching the annual water deficit. The reduction of leaves can be considered as a complete plant mechanism to reduce water loss through transpiration in these plants. Leaf shedding has been reported in response to water shortages occurring sequentially from older to younger leaves and more drought-sensitive genotypes have more leaf shedding (DaMatta and Rena 2001). However, it was proposed that at least senescence may be a result of stress rather than a defense, since sensitive clones of drought, which lose their leaves to a large extent, also show the condition of the most decreased water status in the remaining leaves (DaMatta 2004).

4.3 Leaf Anatomy

The anatomical adaptations of the leaves, such as cell size, number, frequency, and resistance of stomata, and the content of the epicuticular wax were described as basic anatomical indicators to assess the moisture stress in the coconut (Gomes and Prado 2007).

4.3.1 Hydrophobic Waxes

Waxes are involved in the first defense plant against abiotic stresses, in which hydrophobic waxes play a significant role in water retention by reducing nonstomatal water loss (Jenks 2002). Adaptive anatomical features in coconut include leaflets with a higher upper epidermal waxy cuticle thickness of twice the lower epidermis, thicker cuticular edges, thicker xylem tracheids with thick lignification and tracheids with thick scalariform. Furthermore, the water tissue containing thin-walled cells in the upper and lower angles of the straightened leaflet margin and fibrous sheet encircled from seven to eight large vascular bundles in a strong midrib are also appeared in the leaflet lamina. The presence of two layers of hypodermal cells beneath the upper epidermis and a multilayered palisade tissue is tightly packed and is also seen as a contrast to the scanty spongy parenchyma located between the upper and lower hypodermis. Coconut varieties having tolerant behavior have more scalariform thickness on tracheids and large sub-stomatal cavities (Kumar et al. 2007). A negative relationship was observed between the epicuticular waxy content of coconut leaves and the rate of transpiration by demonstrating a high content of epicuticular wax between drought-tolerant and sensitive genotypes (Riedel et al. 2009).

4.3.2 Sclerophylly

Sclerophylly is another adaptation to drought stress, where plants form hard leaves that will not be permanently damaged by wilt and can be restored to full function when natural conditions resume (De Micco et al. 2016). Sclerophylly are widely distributed in arid and semi-arid environments along with seasonal dimorphic species. Sclerophylly has been interpreted as a phenomenon associated with other functions such as protection against pathogens or response to the availability of scarce nutrients (Salleo and Nardini 2000). However, there is evidence that the hard leathery leaves are widespread in species that adapt to the drought that occurs in different environments around the world. Sclerophyllous are characterized by reinforced tissue (such as thick-walled epidermal cells, sclereids, etc.) that prevent the entire structure from collapsing when water is scarce (drought hypothesis), reducing the risk of mechanical damage. Under drought conditions, sclerophyllous leaf slowly reduces volume due to thick cuticle and thick-walled epidermal cells, but thin-walled mesophyll cells shrink sharply, leading to increased intercellular spaces. This allows photosynthesis to remain active also under conditions of extreme water stress when the other leaf types wilt (Shields 1950).

Seasonal dimorphic species are characterized by a seasonal reduction in their transpiring surface. Larger winter mesomorphic leaves that grow on the dolichoblasts (long-twigs) are shed at the onset of arid season and are replaced with smaller summer xeromorphic leaves on the new brachyblasts (short-twigs) (Aronne and De Micco 2001). Both sclerophyllous and summer leaves of seasonal dimorphic plants have many similar traits that allow them to resist drought conditions. These characteristics relate not only to the morphological appearance of the leaves on the axes but also to their anatomical properties both within and at the surface level. It is well accepted that phenomena such as paraheliotropism and acute inclination of leaves and changes in color, due to altered pigment content, are frequent in dry habitats. It helps to reduce solar radiation and thus reduce the heating of leaves and transpiration rates as well as avoid damage to photosystems and phenomenon of inhibition (Arena et al. 2008).

4.3.3 Gas Exchange

Drought adaptation is also achieved by reducing the ratio of external leaf surface to its volume. It is generally accompanied by a compact structure made up of small mesophyll cells with thick cell walls, reduced intercellular spaces and a network of integrated veins. Increased leaf stiffness, despite reduced gas exchange, may reduce water loss, which increases the density of leaf tissue and dry mass per area (Niinemets 2001). Recently, the overall low diffusion conductance across the intercellular spaces from the sub-stomatal cavities to the chloroplasts has been linked not only to reduced aquaporin conductance, but also to the precise anatomical features such as reducing the surface area of the chloroplasts exposed to the intercellular space per unit each leaf area (Tosens et al. 2012). The presence of additional layers of palisade parenchyma at the expense of spongy tissue is also a way to increase the water path across the areas between the cells to reach the stomata, and this will be a strategy to increase the water use efficiency (Lewis 1972). In xeric leaves, additional layers of palisade parenchyma adjacent to the lower epidermis leading to an isobilateral anatomical structure can be developed which, together with the steep leaf inclination, allow for better light interception in the early morning and late afternoon hours. Increased mesophyll thickness enhances photosynthesis ability if accompanied by an increase in the number of exposed chloroplasts near the surface area facing the intercellular spaces (Oguchi et al. 2005). However, other factors, including leaf development stage and light availability, are known to interact with drought in determining the modifications of the differentiation of mesophyll and chloroplast, and ultimately of mesophyll diffusion conductance and photosynthetic capacity (Tosens et al. 2012).

4.3.4 Stomatal Characteristics

Water losses are also strongly affected by the characteristics of stomata. Adaptation to drought involves reducing the size of the stomata, while the density of the stomata shows a more plastic response to environmental changes. In a recent study, stomatal density was found to increase with moderate water shortages, while declines under severe drought (Xu and Zhou 2008). There is a lot of information on how quickly reducing transpiration can be achieved through physiological control of stomatal opening. However, stomatal closure is closely controlled by hydraulic architecture of leaves and stems that regulate critical thresholds for vulnerability to cavitation (Sperry 2000). Plants of arid and semi-arid environments show sunken stomata, often covered by resinous masses and layers of wax or attached to deep crypts of the lamina (Monneveux and Belhassen 1996). These crypts are often covered by wax tubules or trichomes, which may further reduce transpiration. However, the stomata associated with the cuticular structures have also been shown to be adapted to excess water in plants growing in rainforests and cloud forests. These plugs help to maintain photosynthetic activity by preventing the formation of a continuous water film that would inhibit diffusion of carbon oxide into the leaf (Feild et al. 1998). The presence of sunken stomata and hairy leaves is not limited to plants in arid environments and is also associated with other functions such as protection from herbivores (Koster and Baas 1982). Although xeromorphic leaves are usually more hairy than mesomorphic leaves, there is a common agreement on the fact that dead trichomes participate in reducing transpiration when stomata are closed, whereas living trichomes may increase water loss (Shields 1950).

4.3.5 Cuticular Characteristics

Xeromorphic leaves also have thick cuticle. The hydraulic permeability of the cuticle depends on its thickness, chemical composition and crystal forms of cuticular waxes embedded in or deposited over it. The water permeability of the cuticle is less in xeromorphic than in the mesomorphic leaves. The characteristics of this barrier in the interface with the atmosphere have been shown to be genetically controlled (Riederer and Schreiber 2001). Cuticular water permeability also depends on relative humidity, which decreases with the dryness of the environment due to the chemical properties of the cutin and wax domains (Bargel et al. 2004). Moreover, different cuticle components can determine changes in reflection and thus indirectly affect the plate lamina (Monneveux and Belhassen 1996). In fact, the occurrence of cuticular waxes, by increasing the reflection of the leaves at visible and infrared wavelengths, can reduce photoinhibition of photosynthesis and transpiration rates, which has a positive effect on water use efficiency. In addition, ultraviolet radiation is also mitigated by the presence of flavonoids in the cuticle matrix or on the surface of the epicuticular waxes (Bargel et al. 2004).

4.3.6 Phenolic Compounds

Excess radiation can also be filtered by phenolic compounds radiation either in the form of vascular accumulations in glandular leaf hair, in epidermal and parenchyma cells, or linked to membranes through mesophyll cells as also shown in the cortical cells of green photosynthesis branches (De Micco and Aronne 2008). Such phenolics also play an indirect role in drought adaptation, being feeding deterrents and protecting plants from permanent damage caused by grazing and pathogenic attacks. This is critical because the development of spare parts may require additional vital cost, which is often not possible under drought conditions.

The above traits are related to the reduction of transpiration. However, xeromorphic leaves can also be characterized by the presence of multilayered epidermis or parenchyma for water storage. These water storage tissues show less osmotic pressure than photosynthetic cells that, under conditions of low water availability, can get water from water storage cells. The latter is generally thin-walled cells that can easily shrink, although they recover quickly when water becomes available again (Fahn 1964). In xeromorphic leaves, recovery of cell turgidity after shrinking without mechanical damage favors the so-called concertina cells, which include cell walls, which help to rapidly expand when water becomes available again (Aronne and De Micco 2001).

4.4 Stem Anatomy

Stem also has good adaptation strategies against drought stress by reducing water loss and formation of a good water transport system. It also helps in water saving through thick cuticles and thick-walled epidermal cells with suberized sub epidermal layers in young organs or interxylary cork rings in older stems. These structures from hierarchical "series" barriers as hydraulics regulating the exchange of water at plant's atmosphere (De Micco and Aronne 2012). Water is stored through the existence of succulent stems, underground structures such as bulbs, rhizomes living wood fibers and parenchyma tissues comprising reserve materials (Fahn 1964). In addition to analyzing structures that reduce water loss or serve in water storage, most studies have focused on adaptive features that regulate the transfer of water from roots to leaves. Given the metabolic cost of the formation of xylem tissue, the plants that are suitable for xylem formation to suit the expected demand through evapotranspiration of leaves must have an adaptive advantage during evolution (Sperry 2003). In the past few decades, there has been growing interest in studying the relationships between plant anatomy and environmental factors. In general, most anatomical characteristics of plants can be explained on the basis of their functional significance in species survival strategies. They define the hydraulic conductivity and exposure to the cavity (safety) in a particular plant, as well as the biomechanics of the stems and branches (Baas et al. 2004; Hacke and Sperry 2001).

4.4.1 High Water Conductivity

Drought adaptation can be achieved through a kind of compromise between the need to maintain high conductivity when water is available and avoid embolism in dry conditions. The major ecological trends in plant anatomy confirm that when moving from mesic to xeric, plants tend to be less effective in water flow, but are more resistant to embolism and more powerful. In Mediterranean environments, this general trend is often attended by a change in the habit of tree plants to evergreen/deciduous shrubs (De Micco et al. 2008). For many anatomical characteristics of a plant, the presence of direct and indirect proportionality can be traced to the properties of water transport efficiency or safety. These relationships are sometimes vague, keeping in mind that different combinations of different properties can move the balance toward one or other extreme. This balance between efficiency and safety is well achieved by many species of shrubs from semi-arid systems in the Mediterranean ecosystems which specialize in high connectivity when water is available, as well as for safety during droughts (narrow latewood vessels, vasicentric tracheids, etc.) (Carlquist 1989).

4.4.2 Scalariform Perforation Plates

The xeric conditions seem to favor the choice of short elements with simple perforation while those scalariform were generally limited to plant taxa with a mesic or boreal/alpine environment (Baas 1986). Simple perforation plates are the most advanced features than scalariform ones (Wheeler and Baas 1991) and provide the ability to conduct larger amounts of water per unit of time when water is available (Christman and Sperry 2010). In accordance with Sperry (2003), scalariform perforated plates will be useful under drought conditions because they prefer to refill the embolized vessels in a passive manner. More precisely, scalariform perforation would divide large gas bubbles into smaller bubbles that could decompose more quickly than fewer and larger bubbles into filling vessels with simple perforation plates. Xeromorphic plants are generally characterized by an increasing number of narrow vessels. In drought conditions, the redundancy of transport cells allows the transfer of water despite the deactivation of part of the tissue (Baas et al. 1983). The conductivity corresponds to the fourth power of the vessel radius (Zimmermann 1983), narrow vessels allow only a slow flow of water. However, narrow vessels are valuable for safety because they also ensure water transport when large vessels are embolized (Carlquist 1975). In arid and semi-arid environments, the size of the vessel may also be limited by the need to reduce intervessel pitting and embolism by air seeding, a phenomenon that is described as a gas that is pulled through the porous pores (Sperry et al. 2006). The porous membrane porosity between adjacent vessels can be designed to resolve the conflict between functional requirements to reduce vascular resistance, which favors thin films, porosity, and reduce embolism, requiring strong membranes and smaller pores. Indeed, the decrease in the total area of each vessel leads to a decrease in the average size of the membrane pores, thus increasing the safety of the cavity (Wheeler et al. 2005). In light of these considerations, features such as pit membrane permeability, pit membrane area, and pore size can be considered important adaptive traits likely to lead to environmental differences between species (Jansen et al. 2009).

4.4.3 Vessel Grouping

The vessel grouping is another safety-friendly phenomenon common in arid desert and Mediterranean vegetation. More recently, the first empirical evidence of the positive association between vascular aggregation and cavity resistance has been reported in seven species of Acer (Lens et al. 2011). If there is a specific vessel in a group embolises, the surrounding active vessels maintain a three-dimensional connector path. Almost the same role can be played by vasicentric tracheids that are spread in sheaths around vessels, especially in wood from semi-arid ecosystems. These tracheids have been "reinvented" in the corridors that have developed more specialized traps where the primitive tracheids have been replaced by non-conductive fibers (Carlquist 1989). The other type of tracheid reinvention is "vascular tracheids" which are formed only at the end of the growth loop (Carlquist 1989). However, as in very narrow vessels, the rate transmitted through the tracheids is very low, but this is not supposed to be a problem as the transpiration may be expected to decrease and the delivery rates decrease during droughts.

4.4.4 Helical Thickening of Vessels

Another common feature of plants from xeric environment is the occurrence of helical thickness in the vessels. The thickness and density of vessel sculptures in the Acer species have been shown to be associated with cavity resistance (Lens et al. 2011). Apart from the increase in mechanical strength, the helical sculpture has a role in preventing the occurrence and spread of the cavity, because it increases the surface of the wall and then connects the water to the surface (Kohonen and Helland 2009). In plants of xeric environments, helical thickness increases mechanical strength, which can also be ensured by the presence of vessels and imperforate tracheary elements with very thick walls and narrow cavities (Sperry 2003). In addition, other features, such as the length of conduit elements, play an important role in determining the efficiency of water transport and resistance to negative pressures, but because they are associated with the diameter of the vessel cavity, they are less studied. In fact, even within the same plant, specific plant anatomical traits, such as vessel size and cell wall thickness, can vary depending on the season in a type of seasonal dimorphism that leads to the development of safer summer tree rings than the winter tree rings shown in the *Cistus incanus* (De Micco and Aronne 2009). Thus, the high plasticity of specific anatomical features is responsible for their annual variability, which can be linked to environmental fluctuations and used as evidence of growth cycles.

In general, drought adaptation can be achieved by different combinations of wood anatomical characteristics. For example, a larger thickness in any species may compensate for a smaller amount of imperforate tracheary elements when the vessel frequency is too high as described in *Cistus monspeliensis* L. by (De Micco and Aronne 2008). In addition, adaptive plant characteristics can be considered as added (Carlquist 1989) such a feature will explain the distribution of species along the water availability gradients in semi-arid ecosystems such as the Mediterranean types. In these ecosystems, moving from more wet areas to arid zones, there is usually a shift from evergreen to seasonal/seasonal dryness, also associated with increased occurrence of traits in plant anatomy allowing for adaptation to drought conditions (De Micco and Aronne 2009).

4.4.5 Stem Vascular Systems

Plants such as coconut, palm oil, and date have long stalks developed from the apical meristem. Being monocotyledons, vascular bundles like xylem and phloem are appear scattered throughout the ground parenchyma, and there is no distinction between the cortex and the pith in these plants. Vascular bundles have a longitudinal or relative pattern that runs along the stem. The stem anatomy is regulated in a way that reduces the functional segments isolated from the mass flow from the roots to the canopy. The transport capacity of this huge hydraulic system increases with stem diameter, and it works over the lifetime of the plant, which can extend to more than a century. Thus, the ability to withstand water stress in plants such as coconut is supposed to be coordinated with stem processes as well (Couvreur and Baker 2013). This complex process may contribute to the high drought adaptation of tall-coconut genotypes as compared to dwarf species. The stem girth of the tall coconut was found to decrease between dawn and midday before rising again during the afternoon, indicating the organized control of water transport (Passos and da Silva 1991). Variation in stem diameter was also described as a reaction to drought stress in coconut. Stress caused by drought, floods, lack of minerals, and some diseases can cause a decrease in stem diameter, which declines to normal when environmental conditions improve. Therefore, the girth and form of the stem make a good record in the history of stress in palm plants. The phenological changes in the drought sensitivity of the oil palm indicate that palm tree plants are more sensitive because they do not have a large root system and a large stem as in palm tree plants (Legros et al. 2009).

4.5 Root Anatomy

A well-known response of plants to drought is to modify the ratio of roots to shoot dry biomass that is recommended in favor of the former. This governs the increased root density on the leaf area, a phenomenon that brings advantages already discussed for the seedlings. Adaptive approaches for drought are also based on many other features that are considered both root structure and anatomy.

4.5.1 Deep Root System

At the start of the twentieth century, there was a common idea that the roots of arid areas must grow deep in the soil to withstand prolonged periods of severe drought. It has been proven later that there is a variety of adaptive strategies. Of these species, three types of root structures were described for perennial species. The first type, the typical species of succulent, is made of shallow roots, does not grow deeper than about 20 cm. The second type includes both long roots that grow parallel to each other and a few meters deep, shallow, and adventitious roots designed for rapid absorption of water after short short-term precipitation. The third system has several lateral roots of one meter long, with very long taproots that can be tens of meters long (Kummerow 1981). Such roots reach very deep layers of soil where phreatic water is always present, thus avoiding seasonal fluctuations in water availability. In some species, like Agave deserti Engelm, the root system consists of rain-induced and roots that are produced on established roots within a few hours of rain and are flattened when the soil is dry (Hunt Jr. et al. 1987). Although it represents an additional production cost for the plant, rain-induced roots provide an adaptive advantage in desert environments with a higher hydraulic conductivity than established ones. In semi-arid environments, root systems with deep and shallow roots coexist (Kummerow 1981). Furthermore, in the Chilean matorral and Californian reports, some species have been reported to be interrelated (Kummerow 1981). Though the main types of root structure are described, it is also common that the distribution of roots throughout the soil is largely influenced by the moisture content of the superficial layers more than the deeper layers.

4.5.2 Rhizodermis and Small Root Diameter

Water loss is controlled by specialized tissue such as rhizodermis with thick outer cell walls, a well-developed suberized exodermis, often accompanied by many layers of thick or thin suberized cells. The presence of suberized layers of cells in the vicinity of the root is an important mechanism not only in the selection of nutrient uptake, but also in particular because it regulates the inverse flux of water which, in extreme drought conditions, can be transferred from root to soil (Hose et al. 2001). It has been shown experimentally that the restriction on radial hydraulic conductivity of the root through suberized layers increases during the development of roots and during soil drying (North and Nobel 1995). With respect to the internal structure, the adaptive function of specific root traits is generally acceptable even if there are steady evolving trends between plant groups and along the mesic-xeric gradients less investigated than the stem. The Xeromorphism at the root level depends on the presence of features related to the regulation of water absorption, avoiding loss of water and formation of water storage tissues. The presence of small-diameter roots under low water availability is a strategy to increase absorptive surfaces and thus increase water and nutrient uptake rates (Eissenstat 1992).

4.6 Less Cortical Layers and Lacunae

After crossing, the epidermis and the exodermis, the water must pass through the cortical parenchyma. The reduced number of cortical layers is an adaptive advantage in drought conditions because it shortens the path between the soil and the stele which favors the rapid transmission of water (Fahn 1964). However, given the inverse flux of water during drought conditions, cortical rupture is a strategy to create cortical lacunae that cut the pathway of water from stele to soil (Robards et al. 1979). In the Agave deserti Engelm, the development of cortical lacunae between cells has been shown to decrease radial conductivity across the root cortex in response to soil drying (North and Nobel 1995). The opening of large fractions, especially in external cortical layers, can also open a pathway for water release as readily available in *Opuntia ficus-indica* (L.) Miller under dry soil conditions (North and Nobel 1996). Recently, it has also been predicted that the presence of cortical lacunae is beneficial in drought conditions because it decreases the costs of root metabolism by converting living cortical cells into air volume (Zhu et al. 2010). However, the development of cortical lacunae can decline the strength of mechanical roots, making them more vulnerable to soil exposed to cycles of swelling shrinkage caused by frequent fluctuations in water availability (Striker et al. 2007). This weakness in root strength can be balanced by the overall result of further development of lignified tissue with cells characterized by thicker walls in various structures (Mostajeran and Rahimi-Eichi 2008). The shrinkage of cortical parenchyma cells in developing roots under droughts is a common result however, this phenomenon can be considered reversible especially when cell walls are enhanced by deposition of suberin as described in *Lygeum* spp. (Peña-Valdivia et al. 2010).

4.7 Thick Endodermis and Root Hydraulic Conductance

The development of the endodermis with thick cell walls and the formation of additional layers of cells with suberized walls around the stele is seen as a way to prevent drying of meristematic tissues such as pericycle and other tissues within the stele (North and Nobel 1992). The unavailability of water results in the formation of a larger number of endodermal cells with the Casparian bands closer to the root tip in *Opuntia ficus-indica* (North and Nobel 1996). This accompanied by other anatomical changes, including the formation of more suberized peridermal layers, which decrease water permeability of the cortex as measured in other species (Schönherr and Ziegler 1980). The importance of apoplastic barriers in water flow control depends on the stage of root growth, in young unstressed roots, most of the radial hydraulic resistance is evenly distributed on the cortical layers, whereas in the old water-stressed plants it is mainly due to the exodermis and endodermis (Steudle 2000). Once the water has entered the stele, it must be transported efficiently throughout the plant to maintain the continuous water supply of the leaves. Under xeric conditions, plants have developed morpho-functional traits that alter root hydro-logical conductance as a mechanism for regulating the transpiration of those that are played by aerial organs (Trubat et al. 2006). It is recognized that water transport efficiency and hydrological safety of the vascular system are essential for the survival of plants in arid environments. In fact, in the context of climate change, given the increased frequency of drought and generally high temperatures, vegetation in arid and semi-arid environments must adapt to increased xylem vessel cavitation. Within this scenario, in these environments, adaptive capacity of species is closely related to the characteristics of their xylem at the root, as in the stem. In fact, within the same plant, conductivity shows a strong variation between different root types that grow with different directions and at different depths, a phenomenon that must improve water flow according to the availability of changing water (De Micco and Aronne 2010).

As a final test, we may highlight that plants that show different root structures and different combinations of anatomical features that allow adaptation to drought coexist in the same arid and semi-arid regions. In addition, it has been shown that water stress leads to a wide range of morphological and anatomical root responses, where different traits are modified strongly and different trends in different species and even in different cultivars within species (Peña-Valdivia et al. 2010). The comprehensive understanding of different mechanisms to adapt the root of water deficit is a valuable target because the roots can be considered sensors that detect changes in the availability of water in the soil and affect drought resistance at the plant level as a whole.

5 Physiological Adaptations

5.1 Behavioral Adaptations

Occurrence of annual natural defoliation known as "wintering" at the beginning of winter and lasts for approximately 4–6 weeks during the season (Vinod et al. 1996). The flushing of leaves occurs at the end of the winter period and before the arrival of rainfall to the extent that coincides with the months of drought. Flushing requires the flow of large amounts of water to develop and expand the leaves (Williams et al. 2008). Root zone activities show significant absorption of deepwater roots during winter and flushing of plants (Guardiola-Claramonte et al. 2008). The large decrease in transpiration and the increase in the absorption of root water helps trees maintain the stem water potential needed to flush the leaves later (Priyadarshan and Clément-Demange 2004). Plants can preserve extracted water without releasing it into the atmosphere until new leaves are developed (Guardiola-Claramonte et al. 2010). The differential adaptation of drought between the main sub-species of Arab coffee plants, Arabica and Robusta appears to be governed by changes in water use rates and/or soil water extraction efficiency (Pinheiro et al. 2004). During the long dry period, reducing the leaf area and the alternative assimilates partitioning to the new leaves with the main adaptation mechanism for survival in Arab coffee plants (DaMatta et al. 2003). Behavioral differences between tolerant and susceptible genotypes were revealed in relative drought tolerance, governed by deep root adaptation, hydraulic conductivity of the plant, and control of stomata in water loss (Pinheiro et al. 2005). Consequently, a better adaptation of drought between genetic patterns was reported with deeper root systems among the Robusta coffee plants (Pinheiro et al. 2004). The depth of rooting is also reported to be similar to the effect of drought in tea where shallow roots clones were found to be drought-prone as compared to deep-rooted clones. Moreover, drought tolerance was found to enhance with the depth of rooting in shallow-rooted clones, while no significant differences were found in deep-rooted clones (Nagarajah and Ratnasuriya 1981).

5.2 Canopy Architecture

Canopy architecture is found to play an important role in adapting to drought in plants. The structure of the canopy partially determines the hydraulic architecture of the plant (Herzog et al. 1998). Dwarf cultivars with dense crowns are more resilient to drought by delaying dehydration as compared to open-crown varieties. The canopy compactness is also accomplished by reducing the size of the leaves and changing the shape of the crown, resulting in better energy dissipation with less transpiration (Kozlowski and Pallardy 2002). In some plants, the compact canopy is useful for long periods of drought. This prevents high light density from reaching the lower canopy leaves and blocks to penetrate the radiation from access to the plantation floor that can heat the surface soil, change the vapor pressure deficit (VPD) and change the micro climate leading to evapotranspiration loss. The compact canopy has been implicated as an ideal phenotype for drought tolerance (Prakash-MBM and Jacob 1999). The loss associated with microclimate was suggested as one of the possible causes of crop failure, despite adequate supplementary irrigation, in sites with high evaporation requirements for cultivars with open crowns (DaMatta 2004).

5.3 Stomatal Regulation and Signaling

Closing stomata is the first fast-response event of plants for water shortage. The stomatal closure is closely related to the soil moisture content of the leaf water status and is mainly controlled by chemical signals such as ABA, which is produced in dehydration. A direct relationship between xylem ABA content and stomatal conductance has been demonstrated. Changes in plant hydraulic conductivity, plant nutritional status, xylem sap pH, farnesyl transferase activity, lack of leaf vapor pressure to air, and relative water content reduction are other factors that work in stomatal regulation. Although the assimilation of CO_2 and net photosynthesis decreases due to the closure of the stomata, achieving low transpiration rates and preventing water loss from leaves is a good trade-off against growth for survival. Stomata completely close from mild to severe stress depending on plant species control stomata opening, allowing some fixation of carbon and improving water use efficiency. Increasing the resistance of stomata under stress levels indicates the efficiency of species to conserve water. As a reliable indicator of stomatal performance, stomatal conductance has been used to assess the sensitivity of drought, because stomatal conductivity is highly correlated with transpiration rate and photosynthesis (Carr 2011). Stomatal conductance was used as an early indicator of stress in Arabica coffee, because a reduction in stomatal conductance was found to occur even in one-third of the available soil water depletion. Evidence suggests that during short-term water shortages, low yield in coffee genotypes may be associated with stomatal conductance and net carbon assimilation (Nunes 1976). Although weak stomatal control was found during drought, mechanisms that lead to drought tolerance in the Robusta coffee plant are still largely unknown (Pinheiro et al. 2004). In coconut plant, solid evidence of stomatal regulation of plant water status has been recognized at mild to moderate drought (Gomes et al. 2008) making it a useful parameter to distinguish drought-sensitive and tolerant genotypes in combination with tissue water potential (Lakmini et al. 2006). The presence of sensitive VPD in the cocoa plant was reported, where the leaves showed low water efficiency at high VPD, resulting in a decrease in tissue water under limited soil water supply (Gomes et al. 1987).

5.3.1 Hydraulic Signaling

Regulation of stomata in plants occurs in response to signals originating primarily from roots. These signals lead to a series of events such as the induction of differential gene expression, changes in cell metabolism and the development of defense systems in the above-ground organs (Kholodova et al. 2006). Systemic signals are primarily hydraulic in nature and propagate along the xylem tissue that coordinates physiological responses especially in the leaves (Jackson 2002). As soil moisture decreases, this results in a gradual decrease in the water of the shoot, leading to direct hydraulic signals to the leaves. Several studies point to the accumulation of ABA as a primary candidate for hydraulic signals in plants (Kholodova et al. 2006). Root tissue produces ABA as an initial response to decreasing soil moisture availability. The hydraulic signals are immediately spread through the continuous water phase within the plants, culminating in the regulation of the stomata.

5.3.2 Non-hydraulic Signaling

Several studies have shown that moisture-depletion signals reach the leaves from the roots, without assistance from the water status, suggesting a non-hydraulic pathway for signaling (Croker et al. 1998). Although the precise mechanisms involved in the non-hydraulic root for signal capture are still unclear, although they are an important component of plant response to drought (Davies et al. 1994). Recently, non-hydraulic chemical mediation signals have been reported in the coconut plant, in which the chloride ions are involved in sensing the soil water depletion in the root zone and sending the stomata closure signals to the leaves. In coconuts, chloride ions have been found to play two important functions in regulating water balance. First, they regulate the stomatal closure by coordinating the flow of water between six adjacent cells (two guard cells and four subsidiary-cells) of the stomata and secondly, chloride ions increase osmoregulation capacity under water stress (Gomes and Prado 2007).

5.4 *Photosynthesis*

Variation in photosynthesis and associated systems has also been used as indicators of drought tolerance in crop cultivation. To a large extent, the photosynthesis performance of plants is determined by environmental variables under field conditions. Photosynthesis is closely related to the function of stomata, where gases are exchanged in plants through stomata. The drought-induced stomatal closure limits the diffusion of carbon dioxide from the atmosphere to the intercellular spaces, resulting in reduced photosynthesis (Repellin et al. 1997). Plant genetic changes in rates of recovery of gas exchange and internal dehydration have been reported on drought exposure. In dwarf coconut cultivars, photosynthetic acclimation was observed after repeated cycles of drying and recovery (Gomes et al. 2008). Use of photosynthesis and efficient water use where promising reference parameters reported in plants (Nainanayake and Morison 2007). Chlorophyll fluorescence transients have been implicated in the identification and screening of coconut seedlings that can adapt to water stress conditions (Vinod 2012).

5.4.1 Photosynthetic Adaptive Response

Drought stress is known to decrease photosynthesis by reducing both leaf area and the photosynthesis rate per unit of leaf. Low photosynthetic rate is mainly by closing stomata or metabolic impairment (Tezara et al. 1999). Continuous light reactions during drought stress under a limited intercellular CO_2 concentration lead to the accumulation of low photosynthetic electron transport components, resulting in ROS production (Anjum et al. 2017b). ROS can cause severe damage to photosynthetic apparatus (Lawlor and Cornic 2002). Adaptive responses developed by plants to reduce drought caused by photosynthesis include thermal dissipation of light energy, xanthophyll cycle, water cycle, dissociation of light-harvesting complexes from photosynthetic reaction centers. The cause of poor metabolism during drought stress is mainly changes in photosynthetic carbon metabolism. The biochemical efficiency of photosynthesis under drought stress is mainly based on the regeneration of RuBP-5-phosphate (RuBP) and the activity of ribulose-1,5-bisphosphate carboxylase/oxygenase (RuBisCO) (Lawlor 2002). Considerable progress has been made in improving CO_2 diffusion, photosynthetic light reaction, and metabolic changes, including the expression of genes associated with photosynthesis to regulate photosynthesis under drought to improve grain yield (Chaves et al. 2009).

5.4.2 Carbon Fixation Pathway

The C_4 path of carbon uptake has been proposed to be the main adaptation of the C_3 pathway to reduce water loss, reduce photorespiration, and improve photosynthetic efficiency under drought stress (Edwards and Walker 1983). However, several important crops including rice, wheat, soybeans, and potato are using the C_3 path for photosynthesis. Although changing C_4 to C_3 crops is ongoing, its contribution to increasing productivity is very limited (Gowik and Westhoff 2011). Photosynthetic adaptation of plants contains a complex interaction of hormones, ROS, sugars, and other metabolic events to drought stress (Pinheiro and Chaves 2010). Recent studies have used the gene network for rice to determine a TF termed Higher Yield Rice (HYR), which has been strongly associated with primary carbon metabolism and overexpression in the process of improving photosynthesis for rice in normal conditions as well as under drought and high-temperature stress (Ambavaram et al. 2014). HYR regulates many morphophysiological processes leading to increased yield under normal and environmental stress conditions. The study showed that HYR is a key regulator of photosynthesis, directly activating photosynthesis genes, cascades of TFs, and other genes in the downstream involved in photosynthesis, which improves yield.

5.5 Transpiration and Stomatal Conductance

The water absorbed by plants roots is lost through transpiration from the leaves. The proportions vary widely among plant groups: from 10 to 0.1 g of water $dm^{-2}\ h^{-1}$ in hygrophytes species and xerophytes, respectively (Monneveux and Belhassen 1996). The immediate response of plants to exposure to drought stress is to close the stomata. However, closing the stomata not only reduces water loss through transpiration but also reduces the absorption of CO_2 and nutrients, thus altering metabolic pathways such as photosynthesis (Xiong and Zhu 2002). Plants appear in arid and semi-arid environments with xeromorphic properties designed to minimize transpiration under drought conditions. A decrease in transpiration can also be achieved under conditions of drought stress caused by shedding leaves (i.e. deciduous species in drought), as well as reduce number of leaves, leaf size, and branching. In extreme conditions, shedding of branches or twigs or larger parts of plants also occur. In plants that appear leafless under drought conditions, photosynthesis is achieved by green branches, stalks (stem) or petioles (Fahn 1964). Folding the lamina and rolling of leaf is a valuable mechanism to reduce transpiration. In response to drought stress, grasses adapt a special mechanism of leaf rolling due to the loss of bulliform cells that occur in the epidermis or other specialized mesophyll cells (Shields 1950). Recent research has revealed that the low stomatal conductance in response to stress due to drought is not only associated with the low expression of aquaporin genes but also to the anatomical characteristics leading to reduced surface chloroplast surface area exposed to intercellular space per unit of leaf area (Tosens et al. 2012). There are many other factors, including leaf growth stages and the availability of light, which are also known to interact with drought in modulating mesophyll and chloroplasts, ultimately affecting conductivity and photosynthesis (Tosens et al. 2012). Reducing the size of stomata and numbers when exposed to drought is another adaptation to survival under drought conditions. Previous studies have reported that while there is an increase in stomatal density under mild drought stress, there is a decline during severe drought (Xu and Zhou 2008). Thus, all these adaptations in plants reduce the negative effects of drought stress on photosynthesis and thus have a positive effect on Water Use Efficiency (WUE), which in turn will result in high productivity and high yield (Blum 2005). Therefore, the above features represent adaptive mechanisms in plants to survive under drought stress without loss of yield or productivity (Karaba et al. 2007).

6 Biochemical Adaptations

6.1 Osmoregulation

Previous investigations described that one of the key mechanisms by which plants cope with limited water is osmotic adjustment (OA). It is defined as the process of solute accumulation in dividing cells when water potential is low, thus helping to maintain turgor (Chaves and Oliveira 2004). Cell expansion and plant growth depend heavily on the availability of water and help to maintain turgor. Measurement of turgor in growing regions of plants, especially leaves and stems, shows little or no reduction, although cell expansion is inhibited during drought stress and is believed to be due to OA (Serraj and Sinclair 2002). Under drought stress conditions, osmotic adjustment has been involved in maintaining stomatal conductance (Turner 1996), photosynthesis, leaf water volume, and growth (Chaves and Oliveira 2004). In times of drought stress, in addition to reduced water content, there are also other related changes such as increased salt concentration and mechanical resistance (Sauter et al. 2001). These adjustments maintain a positive cell turgor through the active accumulation of compatible solutes. It occurs in cells in response to drought stress signals, where the osmolyte buildup occurs to prevent cellular dehydration. Osmolytes are low molecular weight metabolites capable of compensating high osmotic pressure without interfering with plant metabolism, even at high concentrations. They include inorganic cations, polysaccharides, polyols, amino acids, organic acids, carbohydrates, free amino acids, and quaternary ammonium compounds, known as dominant solutes that accumulate in response to drought stress. Proline is a typical osmolyte, synthesized in plants under various stress conditions. Despite drought sensitivity, plants show effective osmotic adjustment when exposed to drying environments, but suffer yield under stress (Almeida and de Valle 2007). Previous studies revealed that drought-resistant wheat varieties, with yield stability under drought stress, have greater osmoregulation capacity than low resistant varieties (Serraj and Sinclair 2002). The accumulation of compatible solutes such as proline and glycine betaine helps to protect the plants from the harmful effects of drought not only by osmotic adjustment but also by detoxification of ROS, protecting membrane integrity, and stabilizing enzymes or proteins (Ashraf and Foolad 2007). It has been shown that enzymes such as betaine aldehyde dehydrogenase (BADH), pyrroline-5-carboxylate reductase (P5CR), and ornithine ornithine δ-aminotransferase (OAT) play major roles in osmotic adjustment. Osmotic adjustments and stomatal regulation have been reported as one of the mechanisms employed in drought-tolerant coconut plants (Rajagopal and Ramadasan 1999). Traditionally, the analysis of metabolic responses to drought stress has been limited to the analysis of one or two categories of compounds that are considered to be a "key player" in the development of tolerance.

6.2 Metabolomic Profiling

The application of the metabolic approach provides a lower perspective for the metabolic appearance of the response and also helps to discover new metabolic phenotypes. GC-MS metabolomic profiling showed in *Eucalyptus* that drought stress altered more number of leaf metabolites than previously reported in the target analysis. The accumulation of shikimic acid and two cyclohexanepentol stereoisomers was described in response to drought stress for the first time in Eucalyptus. Also, the volume of metabolic adjustments is associated with a response to water stress with the sensitivity/tolerant phenotype. Drought affected approximately 30–40% of the measured metabolites in *Eucalyptus domosa* (a sensitive specie) compared to 10–15% in *Eucalyptus pauciflora* (a drought-tolerant specie) (Warren et al. 2012). Similarly, critical differences were observed in metabolic responses during drought analysis in soybean tolerant (NA5009RG) and sensitive varieties (DM50048) through 1H NMR-based metabolomics. Interestingly, an enhanced accumulation of conventional osmoprotectants, such as proline and soluble sugars such as sucrose or myo-inositol, organic acids or other amino acids (except aspartate), has not been detected in leaves of any genotypes during water stress. In contrast, the levels of 2-oxoglutaric acid, pinitol, and allantoin were affected differentially in genotypes when drought was imposed, indicating possible roles such as osmoprotectants (Silvente et al. 2012).

In contrast to soybeans, amino acid levels, including proline, tryptophan, leucine, isoleucine and valine, were enhanced under drought stress in three different varieties of wheat (*Triticum aestivum*) analyzed for 103 metabolites in the targeted GC-MS approach (Bowne et al. 2012). Metabolic adjustments in response to adverse conditions are temporary and depend on stress severity. In a 17-day time trial in maize (*Zea mays*) subjected to drought stress, GC-MS metabolic analysis showed changes in concentrations of 28 metabolites. The accumulation of soluble carbohydrates, proline and eight other amino acids, shikimate, serine, glycine, and aconitase, was accompanied by reducing leaf starch, malate, fumarate, 2-oxoglutarate and seven amino acids during the dry cycle. However, with the increase in negative water negative potential between days 8 and 10, changes in some metabolites were more dramatic, indicating their dependence on the intensity of stress (Sicher and Barnaby 2012). Accumulation of compatible solutes is an evolutionary characteristic preserved in bacteria, plants, animal cells, and marine algae. After 2 weeks of physiological drought stress, 26 metabolites were affected differentially in gametophores, including altrose, maltitol, L-proline, maltose, isomaltose, and butyric acid, compared to metabolic adjustments previously reported in the *Arabidopsis* stressed leaves. A study of metabolic GC-MS, together with the analysis of primary metabolic fluxes of cell cultures and A. thaliana roots treated with oxidative stressor menadione, showed similarities and differences in metabolic adjustments that triggered in both culture systems.

6.3 Antioxidant System

Various drought adaptation processes include a stress-induced cascade of reactions in plants, including the scavenging of ROS produced during oxidative stress. To prevent oxidative damage, cells contain antioxidants that reveal free radicals (Tanveer and Shabala 2018). Phenolic compounds are cellular compounds with antioxidant properties, and many studies have shown that the production of compounds with effective antioxidant structures, such as the extra hydroxyl groups on the B ring of the flavonoid skeleton, are accelerated during drought stress (Ryan et al. 1998). Plant also contains enzymes, i.e., superoxide dismutase (SOD) and catalase, which protect them by removing superoxide particles and hydrogen peroxide, respectively (Takeuchi et al. 1996). The increase in intercellular levels of ROS is a common result of adverse growth conditions. An imbalance between ROS synthesis and scavenging occurs independently of the nature of stress, resulting from both abiotic and biotic stresses. ROS toxic concentrations severely damage protein structures, inhibit the activity of multiple enzymes from important metabolic pathways, and lead to the oxidation of macromolecules, including lipids and DNA. All these adverse events threaten cellular integrity and may lead to cell death (Kar 2011). Normal cellular metabolic activity also produces ROS in normal growth conditions. Thus, cells perceive an uncontrolled increase in ROS and use it as a signaling mechanism to activate preventive responses (Møller and Sweetlove 2010). In this context, plants have developed effective mechanisms to remove toxic concentrations from ROS. The antioxidant system consists of protective enzymes (e.g., superoxide dismutase, catalase, peroxidase, reductase, and reoxin) and radical scavenger metabolites (mainly GSH and ascorbate). GSH is an important component of an antioxidant system that donates an electron to unstable molecules such as ROS to make it less reactive. It can also act as redox buffer in recycling ascorbic acid from its oxidized form to its reduced form by dehydroascorbate reductase enzyme (Jozefczak et al. 2012). Regeneration of metabolic networks is a critical response that gives cells the best chance of escaping the oxidative challenge. In *A. thaliana*, oxidative therapy with methyl viologen leads to a reduction in the regulation of photosynthesis related genes and the associated discontinuation of starch and sucrose synthesis pathways, and at the same time catabolic pathways are activated. These metabolic adjustment avoid waste of energy used in non-defensive processes and mobilize carbon reserves for emergency relief work such as the accumulation of maltose, a protein structure stabilizer molecule (Scarpeci and Valle 2008). In tea, high polyphenol content was reported as an indicator of drought tolerance (Cheruiyot et al. 2007). Drought-tolerant clones of tea have the highest activity of catalase to decompose the hydrogen peroxide formed in the respiratory pathway (Jeyaramraja et al. 2003). Additionally, drought-induced reduction in catechin content has been reported in tea clones (Singh et al. 2009). This behavior is due either to the instability of catechins under drought or to the potential loss of catechins due to improved cellular injury or to down regulated pathways, thereby reducing the availability of precursor molecules (Singh et al. 2009; Sharma and Kumar 2005).

6.4 TCA Inhibition

Tricarboxylic acid cycle (TCA) inhibition is accompanied by the accumulation of pyruvate and citrate from a decrease in the concentrations of malate, succinate, and fumarate. This early response (0.5 h) was observed in both systems. The inhibition of the TCA cycle simultaneously decreases glutamate and aspartate concentrations due to inhibition of precursor synthesis associated with TCA 2-oxoglutarate and oxaloacetate, respectively. Another early redistribution of metabolic metabolism is the redirection of the carbon flux from glycolysis to the Oxidative Pentose Phosphate (OPP) pathway. This is also reflected by a decrease in glycolytic pools of 6-P and glucose-6 phosphate and fructose 6-P, and an increase in OPP pathway and ribulose 5-phosphate and ribose 5-phosphate. Improved carbon flux through the OPP pathway may provide reducing power (via nicotinamide adenine dinucleotide phosphate, NADPH) for antioxidant activity, because oxidative stress reduces the levels of GSH, ascorbate, and NADPH reduction. After 2 and 6 h of stress development, metabolic adjustment differs in response to oxidative stress in the roots than in cellular suspension cultures. At the roots, TCA cycle intermediates and amino acids are recovered. In contrast, in cell cultures, concentrations of these metabolites remain low throughout the time course, indicating higher levels of oxidative stress in cell cultures. At the end of the treatment period (6 h), 39 metabolites, including GABA, aromatic amino acids (tryptophan, phenylalanine, tyrosine), proline, and other amino acids, were changed dramatically in the roots. These results showed a wide range of metabolic modifications obtained in response to oxidative stress and the impact of the biological system analyzed (Lehmann et al. 2009). The carbon flux reorientation from glycolysis was also reported through the OPP pathway and subsequent increase in NADPH levels in rice cell cultures treated with menadione. The CE-MS analysis of these rice cultures indicated that most of the sugar phosphate produced by glycolysis (pyruvite, 3-phosphoglyceride, dihydroxyacitone phosphate, fructose 6-phosphate, glucose-1-phosphate (G1P), G6P, G3P, phosphoenolpyruvate) and TCA organic acids (2-oxoglutarate, aconitate, citrate, fumarate, isocitrate, malate, succinate) and increases in the levels of OPP pathway intermediates (6-phosphogluconate, ribose 5-phosphate, ribulose 5-phosphate). Gradual increases in the biosynthesis of GSH and internediates (O-acetylL-serine, cysteine and- γ-glutamyl-L-cysteine) were observed in rice cell cultures treated with menadione (Ishikawa et al. 2009).

6.5 Other Biochemical Indicators

Late embryogenesis (LEA) is a distinct group of proteins caused by dehydration stress due to high temperatures, drought, salinity, and some developmental events such as seed maturation (Close 1997). LEA proteins are thought to be a subgroup of dehydrins that are hydrophilic in nature and soluble at high temperatures. LEA proteins are supposed to act as one of the molecular novel forms to help prevent the formation of damaging protein aggregates during water stress (Goyal et al. 2005). Dehydrins is generally a structural stabilizer that protects the nuclear, cytoplasmic, and membrane macromolecules from damage caused by dehydration, thereby maintaining cell structure and integrity. In crops, dehydrin like protein has been identified in coffee and eucalyptus (Hinniger et al. 2006). Another important and very diverse group of proteins involved in drought stress are heat shock proteins (HSPs). HSPs also act as molecular assemblers that reduce protein aggregation and target aggregated protein degradation, while helping to protein folding, aggregation and transport.

7 Hormonal Adaptation

The main plant hormones, such as ABA, CK, auxin, GA and ethylene, regulate various processes that enable the plant to adapt to drought stress (Wilkinson et al. 2012). Plant hormones, ABA and ethylene, play an important role in plant adaptation to environmental stress. The water-deficit sensor and the activation of defense mechanisms come from chemical signals in which abscisic acid plays a vital role. Abscisic acid accumulates in the tissues of plants exposed to water stress and promotes the reduction of transpiration by closing the stomata. Through this mechanism, plants reduce water loss and reduce stress injury. Abscisic acid regulates the expression of many stress-responsive genes, including late embryogenesis abundant proteins (LEA), leading to enhanced drought tolerance in plants (Aroca et al. 2008). When exposed to drought stress, ABA is the main hormone that is synthesized in the roots and transported to leaves to begin adapting plants to drought stress by closing stomata and decreasing plant growth (Wilkinson and Davies 2010). However, modulating the ABA-induced drought adaptation of plants for improved productivity remains a greater challenge because of the potential unintended decrease in carbon gain when ABA-induced senescence, especially if drought occurs in the reproductive stage (Xuemei et al. 2011). There are ABA signaling genes, such as OsNAP, OsNAC5, and DSM2, which help in enhancing yield under reproductive drought (Chen et al. 2014). This nonstomatal ABA-induced adaptations of plants can be exploited under drought stress to improve yield under reproductive drought.

During drought stress, CK is known to delay the premature senescence of leaf and death, adaptive traits are extremely useful for increasing grain yields. The increase in endogenous-levels of CK through the expression of isopentenyl transferase (IPT), a CK biosynthetic pathway gene, leads to adaptation to stress by delaying senescence caused by drought and increased productivity (Peleg and Blumwald 2011). Decreased content of CK and activity due to reduced biosynthesis or improved degradation were observed in drought-stressed plants (Pospíšilová et al. 2000). In alfalfa, reduced CK content during drought resulted in accelerated aging senescence (Goicoechea et al. 1995). Cytokinins are known to delay senescence, and the increase in cytokines' endogenous-levels by overexpression of the *ipt* gene involved in CK biosynthesis has led to adaptation to stress by delaying drought-induced senescence (Peleg and Blumwald 2011). Cytokinin also negatively regulates roots growth and branching, and the degradation of CK in roots has contributed to primary root growth and branching due to drought stress, thus increasing drought tolerance in plants (Werner et al. 2010).

Auxins have been recognized to negatively regulate adaptation to drought in plants. The reduction in the content of indole-3-acetic acid (IAA) has been shown to be associated with increased regulation of genes that encode late embryogenesis abundant proteins (LEA), leading to drought adaptation of plants (Zhang et al. 2009). Recently, DEEPER ROOTING 1 (DRO1), gene determines the location of QTL controlling the angle of root growth, has shown that it regulates negative by auxin. The higher expression of DRO1 in the shallow-rooted rice cultivar resulted to avoid drought and high productivity under drought (Uga et al. 2013). Downregulation of IAA was considered to facilitate the accumulation of LEA mRNA, (LEA), leading to drought adaptation in rice plant (Zhang et al. 2009). In wheat leaves, drought-induced tolerance was accompanied by a reduction in the content of IAA (Xie et al. 2003). However, there is evidence of a transient increase in IAA content in maize leaves during the early stages of exposure to water stress, which subsequently decreases sharply with plant adaptation to water stress (Wang et al. 2008).

Gibberellic acid is proposed to regulate plant adaptation positively with drought stress. A rapid decrease in the levels of endogenous-produced GA in plants exposed to drought stress was observed, which leads to growth inhibition (Wang et al. 2008). Similarly, decreased levels of endogenous zeatin and GA were also observed in corn leaves exposed to water stress, which are associated with higher levels of cell damage and inhibition of plant growth. Thus, the role of the GA in regulating yields of crop plants is an important area that can be further explored.

Ethylene is also a negative regulator of drought stress response by stimulating leaf senescence, inhibition of growth and development of roots, expansion of leaves/shoot, and photosynthesis (Fukao et al. 2006). Ethylene can also directly affect yield by increasing embryo and grain abortion and reducing grain fill rates (Yang et al. 2004). In addition to the major hormones, other hormones such as brassinosteroids, jasmonic acid (JA), salicylic acid (SA) and strigolactone also play an important role in plant growth and development. However, their function under drought stress is comparatively less. It has been suggested that tillering in rice is the result of the interaction between three hormones, CK, auxin, and strigolactone, with CK promoting branching and the other two inhibiting it, suggesting that all hormones do not behave in isolation but instead interact and modulate each other biosynthesis and responses. Therefore, the net result of drought stress response is regulated by the balance of hormones that promote and those that inhibit the traits, rather than individual hormones (Xing and Zhang 2010).

8 Molecular Adaptations

The physiology of plant-wide drought response at plant level is very complex and involves adverse and/or adaptive changes. This complexity is due to factors such as plant species and varieties, environmental dynamics, duration and intensity of soil water depletion, changes in the demand for water from the atmosphere, as well as plant growth and the phenology in which water deficit developed. These complex plant responses include many biochemical and molecular mechanisms. Consecutively, they sense signals, perceieve, and transduce by osmosensors such as AtHK1, kinases, and phospholipases, as well as secondary messengers. Control of transcription by transcription factors such as dehydration responsive transcription factors (DREB); activation of stress response mechanisms such as ROS detoxification by enzymes such as SOD and CAT; osmoprotection by compatible solutes and free radicals such as glutathione and proline and water and ion balance through aquaporins and ion carriers. The results of these response paths are the re-establishment of the cellular homeostasis and the functional and structural protection that enables the plant to adapt to drought conditions.

Plants facing the drought challenge are subject to many adaptive mechanisms at the molecular level to modify the water balance. Plants alter their gene expression in response to drought stress, which results in the up and downregulation of many genes. By this way, stress proteins accumulate in the cells to cope with this situation. (Kavar et al. 2008). A large number of stress-responsive genes have been identified in plants. Generally, two types of responsive genes are identified. First type is the functional genes, which encodes for enzymes and functional or metabolic proteins. The second type of genes is the regulatory one, which encodes for the regulatory proteins. These include the transcription factors and the protein kinases, which regulate the gene expression and signal transduction (Shao et al. 2015). There are many species-specific and conserved genes in response to drought stress, have been identified. These genes include membrane-stabilizing proteins and LEA proteins (late embryogenic abundant proteins), which perform their function in water channel by the means of increase cells' water-binding capacity. Additionally, several transcription factors that also regulate and provide adaptive response under drought stress were identified, including myeloblastosis (MYB), dehydration responsive element binding (DREB), C-repeat binding factor (CBF), abscisic acid responsive elements binding factor (ABF), ABRE binding (AREB), (NAM, ATAF1/2, and CUC2 containing proteins) (NAC), WRKY, and SNF1-related kinase 2 (SnRK2) (Zenda et al. 2019). A significant increase was observed in CDSP 32 (chloroplastic drought-induced stress protein), observed under drought, which protects chloroplast from oxidative damage caused by drought (Broin et al. 2000). Aquaporins are an important group of intrinsic membrane proteins that can help in the passive exchange of water through membranes; this is an increase of 10–20 times in water permeability by regulating the hydraulic conductivity of membranes (Tyerman et al. 2002).

Several dehydration responsive element binding genes for drought response factors are involved in signaling pathways in response to abiotic stresses, including drought (Agarwal et al. 2010). The hydration responsive element/C-repeat (DRE/CRT) cis-acting element and its DNA binding protein is a major transcription system that modulates ABA independent gene expression in response to drought and contains dehydration responsive binding proteins (DREB)/C-repeat binding factor (CBF) family of Proteins. The DREB2 subgroup of DREB/CBF family proteins is expressed under drought to demonstrate the genes responsible for stress tolerance (Seki et al. 2003). Signal transduction pathways are also induced under drought to regulate growth in plants. Heat shock proteins (HSPs), which play a major role in stabilizing protein structure. These HSPs are involved in unwinding of folded proteins and prevention of protein denaturation due to abiotic stress (Zenda et al. 2019). Moreover, there is an early warning response mechanism at plant roots to activate the hydrogen pump ATPase (H^+-ATPase) protein on the plasma membrane of the root hairs before a significant reduction in the RWC of plant. Activation of the plasma membrane of the root hair cell H^+-ATPase led to amplification of the biosynthesis process of major osmolytes such as leaf proline and GB to maintain the water budget of the plants. In addition, there may be interspecific and intraspecific differences in timing of early responses and drought-tolerant varieties to initiate warming responses earlier than many sensitive species (Gong et al. 2010). Plant adaptation was associated with plant response to drought through signaling, as well as its role in responding to many other stresses (Bae et al. 2008). Ornithine decarboxylase (TcODC), arginine decarboxylase (TcADC), S-adenosylmethionine decarboxylase (TcSAMDC), spermidine synthase (TcSPDS), and spermine synthase (TcSPMS) are the expression patterns of genes that encode the enzymes involved in the adaptation of plants in cacao (*Theobroma cacao* L.) leaves. The expression of TcODC, TcADC, and TcSAMDC is induced at the onset of drought, which regulates the stomatal conductance, photosynthesis, photosystem II efficiency, and leaf water potential.

TcSAMDC induction in leaves is closely related to changes in water potential. The first measured drought responses detected in cocao leaves 13 days after the onset of drought were enhanced expression of TcADC and TcSAMDC at the roots with decreased stomatal conductance, photosynthesis and PS II efficiency due to elevated levels of putrescine, spermidine, and Spermine (Bae et al. 2008).

Increased Arabidopsis EDT1/HDG11 expression has been shown to improve drought tolerance for poplar and cotton by increasing the accumulation of solutes such as proline and soluble sugars as well as increasing cotton yield in the field (Yu et al. 2016). However, there are some plants where sugars are the main osmolytes that play an important role in osmotic adjustment, including sucrose, trehalose, glucose, and fructose. Previous studies have shown that over-expression of the fructan-6-fructosyltransferase (6-SFT) gene from *Psathyrostachys huashanica* in tobacco and the trehalose-6-phosphate phosphatase gene OsTPP1 in the rice confers abiotic stress tolerance (He et al. 2015). The researchers also identified QTL for OA on chromosome 8 in rice which is similar to part of chromosome 7 in wheat (Ahn et al. 1993). The transcriptome file of the latex rubber tree contains many genes related to water stress, whose role in stress defense has not yet been known. The transcripts of two genes encoding for *Hevea brasiliensis* Abscisic Acid, Stress, and Ripening (ASR) like proteins such as HbAS-RLP1 and HbASRLP2, were the most abundant in addition to the Rubber Elongation Factor (REF) and the Small Rubber Particle Protein (SRPP) in the latex. These genes are similar to the family of ASR genes, and the presumed proteins encoded by these genes have a similar domain to those of water stress, ABA stress and ripening (Rossi et al. 1996). Furthermore, the gene family of HbRLPs (REF-like proteins) was also expressed at higher levels in latex rubber. HbRLPs and SRPP are structurally closer to stress-related proteins (Ko et al. 2003).

9 Conclusion

Plants are very vulnerable when it comes to water scarcity. Drought will influence on plant growth, development, productivity, and survival. However, the plants have built some protection against drought. A number of improvements have been made in our understanding of how the plant responds to drought stress. Adaptation to drought is seen to include metabolic and morphological changes that prevent plants from dehydration injury. Since they do not conflict with each other, different combinations of traits lead to different adaptive strategies. Such coordination is particularly strong under stressful conditions, when structural investment is regulated to reach optimal carbon allocation to ensure growth, survival, and reproduction. There is wide variation between the morphological, physiological, and anatomical traits of different plant ecosystems in a sort of coordination between plant response and environmental constraints to improve adaptation. Moreover, different degrees of adaptation can coexist, allowing for completely different biological models to share the same environment. They also have some internal defenses that are activated in an attempt to reduce water loss when they realize that water has become scarce. These morphological-physiological and anatomical changes are the molecular mechanisms that regulate the expression of genes involved in different adaptive processes. Although much is now known about the different types of stress sensors, the secondary signal molecules involved and the specific signaling pathways are not deciphered, largely because of the intersection between the different stress signaling pathways. The expression of stress response gene is greatly regulated by transcription factors, which in turn are subject to very complex regulation at chromatin level, RNA level and protein level. Stress-induced chromatin can be re-mediated in adaptation responses and helps the plant adapt better to subsequent stress. Silencing gene connected by microRNA from stress response TFs appeared under non-stress conditions and activated by downregulation of miRNA expression have emerged as another important way to regulate gene expression in stress response. Understanding stress management strategies using model plants and testing them in crop genotypes that show adaptation to stress appears to be a useful approach for improving crop tolerance for drought. Knowing these genes and how they are involved in plant protection from drought provides humankind with a hope to produce drought-resistant GM crops.

References

Agarwal P, Agarwal PK, Joshi AJ, Sopory SK, Reddy MK (2010) Overexpression of PgDREB2A transcription factor enhances abiotic stress tolerance and activates downstream stress-responsive genes. Mol Biol Rep 37:1125

Ahn S, Anderson JA, Sorrells ME, Tanksley SD (1993) Homoeologous relationships of rice, wheat and maize chromosomes. Mol Gen Genet 241:483–490

Akinci S (1997) Physiological responses to water stress by *Cucumis sativus* L. and related species. University of Sheffield

Almeida AAF, de Valle RR (2007) Ecophysiology of the cacao tree. Braz J Plant Physiol 19(4):425–448

Ambavaram MMR, Basu S, Krishnan A, Ramegowda V, Batlang U, Rahman L, Pereira A (2014) Coordinated regulation of photosynthesis in rice increases yield and tolerance to environmental stress. Nat Commun 5:5302

Anjum SA, Ashraf U, Zohaib A, Tanveer M, Naeem M, Ali I, Nazir U (2017a) Growth and development responses of crop plants under drought stress: a review. Zemdirbyste 104(3):267–276

Anjum SA, Ashraf U, Tanveer M, Khan I, Hussain S, Zohaib A Wang L (2017b) Drought tolerance in three maize cultivars is related to differential osmolyte accumulation, antioxidant defense system, and oxidative damage. Front Plant Sci 8:69

Arena C, Vitale L, De Santo AV (2008) Paraheliotropism in Robinia pseudoacacia L.: an efficient strategy to optimise photosynthetic performance under natural environmental conditions. Plant Biol 10(2):194–201

Aroca R, Vernieri P, Ruiz-Lozano JM (2008) Mycorrhizal and non-mycorrhizal Lactuca sativa plants exhibit contrasting responses to exogenous ABA during drought stress and recovery. J Exp Bot 59(8):2029–2041

Aronne G, De Micco V (2001) Seasonal dimorphism in the Mediterranean Cistus incanus L. subsp. incanus. Ann Bot 87(6):789–794

Aronne G, De Micco V (2004) Hypocotyl features of Myrtus communis (Myrtaceae): a many-sided strategy for possible enhancement of seedling establishment in the Mediterranean environment. Bot J Linnean Soc 145(2):195–202

Aronne G, Wilcock C (1997) Reproductive phenology in Mediterranean macchia vegetation. Lagascalia 19(1–2):445–454

Ashraf M, Mr Foolad (2007) Roles of glycine betaine and proline in improving plant abiotic stress resistance. Environ Exp Bot 59(2):206–216

Baas P (1986) Ecological patterns in xylem anatomy. In: On the economy of plant form and function: proceedings of the Sixth Maria Moors Cabot symposium, evolutionary constraints on primary productivity, adaptive patterns of energy capture in plants, Harvard Forest, August 1983. Cambridge University Press, Cambridge [Cambridgeshire]

Baas P, Werker E, Fahn A (1983) Some ecological trends in vessel characters. Iawa Bull 4:141–159

Baas P, Ewers FW, Davis SD, Wheeler EA (2004) Evolution of xylem physiology. In: The evolution plant physiology. Elsevier, pp 273–295

Bae H, Kim SH, Kim M, Sicher RC, Lary D, Strem MD, Bailey BA (2008) The drought response of Theobroma cacao (cacao) and the regulation of genes involved in polyamine biosynthesis by drought and other stresses. Plant Physiol Biochem 46(2):174–188

Bargel H, Barthlott W, Koch K, Schreiber L, Neinhuis C (2004) Plant cuticles: multifunctional interfaces between plant and environment. In: The evolution plant physiology. Elsevier, p 171-III

Battipaglia G, De Micco V, Brand WA, Linke P, Aronne G, Saurer M, Cherubini P (2010) Variations of vessel diameter and δ13C in false rings of Arbutus unedo L. reflect different environmental conditions. New Phytol 188(4):1099–1112

Blilou I, Xu J, Wildwater M, Willemsen V, Paponov I, Friml J Scheres B (2005) The PIN auxin efflux facilitator network controls growth and patterning in Arabidopsis roots. Nature 433(7021):39

Blum A (2005) Drought resistance, water-use efficiency, and yield potential—are they compatible, dissonant, or mutually exclusive? Aust J Agric Res 56(11):1159–1168

Blum A (2011) Genetic resources for drought resistance. In: Plant breeding for water-limited environments. Springer, pp 217–234

Bowne JB, Erwin TA, Juttner J, Schnurbusch T, Langridge P, Bacic A, Roessner U (2012) Drought responses of leaf tissues from wheat cultivars of differing drought tolerance at the metabolite level. Mol Plant 5(2):418–429

Brodribb TJ, Feild TS (2000) Stem hydraulic supply is linked to leaf photosynthetic capacity: evidence from New Caledonian and Tasmanian rainforests. Plant Cell Environ 23(12):1381–1388

Broin M, Cuiné S, Peltier G, Rey P (2000) Involvement of CDSP 32, a drought-induced thioredoxin, in the response to oxidative stress in potato plants. FEBS Lett 467(2–3):245–248

Carlquist SJ (1975) Ecological strategies of xylem evolution. Univ California Press

Carlquist S (1989) Adaptive wood anatomy of chaparral shrubs. The California Chaparral: paradigms re-examined. Los Angeles Country Museum of Natural History Contributions, Los Angeles, pp 25–35

Carr MKV (2011) The water relations and irrigation requirements of coconut (Cocos nucifera): a review. Exp Agric 47(1):27–51

Chaves MM, Oliveira MM (2004) Mechanisms underlying plant resilience to water deficits: prospects for water-saving agriculture. J Exp Bot 55(407):2365–2384

Chaves MM, Flexas J, Pinheiro C (2009) Photosynthesis under drought and salt stress: regulation mechanisms from whole plant to cell. Ann Bot 103(4):551–560

Chen JW, Zhang Q, Li XS, Cao KF (2009) Independence of stem and leaf hydraulic traits in six Euphorbiaceae tree species with contrasting leaf phenology. Planta 230(3):459–468

Chen H, Li Z, Xiong L (2012) A plant microRNA regulates the adaptation of roots to drought stress. FEBS Lett 586(12):1742–1747

Chen X, Wang Y, Lv B, Li J, Luo L, Lu S, Ming F (2014) The NAC family transcription factor OsNAP confers abiotic stress response through the ABA pathway. Plant Cell Physiol 55(3):604–619

Cheruiyot EK, Mumera LM, Ng'etich WK, Hassanali A, Wachira F (2007) Polyphenols as potential indicators for drought tolerance in tea (*Camellia sinensis* L.). Biosci Biotech Biochem 71(9):2190–2197
Christman MA, Sperry JS (2010) Single-vessel flow measurements indicate scalariform perforation plates confer higher flow resistance than previously estimated. Plant Cell Environ 33(3):431–443
Close TJ (1997) Dehydrins: a commonalty in the response of plants to dehydration and low temperature. Physiol Plant 100(2):291–296
Couvreur TLP, Baker WJ (2013) Tropical rain forest evolution: palms as a model group. BMC Biol 11(1):48
Croker JL, Witte WT, Augé RM (1998) Stomatal sensitivity of six temperate, deciduous tree species to non-hydraulic root-to-shoot signalling of partial soil drying. J Exp Bot 49(321):761–774
DaMatta FM (2004) Exploring drought tolerance in coffee: a physiological approach with some insights for plant breeding. Braz J Plant Physiol 16(1):1–6
DaMatta FM, Rena AB (2001) Tolerância do café à seca. Tecnologias de Produção de Café Com Qualidade 65–100
DaMatta FM, Chaves ARM, Pinheiro HA, Ducatti C, Loureiro ME (2003) Drought tolerance of two field-grown clones of Coffea canephora. Plant Sci 164(1):111–117
Davies WJ, Tardieu F, Trejo CL (1994) How do chemical signals work in plants that grow in drying soil? Plant Physiol 104(2):309
De Micco V, Aronne G (2008) Twig morphology and anatomy of Mediterranean trees and shrubs related to drought tolerance. Bot Helv 118(2):139–148
De Micco V, Aronne G (2009) Seasonal dimorphism in wood anatomy of the Mediterranean *Cistus incanus* L. subsp. incanus. Trees 23(5):981–989
De Micco V, Aronne G (2010) Root structure of Rumex scutatus growing on slopes. IAWA J 31(1):13–28
De Micco V, Aronne G (2012) Anatomy and lignin characterisation of twigs in the chaparral shrub Rhamnus californica. IAWA J 33(2):151–162
De Micco V, Aronne G, Baas P (2008) Wood anatomy and hydraulic architecture of stems and twigs of some Mediterranean trees and shrubs along a mesic-xeric gradient. Trees 22(5):643–655
De Micco V, Battipaglia G, Balzano A, Cherubini P, Aronne G (2016) Are wood fibres as sensitive to environmental conditions as vessels in tree rings with intra-annual density fluctuations (IADFs) in Mediterranean species? Trees 30(3):971–983
Deak KI, Malamy J (2005) Osmotic regulation of root system architecture. Plant J 43(1):17–28
Dixit S, Biswal AK, Min A, Henry A, Oane RH, Raorane ML Vardarajan AR (2015) Action of multiple intra-QTL genes concerted around a co-localized transcription factor underpins a large effect QTL. Sci Rep 5:15183
Edwards G, Walker D (1983) *C3, C4*: mechanisms, and cellular and environmental regulation, of photosynthesis. Univ of California Press
Eissenstat DM (1992) Costs and benefits of constructing roots of small diameter. J Plant Nutr 15(6–7):763–782
Fahn A (1964) Some anatomical adaptations of desert plants. Phytomorphology 4:93–101
Feild TS, Zwieniecki MA, Donoghue MJ, Holbrook NM (1998) Stomatal plugs of Drimys winteri (Winteraceae) protect leaves from mist but not drought. Proc Natl Acad Sci 95(24):14256–14259
Fenner M (1999) Seed and seedling ecology. In: Handbook of functional plant ecology, pp 589–648
Flowers TJ (1989) Plants under stress: biochemistry, physiology and ecology and their application to plant improvement, vol 39. Cambridge University Press
Fukao T, Xu K, Ronald PC, Bailey-Serres J (2006) A variable cluster of ethylene response factor–like genes regulates metabolic and developmental acclimation responses to submergence in rice. Plant Cell 18(8):2021–2034
Goicoechea N, Dolezal K, Antolín MC, Strnad M, S´nchez-Díaz M (1995) Influence of mycorrhizae and Rhizobium on cytokinin content in drought-stressed alfalfa. J Exp Bot 46(10):1543–1549
Gomes FP, Prado CHBA (2007) Ecophysiology of coconut palm under water stress. Braz J Plant Physio 19(4):377–391

Gomes ARS, Kozlowski TT, Reich PB (1987) Some physiological responses of Theobroma cacao var. catongo seedlings to air humidity. New Phytol 107(3):591–602

Gomes FP, Oliva MA, Mielke MS, de Almeida AAF, Leite HG, Aquino LA (2008) Photosynthetic limitations in leaves of young Brazilian Green Dwarf coconut (*Cocos nucifera* L.'nana') palm under well-watered conditions or recovering from drought stress. Environ Exp Bot 62(3):195–204

Gong DS, Xiong YC, Ma BL, Wang TM, Ge JP, Qin XL, Li FM (2010) Early activation of plasma membrane H^+-ATPase and its relation to drought adaptation in two contrasting oat (*Avena sativa* L.) genotypes. Environ Exp Bot 69(1):1–8

Gowik U, Westhoff P (2011) The path from C3 to C4 photosynthesis. Plant Physio 155(1):56–63

Goyal K, Walton LJ, Tunnacliffe A (2005) LEA proteins prevent protein aggregation due to water stress. Biochem J 388(1):151–157

Guan YS, Serraj R, Liu SH, Xu JL, Ali J, Wang WS, Li ZK (2010) Simultaneously improving yield under drought stress and non-stress conditions: a case study of rice (*Oryza sativa* L.). J Exp Bot 61(15):4145–4156

Guardiola-Claramonte M, Troch PA, Ziegler AD, Giambelluca TW, Durcik M, Vogler JB, Nullet MA (2010) Hydrologic effects of the expansion of rubber (*Hevea brasiliensis*) in a tropical catchment. Ecohydrology 3(3):306–314

Guardiola-Claramonte M, Troch PA, Ziegler AD, Giambelluca TW, Vogler JB, Nullet MA (2008) Local hydrologic effects of introducing non-native vegetation in a tropical catchment. Ecohydrol: Ecosyst Land Water Process Interact Ecohydrogeomorphol 1(1):13–22

Hacke UG, Sperry JS (2001) Functional and ecological xylem anatomy. Persp Plant Eco Evol Syst 4(2):97–115

He X, Chen Z, Wang J, Li W, Zhao J, Wu J, Chen X (2015) A sucrose: Fructan-6-fructosyltransferase (6-SFT) gene from Psathyrostachys huashanica confers abiotic stress tolerance in tobacco. Gene 570(2):239–247

Henry A, Cal AJ, Batoto TC, Torres RO, Serraj R (2012) Root attributes affecting water uptake of rice (*Oryza sativa*) under drought. J Exp Bot 63(13):4751–4763

Herzog KM, Thum R, Kronfus G, Heldstab HJ, Häsler R (1998) Patterns and mechanisms of transpiration in a large subalpine Norway spruce (*Picea abies* (L.) Karst.). Eco Res 13(2):105–116

Hinniger C, Caillet V, Michoux F, Ben Amor M, Tanksley S, Lin C, Mccarthy J (2006) Isolation and characterization of cDNA encoding three dehydrins expressed during Coffea canephora (Robusta) grain development. Ann Bot 97(5):755–765

Hose E, Clarkson DT, Steudle E, Schreiber L, Hartung W (2001) The exodermis: a variable apoplastic barrier. J Exp Bot 52(365):2245–2264

Hunt Jr ER, Zakir NJD, Nobel PS (1987) Water costs and water revenues for established and rain-induced roots of Agave deserti. Funct Ecol 125–129

Hussain M, Farooq S, Hasan W, Ul-Allah S, Tanveer M, Farooq M, Nawaz A (2018) Drought stress in sunflower: physiological effects and its management through breeding and agronomic alternatives. Agric Water Manag 201:152–166

Ishikawa T, Takahara K, Hirabayashi T, Matsumura H, Fujisawa S, Terauchi R, Kawai-Yamada M (2009) Metabolome analysis of response to oxidative stress in rice suspension cells overexpressing cell death suppressor Bax inhibitor-1. Plant Cell Physiol 51(1):9–20

Jackson MB (2002) Long-distance signalling from roots to shoots assessed: the flooding story. J Exp Bot 53(367):175–181

Jansen S, Choat B, Pletsers A (2009) Morphological variation of intervessel pit membranes and implications to xylem function in angiosperms. Am J Bot 96(2):409–419

Jenks MA (2002) Critical issues with the plant cuticle's function in drought tolerance. Biochem Mol Resp Plants Environ 97:127

Jeyaramraja PR, Kumar RR, Pius PK, Thomas J (2003) Photoassimilatory and photorespiratory behaviour of certain drought tolerant and susceptible tea clones. Photosynthetica 41(4):579–582

Jones MM (1981) Mechanisms of drought resistance. In: The physiology and biochemistry of drought resistance in plants

Jongdee B, Fukai S, Cooper M (2002) Leaf water potential and osmotic adjustment as physiological traits to improve drought tolerance in rice. Field Crops Res 76(2–3):153–163

Jozefczak M, Remans T, Vangronsveld J, Cuypers A (2012) Glutathione is a key player in metal-induced oxidative stress defenses. Int J Mol Sci 13(3):3145–3175

Kar RK (2011) Plant responses to water stress: role of reactive oxygen species. Plant Signal Behav 6(11):1741–1745

Karaba A, Dixit S, Greco R, Aharoni A, Trijatmiko KR, Marsch-Martinez N, Pereira A (2007) Improvement of water use efficiency in rice by expression of HARDY, an Arabidopsis drought and salt tolerance gene. Proc Natl Acad Sci USA 104(39):15270–15275

Kavar T, Maras M, Kidrič M, Šuštar-Vozlič J, Meglič V (2008) Identification of genes involved in the response of leaves of Phaseolus vulgaris to drought stress. Mol Breed 21(2):159–172

Kholodova VP, Meshcheryakov AB, Rakitin VY, Karyagin VV, Kuznetsov VV (2006) Hydraulic signal as a "primary messenger of water deficit" under salt stress in plants. In: Doklady biological sciences, vol 407. Springer, pp 155–157

Ko JH, Chow KS, Han KH (2003) Transcriptome analysis reveals novel features of the molecular events occurring in the laticifers of Hevea brasiliensis (para rubber tree). Plant Mol Biol 53(4):479–492

Kohonen MM, Helland Å (2009) On the function of wall sculpturing in xylem conduits. J Bionic Eng 6(4):324–329

Koster J, Baas P (1982) Alveolar cuticular material in Myristicaceae. In: Cutler DF, Alvin KL, Price CE (eds) The plant cuticle. London

Kozlowski TT, Pallardy SG (2002) Acclimation and adaptive responses of woody plants to environmental stresses. Bot Rev 68(2):270–334

Kramer PJ (1980) Drought, stress, and the origin of adaptations. In: Drought, stress, and the origin of adaptations, pp 7–20

Kubiske ME, Abrams MD, Mostoller SA (1996) Stomatal and nonstomatal limitations of photosynthesis in relation to the drought and shade tolerance of tree species in open and understory environments. Trees 11(2):76–82

Kumar SN, Rajagopal V, Karun A (2007) Leaflet anatomical adaptations in coconut cultivars for drought tolerance

Kummerow J (1981) Structure of roots and root systems. Ecosystems of the World

Lakmini WGD, Nainanayake N, De Costa W (2006) Physiological responses for moisture stress and development of an index for screening coconut (*Cocos nucifera* L.) genotypes for drought. Trop Agric Res Ext 9:17–26

Lawlor DW (2002) Limitation to photosynthesis in water-stressed leaves: stomata vs. metabolism and the role of ATP. Ann Bot 89(7):871–885

Lawlor DW, Cornic G (2002) Photosynthetic carbon assimilation and associated metabolism in relation to water deficits in higher plants. Plant Cell Environ 25(2):275–294

Legros S, Mialet-Serra I, Caliman JP, Siregar FA, Cle´ment-Vidal A Dingkuhn M (2009) Phenology and growth adjustments of oil palm (*Elaeis guineensis*) to photoperiod and climate variability. Ann Bot 104:1171–1182

Lehmann M, Schwarzländer M, Obata T, Sirikantaramas S, Burow M, Olsen CE, Fernie AR (2009) The metabolic response of Arabidopsis roots to oxidative stress is distinct from that of heterotrophic cells in culture and highlights a complex relationship between the levels of transcripts, metabolites, and flux. Mol Plant 2(3):390–406

Lens F, Sperry JS, Christman MA, Choat B, Rabaey D, Jansen S (2011) Testing hypotheses that link wood anatomy to cavitation resistance and hydraulic conductivity in the genus Acer. New Phytol 190(3):709–723

Levitt J (1980a) Response of plants to environmental stresses: chilling, freezing, and high temperature stresses. Physiol Ecol: Ser Monogr Texts Treatises 1:23–64

Levitt J (1980b) Responses of plants to environmental stresses (physiological ecology): chilling, freezing, and high temperature stresses. Academic Press. ISBN 124455018

Lewis MC (1972) The physiological significance of variation in leaf structure. Sci Prog (1933–):25–51

Liu R, Wang L, Tanveer M, Song J (2018) Seed heteromorphism: an important adaptation of halophytes for habitat heterogeneity. Front Plant Sci 9

Moles AT, Westoby M (2004) What do seedlings die from and what are the implications for evolution of seed size? Oikos 106(1):193–199

Møller IM, Sweetlove LJ (2010) ROS signalling–specificity is required. Trends Plant Sci 15(7):370–374

Monneveux P, Belhassen E (1996) The diversity of drought adaptation in the wide. In: Drought tolerance in higher plants: genetical, physiological and molecular biological analysis. Springer, pp 7–14

Morgan JM (1984) Osmoregulation and water stress in higher plants. Ann Rev Plant Physiol 35(1):299–319

Mostajeran A, Rahimi-Eichi V (2008) Drought stress effects on root anatomical characteristics of rice cultivars (*Oryza sativa* L.). Pak J Biol Sci 11(18):2173–2183

Nagarajah S, Ratnasuriya GB (1981) Clonal variability in root growth and drought resistance in tea (*Camellia sinensis*). Plant Soil 60(1):153–155

Nainanayake AD, Morison JIL (2007) A case study on physiology-based drought screening of coconut with selected accessions

Niinemets Ü (2001) Global-scale climatic controls of leaf dry mass per area, density, and thickness in trees and shrubs. Ecology 82(2):453–469

North GB, Nobel PS (1992) Drought-induced changes in hydraulic conductivity and structure in roots of Ferocactus acanthodes and Opuntia ficus-indica. New Phytol 120(1):9–19

North GB, Nobel PS (1995) Hydraulic conductivity of concentric root tissues of Agave deserti Engelm. under wet and drying conditions. New Phytol 130(1):47–57

North GB, Nobel PS (1996) Radial hydraulic conductivity of individual root tissues of *Opuntia ficus-indica* (L.) Miller as soil moisture varies. Ann Bot 77(2):133–142

Nunes MA (1976) Water relations in coffee. Significance of plant water deficit to growth and yield; a review. J Coffee Res (India) 6(1):4–21

Oguchi R, Hikosaka K, Hirose T (2005) Leaf anatomy as a constraint for photosynthetic acclimation: differential responses in leaf anatomy to increasing growth irradiance among three deciduous trees. Plant Cell Environ 28(7):916–927

Osakabe Y, Osakabe K, Shinozaki K, Tran LSP (2014) Response of plants to water stress. Front Plant Sci 5:86

Padilla FM, Pugnaire FI (2007) Rooting depth and soil moisture control Mediterranean woody seedling survival during drought. Funct Ecol 21(3):489–495

Padilla FM, de Dios Miranda J, Pugnaire FI (2007) Early root growth plasticity in seedlings of three Mediterranean woody species. Plant Soil 296(1–2):103–113

Passos EEM, da Silva JV (1991) Détermination de l'état hydrique du cocotier par la méthode dendrométrique. Oleagineux 46:233–238

Passos EE MP, Passos EEM, Prado C (2006) Comportamento sazonal do potencial hídrico e das trocas gasosoas de quatro variedades de coqueiro anão. SciELO, Brasil

Peleg Z, Blumwald E (2011) Hormone balance and abiotic stress tolerance in crop plants. Curr Opin Plant Biol 14(3):290–295

Peña-Valdivia CB, Sánchez-Urdaneta AB, Meza Rangel J, Juárez Muñoz J, García-Nava R, Celis Velázquez R (2010) Anatomical root variations in response to water deficit: wild and domesticated common bean (*Phaseolus vulgaris* L). Biol Res 43(4):417–427

Pereira JS, Chaves MM (1995) Plant responses to drought under climate change in Mediterranean-type ecosystems. In: Global change and Mediterranean-type ecosystems. Springer, pp 140–160

Pinheiro C, Chaves MM (2010) Photosynthesis and drought: can we make metabolic connections from available data? J Exp Bot 62(3):869–882

Pinheiro HA, DaMatta FM, Chaves ARM, Fontes EPB, Loureiro ME (2004) Drought tolerance in relation to protection against oxidative stress in clones of Coffea canephora subjected to long-term drought. Plant Sci 167(6):1307–1314

Pinheiro HA, DaMatta FM, Chaves ARM, Loureiro ME, Ducatti C (2005) Drought tolerance is associated with rooting depth and stomatal control of water use in clones of Coffea canephora. Ann Bot 96(1):101–108

Pospíšilová J, Synková H, Rulcová J (2000) Cytokinins and water stress. Biol Plant 43(3):321–328

Prakash-MBM ASDPG, Jacob SJ (1999) Drought alters the canopy architecture. Trées 13:161–167

Priyadarshan PM, Clément-Demange A (2004) 3 breeding hevea rubber: formal and molecular genetics. Adv Genet 52:51–116

Rajagopal V, Ramadasan A (1999) Advances in plant physiology and biochemistry of coconut palm. APCC, Jakarta

Rajagopal V, Kasturibai KV, Voleti SR (1990) Screening of coconut genotypes for drought tolerance. Oleagineux 45(5):215–223

Reader RJ, Jalili A, Grime JP, Spencer RE, Matthews N (1993) A comparative study of plasticity in seedling rooting depth in drying soil. J Ecol 543–550

Ren H, Yang L, Liu N (2008) Nurse plant theory and its application in ecological restoration in lower subtropics of China. Prog Nat Sci 18(2):137–142

Repellin A, Daniel C, Zuily Fodil Y (1994) Merits of physiological tests for characterizing the performance of different coconut varieties subjected to drought. Oleagineux (France)

Repellin A, Braconnier S, Laffray D, Daniel C, Zuily-Fodil Y (1997) Water relations and gas exchange in young coconut palm (*Cocos nucifera* L.) as influenced by water deficit. Can J Bot 75(1):18–27

Reynolds M, Tuberosa R (2008) Translational research impacting on crop productivity in drought-prone environments. Curr Opin Plant Biol 11(2):171–179

Riedel M, Riederer M, Becker D, Herran A, Kullaya A, Arana-López G, Rohde W (2009) Cuticular wax composition in Cocos nucifera L.: physicochemical analysis of wax components and mapping of their QTLs onto the coconut molecular linkage map. Tree Genet Genomes 5(1):53

Riederer M, Schreiber L (2001) Effects of environmental factors on the water permeability of plant cuticles. J Exp Bot 52:2023–2033

Robards AW, Clarkson DT, Sanderson J (1979) Structure and permeability of the epidermal/hypodermal layers of the sand sedge (*Carex arenaria*, L.). Protoplasma 101(4):331–347

Rossi M, Lijavetzky D, Hopp HE, Iusem N, Bernacchi D (1996) Asr genes belong to a gene family comprising at least three closely linked loci on chromosome 4 in tomato. Mol Gen Genet 252(4):489–492

Ryan KG, Markham KR, Bloor SJ, Bradley JM, Mitchell KA, Jordan BR (1998) UVB radiation induced increase in quercetin: kaempferol ratio in wild-type and transgenic lines of Petunia. Photochem Photobiol 68(3):323–330

Salleo S, Nardini A (2000) Sclerophylly: evolutionary advantage or mere epiphenomenon? Plant Biosys 134(3):247–259

Sánchez-Gómez D, Zavala MA, Valladares F (2006) Seedling survival responses to irradiance are differentially influenced by low-water availability in four tree species of the Iberian cool temperate–Mediterranean ecotone. Acta Oecolog 30(3):322–332

Sangsing K, Kasemsap P, Thanisawanyangkura S, Sangkhasila K, Gohet E, Thaler P, Cochard H (2004) Xylem embolism and stomatal regulation in two rubber clones (Hevea brasiliensis Muell. Arg.). Trees 18(2):109–114

Sauter A, Davies WJ, Hartung W (2001) The long-distance abscisic acid signal in the droughted plant: the fate of the hormone on its way from root to shoot. J Exp Bot 52(363):1991–1997

Scarpeci TE, Valle EM (2008) Rearrangement of carbon metabolism in Arabidopsis thaliana subjected to oxidative stress condition: an emergency survival strategy. Plant Growth Reg 54(2):133–142

Schönherr J, Ziegler H (1980) Water permeability of Betula periderm. Planta 147(4):345–354

Schupp EW (1995) Seed-seedling conflicts, habitat choice, and patterns of plant recruitment. Am J Bot 82(3):399–409

Seki M, Kamei A, Yamaguchi-Shinozaki K, Shinozaki K (2003) Molecular responses to drought, salinity and frost: common and different paths for plant protection. Curr Opin Biotech 14(2):194–199

Sengupta D, Reddy AR (2011) Water deficit as a regulatory switch for legume root responses. Plant Signal Behav 6(6):914–917

Seo PJ, Park CM (2009) Auxin homeostasis during lateral root development under drought condition. Plant Signal Behav 4(10):1002–1004

Serraj R, Sinclair TR (2002) Osmolyte accumulation: can it really help increase crop yield under drought conditions? Plant Cell Environ 25(2):333–341

Shao H, Wang H, Tang X (2015) NAC transcription factors in plant multiple abiotic stress responses: progress and prospects. Front Plant Sci 6:902. https://doi.org/10.3389/fpls.2015.00902

Sharma P, Kumar S (2005) Differential display-mediated identification of three drought-responsive expressed sequence tags in tea [*Camellia sinensis* (L.) O. Kuntze]. J Biosci 30(2):231–235

Sharp RE, Poroyko V, Hejlek LG, Spollen WG, Springer GK, Bohnert HJ, Nguyen HT (2004) Root growth maintenance during water deficits: physiology to functional genomics. J Exp Bot 55(407):2343–2351

Shields LM (1950) Leaf xeromorphy as related to physiological and structural influences. Bot Rev 16(8):399–447

Sicher RC, Barnaby JY (2012) Impact of carbon dioxide enrichment on the responses of maize leaf transcripts and metabolites to water stress. Physiol Plant 144(3):238–253

Silvente S, Sobolev AP, Lara M (2012) Metabolite adjustments in drought tolerant and sensitive soybean genotypes in response to water stress. PLoS ONE 7(6):e38554

Singh K, Kumar S, Ahuja PS (2009) Differential expression of Histone H3 gene in tea (*Camellia sinensis* (L.) O. Kuntze) suggests its role in growing tissue. Mol Biol Rep 36(3):537–542

Sperry JS (2000) Hydraulic constraints on plant gas exchange. Agric For Meteorol 104(1):13–23

Sperry JS (2003) Evolution of water transport and xylem structure. Int J Plant Sci 164(S3):S115–S127

Sperry JS, Tyree MT (1988) Mechanism of water stress-induced xylem embolism. Plant Physiol 88(3):581–587

Sperry JS, Hacke UG, Pittermann J (2006) Size and function in conifer tracheids and angiosperm vessels. Am J Bot 93(10):1490–1500

Steudle E (2000) Water uptake by roots: effects of water deficit. J Exp Bot 51(350):1531–1542

Striker GG, Insausti P, Grimoldi AA, Vega AS (2007) Trade-off between root porosity and mechanical strength in species with different types of aerenchyma. Plant Cell Environ 30(5):580–589

Takahashi N, Yamazaki Y, Kobayashi A, Higashitani A, Takahashi H (2003) Hydrotropism interacts with gravitropism by degrading amyloplasts in seedling roots of Arabidopsis and radish. Plant Physiol 132(2):805–810

Takeuchi Y, Kubo H, Kasahara H, Sakaki T (1996) Adaptive alterations in the activities of scavengers of active oxygen in cucumber cotyledons irradiated with UV-B. J Plant Physiol 147(5):589–592

Tanveer M, Shabala S (2018) Targeting redox regulatory mechanisms for salinity stress tolerance in crops. In: Salinity responses and tolerance in plants, vol 1. Springer, Cham, pp 213–234

Tanveer M, Shahzad B, Sharma A, Khan EA (2018) 24-Epibrassinolide application in plants: an implication for improving drought stress tolerance in plants. Plant Physiol Biochem 135:295–303

Tausend PC, Goldstein G, Meinzer FC (2000) Water utilization, plant hydraulic properties and xylem vulnerability in three contrasting coffee (*Coffea arabica*) cultivars. Tree Physiol 20(3):159–168

Tezara W, Mitchell VJ, Driscoll SD, Lawlor DW (1999) Water stress inhibits plant photosynthesis by decreasing coupling factor and ATP. Nature 401(6756):914

Tosens T, Niinemets U, Vislap V, Eichelmann H, Castro Diez P (2012) Developmental changes in mesophyll diffusion conductance and photosynthetic capacity under different light and water availabilities in Populus tremula: how structure constrains function. Plant Cell Environ 35(5):839–856

Traveset A, Riera N, Mas RE (2001) Ecology of fruit-colour polymorphism in Myrtus communis and differential effects of birds and mammals on seed germination and seedling growth. J Ecol 89(5):749–760
Trubat R, Cortina J, Vilagrosa A (2006) Plant morphology and root hydraulics are altered by nutrient deficiency in *Pistacia lentiscus* (L.). Trees 20(3):334
Turner NC (1996) Further progress in crop water relations. In: Advances in agronomy, vol 58). Elsevier, pp 293–338
Tyerman SD, Niemietz CM, Bramley H (2002) Plant aquaporins: multifunctional water and solute channels with expanding roles. Plant Cell Environ 25(2):173–194
Uga Y, Sugimoto K, Ogawa S, Rane J, Ishitani M, Hara N, Kanno N (2013) Control of root system architecture by DEEPER ROOTING 1 increases rice yield under drought conditions. Nat Genet 45(9):1097
Vinod KK (2012) Stress in plantation crops: adaptation and management. In: Crop stress and its management: perspectives and strategies. Springer, pp 45–137
Vinod KK, Meenattoor JR, Pothen J, Krishnakumar AK, Sethurai MR (1996) Performace analysis for wintering pattern in Hevea brasiliensis clones. Indian J Nat Rub R 1:44–47
Wang C, Yang A, Yin H, Zhang J (2008) Influence of water stress on endogenous hormone contents and cell damage of maize seedlings. J Integr Plant Biol 50(4):427–434
Warren CR, Aranda I, Cano FJ (2012) Metabolomics demonstrates divergent responses of two Eucalyptus species to water stress. Metabolomics 8(2):186–200
Watkins K (2006) Human development report 2006-beyond scarcity: power, poverty and the global water crisis
Werner T, Nehnevajova E, Köllmer I, Novák O, Strnad M, Krämer U, Schmülling T (2010) Root-specific reduction of cytokinin causes enhanced root growth, drought tolerance, and leaf mineral enrichment in Arabidopsis and tobacco. Plant Cell 109
Wheeler EA, Baas P (1991) A survey of the fossil record for Dicotiledonous wood and its significance for evolutionary and ecological wood anatomy. IAWA J 12(3):275–318
Wheeler JK, Sperry JS, Hacke UG, Hoang N (2005) Inter-vessel pitting and cavitation in woody Rosaceae and other vesselled plants: a basis for a safety versus efficiency trade-off in xylem transport. Plant Cell Environ 28(6):800–812
Wilkinson S, Davies WJ (2010) Drought, ozone, ABA and ethylene: new insights from cell to plant to community. Plant Cell Environ 33(4):510–525
Wilkinson S, Kudoyarova GR, Veselov DS, Arkhipova TN, Davies WJ (2012) Plant hormone interactions: innovative targets for crop breeding and management. J Exp Bot 63(9):3499–3509
Williams LJ, Bunyavejchewin S, Baker PJ (2008) Deciduousness in a seasonal tropical forest in western Thailand: interannual and intraspecific variation in timing, duration and environmental cues. Oecologia 155(3):571–582
Wilson BF (1995) Shrub stems: form and function. In: Plant stems. Elsevier, pp 91–102
Xie Z, Jiang D, Cao W, Dai T, Jing Q (2003) Relationships of endogenous plant hormones to accumulation of grain protein and starch in winter wheat under different post-anthesis soil water statusses. Plant Growth Reg 41(2):117–127
Xing Y, Zhang Q (2010) Genetic and molecular bases of rice yield. Ann Rev Plant Biol 61:421–442
Xiong L, Zhu J (2002) Molecular and genetic aspects of plant responses to osmotic stress. Plant Cell Environ 25(2):131–139
Xiong L, Wang RG, Mao G, Koczan JM (2006) Identification of drought tolerance determinants by genetic analysis of root response to drought stress and abscisic acid. Plant Physiol 142(3):1065–1074
Xu Z, Zhou G (2008) Responses of leaf stomatal density to water status and its relationship with photosynthesis in a grass. J Exp Bot 59(12):3317–3325
Xuemei J, Dong B, Shiran B, Talbot MJ, Edlington JE, Trijntje H, Dolferus R (2011) Control of ABA catabolism and ABA homeostasis is important for reproductive stage stress tolerance in cereals. Plant Physiol 111

Yang JC, Zhang JH, Ye YX, Wang ZQ, Zhu QS, Liu LJ (2004) Involvement of abscisic acid and ethylene in the responses of rice grains to water stress during filling. Plant Cell Environ 27(8):1055–1064

Yordanov I, Velikova V, Tsonev T (2000) Plant responses to drought, acclimation, and stress tolerance. Photosynthetica 38(2):171–186

Yu L, Wu S, Peng Y, Liu R, Chen X, Zhao P, Pei Y (2016) Arabidopsis EDT 1/HDG 11 improves drought and salt tolerance in cotton and poplar and increases cotton yield in the field. Plant Biotechol J 14(1):72–84

Zenda T, Liu S, Wang X, Liu G, Jin H, Dong A, Duan H (2019) Key maize drought-responsive genes and pathways revealed by comparative transcriptome and physiological analyses of contrasting inbred lines. Int J Mol Sci 20(6):1268. https://doi.org/10.3390/ijms20061268

Zhang SW, Li CH, Cao J, Zhang YC, Zhang SQ, Xia YF, Sun Y (2009) Altered architecture and enhanced drought tolerance in rice via the down-regulation of indole-3-acetic acid by TLD1/OsGH3. 13 activation. Plant Physiol 151(4):1889–1901

Zhu J, Brown KM, Lynch JP (2010) Root cortical aerenchyma improves the drought tolerance of maize (*Zea mays* L.). Plant Cell Environ 33(5):740–749

Zimmermann MH (1983) Xylem structure and ascent of sap. Springer, Berlin

Special Anatomical Features of Halophytes: Implication for Salt Tolerance

Rizwana Nawaz, Zeshan Ali, Tayyaba Andleeb and Umar Masood Qureshi

Abstract Soil salinity has become a key factor restricting agricultural production. Thus, plants, especially salt-tolerant crops, will have important economic significance. The adaptation or tolerance of plants to salinity stress involves a series of physiological, metabolic, and molecular mechanisms. Halophyte organisms acquire special salt-tolerant mechanisms to cope with a lock of salt and ensure normal growth and development under saline conditions during their long evolutionary adaptation, so understanding halophyte salinity pressure will provide us with methods and means to cultivate and develop salt-tolerant crop varieties. In physiological and molecular strategies adopted by halophytes are various including photosynthetic and transpiration rate of change, the sequestration of Na^+ to cells or vacuole, the regulation of stomatal density and stomatal aperture, accumulation and hormone synthesis and gene expression related to the potential of these physical characteristics, like stress signal transduction, regulation of transcription factor transporter gene activation and expression of synthetase activation or inhibition, etc. This chapter reviews the research progress on physiological and molecular regulation mechanisms and special anatomical features of halophytes and reveals the tolerance and adaptability of halophytes to salt stress.

1 Introduction

All plants face profound stresses during their life cycle, including both abiotic and biotic stresses. Salt stress is the major abiotic stress, currently reducing the plant's productivity specifically crops (Majeed et al. 2010). According to an estimate, salt stress affects 2.1% agriculture of the global dry land (FAO 2003). In semi-arid and arid regions, salinity effects are more pronounced. While elevated temperature, high

R. Nawaz (✉) · T. Andleeb · U. M. Qureshi
Department of Plant Sciences, Quaid-e-Azam University, Islamabad, Islamabad, Pakistan
e-mail: rizwananawaz572@gmail.com

Z. Ali
Plant Physiology Program, National Agricultural Research Centre, Crop Sciences Institute, Park Road, Islamabad 45500, Pakistan

M. Hasanuzzaman and M. Tanveer (eds.), *Salt and Drought Stress Tolerance in Plants*, Signaling and Communication in Plants, https://doi.org/10.1007/978-3-030-40277-8_5

evapotranspiration, and little rainfall along with low water and faulty soil management techniques are the vital causative agents of salt stress (de Azevedo Neto et al. 2006). Earth being a salty planet with its water containing 3% NaCl renders the land salty. According to an estimate, salt stress affects about 900 Mha dry land area considerably posing a serious ultimatum to agricultural productivity (Flowers and Yeo 1995). Ultimately crop growth is adversely affected by high saline conditions. Hence, the prevailing salt stress is a major challenge for food security (Tanveer and Shah 2017).

Plant physiology such as ion toxicity, increased rate of respiration, growth changes, membrane instability and mineral distribution (Sudhir and Murthy 2004) decreased the rate of photosynthesis and membrane permeability is affected by salt stress (Hasegawa et al. 2000; Gupta et al. 2002; Tanveer et al. 2018). Plant physiology at the whole plant level and even at the cellular level is affected by salt stress through ionic and osmotic imbalances (Hasegawa et al. 2000). Ion imbalances caused by salt stress impair root membrane selectivity and induce potassium deficiency (Gadallah 2000). Increased NaCl concentrations in plant chloroplasts during salt stress affect growth rate and cause a reduction in electron transport schemes of photosynthesis (Kirst 1990). In several essential grain legumes, salt stress caused plant growth reduction (Tejera et al. 2006).

The harmful consequences of salt stress are apparent on the whole plant and are evident at any developmental stage like germination, seedling development or physiological maturity. Stress most of the time can appear at a particular stage in the plant life cycle and become adverse at another developmental stage. The ability of plants to respond to particular stress varies according to the stage of development and from species to species (Liu et al. 2018). Stress tolerance in plants is a multiplex trait and is regulated by different mechanisms like regulation of several biochemical pathways leading to the synthesis of free radicals, osmotically active metabolites and specific proteins for maintenance of osmotic and ionic flux. In the same way, histo-anatomical features of plants have physicochemical specificities with the environment, as structures are best suited according to the function they perform. Histo-anatomical features are considered as adaptive features because these justify the position of a plant in a specific ecological group (Grigore and Toma 2006). During evolution, environmental factors had a major, long and shaping influence on the structures of plants. Salt stress, therefore, has pronounced formative abilities on halophyte's anatomy rendering them fit in the environment. There is a need to investigate significant histo-anatomical adaptations in halophytes as implications to salt stress tolerance because it makes sense of the underlying specific mechanisms in which a plant reacts at the environment's requirements. A consolidated anatomical approach is presented in this chapter as halophyte's tendency towards salt stress tolerance. This approach aims to coalesce structure with function of halophytes under salt stress and is a useful indication for plant breeders in the development of stress-tolerant crops.

2 Halophytes-Definition

The identification of halophytes has been made since the time of Goethe but, Schimper and especially Warming brought them to scientific attention. Halophyte's definitions remain ambiguous although their identification was made hundreds of years ago.

The definitions of halophytes are multitudinous; Halophytes are the native plants of saline environments, having an ability to tolerate salt concentrations that are generally harmful to other species. These are the unusual plants that can tolerate salt shock. Plants capable of completing their life cycle in around 200 mM NaCl concentrations and constituting about 1% flora are termed as halophytes. Halophytes are the plants growing in highly saline habitats (Kefu et al. 1995). Halophytes are the natural flora of salty habitats (Jennings 1976).

3 Classification of Halophytes

Halophytes are polyphyletic in origin and mostly belong to higher plant families (Glenn et al. 1999). Chenopodiaceae includes the most significant number of halophytes, i.e., about 550 species (Fig. 1). Poaceae, Fabaceae, and Asteraceae are the other families including halophytes (Aronson and Whitehead 1989).

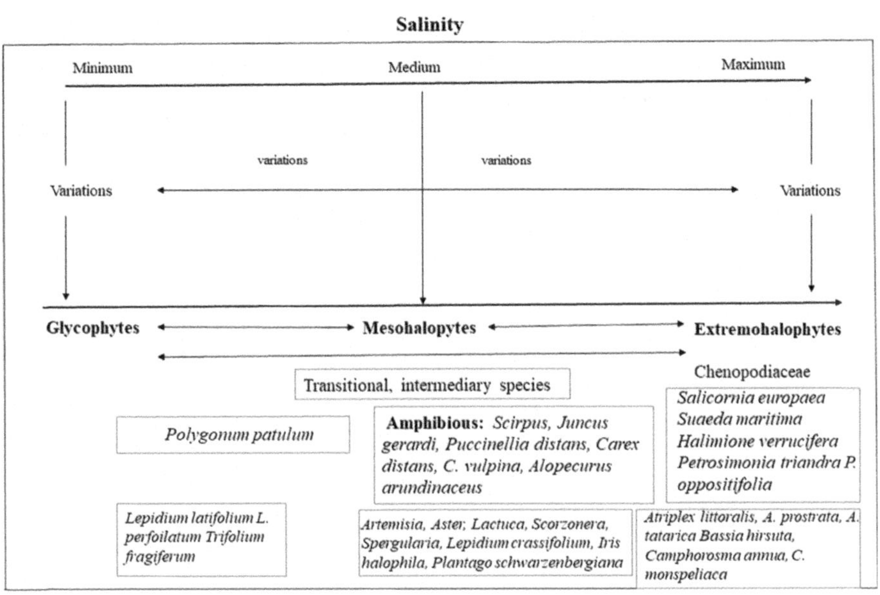

Fig. 1 Proposal of halophyte classification based on anatomical features and their ecological implications (modified from Grigore and Toma 2008)

Classification of halophytes is diverse based on distribution and general ecological behavior, plant growth response to the amount of salt intake and salt stress (Liphschitz and Waisel 1974; Fig. 1). Halophytes grow along the salt swamps are classified as physiotypes. The existence or lack of salt glands can also make a classification, for example, black mangroves possessing highly developed salt glands and red mangroves completely devoid of salt glands (Popp et al. 1993).

3.1 Halophytes Based on Internal Salt Content of the Plant

According to Steiner (1934), halophytes are classified based on their response to internal salt concentration. So, salt marsh halophytes are salt accumulators and salt regulators. Based on internal salt concentration, halophytes are also includers versus excluders (Ashraf et al. 2008).

3.2 Halophytes Based on Morphology

Based on their morphology, halophytes are classified as succulents and excretives.

Halophytes having the ability to excrete excess salt from their bodies are termed as excretives. These types of halophytes possess glandular cells on the leaf surface that help in the elimination of excess salt from the plant body (Engels and Marschner 1995). Succulent halophytes have the presence of a salt bladder on the surface of their leaf. They alleviate the toxicity of salt by reserving a large amount of water within their body (Weber 2009). Desert halophytes belong to this designation.

3.3 Halophytes Based upon Salt Demand

Salt halophytes are categorized as obligate and facultative based on their tolerance for sodium salts (Sabovljevic and Sabovljevic 2007). Obligate halophytes are true halophytes as they require salt for their growth and thrive under 0.5–1% NaCl concentrations (Ungar 1991). Their growth is augmented in increasing salt concentrations (Al Hassan et al. 2017). While facultative halophytes thrive best both in high and low salt concentrations, these include members of Juncaceae, Cyperaceae, Gramineae and a large number of dicots (Sabovljevic and Sabovljevic 2007).

3.4 Habitat Based Halophytes

Habitat or geographical distribution involves two main types of halophytes, i.e., hydrohalophytes and xerohalophytes. Hydrohalophytes grow in moist conditions and include salt marsh species and mangroves along coastal lines. On the other hand, xerophytes grow in saline soils facing water deficit and thus are succulents (Youssef 2009).

3.5 Habitat-Indifferent Halophytes

Habitat indifferent halophytes can cope with the salt stress, but they are habitat insensitive. These plants thrive in salt-stressed as well as non-saline soil conditions. The plants living in saline, as well as non-saline soils, may have some genetic differences rendering them fit the situation. These halophytes include *Agrostis stolonifera*, Juncusbufonius, and Festucarubra (Sabovljevic and Sabovljevic 2007).

4 General Morphological and Anatomical Adaptations in Halophytes

Salinity tolerance in plants is a complex phenomenon and involves diverse mechanisms. According to Colmer and Flowers (2008), it is the potential of plants to complete their life cycle with a sustainable yield and growth. Ion toxicity, uptake, and translocation of nutrients and in turn calcium and potassium ion imbalances and water deficit are the three main factors affecting plant growth under salt stress (Table 1). Interruption of cellular processes and damage of vital physiological processes, i.e., respiration and photosynthesis are the consequences of excessive salt uptake (Marschner et al. 1996). Plants prevent the harmful effects of salt stress by adopting broad-spectrum tolerance and adaptive mechanisms.

The broad spectrum tolerance to saline environments ranges from glycophytes (salt-sensitive species) to halophytes (salt-tolerant species) and is of hereditary nature (Niknam and McComb 2000). Halophytes are the salt-tolerant plants adopting multiple tolerance mechanisms to thrive under high salt stress. These tolerance strategies involve ion control mechanisms by the reduction or exclusion of ion uptake at the root level and hence minimize ion translocation to the shoot (Flowers and Colmer 2008). Plants that thrive best under saline environments have outstanding adaptive as well as tolerance strategies. These include morphological, anatomical, and physiological mechanisms that prevent salt uptake and its translocation in plants.

Plant roots prevent salt uptake by morphological features. Physiological, morphological, and anatomical features impede the detrimental effects of salt stress in halophytes (Hameed et al. 2009). Two general salt tolerance mechanisms in plants

Table 1 Na^+, Cl^-, K^+ and Glycine betaine concentrations expressed on a water basis (mM) in shoot parts and the contribution of Na^+, Cl^- and K^+ to the osmotic potential of the shoot of *Salicornia europaea* hydroponically grown at 50–500 mM NaCl in the root medium for 40 d

Treatment	Shoot			Na^+, Cl^- and K^+ contribution to Ψs (%)	Glycine betaine concentration (mM)	
Salinity (mM)	Na^+	Cl^-	K^+		Shoot	Root
50	330	270	80	77	40	30
100	321	300	60	70	32	27
150	305	302	50	59	31	27
200	350	315	40	65	34	30
250	360	320	35	67	37	30
300	390	340	30	67	34	31
400	510	460	40	81	35	35
500	580	638	35	80	46	37

The molarities of Na^+, Cl^- and K^+ were calculated from the ion data. Values are mean ± S.E. of 5–9 replicate. ANOVA p-values were for: shoots $p = 0.054$ and for roots $p > 0.05$

are the avoidance and the compartmentalization of cumulated salts in plant cell vacuoles. Avoidance strategies involve the prevention of salt uptake by metabolically active tissues (Munns and Tester 2008). These include active ion expulsion by ion pumps, passive ion exclusion by a permeable membrane or by ion dilution in tissues of plants (Allen et al. 1994). Compartmentalization is also a defensive strategy in plants against salt stress. These important mechanisms in plants prevent damage of plant tissues in excess salt concentrations and thus ensure plant survival under highly saline environments. Salt glands present on the leaf and extensive root system of plants are important tolerance mechanisms in plants (Naz et al. 2009; Tanveer and Shah 2017). Moreover, controlled or reduced salt uptake, tissue tolerance, salt accumulation in vacuoles, K^+, Na^+, Cl^- and SO_4^{2-}, i.e., ion discrimination and low molecular weight osmolytes production are the other salt tolerance mechanisms in plants (Munns and Tester 2008). So, plants vary in their salt tolerance behaviors due to these different mechanisms.

Small-sized leaves, lesser stomata per unit leaf area, increased the thickness of leaf cuticle, increased succulence, and wax deposition are the most essential morphological adaptations of plants from salt-stressed arid regions (Gale 1975). These adaptations are involved in water conservation for maximum plant growth under salt-stressed conditions. In *Cenchrus ciliaris*, a considerable ecotypic variability enables the grass to withstand extreme conditions unusually severe drought conditions (Mansoor et al. 2002). Conspicuous anatomical adaptations at leaf level include appendages that secrete excess salts from the plant body (Dickison 2000). These salt secreting appendages include salt secretory trichomes for example in *Atriplex* spp. and multicellular salt glands in flowering plants native of coastal and desert habitats. Salt glands occur in families including Aveceniaceae, Frankiaceae, Tamaricaceae, Plumbaginaceae, and Poaceae. Succulent stems having a well-developed water-storing tissue in pith and cortex are the characteristics of desert plants (Dickison 2000). The epidermis is thick walled and many-layered with thick cuticle and deposition of wax, e.g., *Anbasis* spp. has eight to eleven layered epidermis and a stomatal crypt having stomata at the base (Wahid 2003). Contrastingly, the stem of halophytic plant *Salicornia fruiticosa* has a simple cortex and single-layered thin-walled epidermis. Roots of saline plants have the adaptation of reduced cortex to lessen the distance between stele and epidermis. Highly arid and salt-stressed plants have wider Casparian strip in comparison to mesophytes. In plants of saline areas, the exodermis and endodermis are the barriers of varying resistance to the movement of ions and water from cortex to stele under induced conditions (Hose et al. 2001).

Special halophytic features offloading the excess of ions in arid regions are the salt secretory trichomes and salt glands (Tanveer and Shah 2017). Salt secretory trichomes apparent in *Atriplex* spp. are the bladder-like hairy projections of leaf surface. These have a large bladder cell at the top and a unicellular or sometimes multicellular stalk (Dickison 2000). Dictyosomes, mitochondria, ribosomes, a large nucleus, and endoplasmic reticulum are present in these cells. These have partially developed or rudimentary chloroplasts. In bladder cells, there is a single large vacuole in contrast to stalk cells containing many small vacuoles (Osmond et al. 1969). Ionic movement from mesophyll cells to the bladder cells takes place through a symplastic continuum. Bladder and stalk cells have external cutinized walls while internal primary walls are not cutinized (Thomson and Platt-Aloia 1979). Offloading of excess salts takes place by leaf shedding. Our experiment on *S. europaea* under different salt treatments showed variation in leaf anatomy. Epidermal cells are tangential in treatment two and radially mixed with tangentially elongated cells in treatment one and three, covered with thick, warty cutin. Cortex consists of 3–4 layers of chlorenchyma cells followed by 1–2 layers of parenchyma. Pericycle consists of parenchymatous cells. Vascular cylinder is eustele, composed of 15–16 bundles in *S. europaea* from treatment one, 15–17 in treatment two and 18–20 in treatment three, each with will defined patches of phloem and well-defined xylem vessels. The medullary rays are wide. Pith is wide and homogenous, consists of thin-walled round to polygonal parenchymatous cells. Schizogenous canals are recorded in cortex and pith (Fig. 2). On the other hand, salt glands are more conspicuous, multicellular and elaborate structures. The glands are generally bicelled in grasses having an outer cap cell and a basal cell and may be sunken, sub-sunken or exist as extensions of the epidermis (Marcum and Murdoch 1994). Glands may lie recumbent to the leaf surface (Marcum et al. 1998). In dicots, multicellular salt glands consisting of basal and secretory cells are present. Different genera differ in the number of cells ranging from 6 to 40 (Faraday and Thomson 1986). *Tamarix* spp. has salt glands consisting of two basal collecting cells and outer cytoplasmic secretory cells that are six in number. However, Glaux and Avicennia glands have many secretory cells on a single disc-shaped basal cell (Rozema et al. 1977). Glands may be protruding out of the abaxial surface of leaf-like trichomes (Avicennia), lateral to the epidermis (Tamarix) or present in pits or depressions of the epidermis (Glaux) (Rozema et al. 1977).

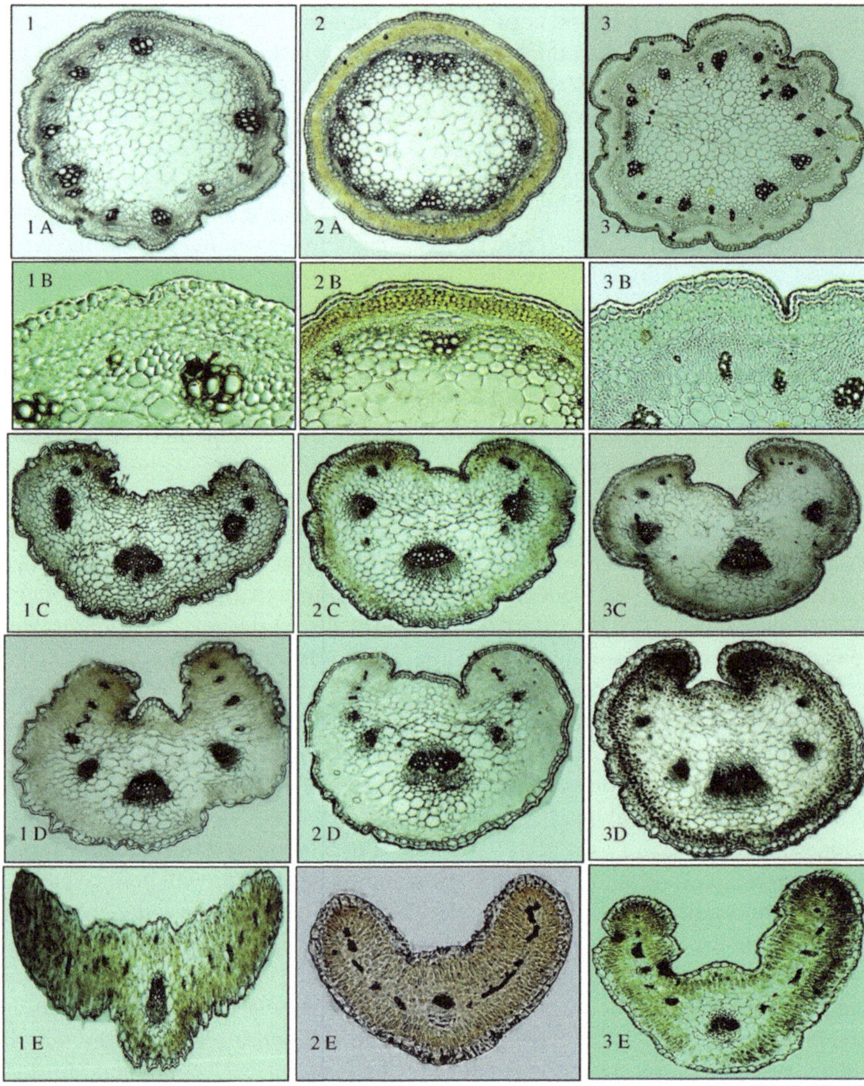

Fig. 2 Anatomical structure of *Salicornia europaea* from different treatments. Figure 1: from 200 mM; Fig. 2: from 400 mM; Fig. 3: from 500 mM; A = outline of stem; B = sector of stem; C = petiole outline; D = raches of blade; E = lobe of blade

5 Halophytes and Salt Stress

Salinity disrupts ion distribution and water potential. Homeostasis is disrupted at both the whole-plant as well as cellular levels. Consequently, there is a condition of arrested growth, molecular damage, and death of the plant body. Three major interconnected outlooks to achieve salt tolerance in plants are present. First is the damage prevention or alleviation. Second is the re-establishment of homeostatic conditions in the new stressed conditions. The third is the growth resumption (Zhu 2001).

5.1 Detoxification

Damage imposed by high salt concentrations on plants is not completely obvious. Disruption of activities of various enzymes, acquisition of nutrients, the functioning of photosynthetic apparatus and cell membrane integrity are the major consequences of high salt conditions. ROS are the major causative agents of these losses in salt stress (Tanveer and Shabala 2018). Complex molecular responses including the production of accordant osmolytes and stress proteins production are displayed by the plants subjected to high salt stress (Zhu et al. 1997). These stress proteins and osmolytes probably have a detoxification function by scavenging reactive oxygen species (ROS).

This detoxification strategy is essential for the achievement of transgenic improvements in plant tolerance. It is considerable in transgenic plants having overexpression of enzymes involved in protection against oxidative damage, such as superoxide dismutase, ascorbate peroxidases, glutathione reductases and glutathione peroxidases (Allen et al. 1997). A notable example in this regard is of protein NPK1, a MAP kinase (mitogen-activated protein kinase) that mediates oxidative stress response (Kovtun et al. 2000). Another importance in tolerance of salt stress also comes from the characterization of an *Arabidopsis* mutant pst1, having a mutation in a yet unidentified negative regulatory agent of oxidative stress responses (Tsugane et al. 1999). The ability of pst1 mutant plants to tolerate high salt concentrations is correlated with the increased ability to tolerate oxidative stress. Oxidative detoxification also involves engineering osmolytes like fructans, trehalose, proline, mannitol, ectoine, and glycine betaine. These osmolytes actively scavenge ROS. Another critical strategy involved in better protection is the targeted osmolytes production in chloroplast (Shen et al. 1997). This is by the notion that chloroplasts are the primary ROS producing sites. Another support in the osmolytes mode of action in this way is that the osmolytes levels are too low for their significance in the osmotic adjustment of transgenic plants. So, osmolytes production in transgenic plants is an essential aspect towards their improved tolerance in salt-stressed conditions and also in other stresses like freezing, chilling, drought and heat leading to the production of ROS (Hayashi et al. 1998).

5.2 Homeostasis

Homeostasis is another important strategy for the achievement of greater salt tolerance in plants. The restoration of both osmotic and ionic homeostasis is essential. Many ion transporters determine ionic homeostasis. High sodium accumulation in the cytoplasm or organelles other than vacuole is prevented because it causes inhibition of essential enzymes. So, sodium entry should be alleviated or prevented. This sodium entry into plant cells is mediated by NSCs (Nonselective cation channels) although they have no molecular identification yet (Amtmann and Sanders 1998). LCT1 and HKT1 are the transporters that have been shown to be permeable to sodium ions in yeasts and oocytes (Rubio et al. 1995). So, these gene products which were spotted as potassium ion transporters may be involved in the mediation of sodium influxes in plant cells. Salt tolerance studies also aim at the establishment of the functions of the transporters involved in the influx of sodium ions into plant cell so that influx of sodium ions may be blocked for the achievement of increased salt tolerance. Functional characterization of sodium ions influx mechanisms also involves the screening of accessory site mutations that cause partial suppression of the salt-hypersensitive phenotypes of *SOS*3 or *SOS*2. Mutations in sodium ions influx transporters might be necessary for the alleviation of their salt hypersensitivity because there is an increased sodium accumulation in *SOS*3 and *SOS*2 *Arabidopsis* mutants (Zhu et al. 1998). Any sodium influx in the cell is exported out of the cell or is stored in the vacuole. Sodium toxicity in the cytosol is prevented by sodium compartmentation because sodium ions are important as an osmolytes in the vacuole for the maintenance of osmotic homeostasis. This strategy is vital in many halophytes (Flowers et al. 1977). In tonoplast vesicle preparations, sodium hydrogen antiport activities were detected (Blumwald and Poole 1985).

Sodium ion efflux is also of considerable significance in the maintenance of a low sodium ion concentration in the cytoplasm. In contrast to animal cells which have sodium–potassium ATPases, plant cells do not have sodium ATPases. However, in plasma membrane enriched membrane vesicles there is an indication of sodium hydrogen antiport activities (Blumwald et al. 2000). Recently, a putative plasma membrane sodium hydrogen antiporter is reported to have been encoded by the *SOS1* gene (Shi et al. 2000). Due to any mutation in SOSI, Arabidopsis plants become intensely sensitive to sodium stress. Plants under stress also establish osmotic homeostasis in the cytosol along with the maintenance of ionic homeostasis. There is an accumulation of many compatible osmolytes in the plant cytosol sustaining water uptake from saline soils and reducing the osmotic potential (Zhu et al. 1997). Some organic osmolytes are of considerable significance in the protection of cellular structures by the detoxification of reactive oxygen species (ROS). The water flux speed across cell membranes under stress is controlled by water channel proteins (Chrispeels et al. 1999).

5.3 Growth Regulation

Like other abiotic stresses, plant growth is inhibited under salt stress. An adaptive feature for the survival of plants under salt stress is the reduction in growth, enabling the plant to rely on multiple resources, e.g., energy and building blocks to overcome stress. Usually, the extent of growth is inversely related to the extent of drought or salt tolerance. Inadequate photosynthetic rate due to stomatal closure and consequently reduced CO_2 uptake is one major causes of the reduction in growth rate. There might be an inhibition in cell division and expansion under stress conditions. There is a need to better understand the interconnection of stress signaling and cell division and expansion control. There is a reduction in growth and a significant loss of productivity in plants even in mild stresses. Some plants are incredibly responsive to stresses that there is a cessation of growth even in mild stress. Contrastingly, some plants are not so responsive to stress conditions. To improve productivity under drought or salt stress, fine-tuning of this responsiveness is of considerable significance (Zhu 2001). By the induction of ICK1 by abscisic acid in *Arabidopsis*, there is the discovery of a potentially significant link between cell division and stress (Wang et al. 1998). ICK1 is a cyclin-dependent protein kinase inhibitor, and it might cause hindrance in cell division by the reduction of cyclin-dependent protein kinases activities that manage cell cycle. Cell division is inhibited by the accumulation of abscisic acid in salt and water stress conditions causing the induction of ICK1. These stress conditions also affect cell division the post-transcriptional or transcriptional regulation of other cell cycle machinery components. There is no careful examination of the link between cell expansion control and water or salt stress. The inhibition of cell expansion by stress by the reduction of the concentration of growth-promoting hormones (auxin, brassinolides, and gibberellins) is not surprising, because several hormones are considered significant in the regulation of cell elongation.

There are also implications of interconnection between growth regulation and stress in studies with transgenic overexpression of the components of salt stress.

6 Histo-anatomical Features of Halophytes Under Salt Stress

Histo-anatomical features are one of the ambient factors considered as an adaptation of the plants to salt stress. Different histo-anatomical features of the halophytes are observed and discussed below.

There is a typical secondary structure in Spergularia media root and *Plantago schwarzenbergiana* root and rhizome. Lateral meristems, phellogen, and cambium also play a role in this architecture. Another common histo-anatomical adaptation in *Juncus gerardi*, *P. schwarzenbergiana* and Spergularia media is the presence of auriferous cavities. Well-developed endodermis is also an important histo-anatomical feature present in halophytes. In the *J. gerardi* rhizome, the presence of a tertiary endodermis is also evidenced. There is the presence of a central cylinder in the aerial stem of *J. gerardi* and a thick sclerenchyma ring containing vascular bundles and a pericycle in the stem of Spergularia media with the pericycle being highly lignified and sclerenchymatous. The other species have higher lignification of xylem and schlerenchymatic pericycle at the lower level of the stem. There is the presence of calcium oxalate crystals in the stem cortical parenchyma and mesophyll of Spergularia media. There is evidence of the presence of biperygen type stomata in lower epidermis of *J. gerardi* vagina considering the leaf as a lateral vegetative organ. At the inferior level of the vagina, there is the presence of lower epidermis with tangentially elongated and small cells having thicker external walls covered by a cuticle and the upper epidermis consisting of large cells (Grigore and Toma 2008).

7 Potential Use of Halophytes Under Saline Conditions

Due to the rapidly changing climate, the area affected by salinity is at a continuous pace. Recently, the development of highly tolerant crops is the utmost requirement of the time to cope with the underlying challenges. Halophytes have the potential of high yield under salt stress conditions. So, halophytes can be utilized for ecological, industrial, or agricultural purposes (Fig. 3). In agronomic field trials, halophytes have been analyzed as forage, oilseed crops, and vegetables. Seawater irrigated species have the most productive yield of 10–20 tons per hectare biomass that is equivalent to

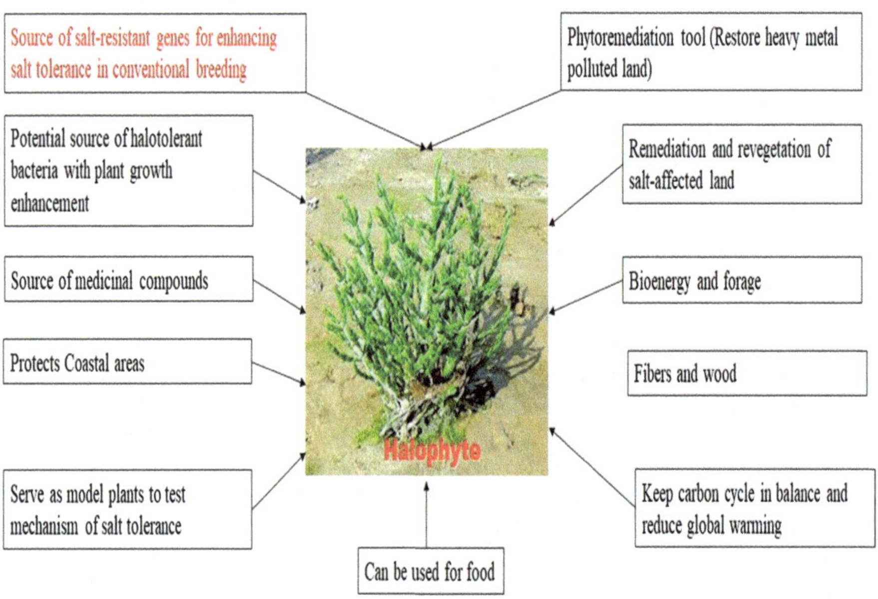

Fig. 3 Potential application of halophytes

conventional crops. An oilseed halophyte- *S. bigelovii* yield 2 tons per hectare of seed containing 31% protein and 28% oil which is comparable to yield and seed quality of soybean (Glenn et al. 1999). Several plant species have been traditionally used as vegetables and herbs and so, rediscovering the potentials of several halophytes to be framed as leafy vegetables is at an increased pace for a couple of decades (Ventura and Sagi 2013). Some halophyte species have the potential use as a good fodder and can be used for feeding animals in salt stress areas. Contrastingly, due to excessive salt concentrations and antinutritional compounds, several halophytes are nutritional barriers (Khan et al. 2009). Salt-affected soils reclamation is not always cost-effective and also not feasible; there is a need to explore biosaline agriculture and thus a better understanding of the halophytes to manage salts. Halophytes study is of major significance because of the three major perspectives (Glenn et al. 1999).

The mechanisms of halophytes survival and the maintenance of productivity in saline areas can be used in the development of salt-tolerant varieties in conventional crops (Zhu et al. 1997). The overall feasibility of highly saline agriculture can be evaluated by the halophytes grown in an agronomic setting (Glenn et al. 1997). Halophytes can become a potential source of new crops (Hamdy 1996).

On the other hand, in some situations, the performance of halophytes is affected. For example, halophytes may have low biomass comparable to several glycophytes (Zaier et al. 2010). Halophytes have the potential of salt accumulation from the saline areas as well as the remediation of toxic metals. So, halophytes are well adapted in metal regions affected in contrast to glycophytes, thus making them eco-friendly and as a cleanup for contaminated coastal environments (Anjum et al. 2014).

8 Conclusion and Future Outlook

It is clear from the above discussion that salinity has become a major problem all over the world and is increasing at a rapid pace. The problem is more drastic in arid and semi-arid areas (Munns 2002) and curb plant growth and productivity all over the world. Specific changes including morphological, physiological, and anatomical are induced at cell, tissue and organ level due to salinity (Fig. 2). Anatomical alterations such as reduced stomatal number, leaf thickness, epidermal cell number and distance between vascular bundles are evidenced in salinity stress (Kiliç 2007). Anatomical alterations of halophytes are of significance as an adaptive response to the native environment of a species(Chrispeels and Sadava 2003; Grigore and Toma 2007). According to literature, plants have plastic behavior for many ecologically significant traits ranging from anatomy, physiology, and morphology to patterns of offspring development and breeding system (Al-Taisan 2010).

Despite this, our knowledge about the response of different plant species to variable salinities is limited. There is a need for a better understanding of halophytes to cope with the underlying salinity-induced challenges. The underlying physiological and biochemical mechanisms must be revealed for the complete understanding of these salinity-induced strategies. At the molecular level, response and tolerance mechanisms are essential in engineering plants that are salt tolerant with high economic gains. Phytoremediation of highly saline soils by halophytes can be improved as an environmental friendly and economically important technology. There is a need to develop hyper-accumulators that have diverse salt accumulating capacity in an economically beneficial way. Novel salt-tolerant gene identification and the production of transgenic plants are also other areas of research. Briefly, the physiological, biochemical and molecular studies on halophytes open their salt-tolerant door, therefore we are capable to acquire more strategies to develop crops with improved salt tolerance permitting the crops to be able to develop usually in saline lands in future.

References

Al Hassan M, Estrelles E, Soriano P, López-Gresa MP, Bellés JM, Boscaiu M, Vicent O (2017) Unraveling salt tolerance mechanisms in halophytes: a comparative study on four Mediterranean Limonium species with different geographic distribution patterns. Front Plant Sci 8:1438

Allen JA, Chambers JL, Stine M (1994) Prospects for increasing the salt tolerance of forest trees: a review. Tree Physiol 14:843–853

Allen RD, Webb RP, Schake SA (1997) Use of transgenic plants to study antioxidant defenses. Free Radical Biol Med 23:473–479

Al-Taisan W (2010) The relation between phenotypic plasticity of *Senecio glaucus* L. and some soil factors. Aust J Basic Appl Sci 4:1369–1375

Amtmann A, Sanders D (1998) Mechanisms of Na+ uptake by plant cells. In: Advances in botanical research. Elsevier, pp 75–112

Anjum NA, Ahmad I, Válega M, Mohmood I, Gill SS, Tuteja N, Duarte AC, Pereir E (2014) Salt marsh halophyte services to metal–metalloid remediation: assessment of the processes and underlying mechanisms. Crit Rev Environ Sci Technol 44:2038–2106

Aronson JA, Whitehead EE (1989) HALOPH: a data base of salt tolerant plants of the world

Ashraf M, Hameed M, Arshad M, Ashraf Y, Akhtar K (2008) Salt tolerance of some potential forage grasses from Cholistan desert of Pakistan. In: Ecophysiology of high salinity tolerant plants. Springer, pp 31–54

Blumwald E, Poole RJ (1985) Na+/H+ antiport in isolated tonoplast vesicles from storage tissue of *Beta vulgaris*. Plant Physiol 78:163–167

Blumwald E, Aharon GS, Apse MP (2000) Sodium transport in plant cells. Biochim Biophys Acta Biomembr 1465:140–151

Chrispeels MJ, Sadava DE (2003) Plants, genes, and crop biotechnology. Jones & Bartlett Learning

Chrispeels MJ, Crawford NM, Schroeder JI (1999) Proteins for transport of water and mineral nutrients across the membranes of plant cells. Plant Cell 11:661–675

de Azevedo Neto AD, Prisco JT, Enéas-Filho J, de Abreu CEB, Gomes-Filho E (2006) Effect of salt stress on antioxidative enzymes and lipid peroxidation in leaves and roots of salt-tolerant and salt-sensitive maize genotypes. Environ Exp Bot 56:87–94

Dickison W (2000) Integrative plant anatomy. Harcourt Academic Press, San Diego, CA (Calif Google Scholar)

Engels C, Marschner H (1995) Plant uptake and utilization of nitrogen. Nitrogen Fertil Environ 41–81

Faraday CD, Thomson WW (1986) Functional aspects of the salt glands of the Plumbaginaceae. J Exp Bot 37:1129–1135

Flowers TJ, Colmer TD (2008) Salinity tolerance in halophytes. New Phytol 179:945–963

Flowers T, Yeo A (1995) Breeding for salinity resistance in crop plants: where next? Funct Plant Biol 22:875–884

Flowers T, Troke P, Yeo A (1977) The mechanism of salt tolerance in halophytes. Ann Rev Plant Physiol 28:89–121

Food FAO (2003) Agriculture organization. Gender and access to land, FAO land tenure studies, 4

Gadallah M (2000) Effects of acid mist and ascorbic acid treatment on the growth, stability of leaf membranes, chlorophyll content and some mineral elements of *Carthamus tinctorius*, the safflower. Water Air Soil Pollut 118:311–327

Gale J (1975) Water balance and gas exchange of plants under saline conditions. In: Plants in saline environments. Springer, pp168–185

Glenn E, Miyamoto S, Moore D, Brown JJ, Thompson TL, Brown P (1997) Water requirements for cultivating *Salicornia bigelovii* Torr. with seawater on sand in a coastal desert environment. J Arid Environ 36:711–730

Glenn EP, Brown JJ, Blumwald E (1999) Salt tolerance and crop potential of halophytes. Crit Rev Plant Sci 18:227–255

Grigore MN, Toma C (2006) Evidencing the successive cambia phenomenon on some halophylous representatives among Chenopodiaceae and its possible ecological-adaptive implications. Stud Com Complexul Muzeal Şt Nat Ion Borcea Bacău 21:87–93

Grigore MN, Toma C (2007) Histo-anatomical strategies of Chenopodiaceae halophytes: adaptive, ecological and evolutionary implications. WSEAS Trans Biol Biomed 4:204–218

Grigore M, Toma C (2008) Ecological anatomy investigations related to some halophyte species from Moldavia. Rom J Biol Plant Biol 53:23–30

Gupta N, Meena S, Gupta S, Khandelwal S (2002) Gas exchange, membrane permeability, and ion uptake in two species of Indian jujube differing in salt tolerance. Photosynthetica 40:535–539

Hamdy A (1996) Saline irrigation: assessment and management techniques. In: Chouckr-Allah R, Malcolm CV, Hamdy A (eds) Halophytes and biosaline agriculture. Marcel Dekker Publ

Hameed M, Ashraf M, Naz N (2009) Anatomical adaptations to salinity in cogon grass [*Imperata cylindrica* (L.) Raeuschel] from the Salt Range, Pakistan. Plant Soil 322:229–238

Hasegawa PM, Bressan RA, Zhu JK, Bohnert HJ (2000) Plant cellular and molecular responses to high salinity. Annu Rev Plant Biol 51:463–499

Hayashi H, Sakamoto A, Murata N (1998) Enhancement of the tolerance of Arabidopsis to high temperatures by genetic engineering of the synthesis of glycinebetaine. Plant J 16:155–161

Hose E, Clarkson D, Steudle E, Schreiber L, Hartung W (2001) The exodermis: a variable apoplastic barrier. J Exp Bot 52:2245–2264

Jennings BD (1976) The effects of sodium chloride on higher plants. Biol Rev 51:453–486

Kefu Z, Zi-Yi C, Shou-Jin F, Giang H, Hai F, Zeng L, Harris P (1995) Halophytes in China. Biology of salt tolerant plants. Department of Botany, University of Karachi, Pakistan, pp 284–293

Khan MA, Ansari R, Ali H, Gul B, Nielsen BL (2009) *Panicum turgidum*, a potentially sustainable cattle feed alternative to maize for saline areas. Agric Ecosyst Environ 129:542–546

Kiliç S (2007) Effects of 24-epibrassinolide on salinity stress induced inhibition of seed germination, seedling growth and leaf anatomy of barley. http://dergipark.gov.tr/sdufeffd 2

Kirst G (1990) Salinity tolerance of eukaryotic marine algae. Ann Rev Plant Biol 41:21–53

Kovtun Y, Chiu WL, Tena G, Sheen J (2000) Functional analysis of oxidative stress-activated mitogen-activated protein kinase cascade in plants. Proc Natl Acad Sci USA 97:2940–2945

Liphschitz N, Waisel Y (1974) Existence of salt glands in various genera of the Gramineae. New Phytol 73:507–513

Liu R, Wang L, Tanveer M, Song J (2018) Seed heteromorphism: an important adaptation of halophytes for habitat heterogeneity. Front Plant Sci 9

Majeed A, Nisar MF, Hussain K (2010) Effect of saline culture on the concentration of Na+, K+ and Cl-in *Agrostis tolonifera*. Curr Res J Biol Sci 2:76–82

Mansoor U, Hameed M, Wahid A, Rao AR (2002) Ecotypic variability for drought resistance in Cenchrus ciliaris L. germplasm from Cholistan Desert in Pakistan. Inter J Agric Biol 4:392–397

Marcum KB, Murdoch CL (1994) Salinity tolerance mechanisms of six C4 turfgrasses. J Amer Soc Horti Sci 119:779–784

Marcum KB, Anderson SJ, Engelke M (1998) Salt gland ion secretion: a salinity tolerance mechanism among five zoysiagrass species. Crop Sci 38:806–810
Marschner H, Kirkby E, Cakmak I (1996) Effect of mineral nutritional status on shoot—root partitioning of photoassimilates and cycling of mineral nutrients. J Exp Bot 1255–1263
Munns R (2002) Comparative physiology of salt and water stress. Plant Cell Environ 25:239–250
Munns R, Tester M (2008) Mechanisms of salinity tolerance. Annu Rev Plant Biol 59:651–681
Naz N, Hameed M, Wahid A, Arshad M, Aqeel Ahmad MS (2009) Patterns of ion excretion and survival in two stoloniferous arid zone grasses. Physiol Planta 135:185–195
Niknam S, McComb J (2000) Salt tolerance screening of selected Australian woody species—a review. For Eco Manag 139:1–19
Osmond C, Lüttge U, West K, Pallaghy C, Shacher-Hill B (1969) Ion absorption in Atriplex leaf tissue II. Secretion of ions to epidermal bladders. Aust J Biol Sci 22:797–814
Popp M, Polania J, Weiper M (1993) Physiological adaptations to different salinity levels in mangrove. In: Towards the rational use of high salinity tolerant plants. Springer, pp 217–224
Rozema J, Riphagen I, Sminia T (1977) A light and electron-microscopical study on the structure and function of the salt gland of *Glaux Maritima* L. New Phytol 79:665–671
Rubio F, Gassmann W, Schroeder JI (1995) Sodium-driven potassium uptake by the plant potassium transporter HKT1 and mutations conferring salt tolerance. Science 270:1660–1663
Sabovljevic M, Sabovljevic A (2007) Contribution to the coastal bryophytes of the Northern Mediterranean: are there halophytes among bryophytes. Phytol Balcan 13:131–135
Shen B, Jensen RG, Bohnert HJ (1997) Increased resistance to oxidative stress in transgenic plants by targeting mannitol biosynthesis to chloroplasts. Plant Physiol 113:1177–1183
Shi H, Ishitani M, Kim C, Zhu JK (2000) The *Arabidopsis thaliana* salt tolerance gene SOS1 encodes a putative Na+/H+ antiporter. Proc Natl Acad Sci USA 97:6896–6901
Sudhir P, Murthy S (2004) Effects of salt stress on basic processes of photosynthesis. Photosynthetica 42:481–486
Tanveer M, Shabala S (2018) Targeting redox regulatory mechanisms for salinity stress tolerance in crops. Salinity responses and tolerance in plants, vol 1. Springer, Cham, pp 213–234
Tanveer M, Shah AN (2017) An insight into salt stress tolerance mechanisms of *Chenopodium album*. Environ Sci Pollut Res 24:16531–16535
Tanveer M, Shahzad B, Sharma A, Biju S, Bhardwaj R (2018) 24-Epibrassinolide; an active brassinolide and its role in salt stress tolerance in plants: a review. Plant Physiol Biochem 130:69–79
Tejera N, Ortega E, Rodes R, Lluch C (2006) Nitrogen compounds in the apoplastic sap of sugarcane stem: Some implications in the association with endophytes. J Plant Physiol 163:80–85
Thomson W, Platt-Aloia K (1979) Ultrastructural transitions associated with the development of the bladder cells of the trichomes of Atriplex. Cytobios 25:105–114
Tsugane K, Kobayashi K, Niwa Y, Ohba Y, Wada K, Kobayashi H (1999) A recessive *Arabidopsis* mutant that grows photoautotrophically under salt stress shows enhanced active oxygen detoxification. Plant Cell 11:1195–1206
Ungar IA (1991) Ecophysiology of vascular halophytes. CRC Press
Ventura Y, Sagi M (2013) Halophyte crop cultivation: the case for *Salicornia* and *Sarcocornia*. Environ Exp Bot 92:144–153
Wahid A (2003) Physiological significance of morpho-anatomical features of halophytes with particular reference to Cholistan flora. Inter J Agric Biol 5:207–212
Wang H, Qi Q, Schorr P, Cutler AJ, Crosby WL, Fowke LC (1998) ICK1, a cyclin-dependent protein kinase inhibitor from *Arabidopsis thaliana* interacts with both Cdc2a and CycD3, and its expression is induced by abscisic acid. Plant J 15:501–510
Weber D (2009) Adaptive mechanisms of halophytes in desert regions. In: Salinity and water stress. Springer, pp 179–185
Youssef AM (2009) Salt tolerance mechanisms in some halophytes from Saudi Arabia and Egypt. Res J Agric Biol Sci 5:191–206

Zaier H, Ghnaya T, Lakhdar A, Baioui R, Ghabriche R, Mnasri M, Sghair S, Lutts S, Abdelly C (2010) Comparative study of Pb-phytoextraction potential in *Sesuvium portulacastrum* and *Brassica juncea*: tolerance and accumulation. J Hazardous Mater 183:609–615
Zhu J-K (2001) Plant salt tolerance. Trends Plant Sci 6:66–71
Zhu JK, Hasegawa PM, Bressan RA, Bohnert HJ (1997) Molecular aspects of osmotic stress in plants. Crit Rev Plant Sci 16:253–277
Zhu JK, Liu J, Xiong L (1998) Genetic analysis of salt tolerance in *Arabidopsis*: evidence for a critical role of potassium nutrition. Plant Cell 10:1181–1191

Plant Roots—The Hidden Half for Investigating Salt and Drought Stress Responses and Tolerance

B. Sánchez-Romera and Ricardo Aroca

Abstract Plant roots are not just a mere organ involved in soil anchoring and nutrient absorption. But it is a tool used by plants to survive against adverse conditions that surround them since plants are sessile organisms which cannot move to escape stressful conditions. This chapter will focus on exposing the strategies that plants use to cope with two stress situations, namely drought and salinity. These strategies range from changes in the root external morphology as changes at the cellular level, where the permeability of the membranes and gene expression are changed.

1 Introduction

Arable areas are being increased in recent decades, given the demand from the growing population. As a result, arable soils are being degraded by intensive agriculture and the application of fertilizers in excess, increasing the salt content in the soil and reducing the water availability (Lichtfouse 2013). To this fact, we must add the increase in temperature and the reduction of rainfall caused by climate change (Hatfield and Dold 2019), making drought and salinity stresses the most important causes of the biggest economic losses in agriculture. For that reason, both stresses have to be deeply studied and it is required to find solutions to face them and minimize the economic losses.

The roots of the plants are organs with a great adaptive capacity, being able to grow and carry out their development under different types of substrate, hydric and nutritional conditions and under adverse environmental conditions (Babé et al. 2012; Atkinson et al. 2014; Julkowska et al. 2014; Koevoets et al. 2016; Li et al. 2016c; Gray and Brady 2016). But the molecular routes by which roots are able to grow under such conditions are being elucidated at present.

The radical system of a majority of plants consists mainly of two parts. The main or primary root which is formed embryonically (Scheres et al. 1994), and lateral roots

B. Sánchez-Romera · R. Aroca (✉)
Estación Experimental del Zaidin, CSIC, Granada, Spain
e-mail: ricardo.aroca@eez.csic.es

M. Hasanuzzaman and M. Tanveer (eds.), *Salt and Drought Stress Tolerance in Plants*, Signaling and Communication in Plants, https://doi.org/10.1007/978-3-030-40277-8_6

and adventitious roots which are included as secondary roots and are formed post-embryonically (Verstraeten et al. 2014; Bellini et al. 2014). There are also differences between monocot and dicot roots. For example, dicots present a tap root system with a well-developed main root from which lateral roots come out, whereas monocots have a fibrous root system, formed by the crown or adventitious roots (Coudert et al. 2013). All this root diversity has to be taken into account to study the role of a given root system (Olatunji et al. 2017).

Nevertheless, it is known that the roots of the plants, in addition to having a nutritional function, are responsible for keeping the plants anchored to the ground (Koevoets et al. 2016; Lynch 2018), developing a strategy of defence against underground pathogens (Ray et al. 2018; Elhady et al. 2018), establishing symbiotic relationships with beneficial microorganisms of the soil to improve their nutrition (Dodd and Ruiz-Lozano 2012; Santander et al. 2017), and even to explore the soil in search of water and nutrients when their availability is low (Li et al. 2016c).

The roots have been extensively studied throughout history, provided their internal structure and external anatomy are known (Fitter 1986; Lynch 2011, 2019; Fitters et al. 2017). All this information has made it possible to publicize the ability of plant adaptation to living in extreme climates. Accordingly, specialized plants of certain regions have developed strategies that allow them to live in extreme areas like aquatic environments and desert areas (Nobel 1984; Nie et al. 2015; Méndez-Alonzo et al. 2016).

In this chapter, we will expose the adaptations at the morphological and molecular levels of the plant roots grown under drought and salinity conditions. In order to identify genes that may be involved in the tolerance to both stresses and develop tolerant plants against such stresses, such genes will be highlighted.

2 Root Morphological Responses to Drought and Salt Stresses

The root system architecture (RSA) plays an important role in dealing with stresses like salinity and drought. For this, plants modify their root morphology to counteract the limited water availability and high salt content in the soil.

Some of these modifications are specific to each stress, while others are present in both osmotic stresses. However, there is much more information available about the changes presented by the roots subjected to drought than about the changes caused by salinity. In this section, the most important characteristics acquired by the roots in response to each of the stresses will be described.

2.1 Drought

Drought is a stress which is based on a reduction of water in the surface layers of the soil, caused either by high temperatures, high force winds or rain scarcity. Among the adaptations carried out by plant roots to cope with drought, these are the most important.

2.1.1 Root Angle

The normal root growth is due to gravitropism. This process, which is regulated by polar auxin transport (Aloni et al. 2006; Geisler et al. 2014), makes that plants develop their roots following a vertical growth. Under drought conditions, the roots maintain their vertical growth but increase the length of the root to get the soil water located in the deeper layer of soil (Alsina et al. 2011; Comas et al. 2013; Fenta et al. 2014; Koevoets et al. 2016). Nevertheless, this angle can be modified by plants to avoid low water availability areas, changing the direction of their main roots towards areas with high water potential. The root cap situated in the root extreme acts as a sensor to detect different water potentials. This phenome is known as hydrotropism and it has been described in several species like maize, cucumber, *Arabidopsis* and pea (Miyazawa et al. 2007; Takahashi et al. 2009; Cassab et al. 2013; Eapen et al. 2017; Nakajima et al. 2017; Tanaka-Takada et al. 2019).

2.1.2 Length and Branched Root

Drought induces growth of the main root and ramified roots to improve soil prospection to find water deposits, where branching roots are more efficient in taking up water (Osmont et al. 2007; Meister et al. 2014; Jaganathan et al. 2015; Salazar-Henao and Schmidt 2016). For instance, sugar beet plants produce root proliferation only when the water of superficial layers of the soil is reduced (Fitters et al. 2017). However, the number of lateral roots in barley and maize plants was reduced by drought (Babé et al. 2012).

Recently, a new process has been described and it is called hydropatterning. In it, the plants develop lateral roots on the main root in the function of the spatial disposition of water in the soil (Bao et al. 2014).

In addition, crown root number is a characteristic of some plants like maize and is the number of belowground nodal whorls and the number of roots per whorl (Saengwilai et al. 2014). Under drought conditions, a reduction of crown root number makes those plants develop deeper roots, improving the drought tolerance (Gao and Lynch 2016).

2.1.3 Root Diameter

The diameter of the root is an important factor involved in the soil penetration capacity of the root. It has been observed that thicker roots can reach deeper layer soil (Yu et al. 1995; Zheng et al. 2000). Concretely, two transgenic rice plants were more drought tolerant because of increase in the root diameter (Redillas et al. 2012; Lee et al. 2017).

2.1.4 Aerenchyma

The formation of aerenchyma has not only been related to aquatic plants and plants that grow in flooded soil, but it has also been observed in plants that grow under other stress conditions. The aerenchyma is an important adaptation because it improve the gas exchange. For instance, maize plants reduce the proportion of root cortical aerenchyma under drought stress conditions but it was developed in the root maturation zone (Díaz et al. 2018). On the other hand, transgenic rice plants were more tolerant against drought stress probably because their aerenchyma was increased (Redillas et al. 2012).

2.1.5 Apoplastic Barriers

It has been observed that plants grown under drought conditions showed the accumulation of suberin and lignin deposition. Specifically, plants of rice subjected to drought increased the suberization of the endodermis to hold water, making these plants more drought tolerant (Henry et al. 2012).

2.2 *Salinity*

Plants are subjected to salinity stress when the concentration of salts in the soil makes the soil water inaccessible for plants because of low water potentials. For that, plants have developed some root morphology changes to cope with it.

2.2.1 Root Angle

Plants, which are grown under salinity conditions, show a horizontal root growth, developing radical growth parallel to the soil. This effect is dependent on salt concentration. When the stress is moderate, the root growth follows a horizontal root growth, but when the plant is subjected to high salt concentrations the growth is vertical (gravitropism) (Shelef et al. 2010). In addition, similar to hydrotropism, the root plant can change the direction of their growth to avoid high salt concentrations in the soil. That process is called halotropism and it has been observed in plants like Arabidopsis, tomato and sorghum (Galvan-Ampudia et al. 2013). This movement is regulated by auxin distribution on the root which is a specific response to Na^+ ions (Galvan-Ampudia et al. 2013; Pierik and Testerink 2014).

2.2.2 Length and Branched Roots

It has been described that main root growth and lateral root are inhibited by salinity in plants of rice, rye, and maize (Rodriguez et al. 1997; Rahman et al. 2001; Ogawa et al.2006; Julkowska et al. 2014). *Arabidopsis* and maize plants present a quiescent phase when they are subjected to high salinity (Rodriguez et al. 1997). In rye, the inhibition of root growth is due to a decrease in cell division and an increase in cell death (Ogawa et al. 2006). The quiescent phase in *Arabidopsis* is temporary, recovering the root growth properties after salinity stress relief (West 2004). Sebastian et al. (2016) postulate that crown roots act as a water availability sensor. This reduction has been seen in several plants like *Setaria viridis* (Sebastian et al. 2016) and maize (Gao and Lynch 2016). Li et al. (2019) observed that *Solanum lycopersycum* plants treated with salt had a reduction of total root length, surface area, volume and number of forks but these effects were offset with good soil aeration.

2.2.3 Root Diameter

Salinity increases the root diameter of the tip and middle segments compared to upper zones (Barzegargolchini et al. 2017).

2.2.4 Aerenchyma

Compared to drought, plants subjected to salinity show an increase in the percentage of aerenchyma in the root to improve the gas exchange, since in many cases, plants subjected to high concentrations of salt are usually flooded (Tong et al. 2014; Naz et al. 2018).

2.2.5 Apoplastic Barriers

In plants of *Aeluropus littoralis* has been observed an increase of endodermis cell wall thickness and lignification of protoxylem in undifferentiated root tips to improve the tolerance in early stages of salinity (Barzegargolchini et al. 2017).

A summary of this section is represented in Fig. 1.

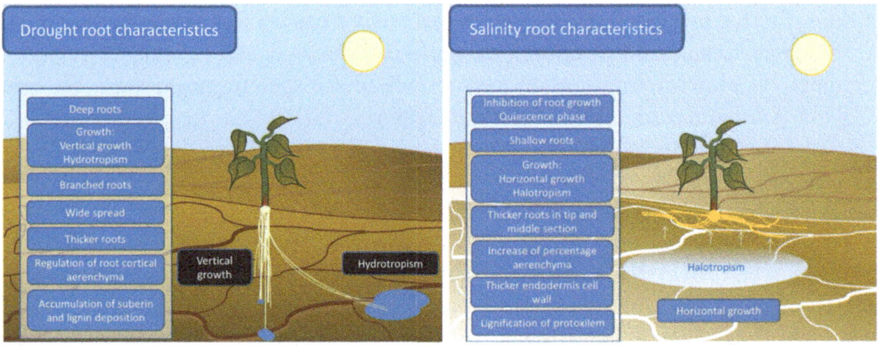

Fig. 1 Morphological changes that occur in plant roots caused by drought and salinity are shown schematically in this picture

3 Root Water Uptake Under Drought and Salt Stresses

The plants adapt to the environment that is not only constantly changing through changes in the anatomy of the roots, but also making changes at the cellular and molecular level that favour the adaptation of the plant. In this section, we are going to focus on the molecular changes related to water uptake.

The direction of water flow inside plants follows the soil–plant–atmosphere continuum (SPAC). Water penetrates into the roots due to an osmotic or hydrostatic gradient which is known as radial transport (Doussan et al. 1998; Steudle and Peterson 1998; Knipfer and Fricke 2011). This transport is composed of the sum of three types of routes: apoplastic, symplastic and transcellular. In the apoplastic path, water circulates across the pores of the cell walls and intercellular spaces. In the symplastic path, water is moving by plasmodesmata which connect the cytoplasm of adjacent cells and in the transcellular path, water flows across the cellular membranes through aquaporins, which are intrinsic membrane proteins that act as water channels. The sum of symplastic and transcellular paths is known as cell-to-cell path because it is not possible to quantify them separately (Steudle and Peterson 1998). The three pathways act simultaneously. Nevertheless, the kind of plants and environmental conditions can favour one route more than the others like an adaptation strategy (Steudle 2000a). Under unstressed conditions the dominant route is the apoplastic one, nevertheless, there are some species in which the predominant route is other different (Steudle and Peterson 1998; Steudle 2000b).

Root hydraulic conductance (L) is a parameter which provides information on the predominant route that is used in water transport (depending on the method used) and the root water transport capacity (Calvo-Polanco et al. 2014; Sánchez-Romera et al. 2014). For example, under drought and salt stresses, plants reduce L and transpiration rate to avoid water loss (Aroca et al. 2006, 2008; Gao et al. 2010). In these cases, the apoplastic route is low, the percentage of water flowing from cell-to-cell path having increased (Steudle and Peterson 1998; Javot and Maurel

2002). Nevertheless, contradictory results have been observed. Concretely, poplar plants showed a reduction of L caused by drought, where the apoplastic path was predominant (Siemens and Zwiazek 2003, 2004).

The root water permeability can be modified due to several factors, such as suberin and lignin depositions, aquaporins and plant hormones. Suberin and lignin depositions, as was shown in the previous section, are a strategy to avoid the loss of water when the plants are subjected to osmotic stresses. It is known that these accumulations reduce water transport by the apoplastic pathway (Schreiber et al. 2005; Vandeleur et al. 2008; Ranathunge et al. 2011; Krishnamurthy et al. 2011) although some exceptions have also been found (Ranathunge and Schreiber 2011).

Aquaporins are water channels located in cell membranes. There are different types of aquaporins according to their amino acid sequence and predominant location: plasma membrane intrinsic proteins (PIP), tonoplast intrinsic proteins (TIP), nodulin-like intrinsic proteins (NIP), unrecognized intrinsic proteins (XIP) and small and basic intrinsic proteins (SIP) (Maurel 2007). But there is mobility of aquaporins between different organelles, so there is traffic of aquaporins between the membranes of different organelles, allowing plants to modify the permeability of membranes according to environmental conditions (Boursiac 2005; Boursiac et al. 2008b).

At the same time, aquaporins can be subdivided in different isoforms. Specifically, PIP2s aquaporins are known to have higher water transport capacity than PIP1s proteins. In addition, it is known that aquaporins need to be phosphorylated to be active, so the plant also controls their phosphorylation in order to adapt to external water conditions (Prak et al. 2008; Zhang et al. 2019).

Many studies have been carried out to investigate the changes in the expression and abundance of aquaporins under drought and salt conditions. Nevertheless, the results obtained did not display a clear pattern, since it depends on many factors such as the plant species, the severity of the stress and the growing conditions.

In general, stress conditions tend to reduce the abundance and expression of aquaporins. However, specific aquaporins have been identified to confer stress tolerance (Perrone et al. 2012; Ma et al. 2019). For example, it has been seen that there are isoforms that are overexpressed in plants subjected to osmotic stress. Concretely, it has been observed that plants of soybean increase the expression of *GmPIP2;9* aquaporin under drought conditions. Then, transgenic plants which overexpress that gen showed an increase of drought tolerance in comparison with wild-type plants (Lu et al. 2018). Similarly, overexpression of *SlPIP2;1*, *SlPIP2;7* or *SlPIP2;5* in transgenic *Arabidopsis* and tomato plants caused an increase of hydraulic conductivity in plants under unstressed and drought conditions, suggesting their important implication in enhancing plant water content and osmotic balance (Li et al. 2016b), whereas PIP1b overexpression in tobacco plants caused a rapid wilting under drought stress (Refael Aharon et al. 2003). Perrone et al. (2012) studied that the effect of *VvPIP2;4* overexpression in grapevine roots and observed an enhance of gas exchange, shoot growth and L under unstressed conditions, but had a few effects over L under drought (Perrone et al. 2012).

In relation to salinity, it has been found that salt stress causes mobilization of aquaporins inside cells; concretely, PIPs were accumulated in intracellular membranes after salt application which caused a decrease of L (Boursiac et al. 2008a). Prak et al. (2008) noticed that PIPs lacked phosphorylation of Ser-283 under salt conditions, indicating that this phosphorylation also plays a role in regulating the aquaporin trafficking among membranes under salt treatment (Prak et al. 2008). Later, Calvo-Polanco et al. (2014) observed that bean plants treated with salt showed a recovery of L after 6 days of treatment. This increment of L was accompanied by a change in the location of PIPs aquaporins from cortex cells to the epidermis and in cells surrounding the xylem vessels (Calvo-Polanco et al. 2014).

Recent studies have shown the important role of aquaporins to cope with drought and salinity. For example, plants of maize grown under both (no simultaneously) stress conditions increased the expression of *ZmPIP1;1* in root and leaves. Moreover, in transgenic *Arabidopsis* plants which overexpressed that aquaporin (*ZmPIP1;1*), enhanced drought and salt stress tolerance has been found because it reduce the oxidative damage in these plants (Zhou et al. 2018). Similarly, *ScPIP1* overexpressed Arabidopsis plants grown under both stresses showed longer roots than wild-type plants because of decreased membrane damage and improved osmotic adjustment and decreased malonaldehyde (MDA) (stress indicator compound) and increased proline contents (osmoprotective compound) (Wang et al. 2019). Ectopic expression of *SpPIP1* as well as the drought tolerance in *Arabidopsis thaliana* increase (Chen et al. 2018).

In the same way, *PgTIP1 Arabidopsis* and soybean transgenic plants were more tolerant against drought and salinity stresses because of increased expression of genes related with stress such as ascorbate peroxidase, catalase, SOS1 (a transporter of Na^+), and synthesis of ABA and proline (An et al. 2017). However, Peng et al. (2007) discovered that *PgTIP1* overexpression in *Panax Ginseng* increased drought tolerance when the plants were grown under 45 cm deep pots but decreased when they were grown in 10 cm deep pots (Peng et al. 2007) because transgenic plants depleted the water faster from the soil.

4 Implication of Plants Hormones in Root System Architecture

Phytohormones are compounds produced by plants and have a key function to overcome biotic and abiotic stresses. The best-known phytohormones are auxin, abscisic acid, jasmonic acid, ethylene, cytokinins, gibberellins and brassinosteroids. All these phytohormones have been thoroughly studied regarding different plant processes including their role in the face of drought and salinity, focusing on the root system.

4.1 Auxin (IAA)

Auxins are phytohormones involved in plant growth and the response to several stresses (Singh et al. 2017). It is produced mainly in young leaves and it is transported by the phloem to the roots. It intervenes in gravitropism and patterning, explained in the previous section. Root morphological adaptations are carried out by a gradient of the endogenous auxin concentration. Its ability to shape the roots according to the conditions of the rhizosphere makes this hormone important to cope with drought and salinity stresses. Specifically, rice plants overexpressing *OsIAA6* gene, one AUX/IAA gene involved in auxin signal, were more drought tolerant (Jung et al. 2015). Similarly, potato and poplar plants increased the production of auxin under drought conditions. Although, this hormone has been much studied for its role in the development and growth of lateral roots and the main root; note that its ability to adapt the characteristics of the root according to the conditions of the rhizosphere makes this hormone vital to cope with saline stress and drought.

4.2 Abscisic Acid (ABA)

Abscisic acid is the most studied stress hormone. Its main function to defend the plant from stress is to delay the germination of the seeds under adverse conditions, regulate the closure of the stomata (which would affect the photosynthetic capacity of the plant and also prevent the loss of water), regulate the water uptake and its location in the plants (aquaporins and L are regulated under stress conditions by ABA), reduce the oxidative damage and finally modify the root system architecture features (Harris 2015; Barberon et al. 2016).

Similarly to IAA, endogenous ABA levels are altered in plants in response to the heterogeneous disposition of water in the soil (Puértolas et al. 2015). Therefore, ABA is involved in root local and systemic changes caused by environmental conditions. Under stress conditions, like drought, ABA synthesis is increased to maintain root development (Sharp et al. 1994; Spollen et al. 2000). It has been observed that ABA also regulates the lateral and primary root development (Bennett et al. 2013). Recently, Rosales et al. (2019) found that the effect of ABA on root morphology under drought conditions depends on the ABA internal concentrations. At mild drought stress conditions, ABA promotes root growth, but under more severe drought conditions ABA induces a reduction of root growth (Rosales et al. 2019). Under salt conditions, ABA induces changes in root architecture, carrying out a quiescence period and then, promoting root growth at the stage of recovery after a short period of salt stress (Zhao et al. 2014; Chen et al. 2017a). The study with ABA mutants defective in ABA synthesis showed that the hydrotropism process requires ABA signalling (Takahashi et al. 2002; Eapen et al. 2017) whereas the hydropatterning process is independent of ABA (Bao et al. 2014).

4.3 Brassinosteroids (BRs)

Brassinosteroids are a kind of phytohormones involved in plant growth and development and response against several stresses (Chen et al. 2017a). In particular, exogenous application of brassinosteroids increases waving frequency and torsion in Arabidopsis roots (Lanza et al. 2012). Several studies showed how brassinosteroids have a positive effect on drought tolerance (Kagale et al. 2007; Sahni et al. 2016). Nevertheless, a negative role of BR has also been observed under drought condition, where defective BR plants were more tolerant to drought (Northey et al. 2016; Ye et al. 2017).

4.4 Jasmonic Acid (JA)

Jasmonic acid is another phytohormone widely studied for its defensive role against plant pathogens; however, there are also studies where it has been seen to reduce the negative effects of drought and salinity because it is involved in stomatal regulation, root development and oxidative damage reduction. However, most important is its role in regulating root abundance and aquaporin expression and L to cope with drought stress (Munemasa et al. 2007; Riemann et al. 2015; Sánchez-Romera et al. 2016).

4.5 Ethylene

Ethylene is a gaseous hormone involved in the process like fruit ripening, senescence and response to several stresses among others (Arraes et al. 2015). In particular, ethylene may contribute to reducing the root development under drought and salt stresses (Cao et al. 2007; Liu and Zhang 2017), reducing the cell proliferation at the root apical meristem (Street et al. 2015).

4.6 Hormones Interaction

There is a lot of information about the role of each hormone in the root development; however, there is little information that indicates how all hormones act simultaneously on the same process. It is known that hormones are molecules that regulate several physiological processes in plants by modifying their concentrations (hormone balance) or travelling through the plant. Plants use this hormonal balance as a strategy to regulate root growth and to cope with adverse conditions caused by periods of stress. The interaction between hormones has been widely studied at the level of aerial part, however, in the roots, there is much to discover (Kushwah et al. 2011; Ullah et al. 2018).

ABA and cytokinin (CK) (another phytohormone) interact by reducing the polar transport of auxin to carry out lateral root formation (Shkolnik-Inbar et al. 2013). Other studies observed that property concentration of CK and auxin are essential to root development. Similarly, it has been observed that JA inhibits the root length but is involved in lateral root development because JA can regulate the synthesis of IAA (Cai et al. 2014; Wasternack and Song 2017). With reference to JA, there is a study which observed that it regulates the root hydraulic properties in tomato plants and this effect was due to changes in calcium and ABA concentration in roots (Sánchez-Romera et al. 2014). On the other hand, ABA inhibits ethylene synthesis to promote the root growth under drought conditions (Ghassemian et al. 2000). In salinity, crosstalk between ABA and GA has been found' negative feedback of these hormones regulates root development (Achard et al. 2008).

5 Role of Plant Nutrients in Root Architecture Under Drought and Salt Stresses

The nutritional status of plants is an important factor since plants with an optimal nutritional status can increase harvest and cope with stress conditions (Marschner et al. 1996; Amtmann and Armengaud 2009; George et al. 2011; Li et al. 2016c). Plants absorb the nutrients and water necessary to carry out their metabolic processes through their roots. The nutrients are divided into macronutrients and micronutrients, according to the concentrations required by the plant. Both high levels of nutrients in the soil and their scarcity cause serious physiological disorders in plants. In order to tolerate these concentrations, plants make morphological changes mainly in the roots although the whole plant is affected by them (Santos et al. 2017; dos Santos Araújo et al. 2018; Lynch 2018). The most common problem is the lack of nutrients in the soil due to intensive agriculture and weather effects. Scarce nutrient availability can be due to two factors: their abundance and mobility. In relation to the abundance, plants grown in soil poor of nutrients are smaller and more sensitive to stress conditions, which translate into economic losses (Gojon et al. 2009; Niu et al. 2013). In the case of mobility, it depends on the kind of nutrient and its form available in the soil. Nutrients can be mobile or immobile in the soil. This property is in part determined by the presence of water in the soil since the mobile elements are mainly dissolved in water. On the other hand, immobile elements are usually found on the soil surface and are retained into soil particles (Peret et al. 2014; Saengwilai et al. 2014; Dai et al. 2015).

Drought and salinity stresses, in addition to affecting the decomposition of soil organic matter, affect the mobility of nutrients. Therefore, root anatomy changes, made by plants in response to nutrient availability, are also related to water availability (Thorup-Kristensen and Kirkegaard 2016; Dathe et al. 2016).

Table 1 Effects that the most important nutrients in plant nutrition have on the morphological characteristics of the root when they are subjected to abiotic stresses (nutrient deficiency, drought and salinity)

Nutrient	Stress	References
N	Deficiency: development of deeper and longer roots	Seangwilai et al. (2014)
	Drought: ammonium treatment improves drought tolerance, increasing AQP Ding et al. 2016 expression	Ding et al. (2016)
	Salinity: N treatment improves salt tolerance, increasing the root development	Dai et al. (2015)
P	Deficiency: main root is reduced but increases the proliferation of hair and lateral roots	Bates and Lynch (2000)
	Drought: P regulates the AQP phosphorilation. Increase secondary root development and AQP expressions	Sun et al. (2017)
	Salinity: horizontal root growth and increased lateral root length	De Bauw et al. 2019
	P application improves salt tolerance	Zhou et al. (2018)
K	Deficiency: inhibition of root system	Gruber et al. (2013)
	Drought: K treatment improves drought tolerance. K regulates L and AQP	Galmés et al. (2007)
	Salinity: K application enhances salt tolerance. Regulation of ionic homeostasis and L	García-Martí et al. (2019)
Ca	Deficiency: increment of root fresh weight and low expression of AQP	Gruber et al. (2013)
	Drought and salinity: Ca applications improve tolerance	Cabañero et al. (2006)
B	Deficiency: inhibition of root elongation of principal and lateral roots	Kobayashi et al. (2017)
	Water relation: regulation of aquaporin expression involved in B transport	Tanaka et al. (2008)

In this section, we will be focusing on nitrogen and phosphorus due to their important role as a constituent of genetic material and cellular structures. Moreover, the implications of N and P in root development and stress tolerance have been deeply studied. Other nutrients such as potassium, calcium or boron will be briefly explained (see Table 1).

5.1 Nitrogen (N)

Nitrogen is necessary for numerous physiological processes to be carried out in plants and that is because it is involved in the synthesis of molecules such as amino acids, proteins, coenzymes, nucleic acids and chlorophylls (Marschner et al. 1996; Amtmann and Armengaud 2009; Xu et al. 2012). N is found in the form of nitrate (NO_3^-) or ammonium (NH_4^+) in the soil, as a result of processes of mineralization of soil organic matter, and is absorbed by plants through specific transporters that are found in the roots (Dai et al. 2015; Lynch 2019). In agriculture, urea is used as nitrogen fertilizer. Nitrate is the main source of nitrogen for plants since being a soluble ion is easier to be absorbed by the roots. However, ammonium is more difficult to assimilate by plants because is an insoluble ion. Therefore, it has low mobility and is found in the most superficial layers of the soil (Lynch 2018, 2019).

5.1.1 Nitrogen Deficiency

Plants have the capacity of changing their root anatomy to enhance nutrient uptake. For example, plants of rice and maize grown in soil with low nitrogen concentration showed longer and deeper roots (Ke et al. 2008; Atkinson et al. 2014; Wang et al. 2015; Yu et al. 2015). In addition, several maize genotypes were compared to find the root characteristics which improved the nitrogen absorption. The results showed that genotypes with a deeper root and fewer crowns were the most efficient that uptake nitrogen (Saengwilai et al. 2014).

5.1.2 Nitrogen and Water Relations

As it was explained in the previous section, plants develop longer roots in order to reach water storage in the deepest layers of the soil. This root characteristic is directly related to the uptake of mobile ions like nitrate since mobile ions are dissolved in soil water. Then, the nitrate source may be also found in the deeper water reserves and the same root adaptations enhance the water and nutritional status. This connection between nitrate availability and water uptake has been observed in several kinds of plants like rice, maize, tomato and cucumber which increases the water absorption after N treatment (Gorska et al. 2008; Ishikawa-Sakurai et al. 2014). Aquaporins play an important role in nitrogen metabolism because they are involved in the uptake, mobilization and detoxification of N (Gerbeau et al. 1999; Liu et al. 2003; Loque et al. 2005). Several aquaporin subfamilies like PIP, NIP and TIP are able to transport N compounds (Gerbeau et al. 1999; Liu et al. 2003; Jahn et al. 2004; Bárzana et al. 2014).

Concretely, plants showed an increase of aquaporin expressions after ammonium or nitrate application (Hacke et al. 2010; Ren et al. 2015; Ding et al. 2016), being higher with ammonium than nitrate (Guo et al. 2007a; Ding et al. 2016; Korhonen et al. 2018). Related to up-aquaporin expression after nitrogen treatment, it was also observed that a reduction of root aerenchyma and lignin accumulation explains the improvement of water absorption rate (Wang et al. 2001; Gaspar et al. 2003; Ishikawa-Sakurai et al. 2014; Ren et al. 2015). On the other hand, plants grown under low nitrogen levels showed an increase of aerenchyma formation, and a reduction of L values and root aquaporin gene expression (Ranathunge et al. 2016). Nevertheless, an increase of *Zm*TIP4;4 was found in maize roots and was related to urea mobilization across tonoplast, suggesting that *Zm*TIP4;4-regulated urea transport was essential for unloading vacuolar urea across the tonoplast under N starvation conditions (Gu et al. 2011). The use of nrt2.1 (the high-affinity NO_3^- transporter) mutant plants showed a diminution of NO_3^- content in roots and therefore, a reduction of L and PIP expression, suggesting a positive correlation between N and L (Li et al. 2016a).

Other studies suggest that nitrogen has a different effect on water relations which varies along the time. In short time, nitrogen application induce up expression of aquaporins whereas, after long time of nitrogen exposition, the plant act making modifications in root system architecture, increasing nitrate uptake (Wang et al. 2001; Remans et al. 2006; Walch-Liu et al. 2006).

Several TIPs and NIPs are involved in the N compounds transport across tonoplast (Gaspar et al. 2003; Soto et al. 2008; Gu et al. 2011; Bárzana et al. 2014; Zhang et al. 2016), either as a strategy to detoxify excess nitrogen (Wang et al. 2008), to release nitrogen reserve sources in situations of deficiency (Liu et al. 2003).

Drought

In relation to plants grown in nitrogen-poor soils, besides increasing the root length to improve its absorption, a reduction of enzymes and transporters related to nitrogen metabolism was observed when plants were subjected to drought conditions. For example, drought reduced the expression of some nitrate transporter but increased the expression of two ammonium transporters (AMT1; 2 and AMT4; 2) and two enzymes of nitrate reductase (NR and NRT 2; 5) in *Malus prunifolia* plants. Moreover, plants of *Malus hupehensis* increased NH_4^+/NO_3^- ratio in root and shoot as a strategy to cope with drought stress (Huang et al. 2018a, b). In relation to ammonium role in drought tolerance, it has been observed that ammonium treatment increased aquaporin expressions, improving the root water uptake under drought conditions (Ding et al. 2015a).

Salinity

There are a few studies related to the effect of N in plants subjected to salinity. Probably because N is only limited under reduced water conditions since nitrate is a mobile ion. Nevertheless, it has been seen that cotton plants treated with nitrate were more tolerant of salinity than plants treated with ammonium because they develop bigger roots and accumulated less Na^+ (Dai et al. 2015).

5.2 Phosphorus (P)

Phosphorus is part of RNA and DNA and it is used as an energy storage molecule forming part of ATP and as a constituent of phospholipid membranes. It is an ion with low mobility and is located in the superficial layers of the soil because it comes from the degradation of organic matter (Lynch 2011; Shen et al. 2011).

5.2.1 Phosphorus Deficiency

In the case of Arabidopsis, rice and maize plants grown under P deficiency conditions, the root anatomy presents an increase of lateral root and hair root proliferation to increase the capacity of soil exploration, but the main root is reduced (Kirk and Van Du 1997; Bates and Lynch 2000; Williamson et al. 2001; Postma et al. 2014; Kawa et al. 2016). As well, plants of the Proteaceae family have developed special roots called proteoid or cluster roots. These roots are very branched, covered by a large number of long and densely clustered trichoblasts (Lamont 1972; Neumann et al. 2000).

5.2.2 Phosphorus and Water Relations

P is directly related to aquaporin activity because aquaporins are able to transport water when they are in the phosphorylated state. Therefore, P deprivation decreases the aquaporin activity and water uptake in roots (Carvajal et al. 1996; Wang et al. 2016).

Drought

With respect to plants grown under low phosphorus availability and drought, it has been observed that plants showed the same changes as for low phosphorus availability. Under water limitations, plants showed an increase of secondary root branching and a decrease of nodal thickness. Nevertheless, these changes managed to improve P uptake efficiency and increase the root:shoot ratio (De Bauw et al. 2019).

Molecular changes like an increase in the expression of phosphorus transporter genes, such as *PHT1;7*, *PHT1;12* and *PHT2;1*, has been observed in the root of *Malus domestica* to face drought and starvation of P (Sun et al. 2017).

Other studies have been focused on improving the nutrient status to cope with drought. For example, plants of *Matricaria chamomilla* subjected to severe drought stress were treated with vermicompost and the results showed an increase of N, P and K uptake and it was related to enhancing drought tolerance (Salehi et al. 2016). In the same way, barley plants grown under severe drought stress increased H^+/K^+ ATPase activity and K^+ uptake, this ion balance stimulated drought tolerance (Feng et al. 2015).

Salinity

On the other hand, under salinity conditions plants develop roots that grow parallel to the soil; therefore this root system may favour the absorption of nutrients that are immobile as is the case of phosphorus. Arabidopsis plants, subjected to double stress, salinity and P deprivation, showed an increase of density and length of lateral roots (Kawa et al. 2016). Experiments with maize plants treated with P were more tolerant to salt stress because enhanced P uptake and this resulted in a Na^+ exclusion (Zhou et al. 2018).

5.3 Potassium (K)

K^+ is a plant macronutrient which is known to be involved in processes like enzyme activation, osmotic adjustment, cell expansion, regulation of membrane electric potential and pH homeostasis (Hawkesford et al. 2011; Ragel et al. 2019).

5.3.1 Potassium and Water Relations

It has been observed a positive correlation between K^+ and water uptake (Guo et al. 2007b). Therefore, AQP could play a role as turgor sensors to regulate K^+ channel activities (Hill et al. 2004). K^+ channel inhibitors reduced the expression of aquaporin and K^+ channels in Arabidopsis roots, suggesting that K^+ is involved in the hydraulic conductivity of plasma membranes (Sahr et al. 2005).

Drought

K^+ starving and drought stress caused similar symptoms in plants (Liu et al. 2006). Several studies suggest that aquaporins and K^+ channels can act like osmoregulators to improve tolerance to drought stress. Then K^+ treatment could help to improve plant water status because AQP is activated to mobilized K^+ ions under drought stress (Liu et al. 2006; Galmés et al. 2007).

Salinity

García-Martí et al. (2019) observed that plants treated with an increase in K^+ and Ca^{2+} concentrations in the irrigation solution (higher than recommended values) and subjected to salt and heat stresses improved their biomass production and reduced their oxidative damage than plants watered with Hoagland solution, suggesting the importance of plant nutrition to cope with combined stress situations (García-Martí et al. 2019).

5.4 Calcium (Ca)

Calcium is an essential element involved in the growth and development of plants. Calcium is important in the formation of the cell wall and membrane stability, but its main function is as the second messenger in many plant physiological processes, such as response of plants to abiotic stresses.

5.4.1 Calcium Deficiency

Although calcium deficiency in the soil, is not common, the symptoms are observed often in developing tissue-like young leaves and fruits because calcium shows low mobilization via phloem from old tissues to young ones (White and Broadley 2003; Thor 2019). The studies related to calcium role in roots are few. Nevertheless, it has been observed that Arabidopsis plants subjected to calcium deprivation treatment presented an increase of root fresh weight and a superficial and highly branched root system. The main root was drastically reduced after 100 μM calcium application and moderately reduced after 500 μM calcium. Nevertheless, although the main root was inhibited, the density of first-order lateral root was increased (Gruber et al. 2013).

5.4.2 Drought and Salinity

Calcium application improves drought tolerance in plants of sugar beet as well as increases the plant biomass and sugar root concentration. Salt stress inhibits the aquaporin activity; this fact was related to a reduction of cytosolic calcium in pepper plants (Martínez-Ballesta et al. 2008). In the same way, calcium treatment improves water transport which could be due to aquaporin activity (Carvajal et al. 1996, 2000; Cabañero et al. 2006).

In brief, these results suggest that calcium could act like an aquaporin regulator to respond to environmental stimuli (Steudle and Henzler 1995; Johansson et al. 1998; Vera-Estrella et al. 2004; Luu and Maurel 2005).

5.5 Boron (B)

Boron (B) is one of the essential micronutrients. Although the necessary concentration of B is low, this function is critical to carry out the growth and development of plants. Its function is focused on the structure and function of the plant cell wall (Shireen et al. 2018).

5.5.1 B Deficiency and Toxicity

Boron deficiency is a large-scale problem, leading to losses in the quality and yield of many crops at a global level. B deprivation causes a speed inhibition of primary and lateral root elongation (Kobayashi et al. 2017; Wu et al. 2017; Riaz et al. 2018). This effect is due to alteration of cell division, cell wall component and cell elongation in the root elongation zone (De Cnodder et al. 2000; González-Fontes et al. 2016).

5.5.2 Boron and Water Relations

Aquaporins are able to transport some ions like B and distribute it to all plant tissues (Dordas et al. 2000; Fitzpatrick and Reid 2009). Specifically, AtNIP5;1 is an aquaporin involved in a boric acid transport; its expression is upregulated under B deficiency conditions. Its function is essential for adequate B transport into the roots. Knockout Arabidopsis plants for nip5;1 showed inhibition of root and shoot growth under B deficiency (Takano et al. 2006; Kato et al. 2009). Other aquaporins, like AtNIP6;1 in Arabidopsis and OsNIP3;1 in rice plants, are implicated in uptake and distribution of B in the plant (Hanaoka and Fujiwara 2007; Tanaka et al. 2008).

In barley, it has been observed that a down regulation of HvNIP2;1 expression to induce B toxicity tolerance, reduces the B uptake and accumulation inside the plant (Schnurbusch et al. 2010). Similarly, overexpression of AtTIP5;1 in Arabidopsis plants in the tonoplast membrane was found to store B inside the vacuole and reduce the B toxicity (Pang et al. 2010). Other PIPs aquaporins were also involved in B transport in rice plants to reduce B toxicity (Kumar et al. 2014; Mosa et al. 2016).

6 Root Molecular Responses to Drought and Salt Stresses

This section has been focused on collecting a series of genes which are directly or indirectly involved in the regulation of root system architecture (RSA). Several publications have shown that these genes could play an important role in facing the drought and salinity stresses at the root level. A summary is showed in Table 2.

6.1 DEEPER ROOTING 1 (DRO1)

DEEPER ROOTING 1 (DRO1) is a gene which is implicated in the regulation of length and angle of the root (Uga et al. 2011, 2013). Uga et al. (2013) realized an experiment where one tolerant and another sensitive rice species were compared. Tolerant plants which have a full-length copy of DRO1 showed deeper root and the grain production was not affected by stress conditions. On the other hand, sensitive plants, which have a truncated copy of the gene, showed superficial roots and lower

Table 2 Summary of genes involved in changes that occur in root morphology to improve tolerance to drought and salinity

Gene	Abbreviation	Effect	Drought	Salinity	References
DEEPER ROOT 1	DROB1	Rooting depth	Tolerance	Not reported	Uga et al. (2013), Guseman et al. (2017)
ENHANCED DROUGHT TOLERANCE 1	EDT1	Increase diameters, length and root weight	Tolerance	Not reported	Yu et al. (2013)
HOMEODOMAIN GLABROUS11	HDG11	Increase root length and root dry biomass	Tolerance	Tolerance	Yu et al. (2013, Yu et al. 2016)
GLYCINE SOJA 14-3-3	GF14-3-3	Increase length root	Opposite results	Tolerance	Sun et al. (2014), Chen et al. (1994)
NON-SPECIFIC PHOSPHOLIPASE C5	NPC5	Proliferation of lateral roots	Tolerance	Tolerance	Peters et al. (2014)
No apical meristem (NAM) Arabidopsis transcription activation factor (ATAF1) CUC2 (cup-shaped cotyledon)	NAC	Promote lateral root development	Not reported	Tolerance	Redillas et al. (2012)
Sucrose non-fermenting-1 related protein kinase	SnRK2	Longer primary root and higher relative water content	Tolerance	Tolerance	Zhang et al. (2010)
WRKY transcription factor	WRKY46	Regulate the lateral root formation	Tolerance	Tolerance	Ding et al. (2015b)
TRANSPORT INHIBITOR RESPONSE 1/AFB	TIR1/AFB	Suppression of lateral roots	Tolerance	Tolerance	Shkolnik-Inbar et al. (2014)
ABSCISIC ACID INSENSITIVE 4	ABI4	Inhibition of lateral roots	Not reported	Tolerance	Shkolnik-Inbar et al. (2014)
FASCICLIN-LIKE ARABINOGALACTAN PROTEIN4	FLA4	Involved in cell wall biosynthesis and root development	Not reported	Tolerance	Seifert et al. (2014)

(continued)

Table 2 (continued)

Gene	Abbreviation	Effect	Drought	Salinity	References
Rho-family GTPases	ROP	Increase primary root length and survival rate	Tolerance	Tolerance	Miao et al. (2018)
CALCINEURINB-LIKE PROTEINS	CBL	Improve the root system architecture, survival and germination rate, and reduce water loss rate	Tolerance	Not reported	Cui et al. (2018)
ROOT HAIR DEFECTIVE-3	RHD3	Increase biomass, root hair abundance and branching roots	Tolerance	Not reported	Wong et al. (2018)
PYR1/PYL/RCAR (pyrabactin resistance 1/PYR1-like/regulatory components of ABA receptors)	PYLs	Increase the primary root length	Tolerance	Not reported	Chen et al. (2017a, b), Li et al. (2018)

weight gain and production. Later, it was observed that sensitive plants had not a functional allele of DRO1, DRO2 and DRO3 as well were identified. In addition, DRO3 is involved in RSA only when DRO1 is functional in the plant, suggesting that DRO3 is involved in the DRO1 regulation (Uga et al. 2015). Other studies showed that overexpression of AtDRO1 in Arabidopsis plants decreased the angle of lateral root and narrower lateral roots, and overexpression of PpeDRO1 in *Prunus domestica* increased the length root with respect to control plants (Guseman et al. 2017).

The implication of this gene on drought tolerance has been accepted, whereas no information about its role against salinity stress has been found. This gene could be proposed to be effective to improve salt tolerance.

6.2 *ENHANCED DROUGHT TOLERANCE1/HOMEODOMAIN GLABROUS11 (AtEDT1/HDG11)*

ENHANCED DROUGHT TOLERANCE1/HOMEODOMAIN GLABROUS11 (AtEDT1/HDG11) is a gene involved in the encoding of transcription factors of homeodomain-leucine zipper (HD-ZIP) (Schrick et al. 2004). Specifically,

AtEDT1/HDG11 genes perform on the encoding of cell wall proteins involved in root cell elongation and controlling the extended root system, which is related to drought tolerance (Xu et al. 2014).

Yu et al. (2016) observed that AtHDG11-overexpressing plants of cotton (*Gossypium hirsutum*) and poplar (*Populus tomentosa* Carr.) were more tolerant of salt and drought conditions. In the case of transgenic cotton plants, they were more tolerant against salt and drought stresses because they improved the RSA. These plants produced longer roots and increased root dry biomass. In the same way, transgenic poplar had also a bigger developed root system with longer primary root and more lateral roots. For this reason, they were more resistant than wild-type plants under drought and salt conditions. Moreover, these plants showed better water status and less damage (Yu et al. 2016).

On the other hand, overexpression of AtEDT1 enhances the elongation of roots in *Salvia miltiorrhiza* (Liu et al. 2017) and AtEDT1-overexpression in alfalfa (*Medicago sativa* L.) plants also improved the root system, showing an increase of root length, root weight and root diameters with respect to wild-type plants. It should be noted that these plants were more tolerant to drought, so this gene is of great importance to make plants more resistant (Fan et al. 2017).

These improvements against drought and salt conditions have also been observed in several kinds of (AtEDT1/HDG11) transgenic plants like rice, tobacco, sweet potato, wheat, poplar and cotton (Yu et al. 2013, 2016; Zhu et al. 2016; Chen et al. 2017b; Tariq et al. 2017; Guo et al. 2019). Therefore, these genes could be a good tool to make drought- and salt-tolerant plants.

6.3 14-3-3 Genes

The 14-3-3 proteins are phosphopeptide-binding proteins (Campo et al. 2012). They are characterized because they interact with proteins of phosphoserine/threonine motifs to regulate signal transduction due to their properties, such as their ability to move between different subcellular localizations, stability and affinity to interact with other proteins (Paul et al. 2012; Tan et al. 2016). Specifically, these genes codify proteins which are involved in the regulation of other target proteins related with signalling, transcription activation and defence against stress (Robert et al. 2002; Jaganathan et al. 2015; Cao et al. 2016; Liu et al. 2016).

For example, it has been observed that tobacco plants decreased the expression of T14-3-3 mRNA under salt conditions (Chen et al. 1994; Zhang et al. 2018). However contradictory results have been found in relation to drought. On the one hand, GsGF14 overexpression in *Arabidopsis thaliana* had a negative effect on drought tolerance and plant growth because the plants showed deficits in root hair formation and development and changes in stomata size, and thereby reduced the water intake capacity. Moreover, knockout mutant Arabidopsis (AtGF14) improved drought tolerance and seed germination rate (Sun et al. 2014). Then, it could be explained as an decrease in the expression of this gene or their protein could improve the tolerance against the salt and drought stresses (Tan et al. 2016; Guo et al. 2018). However, opposite results have been obtained because the At14-3-3λ overexpression in maize plants enhanced their drought tolerance, modifying stomatal size and root system architecture (Yan et al. 2004; Campo et al. 2012; He et al. 2015). In addition, wheat plants increase the expression of 14-3-3 gene when they were grown under polyethylene glycol 6000, NaCl, H_2O_2 and abscisic acid treatments. Therefore, this gene also confers tolerance to drought and salt stresses. Transgenic tobaccos showed an increase of length root, relative water content, survival rate, photosynthetic rate and water use efficiency than wild-type plants under drought and salt stress conditions. These results suggest that this gene enhanced the drought and salt tolerance and that it is regulated by ABA signalling (Zhang et al. 2018).

6.4 Non-specific Phospholipase C5

Non-specific phospholipase C5 (NPC5) and its derived lipid mediator diacylglycerol (DAG) have been involved in salt stress tolerance. Concretely, an increase of lateral root development was observed under salt stress in *Arabidopsis thaliana*. In knock-out mutant npc5-1, lateral roots were decreased under mild NaCl stress, whereas overexpression of NPC5 increased it. The increase of lateral root by NPC5 was independent of auxin, being regulated by stress (Peters et al. 2014). On the other hand, no studies in relation to drought tolerance by NPC5 have been observed.

6.5 NAC Family Transcription Factors

NAC is the result of the abbreviations of NAM (no apical meristem), ATAF1-2 (Arabidopsis transcription activation factor) and CUC2 (cup-shaped cotyledon) (Aida et al. 1997). NAC-type transcription factor gene AtNAC2 has been identified in *Arabidopsis thaliana*, mainly expressed in roots. It has been observed that its expression is induced under salinity conditions, increasing abscisic acid (ABA), 1-Aminocyclopropane-1-carboxylic acid (ACC ethylene precursor) and naphthaleneacetic acid (NAA is a synthetic plant hormone in the auxin family) content also. Overexpression of AtNAC2 in transgenic Arabidopsis plants resulted in the promotion of lateral root development. These results suggest that AtNAC2 may be a transcription factor induced by environmental and endogenous conditions to regulate plant lateral root development (He et al. 2005).

Root-specific (RCc3) and constitutive (GOS2) promoters were used to create two lines of transgenic plants (RCc3:OsNAC9 and GOS2:OsNAC9) which overexpressed OsNAC9. The two lines showed an enlarged stele and aerenchyma. Concretely, RCc3:OsNAC9 roots had a greater extent aerenchyma than those of GOS2:OsNAC9 and non-transgenic roots. Therefore, this phenotype could be used to enhance drought resistance (Redillas et al. 2012).

6.6 Sucrose Non-fermenting1-Related Protein Kinase 2

Sucrose non-fermenting1-related protein kinase 2 (SnRK2) is involved in abiotic stress signalling. Overexpression of TaSnRK2.8 in transgenic Arabidopsis improved drought and salt tolerance. These plants showed longer primary roots, higher relative water content, strengthened cell membrane stability, lower osmotic potential, increased chlorophyll content and enhanced PSII activity. Moreover, TaSnRK2.8 overexpressed plants increased the transcript level of ABA biosynthesis, ABA signalling and stress-responsive genes under control and stressed conditions (Zhang et al. 2010).

6.7 WRKY Transcription Factor

WRKY transcription factor (WRKY46) is expressed along lateral root primordia during early lateral root development but this expression is restricted to the stele of the mature LR. Under osmotic and salt stress conditions, lack of WRKY46 (*wrky46* mutants) decreased lateral root formation, whereas overexpression of WRKY46 improved it. Auxin application restores LR development in defective plants (Ding et al. 2015b).

6.8 TRANSPORT INHIBITOR RESPONSE 1/AUXIN SIGNALLING F-BOX PROTEIN

Auxin joins TIR1/AFB receptor proteins to regulate the gene expressions. Two auxin receptors have been found in Arabidopsis mutant plants to induce more stress tolerance, *tir1 afb2* being more resistant against salinity because of enhanced germination rate and root development and decreased oxidative damage (Iglesias et al. 2010). Subsequently, the same authors observed that miR393, which form part of miR393-TIR1/AFB2/AFB3 regulatory module, is involved in suppression lateral root development during salt stress (Iglesias et al. 2014). Other studies showed how overexpression of a miR393-resistant TIR1 gene improved salt stress tolerance, increasing the germination rate and proline and anthocyanin content, reducing water loss and inhibition of root length and delaying senescence (Chen et al. 2015). Other studies observed that TIR1 mutant plants of Arabidopsis did not modify their root angle when the plants were under water deficit conditions, therefore the auxin is essential to suggesting redirected root growth angles downward under drought stress (Rellán-Álvarez et al. 2015).

6.9 ABSCISIC ACID INSENSITIVE4

ABA is a plant hormone involved in stress defence. ABSCISIC ACID INSENSITIVE4 (ABI4) gene which is expressed in the root stele is also implicated in RSA (Shkolnik-Inbar and Bar-Zvi 2010; Shkolnik-Inbar et al. 2013; Rowe et al. 2016). In relation to salinity, Arabidopsis mutants affected in ABI 4 gene showed tolerance to salinity because they stored lower levels of sodium ions. That was due to an increase in the expression of HKT1;1 gene (transporter of Na^{+}). Moreover, these plants showed an increment proline content which reduced the oxidative damage. Therefore, ABI4-overexpression plants were sensitive to salinity, suggesting the important role of ABI4 in the root formation regulation (Shkolnik-Inbar et al. 2013).

6.10 FASCICLIN-LIKE ARABINOGALACTAN PROTEIN 4

FASCICLIN-LIKE ARABINOGALACTAN PROTEIN 4 (At-FLA4) gene is required in the root development of plants subjected to salt conditions. Seifert et al. (2014) observed that the ABA application suppressed the At-FLA4 function in Arabidopsis plants grown under salt stress. They proposed that the At-FLA4 gene is involved in cell wall biosynthesis and root growth but its role is dependent on ABA signalling (Seifert et al. 2014).

6.11 Rho-Like GTPases

Rho-family GTPases are plant-specific molecular switches that are involved in the regulation of processes like cytoskeletal reorganization and vesicular trafficking (Nagawa et al. 2010). Rho-like GTPases from plants (ROPs) are related to plant survival to cope with abiotic stresses. Seventeen novel ROP proteins had been identified and characterized from *Musa acuminata* (MaROPs). Six genes of them were highly expressed in response to cold, salt and drought stress conditions in two genotypes with similar behaviour against these abiotic stresses, MaROP5g being more expressed under salinity conditions. Subsequently, transgenic *Arabidopsis thaliana* plants overexpressing MaROP5g grown under salt stress showed longer primary roots and more survival rate than wild-type *A. thaliana*. Therefore, ROPs genes might be taken into account as a gene involved in tolerance to saline stress (Miao et al. 2018).

6.12 *Calcineurin B-Like Proteins (CBLs)*

Calcineurin B-like proteins (CBLs) are calcium sensors that can interact with a family of protein kinases, known as CBL-interacting protein kinases (CIPKs). Expression analysis in Arabidopsis pants showed that CIPK23 genes are expressed in roots and leaves (Cheong et al. 2007). Transgenic plants, which overexpressed TaCIPK23 gene in wheat and Arabidopsis plants grown under drought conditions, improved survival and germination rate, root system architecture, accumulation of osmolytes and reduced water loss rate (Cui et al. 2018).

6.13 *Root Hair Defective-3 (MaRHD3)*

Wong et al. (2018) observed that transgenic Arabidopsis plants expressing root hair defective-3 (MaRHD3) were more tolerant to drought stress. The MaRHD3 plants increased biomass accumulation, relative water content, chlorophyll content and abundance of root hairs and branching roots than control plants (Wong et al. 2018).

6.14 *RCAR/PYR1/PYL*

RCAR/PYR1/PYL is ABA-binding receptor involved in the ABA signalling pathway. Transgenic plants in which RCAR11–RCAR14 overexpressed improved ABA sensitivity, root length and drought resistance (Li et al. 2018).

In a study carried out with cotton (*Gossypium hirsutum*), 27 predicted PYL proteins were described. Expression determinations showed that nine of GhPYL genes were down-regulated, whereas three of them increased their expression when the plants were subjected to drought stress. Overexpression of GhPYL10/12/26 in Arabidopsis made that transgenic plants bigger and more drought tolerant than wild-type plants, increasing the length of primary roots under unstressed conditions and mannitol stress. These results proposed that PYL genes may be essential in plant response to deal with drought/osmotic stresses (Chen et al. 2017a).

7 Perspectives

Environmental problems such as soil degradation by intensive agriculture, deforestation and an increase in greenhouse gases in the atmosphere are causing plants to develop new strategies that allow them to survive a stressful situation. In addition, if it is considered that the world population is rising, it is necessary that plants of great economic interest (tomato, beans, corn, wheat and rice) would be able to adapt to such unfavourable conditions without greatly affecting their economic production. Throughout this chapter, we have presented information about adaptations that plants suffer in nature to cope with such stresses but we have also talked about genes involved in the regulation of these changes. Given the information collected so far, we propose new ideas to give light to the unknown remaining mechanisms.

In addition, there are many studies carried out to study the effect of drought on plant roots, but little is known about the effect of salinity despite the fact that sometimes both stresses complement each other. Therefore, studies related to salinity should be increased, especially related to the absorption of nutrients and the overexpression of aquaporins. There are some topics which may shed new light on the root role. For example, to study the changes which roots undergo by contrasting a root of dicotyledons and monocotyledons plants, since both roots are embryologically different. Although there are studies where these roots are analyzed separately, it is necessary to include experiments where a comparison of these both kinds of roots will be grown under the same conditions of humidity, soil characteristics and nutrients, because all these factors affect drastically the root development.

On the other hand, I believe that the role of root exudates in the soil is an unexplored target. It could be interesting to study the changes in the composition of exudates against conditions of salinity and drought since they can influence the morphology of the root and therefore plant survival.

In relation to monocotyledons and dicotyledons differences and root exudates, to continue with the use of technologies such as quantitative trait loci (QTL), amplified fragment length polymorphism (AFLP) or clustered regularly interspaced short palindromic repeats (CRISPR) to identify new genes of interest involved in the tolerance of plants to saline stress and drought. Subsequently, these genes will be used to create transgenic plants of economic interest which will show these strategies, will be more resistant and will reduce crop losses.

Finally, many studies that focused on hormones signalling may be carried out to clarify the role of hormones, their receptors, intermediary products and interaction with other signal molecules that occur in the root system, since in truth all of them work together to achieve the desired result by the plant. The ideal would be to understand how a stress-tolerant plant can do so in terms of hormonal balance, root morphology, ability to regulate and store water in its tissues, reduction of oxidative damage and nutrient availability, and thus understand the communication that occurs between the root and aerial part of the plant.

References

Achard P, Renou JP, Berthomé R, Harberd NP, Genschik P (2008) Plant DELLAs restrain growth and promote survival of adversity by reducing the levels of reactive oxygen species. Curr Biol 18:656–660

Aharon Refael, Shahak Y, Wininger S, Bendov R, Kapulnik Y, Galili G (2003) Overexpression of a plasma membrane aquaporin in transgenic tobacco improves plant vigor under favorable growth conditions but not under drought or salt stress. Plant Cell 15:439–447

Aida M, Ishida T, Fukaki H, Fujisawa H, Tasaka M (1997) Genes involved in organ separation in Arabidopsis: an analysis of the cup-shaped cotyledon mutant. Plant Cell 9:841–857
Aloni R, Aloni E, Langhans M, Ullrich CI (2006) Role of cytokinin and auxin in shaping root architecture: regulating vascular differentiation, lateral root initiation, root apical dominance and root gravitropism. Ann Bot 97:883–893
Alsina MM, Smart DR, Bauerle T, De Herralde F, Biel C, Stockert C, Negron C, Save R (2011) Seasonal changes of whole root system conductance by a drought-tolerant grape root system. J Exp Bot 62:99–109
Amtmann A, Armengaud P (2009) Effects of N, P, K and S on metabolism: new knowledge gained from multi-level analysis. Curr Opin Plant Biol 12:275–283
An J, Hu Z, Che B, Chen H, Yu B, Cai W (2017) Heterologous expression of panax ginseng PgTIP1 confers enhanced salt tolerance of soybean cotyledon hairy roots, composite, and whole plants. Front Plant Sci 8:1–15
Aroca R, Ferrante A, Vernieri P, Chrispeels MJ (2006) Drought, abscisic acid and transpiration rate effects on the regulation of PIP aquaporin gene expression and abundance in Phaseolus vulgaris plants. Ann Bot 98:1301–1310
Aroca R, Vernieri P, Ruiz-Lozano JM (2008) Mycorrhizal and non-mycorrhizal Lactuca sativa plants exhibit contrasting responses to exogenous ABA during drought stress and recovery. J Exp Bot 59:2029–2041
Arraes FBM, Beneventi MA, Lisei de Sa ME, Paixao JFR, Albuquerque EVS, Marin SRR, Purgatto E, Nepomuceno AL, Grossi-de-Sa MF (2015) Implications of ethylene biosynthesis and signaling in soybean drought stress tolerance. BMC Plant Biol 15:1–20
Atkinson JA, Rasmussen A, Traini R, Voss U, Sturrock C, Mooney SJ, Wells DM, Bennett MJ (2014) Branching out in roots: uncovering form, function, and regulation. Plant Physiol 166:538–550
Babé A, Lavigne T, Séverin JP, Nagel KA, Walter A, Chaumont F, Batoko H, Beeckman T, Draye X (2012) Repression of early lateral root initiation events by transient water deficit in barley and maize. Philos Trans R Soc B: Biol Sci 367:1534–1541
Bao Y, Aggarwal P, Robbins NE, Sturrock CJ, Thompson MC, Tan HQ, Tham C, Duan L, Rodriguez PL, Vernoux T et al (2014) Plant roots use a patterning mechanism to position lateral root branches toward available water. Proc Natl Acad Sci USA 111:9319–9324
Barberon M, Vermeer JEM, De Bellis D, Wang P, Naseer S, Andersen TG, Humbel BM, Nawrath C, Takano J, Salt DE et al (2016) Adaptation of root function by nutrient-induced plasticity of endodermal differentiation. Cell 164:447–459
Bárzana G, Aroca R, Bienert GP, Chaumont F, Ruiz-Lozano JM (2014) New insights into the regulation of aquaporins by the arbuscular mycorrhizal symbiosis in maize plants under drought stress and possible implications for plant performance. Mol Plant-Microbe Interact 27:349–363
Barzegargolchini B, Movafeghi A, Dehestani A, Mehrabanjoubani P (2017) Increased cell wall thickness of endodermis and protoxylem in *Aeluropus littoralis* roots under salinity: the role of LAC4 and PER64 genes. J Plant Physiol 218:127–134
Bates TR, Lynch JP (2000) Plant growth and phosphorus accumulation of wild type and two root hair mutants of *Arabidopsis thaliana* (Brassicaceae). Am J Bot 87:958–963
Bellini C, Pacurar DI, Perrone I (2014) Adventitious roots and lateral roots: similarities and differences. Ann Rev Plant Biol 65:639–666
Bennett MJ, Duan L, Bhalerao R, Chan PMY, Ng CH, Dinneny JR, Dietrich D (2013) Endodermal ABA signaling promotes lateral root quiescence during salt stress in Arabidopsis seedlings. Plant Cell 25:324–341
Boursiac Y (2005) Early effects of salinity on water transport in Arabidopsis roots. Molecular and cellular features of aquaporin expression. Plant Physiol 139:790–805
Boursiac Y, Boudet J, Postaire O, Luu DT, Tournaire-Roux C, Maurel C (2008a) Stimulus-induced downregulation of root water transport involves reactive oxygen species-activated cell signalling and plasma membrane intrinsic protein internalization. Plant J 56:207–218
Boursiac Y, Prak S, Boudet J, Postaire O, Luu D-T, Tournaire-Roux C, Santoni V, Maurel C (2008b) The response of Arabidopsis root water transport to a challenging environment implicates reactive

oxygen species-and phosphorylation-dependent internalization of aquaporins. Plant Signal Behav 3:18573191

Cabañero FJ, Martínez-Ballesta MC, Teruel JA, Carvajal M (2006) New evidence about the relationship between water channel activity and calcium in salinity-stressed pepper plants. Plant Cell Physiol 47:224–233

Cai XT, Xu P, Zhao PX, Liu R, Yu LH, Bin Xiang C (2014) Arabidopsis ERF109 mediates cross-talk between jasmonic acid and auxin biosynthesis during lateral root formation. Nat Commun 5:1–13

Calvo-Polanco M, Sánchez-Romera B, Aroca R (2014) Mild salt stress conditions induce different responses in root hydraulic conductivity of *Phaseolus vulgaris* over-time. PLoS ONE 9

Campo S, Peris-Peris C, Montesinos L, Peñas G, Messeguer J, San Segundo B (2012) Expression of the maize ZmGF14-6 gene in rice confers tolerance to drought stress while enhancing susceptibility to pathogen infection. J Exp Bot 63:983–999

Cao WH, Liu J, He XJ, Mu RL, Zhou HL, Chen SY, Zhang JS (2007) Modulation of ethylene responses affects plant salt-stress responses. Plant Physiol 143:707–719

Cao H, Xu Y, Yuan L, Bian Y, Wang L, Zhen S, Hu Y, Yan Y (2016) Molecular characterization of the 14-3-3 gene family in Brachypodium distachyon L. reveals high evolutionary conservation and diverse responses to abiotic stresses. Front Plant Sci 7:1099

Carvajal M, Cooke DT, Clarkson DT (1996) Responses of wheat plants to nutrient deprivation may involve the regulation of water-channel function. Planta 199:372–381

Carvajal M, Cerdá A, Martínez V (2000) Does calcium ameliorate the negative effect of NaCl on melon root water transport by regulating aquaporin activity? New Phytol 145:439–447

Cassab GI, Eapen D, Campos ME (2013) Root hydrotropism: an update. Am J Bot 100:14–24

Chen Z, Fu H, Liu D, Chang PL, Narasimhan M, Ferl R, Hasegawa PM, Bressan RA (1994) A NaCl-regulated plant gene encoding a brain protein homolog that activates ADP ribosyltransferase and inhibits protein kinase C. Plant J 6:729–740

Chen Z, Hu L, Han N, Hu J, Yang Y, Xiang T, Zhang X, Wang L (2015) Overexpression of a miR393-resistant form of transport inhibitor response protein 1 (mTIR1) enhances salt tolerance by increased osmoregulation and Na^+ exclusion in *Arabidopsis thaliana*. Plant Cell Physiol 56:73–83

Chen Y, Feng L, Wei N, Liu ZH, Hu S, Li XB (2017a) Overexpression of cotton PYL genes in *Arabidopsis* enhances the transgenic plant tolerance to drought stress. Plant Physiol Biochem 115:229–238

Chen E, Zhang X, Yang Z, Wang X, Yang Z, Zhang C, Wu Z, Kong D, Liu Z, Zhao G et al (2017b) Genome-wide analysis of the HD-ZIP IV transcription factor family in *Gossypium arboreum* and GaHDG11 involved in osmotic tolerance in transgenic Arabidopsis. Mol Genet Genomics 292:593–609

Chen Q, Yang S, Kong X, Wang C, Xiang N, Yang Y, Yang Y (2018) Molecular cloning of a plasma membrane aquaporin in *Stipa purpurea*, and exploration of its role in drought stress tolerance. Gene 665:41–48

Cheong YH, Pandey GK, Grant JJ, Batistic O, Li L, Kim BG, Lee SC, Kudla J, Luan S (2007) Two calcineurin B-like calcium sensors, interacting with protein kinase CIPK23, regulate leaf transpiration and root potassium uptake in Arabidopsis. Plant J 52:223–239

Comas LH, Becker SR, Cruz VMV, Byrne PF, Dierig DA (2013) Root traits contributing to plant productivity under drought. Front Plant Sci 4. https://doi.org/10.3389/fpls.2013.00442

Coudert Y, Le Thi A, Gantet P (2013) Rice: a model plant to decipher the hidden origin of adventitious roots. In: Plant roots: the hidden half, 4th edn, pp 145–154

Cui XY, Du YT, Fu J dong, Yu TF, Wang CT, Chen M, Chen J, Ma YZ, Xu ZS (2018) Wheat CBL-interacting protein kinase 23 positively regulates drought stress and ABA responses. BMC Plant Biol 18:93

Dai J, Duan L, Dong H (2015) Comparative effect of nitrogen forms on nitrogen uptake and cotton growth under salinity stress. J Plant Nutr 38:1530–1543

Dathe A, Postma JA, Postma-Blaauw MB, Lynch JP (2016) Impact of axial root growth angles on nitrogen acquisition in maize depends on environmental conditions. Ann Bot 118:401–414

De Bauw P, Vandamme E, Lupembe A, Mwakasege L, Senthilkumar K, Merckx R (2019) Architectural root responses of rice to reduced water availability can overcome phosphorus stress. Agronomy 9

De Cnodder T, Vissenberg K, Van Der Straeten D (2000) Regulation of cell length in the *Arabidopsis thaliana* root by the ethylene precursor 1-aminocyclopropane-1-carboxylic acid: a matter of apoplastic reactions. New Phytol 168:541–550

Díaz AS, Aguiar GM, Pereira MP, Mauro de Castro E, Magalhães PC, Pereira FJ (2018) Aerenchyma development in different root zones of maize genotypes under water limitation and different phosphorus nutrition. Biol Plant 62:561–568

Ding L, Gao C, Li Y, Li Y, Zhu Y, Xu G, Shen Q, Kaldenhoff R, Kai L, Guo S (2015a) The enhanced drought tolerance of rice plants under ammonium is related to aquaporin (AQP). Plant Sci 234:14–21

Ding ZJ, Yan JY, Li CX, Li GX, Wu YR, Zheng SJ (2015b) Transcription factor WRKY46 modulates the development of Arabidopsis lateral roots in osmotic/salt stress conditions via regulation of ABA signaling and auxin homeostasis. Plant J 84:56–69

Ding L, Li Y, Wang Y, Gao L, Wang M, Chaumont F, Shen Q, Guo S (2016) Root ABA accumulation enhances rice seedling drought tolerance under ammonium supply: interaction with aquaporins. Front Plant Sci 7. https://doi.org/10.3389/fpls.2016.01206

Dodd IC, Ruiz-Lozano JM (2012) Microbial enhancement of crop resource use efficiency. Curr Opin Biotechnol 23:236–242

Dordas C, Chrispeels MJ, Brown PH (2000) Permeability and channel-mediated transport of boric acid across membrane vesicles isolated from squash roots. Plant Physiol 124:1349–1361

dos Santos Araújo G, de Souza Miranda R, Mesquita RO, de Oliveira Paula S, Prisco JT, Gomes-Filho E (2018) Nitrogen assimilation pathways and ionic homeostasis are crucial for photosynthetic apparatus efficiency in salt-tolerant sunflower genotypes. Plant Growth Regul 86:375–388

Doussan C, Pagès L, Vercambre G (1998) Modelling of the hydraulic architecture of root systems: an integrated approach to water absorption—model description. Ann Bot 81:213–223

Eapen D, Martínez-Guadarrama J, Hernández-Bruno O, Flores L, Nieto-Sotelo J, Cassab GI (2017) Synergy between root hydrotropic response and root biomass in maize (*Zea mays* L.) enhances drought avoidance. Plant Sci 265:87–99

Elhady A, Adss S, Hallmann J, Heuer H (2018) Rhizosphere microbiomes modulated by pre-crops assisted plants in defense against plant-parasitic nematodes. Front Microbiol 9. https://doi.org/10.3389/fmicb.2018.01133

Fan C, Wang X, Pang Y, Di S, Zheng G, Xiang C (2017) Over-expression of Arabidopsis EDT1 gene confers drought tolerance in alfalfa (*Medicago sativa* L.). Front Plant Sci 8:1–14

Feng X, Liu W, Zeng F, Chen Z, Zhang G, Wu F (2015) K^+ uptake, H^+-ATPase pumping activity and Ca^{2+} efflux mechanism are involved in drought tolerance of barley. Environ Exp Bot 129:57–66

Fenta B, Beebe S, Kunert K, Barlow K, Burridge J, Lynch J, Foyer C (2014) Field phenotyping of soybean roots for drought stress tolerance. Agron 4:418–435

Fitter A (1986) Spatial and temporal patterns of root activity in a species-rich alluvial grassland. Oecologia 69:594–599

Fitters TFJ, Bussell JS, Mooney SJ, Sparkes DL (2017) Assessing water uptake in sugar beet (*Beta vulgaris*) under different watering regimes. Environ Exp Bot 144:61–67

Fitzpatrick KL, Reid RJ (2009) The involvement of aquaglyceroporins in transport of boron in barley roots. Plant Cell Environ 32:1357–1365

Galmés J, Pou A, Alsina MM, Tomàs M, Medrano H, Flexas J (2007) Aquaporin expression in response to different water stress intensities and recovery in Richter-110 (Vitis sp.): relationship with ecophysiological status. Planta 226:671–681

Galvan-Ampudia CS, Julkowska MM, Darwish E, Gandullo J, Korver RA, Brunoud G, Haring MA, Munnik T, Vernoux T, Testerink C (2013) Halotropism is a response of plant roots to avoid a saline environment. Curr Biol 23:2044–2050
Gao Y, Lynch JP (2016) Reduced crown root number improves water acquisition under water deficit stress in maize (*Zea mays* L.). J Exp Bot 67:4545–4557
Gao Z, He X, Zhao B, Zhou C, Liang Y, Ge R, Shen Y, Huang Z (2010) Overexpressing a putative aquaporin gene from wheat, TaNIP, enhances salt tolerance in transgenic arabidopsis. Plant Cell Physiol 51:767–775
García-Martí M, Piñero MC, García-Sanchez F, Mestre TC, López-Delacalle M, Martínez V, Rivero RM (2019) Amelioration of the oxidative stress generated by simple or combined abiotic stress through the K^+ and Ca^{2+} supplementation in tomato plants. Antioxidants 8:81
Gaspar M, Bousser A, Sissoëff I, Roche O, Hoarau J, Mahé A (2003) Cloning and characterization of ZmPIP1-5b, an aquaporin transporting water and urea. Plant Sci 165:21–31
Geisler M, Wang B, Zhu J (2014) Auxin transport during root gravitropism: transporters and techniques. Plant Biol 16:50–57
George E, Horst WJ, Neumann E (2011) Adaptation of plants to adverse chemical soil conditions. In: Marschner's mineral nutrition of higher plants, 3rd edn, pp 409–472
Gerbeau P, Güçlü J, Ripoche P, Maurel C (1999) Aquaporin Nt-TIPa can account for the high permeability of tobacco cell vacuolar membrane to small neutral solutes. Plant J 18:577–587
Ghassemian M, Nambara E, Cutler S, Kawaide H, Kamiya Y, McCourt P (2000) Regulation of abscisic acid signaling by the ethylene response pathway in arabidopsis. Plant Cell 12:1117–1126
Gojon A, Nacry P, Davidian JC (2009) Root uptake regulation: a central process for NPS homeostasis in plants. Curr Opin Plant Biol 12:328–338
González-Fontes A, Herrera-Rodríguez MB, Martín-Rejano EM, Navarro-Gochicoa MT, Rexach J, Camacho-Cristóbal JJ (2016) Root responses to boron deficiency mediated by ethylene. Front Plant Sci 6:1–6
Gorska A, Zwieniecka A, Michele Holbrook N, Zwieniecki MA (2008) Nitrate induction of root hydraulic conductivity in maize is not correlated with aquaporin expression. Planta 228:989–998
Gray SB, Brady SM (2016) Plant developmental responses to climate change. Dev Biol 419:64–77
Gruber BD, Giehl RFH, Friedel S, von Wirén N (2013) Plasticity of the Arabidopsis root system under nutrient deficiencies. Plant Physiol 163:161–179
Gu R, Chen X, Zhou Y, Yuan L (2011) Isolation and characterization of three maize aquaporin genes. BMB Rep 45:96–101
Guo S, Kaldenhoff R, Uehlein N, Sattelmacher B, Brueck H (2007a) Relationship between water and nitrogen uptake in nitrate- and ammonium-supplied *Phaseolus vulgaris* L. plants. J Plant Nut Soil Sci 170:73–80
Guo S, Shen Q, Brueck H (2007b) Effects of local nitrogen supply on water uptake of bean plants in a split root system. J Int Plant Biol 49:472–480
Guo Y, Jiang Q, Hu Z, Sun X, Fan S, Zhang H (2018) Function of the auxin-responsive gene *TaSAUR75* under salt and drought stress. Crop J. https://doi.org/10.1016/j.cj.2017.08.005
Guo XY, Wang Y, Zhao PX, Xu P, Yu GH, Zhang LY, Xiong Y, Bin Xiang C (2019) AtEDT1/HDG11 regulates stomatal density and water-use efficiency via ERECTA and E2Fa. New Phytol 223:1478–1488
Guseman JM, Webb K, Srinivasan C, Dardick C (2017) DRO1 influences root system architecture in *Arabidopsis* and *Prunus* species. Plant J 89:1093–1105
Hacke UG, Plavcová L, Almeida-Rodriguez A, King-Jones S, Zhou W, Cooke JEK (2010) Influence of nitrogen fertilization on xylem traits and aquaporin expression in stems of hybrid poplar. Tree Physiol 30:1016–1025
Hanaoka H, Fujiwara T (2007) Channel-mediated boron transport in rice. Plant Cell Physiol 48:227
Harris JM (2015) Abscisic acid: Hidden architect of root system structure. Plants 4:548–572
Hatfield JL, Dold C (2019) Water-use efficiency: advances and challenges in a changing climate. Front Plant Sci 10

Hawkesford M, Horst W, Kichey T, Lambers H, Schjoerring J, Møller IS, White P (2011) Functions of macronutrients. Elsevier Ltd
He XJ, Mu RL, Cao WH, Zhang ZG, Zhang JS, Chen SY (2005) AtNAC2, a transcription factor downstream of ethylene and auxin signaling pathways, is involved in salt stress response and lateral root development. Plant J 44:903–916
He Y, Wu J, Lv B, Li J, Gao Z, Xu W, Baluška F, Shi W, Shaw PC, Zhang J (2015) Involvement of 14-3-3 protein GRF9 in root growth and response under polyethylene glycol-induced water stress. J Exp Bot 66:2271–2281
Henry A, Cal AJ, Batoto TC, Torres RO, Serraj R (2012) Root attributes affecting water uptake of rice (*Oryza sativa*) under drought. J Exp Bot 63:4751–4763
Hill AE, Shachar-Hill B, Shachar-Hill Y (2004) What are aquaporins for? J Membr Biol 197:1–32
Huang L, Li M, Shao Y, Sun T, Li C, Ma F (2018a) Ammonium uptake increases in response to PEG-induced drought stress in *Malus hupehensis* Rehd. Environ Exp Bot 151:32–42
Huang L, Li M, Zhou K, Sun T, Hu L, Li C, Ma F (2018b) Uptake and metabolism of ammonium and nitrate in response to drought stress in *Malus prunifolia*. Plant Physiol Biochem 127:185–193
Iglesias MJ, Terrile MC, Bartoli CG, D'Ippólito S, Casalongué CA (2010) Auxin signaling participates in the adaptative response against oxidative stress and salinity by interacting with redox metabolism in *Arabidopsis*. Plant Mol Biol 74:215–222
Iglesias MJ, Terrile MC, Windels D, Lombardo MC, Bartoli CG, Vazquez F, Estelle M, Casalongué CA (2014) MiR393 regulation of auxin signaling and redox-related components during acclimation to salinity in Arabidopsis. PLoS ONE 9
Ishikawa-Sakurai J, Hayashi H, Murai-Hatano M (2014) Nitrogen availability affects hydraulic conductivity of rice roots, possibly through changes in aquaporin gene expression. Plant Soil 379:289–300
Jaganathan D, Thudi M, Kale S, Azam S, Roorkiwal M, Gaur PM, Kishor PBK, Nguyen H, Sutton T, Varshney RK (2015) Genotyping-by-sequencing based intra-specific genetic map refines a "QTL-hotspot" region for drought tolerance in chickpea. Mol Genet Genomics 290:559–571
Jahn TP, Møller ALB, Zeuthen T, Holm LM, Klærke DA, Mohsin B, Kühlbrandt W, Schjoerring JK (2004) Aquaporin homologues in plants and mammals transport ammonia. FEBS Lett 574:31–36
Javot H, Maurel C (2002) The role of aquaporins in root water uptake. Ann Bot 90:301–313
Johansson I, Karlsson M, Shukla VK, Chrispeels MJ, Larsson C (1998) Water transport activity of plasma membrane aquaporin PM28A is regulated by phosphorylation. Plant Cell 10:451–459
Julkowska MM, Hoefsloot HCJ, Mol S, Feron R, de Boer G-J, Haring MA, Testerink C (2014) Capturing Arabidopsis root architecture dynamics with ROOT-fit reveals diversity in responses to salinity. Plant Physiol 166:1387–1402
Jung H, Lee DK, Do Choi Y, Kim JK (2015) OsIAA6, a member of the rice Aux/IAA gene family, is involved in drought tolerance and tiller outgrowth. Plant Sci 236:304–312
Kagale S, Divi UK, Krochko JE, Keller WA, Krishna P (2007) Brassinosteroid confers tolerance in *Arabidopsis thaliana* and *Brassica napus* to a range of abiotic stresses. Planta 225:353–364
Kato Y, Miwa K, Takano J, Wada M, Fujiwara T (2009) Highly boron deficiency-tolerant plants generated by enhanced expression of NIP5;1, a boric acid channel. Plant Cell Physiol 50:58–66
Kawa D, Julkowska M, Montero Sommerfeld H, ter Horst A, Haring MA, Testerink C (2016) Phosphate-dependent root system architecture responses to salt stress. Plant Physiol 172:690–706
Ke S-D, Chen X, Oliver DJ, Liu H-Y, Xu P, Zhu J-K, Wang Y, Xiang C-B, Yu H, Hong Y-Y (2008) Activated expression of an Arabidopsis HD-start protein confers drought tolerance with improved root system and reduced stomatal density. Plant Cell 20:1134–1151
Kirk GJD, Van Du L (1997) Changes in rice root architecture, porosity, and oxygen and proton release under phosphorus deficiency. New Phytol 135:191–200
Knipfer T, Fricke W (2011) Water uptake by seminal and adventitious roots in relation to whole-plant water flow in barley (*Hordeum vulgare* L.). J Exp Bot 62:717–733
Kobayashi K, Kobayashi H, Maesato M, Hayashi M, Yamamoto T, Yoshioka S, Matsumura S, Sugiyama T, Kawaguchi S, Kubota Y et al (2017) Discovery of hexagonal structured Pd–B nanocrystals. Angew Chem Int Ed 56:6578–6582

Koevoets IT, Venema JH, Elzenga JTM, Testerink C (2016) Roots withstanding their environment: exploiting root system architecture responses to abiotic stress to improve crop tolerance. Front Plant Sci 7. https://doi.org/10.3389/fpls.2016.01335

Korhonen A, Lehto T, Heinonen J, Repo T (2018) Whole-plant frost hardiness of mycorrhizal (*Hebeloma* sp. or *Suillus luteus*) and non-mycorrhizal Scots pine seedlings (T Näsholm, ed). Tree Physiol 39

Krishnamurthy P, Ranathunge K, Nayak S, Schreiber L, Mathew MK (2011) Root apoplastic barriers block Na^+ transport to shoots in rice (*Oryza sativa* L.). J Exp Bot 62:4215–4228

Kumar K, Mosa KA, Chhikara S, Musante C, White JC, Dhankher OP (2014) Two rice plasma membrane intrinsic proteins, OsPIP2;4 and OsPIP2;7, are involved in transport and providing tolerance to boron toxicity. Planta 239:187–198

Kushwah S, Jones AM, Laxmi A (2011) Cytokinin interplay with ethylene, auxin, and glucose signaling controls arabidopsis seedling root directional growth. Plant Physiol 156:1851–1866

Lamont B (1972) The morphology and anatomy of proteoid roots in the genus Hakea. Aust J Bot 20:155–174

Lanza M, Garcia-Ponce B, Castrillo G, Catarecha P, Sauer M, Rodriguez-Serrano M, Páez-García A, Sánchez-Bermejo E, Tc M, Leo del Puerto Y et al (2012) Role of actin cytoskeleton in brassinosteroid signaling and in its integration with the auxin response in plants. Dev Cell 22:1275–1285

Lee DK, Yoon S, Kim YS, Kim JK (2017) Rice OsERF71-mediated root modification affects shoot drought tolerance. Plant Signal Behav 12(1):e1268311

Li G, Tillard P, Gojon A, Maurel C (2016a) Dual regulation of root hydraulic conductivity and plasma membrane aquaporins by plant nitrate accumulation and high-affinity nitrate transporter NRT2.1. Plant Cell Physiol 57:733–742

Li R, Wang J, Li S, Zhang L, Qi C, Weeda S, Zhao B, Ren S, Guo YD (2016b) Plasma membrane intrinsic proteins SlPIP2;1, SlPIP2;7 and SlPIP2;5 conferring enhanced drought stress tolerance in Tomato. Sci Rep 6. https://doi.org/10.1038/srep31814

Li X, Zeng R, Liao H (2016c) Improving crop nutrient efficiency through root architecture modifications. J Integr Plant Biol 58:193–202

Li X, Li G, Li Y, Kong X, Zhang L, Wang J, Li X, Yang Y (2018) ABA receptor subfamily III enhances abscisic acid sensitivity and improves the drought tolerance of arabidopsis. Int J Mol Sci 19

Li Y, Niu W, Cao X, Wang J, Zhang M, Duan X, Zhang Z (2019) Effect of soil aeration on root morphology and photosynthetic characteristics of potted tomato plants (*Solanum lycopersicum*) at different NaCl salinity levels. BMC Plant Biol 19:331

Lichtfouse E (ed) (2013) Sustainable agriculture reviews. Springer, Dordrecht

Liu C, Zhang T (2017) Expansion and stress responses of the AP2/EREBP superfamily in cotton. BMC Genom 18:1–16

Liu LH, Ludewig U, Gassert B, Frommer WB, Von Wirén N (2003) Urea transport by nitrogen-regulated tonoplast intrinsic proteins in Arabidopsis. Plant Physiol 133:1220–1228

Liu HY, Sun WN, Su WA, Tang ZC (2006) Co-regulation of water channels and potassium channels in rice. Physiol Plant 128:58–69

Liu Q, Zhang S, Liu B (2016) 14-3-3 proteins: macro-regulators with great potential for improving abiotic stress tolerance in plants. Biochem Biophys Res Commun 477:9–13

Liu Y, Sun G, Zhong Z, Ji L, Zhang Y, Zhou J, Zheng X, Deng K (2017) Overexpression of AtEDT1 promotes root elongation and affects medicinal secondary metabolite biosynthesis in roots of transgenic Salvia miltiorrhiza. Protoplasma 254:1617–1625

Loque D, Ludewig U, Yuan L, Von Wire N (2005) Tonoplast facilitate NH 3 transport into the vacuole 1. Plant Physiol 137:671–680

Lu L, Dong C, Liu R, Zhou B, Wang C, Shou H (2018) Roles of soybean plasma membrane intrinsic protein GmPIP2;9 in drought tolerance and seed development. Front Plant Sci 9. https://doi.org/10.3389/fpls.2018.00530

Luu DT, Maurel C (2005) Aquaporins in a challenging environment: molecular gears for adjusting plant water status. Plant Cell Environ 28:85–96
Lynch JP (2011) Root phenes for enhanced soil exploration and phosphorus acquisition: tools for future crops. Plant Physiol 156:1041–1049
Lynch JP (2018) Rightsizing root phenotypes for drought resistance. J Exp Bot 69:3279–3292
Lynch JP (2019) Root phenotypes for improved nutrient capture: an underexploited opportunity for global agriculture. New Phytol
Ma Q, Xia Z, Cai Z, Li L, Cheng Y, Liu J, Nian H (2019) GmWRKY16 enhances drought and salt tolerance through an ABA-mediated pathway in Arabidopsis thaliana. Front Plant Sci 9
Marschner H, Kirkby EA, Cakmak I (1996) Effect of mineral nutritional status on shoot-root partitioning of photoassimilates and cycling of mineral nutrients. J Exp Bot 47:1255–1263
Martínez-Ballesta MC, Cabañero F, Olmos E, Periago PM, Maurel C, Carvajal M (2008) Two different effects of calcium on aquaporins in salinity-stressed pepper plants. Planta 228:15–25
Maurel C (2007) Plant aquaporins: novel functions and regulation properties. FEBS Lett 581:2227–2236
Meister R, Rajani MS, Ruzicka D, Schachtman DP (2014) Challenges of modifying root traits in crops for agriculture. Trends Plant Sci 19:779–788
Méndez-Alonzo R, López-Portillo J, Moctezuma C, Bartlett MK, Sack L (2016) Osmotic and hydraulic adjustment of mangrove saplings to extreme salinity. Tree Physiol 36:1562–1572
Miao H, Sun P, Liu J, Wang J, Xu B, Jin Z (2018) Overexpression of a novel ROP gene from the banana (MaROP5g) confers increased salt stress tolerance. Int J Mol Sci 19. https://doi.org/10.3390/ijms19103108
Miyazawa Y, Takahashi H, Kaneyasu T, Fujii N, Kobayashi A, Nakayama M (2007) Auxin response, but not its polar transport, plays a role in hydrotropism of Arabidopsis roots. J Exp Bot 58:1143–1150
Mosa KA, Kumar K, Chhikara S, Musante C, White JC, Dhankher OP (2016) Enhanced boron tolerance in plants mediated by bidirectional transport through plasma membrane intrinsic proteins. Sci Rep 6:1–14
Munemasa S, Oda K, Watanabe-Sugimoto M, Nakamura Y, Shimoishi Y, Murata Y (2007) The coronatine-insensitive 1 mutation reveals the hormonal signaling interaction between abscisic acid and methyl jasmonate in Arabidopsis guard cells. Specific impairment of ion channel activation and second messenger production. Plant Physiol 143:1398–1407
Nagawa S, Xu T, Yang Z (2010) RHO GTPase in plants: conservation and invention of regulators and effectors. Small GTPases 1:78–88
Nakajima Y, Nara Y, Kobayashi A, Sugita T, Miyazawa Y, Fujii N, Takahashi H (2017) Auxin transport and response requirements for root hydrotropism differ between plant species. J Exp Bot 68:3441–3456
Naz N, Fatima S, Hameed M, Muhammad A, Naseer M, Ahmad F, Zahoor A (2018) Structural and functional aspects of salt tolerance in differently adapted ecotypes of Aeluropus lagopoides from saline desert habitats. Int J Agric Biol 20:41–51
Neumann G, Massonneau A, Langlade N, Dinkelaker B, Hengeler C, Römheld V, Martinoia E (2000) Physiological aspects of cluster root function and development in phosphorus-deficient white lupin (Lupinus albus L.). Ann Bot 85:909–919
Nie Q, Gao GL, Fan Q jie, Qiao G, Wen XP, Liu T, Peng ZJ, Cai YQ (2015) Isolation and characterization of a catalase gene 'HuCAT3' from pitaya (Hylocereus undatus) and its expression under abiotic stress. Gene 563:63–71
Niu YF, Chai RS, Jin GL, Wang H, Tang CX, Zhang YS (2013) Responses of root architecture development to low phosphorus availability: a review. Ann Bot 112:391–408
Nobel P (1984) Extreme temperatures and thermal tolerances for seedlings of desert succulents. Oecologia 62:310–317
Northey JGB, Liang S, Jamshed M, Deb S, Foo E, Reid JB, Mccourt P, Samuel MA (2016) Farnesylation mediates brassinosteroid biosynthesis to regulate abscisic acid responses. Nat Plants 2:1–7

Ogawa A, Kitamichi K, Toyofuku K, Kawashima C (2006) Quantitative analysis of cell division and cell death in seminal root of rye under salt stress. Plant Prod Sci 9:56–64
Olatunji D, Geelen D, Verstraeten I (2017) Control of endogenous auxin levels in plant root development. Int J Mol Sci 18. https://doi.org/10.3390/ijms18122587
Osmont KS, Sibout R, Hardtke CS (2007) Hidden branches: developments in root system architecture. Annu Rev Plant Biol 58:93–113
Pang Y, Li L, Ren F, Lu P, Wei P, Cai J, Xin L, Zhang J, Chen J, Wang X (2010) Overexpression of the tonoplast aquaporin AtTIP5;1 conferred tolerance to boron toxicity in Arabidopsis. J Genet Genomics 37:389–397
Paul AL, Denison FC, Schultz ER, Zupanska AK, Ferl RJ (2012) 14-3-3 Phosphoprotein interaction networks—does isoform diversity present functional interaction specification? Front Plant Sci 3:1–14
Peng Y, Lin W, Cai W, Arora R (2007) Overexpression of a Panax ginseng tonoplast aquaporin alters salt tolerance, drought tolerance and cold acclimation ability in transgenic Arabidopsis plants. Planta 226:729–740
Peret B, Desnos T, Jost R, Kanno S, Berkowitz O, Nussaume L (2014) Root architecture responses: in search of phosphate. Plant Physiol 166:1713–1723
Perrone I, Gambino G, Chitarra W, Vitali M, Pagliarani C, Riccomagno N, Balestrini R, Kaldenhoff R, Uehlein N, Gribaudo I et al (2012) The grapevine root-specific aquaporin VvPIP2;4n controls root hydraulic conductance and leaf gas exchange under well-watered conditions but not under water stress. Plant Physiol 160:965–977
Peters C, Kim SC, Devaiah S, Li M, Wang X (2014) Non-specific phospholipase C5 and diacylglycerol promote lateral root development under mild salt stress in Arabidopsis. Plant Cell Environ 37:2002–2013
Pierik R, Testerink C (2014) The art of being flexible: how to escape from shade, salt, and drought. Plant Physiol 166:5–22
Postma JA, Dathe A, Lynch JP (2014) The optimal lateral root branching density for maize depends on nitrogen and phosphorus availability. Plant Physiol 166:590–602
Prak S, Hem S, Boudet J, Viennois G, Sommerer N, Rossignol M, Maurel C, Santoni V (2008) Multiple phosphorylations in the C-terminal tail of plant plasma membrane aquaporins. Mol Cell Proteomics 7:1019–1030
Puértolas J, Conesa MR, Ballester C, Dodd IC (2015) Local root abscisic acid (ABA) accumulation depends on the spatial distribution of soil moisture in potato: implications for ABA signalling under heterogeneous soil drying. J Exp Bot 66:2325–2334
Ragel P, Raddatz N, Leidi EO, Quintero FJ, Pardo JM (2019) Regulation of K^+ nutrition in plants. Front Plant Sci 10. https://doi.org/10.3389/fpls.2019.00281
Rahman MS, Matsumuro T, Miyake H, Takeoka Y (2001) Effects of salinity stress on the seminal root tip ultrastructures of rice seedlings (*Oryza sativa* L.). Plant Prod Sci 4:103–111
Ranathunge K, Schreiber L (2011) Water and solute permeabilities of Arabidopsis roots in relation to the amount and composition of aliphatic suberin. J Exp Bot 62:1961–1974
Ranathunge K, Schreiber L, Franke R (2011) Suberin research in the genomics era-New interest for an old polymer. Plant Sci 180:399–413
Ranathunge K, Schreiber L, Bi YM, Rothstein SJ (2016) Ammonium-induced architectural and anatomical changes with altered suberin and lignin levels significantly change water and solute permeabilities of rice (*Oryza sativa* L.) roots. Planta 243:231–249
Ray S, Mishra S, Bisen K, Singh S, Sarma BK, Singh HB (2018) Modulation in phenolic root exudate profile of Abelmoschus esculentus expressing activation of defense pathway. Microbiol Res 207:100–107
Redillas MCFR, Jeong JS, Kim YS, Jung H, Bang SW, Choi YD, Ha SH, Reuzeau C, Kim JK (2012) The overexpression of OsNAC9 alters the root architecture of rice plants enhancing drought resistance and grain yield under field conditions. Plant Biotechnol J 10:792–805

Rellán-Álvarez R, Lobet G, Lindner H, Pradier PL, Sebastian J, Yee MC, Geng Y, Trontin C, Larue T, Schrager-Lavelle A et al (2015) GLO-roots: an imaging platform enabling multidimensional characterization of soil-grown root systems. eLife. https://doi.org/10.7554/elife.07597

Remans T, Nacry P, Pervent M, Filleur S, Diatloff E, Mounier E, Tillard P, Forde BG, Gojon A (2006) The Arabidopsis NRT1.1 transporter participates in the signaling pathway triggering root colonization of nitrate-rich patches. Proc Natl Acad Sci USA 103:19206–19211

Ren B, Wang M, Chen Y, Sun G, Li Y, Shen Q, Guo S (2015) Water absorption is affected by the nitrogen supply to rice plants. Plant Soil 396:397–410

Riaz M, Yan L, Wu X, Hussain S, Aziz O, Jiang C (2018) Boron increases root elongation by reducing aluminum induced disorganized distribution of HG epitopes and alterations in subcellular cell wall structure of trifoliate orange roots. Ecotoxicol Environ Saf 165:202–210

Riemann M, Dhakarey R, Hazman M, Miro B, Kohli A, Nick P (2015) Exploring jasmonates in the hormonal network of drought and salinity responses. Front Plant Sci 6:1–16

Robert M, Salinas J, Collinge D (2002) 14-3-3 Proteins and the response to abiotic and biotic stress. Plant Mol Biol 50:1031–1039

Rodriguez HG, Roberts J, Jordan WR, Drew MC (1997) Growth, water relations, and accumulation of organic and inorganic solutes in roots of maize seedlings during salt stress. Plant Physiol 113:881–893

Rosales MA, Maurel C, Nacry P (2019) Abscisic acid coordinates dose-dependent developmental and hydraulic responses of roots to water deficit. Plant Physiol 180:2198–2211

Rowe JH, Topping JF, Liu J, Lindsey K (2016) Abscisic acid regulates root growth under osmotic stress conditions via an interacting hormonal network with cytokinin, ethylene and auxin. New Phytol 211:225–239

Saengwilai P, Tian X, Lynch JP (2014) Low crown root number enhances nitrogen acquisition from low-nitrogen soils in maize. Plant Physiol 166:581–589

Sahni S, Prasad BD, Liu Q, Grbic V, Sharpe A, Singh SP, Krishna P (2016) Overexpression of the brassinosteroid biosynthetic gene DWF4 in Brassica napus simultaneously increases seed yield and stress tolerance. Sci Rep 6:1–14

Sahr T, Voigt G, Paretzke HG, Schramel P, Ernst D (2005) Caesium-affected gene expression in Arabidopsis thaliana. New Phytol 165:747–754

Salazar-Henao JE, Schmidt W (2016) An inventory of nutrient-responsive genes in Arabidopsis root hairs. Front Plant Sci 7. https://doi.org/10.3389/fpls.2016.00237

Salehi A, Tasdighi H, Gholamhoseini M (2016) Evaluation of proline, chlorophyll, soluble sugar content and uptake of nutrients in the German chamomile (*Matricaria chamomilla* L.) under drought stress and organic fertilizer treatments. Asian Pac J Trop Biomed 6:886–891

Sánchez-Romera B, Ruiz-Lozano JM, Li G, Luu D-T, Martínez-Ballesta M del C, Carvajal M, Zamarreño AM, García-Mina JM, Maurel C, Aroca R (2014) Enhancement of root hydraulic conductivity by methyl jasmonate and the role of calcium and abscisic acid in this process. Plant Cell Environ 37:995–1008

Sánchez-Romera B, Ruiz-Lozano JM, Zamarreño ÁM, García-Mina JM, Aroca R (2016) Arbuscular mycorrhizal symbiosis and methyl jasmonate avoid the inhibition of root hydraulic conductivity caused by drought. Mycorrhiza 26:111–122

Santander C, Aroca R, Ruiz-Lozano JM, Olave J, Cartes P, Borie F, Cornejo P (2017) Arbuscular mycorrhiza effects on plant performance under osmotic stress. Mycorrhiza 27:639–657

Santos EF, Kondo Santini JM, Paixão AP, Júnior EF, Lavres J, Campos M, dos Reis AR (2017) Physiological highlights of manganese toxicity symptoms in soybean plants: Mn toxicity responses. Plant Physiol Biochem 113:6–19

Scheres B, Wolkenfelt H, Willemsen V, Terlouw M, Lawson E, Dean C, Weisbeek P (1994) Embryonic origin of the Arabidopsis primary root and root meristem initials. Development 120:2475–2487

Schnurbusch T, Hayes J, Hrmova M, Baumann U, Ramesh SA, Tyerman SD, Langridge P, Sutton T (2010) Boron toxicity tolerance in barley through reduced expression of the multifunctional aquaporin HvNIP2;1. Plant Physiol 153:1706–1715

Schreiber L, Franke R, Hartmann KD, Ranathunge K, Steudle E (2005) The chemical composition of suberin in apoplastic barriers affects radial hydraulic conductivity differently in the roots of rice (*Oryza sativa* L. cv. IR64) and corn (*Zea mays* L. cv. Helix). J Exp Bot 56:1427–1436
Schrick K, Nguyen D, Karlowski WM, Mayer KFX (2004) START lipid/sterol-binding domains are amplified in plants and are predominantly associated with homeodomain transcription factors. Genome Biol 5:1–16
Sebastian J, Yee M-C, Goudinho Viana W, Rellán-Álvarez R, Feldman M, Priest HD, Trontin C, Lee T, Jiang H, Baxter I et al (2016) Grasses suppress shoot-borne roots to conserve water during drought. Proc Natl Acad Sci USA
Seifert GJ, Acet T, Xue H (2014) The Arabidopsis thaliana FASCICLIN LIKE ARABINOGALACTAN PROTEIN 4 gene acts synergistically with abscisic acid signalling to control root growth. Ann Bot 114:1125–1133
Sharp RE, Wu Y, Voetberg GS, Aab IN, LeNoble ME (1994) Confirmation that abscisic acid accumulation is required for maize primary root elongation at low water potentials. J Exp Bot 45:1743–1751
Shelef O, Lazarovitch N, Rewald B, Golan-Goldhirsh A, Rachmilevitch S (2010) Root halotropism: salinity effects on Bassia indica root. Plant Biosyst 144:471–478
Shen J, Yuan L, Zhang J, Li H, Bai Z, Chen X, Zhang W, Zhang F (2011) Phosphorus dynamics: from soil to plant. Plant Physiol 156:997–1005
Shireen F, Nawaz MA, Chen C, Zhang Q, Zheng Z, Sohail H, Sun J, Cao H, Huang Y, Bie Z (2018) Boron: functions and approaches to enhance its availability in plants for sustainable agriculture. Int J Mol Sci 19:95–98
Shkolnik-Inbar D, Bar-Zvi D (2010) ABI4 mediates abscisic acid and cytokinin inhibition of lateral root formation by reducing polar auxin transport in arabidopsis. Plant Cell 22:3560–3573
Shkolnik-Inbar D, Adler G, Bar-Zvi D (2013) ABI4 downregulates expression of the sodium transporter HKT1;1 in Arabidopsis roots and affects salt tolerance. Plant J 73:993–1005
Siemens JA, Zwiazek JJ (2003) Effects of water deficit stress and recovery on the root water relations of trembling aspen (*Populus tremuloides*) seedlings. Plant Sci 165:113–120
Siemens JA, Zwiazek JJ (2004) Changes in root water flow properties of solution culture-grown trembling aspen (*Populus tremuloides*) seedlings under different intensities of water-deficit stress. Physiol Plant 121:44–49
Singh M, Gupta A, Laxmi A (2017) Striking the right chord: signaling enigma during root gravitropism. Front Plant Sci 8:1–17
Soto G, Alleva K, Mazzella MA, Amodeo G, Muschietti JP (2008) AtTIP1;3 and AtTIP5;1, the only highly expressed Arabidopsis pollen-specific aquaporins, transport water and urea. FEBS Lett 582:4077–4082
Spollen WG, Lenoble ME, Samuels TD, Bernstein N, Sharp RE (2000) Abscisic acid accumulation maintains maize primary root elongation at low water potentials by restricting ethylene production. Plant Physiol 122:967–976
Steudle E (2000a) Water uptake by roots: effects of water deficit. J Exp Bot 51:1531–1542
Steudle E (2000b) Water uptake by plant roots: an integration of views. Plant Soil 226:45–56
Steudle E, Henzler T (1995) Water channels in plants: do basic concepts of water transport change? J Exp Bot 46:1067–1076
Steudle E, Peterson CA (1998) How does water get through roots? J Exp Bot 49:775–788
Street IH, Aman S, Zubo Y, Ramzan A, Wang X, Shakeel SN, Kieber JJ, Eric Schaller G (2015) Ethylene inhibits cell proliferation of the arabidopsis root meristem. Plant Physiol 169:338–350
Sun X, Luo X, Sun M, Chen C, Ding X, Wang X, Yang S, Yu Q, Jia B, Ji W et al (2014) A glycine soja 14-3-3 protein gsgf14o participates in stomatal and root hair development and drought tolerance in arabidopsis thaliana. Plant Cell Physiol 55:99–118
Sun T, Li M, Shao Y, Yu L, Ma F (2017) Comprehensive genomic identification and expression analysis of the phosphate transporter (PHT) gene family in apple. Front Plant Sci 8. https://doi.org/10.3389/fpls.2017.00426

Takahashi N, Goto N, Okada K, Takahashi H (2002) Hydrotropism in abscisic acid, wavy, and gravitropic mutants of *Arabidopsis thaliana*. Planta 216:203–211
Takahashi H, Miyazawa Y, Fujii N (2009) Hormonal interactions during root tropic growth: hydrotropism versus gravitropism. Plant Mol Biol 69:489–502
Takano J, Wada M, Ludewig U, Schaaf G, Von Wirén N, Fujiwara T (2006) The Arabidopsis major intrinsic protein NIP5;1 is essential for efficient boron uptake and plant development under boron limitation. Plant Cell 18:1498–1509
Tan T, Cai J, Zhan E, Yang Y, Zhao J, Guo Y, Zhou H (2016) Stability and localization of 14-3-3 proteins are involved in salt tolerance in Arabidopsis. Plant Mol Biol 92:391–400
Tanaka M, Wallace IS, Takano J, Roberts DM, Fujiwara T (2008) NIP6;1 is a boric acid channel for preferential transpórt of boron to growing shoot tissues in Arabidopsis. Plant Cell 20:2860–2875
Tanaka-Takada N, Kobayashi A, Takahashi H, Kamiya T, Kinoshita T, Maeshima M (2019) Plasma membrane-associated Ca^{2+}-binding protein PCaP1 is involved in root hydrotropism of *Arabidopsis thaliana*. Plant Cell Physiol 0:1–11
Tariq A, Pan K, Olatunji OA, Graciano C, Li Z, Sun F, Sun X, Song D, Chen W, Zhang A et al (2017) Phosphorous application improves drought tolerance of Phoebe zhennan. Front Plant Sci 8. https://doi.org/10.3389/fpls.2017.01561
Thor K (2019) Calcium—nutrient and messenger. Front Plant Sci 10. https://doi.org/10.3389/fpls.2019.00440
Thorup-Kristensen K, Kirkegaard J (2016) Root system-based limits to agricultural productivity and efficiency: the farming systems context. Ann Bot 118:573–592
Tong H-Y, Gu C-S, Yuan H-Y, Ma J-J, Huang S-Z (2014) Effects of salt on the growth, photosyntheticpigments and structure of two halophytes, Iris halophyla and I. lactea Var. chinensis. Fresenius Environ Bull 23:84–90
Uga Y, Okuno K, Yano M (2011) Dro1, a major QTL involved in deep rooting of rice under upland field conditions. J Exp Bot 62:2485–2494
Uga Y, Sugimoto K, Ogawa S, Rane J, Ishitani M, Hara N, Kitomi Y, Inukai Y, Ono K, Kanno N et al (2013) Control of root system architecture by DEEPER ROOTING 1 increases rice yield under drought conditions. Nat Genet 45:1097
Uga Y, Kitomi Y, Ishikawa S, Yano M (2015) Genetic improvement for root growth angle to enhance crop production. Breed Sci 65:111–119
Ullah A, Manghwar H, Shaban M, Khan AH, Akbar A, Ali U, Ali E, Fahad S (2018) Phytohormones enhanced drought tolerance in plants: a coping strategy. Environ Sci Pollut Res 25:33103–33118
Vandeleur RK, Mayo G, Shelden MC, Gilliham M, Kaiser BN, Tyerman SD (2008) The role of plasma membrane intrinsic protein aquaporins in water transport through roots: diurnal and drought stress responses reveal different strategies between isohydric and anisohydric cultivars of grapevine. Plant Physiol 149:445–460
Vera-Estrella R, Barkla BJ, Bohnert HJ, Pantoja O (2004) Novel regulation of aquaporins during osmotic stress. Plant Physiol 135:2318–2329
Verstraeten I, Schotte S, Geelen D (2014) Hypocotyl adventitious root organogenesis differs from lateral root development. Front Plant Sci 5:1–13
Walch-Liu P, Ivanov II, Filleur S, Gan Y, Remans T, Forde BG (2006) Nitrogen regulation of root branching. Ann Bot 97:875–881
Wang YH, Garvin DF, Kochian LV (2001) Nitrate-induced genes in tomato roots. Array analysis reveals novel genes that may play a role in nitrogen nutrition. Plant Physiol 127:345–359
Wang WH, Köhler B, Cao FQ, Liu LH (2008) Molecular and physiological aspects of urea transport in higher plants. Plant Sci 175:467–477
Wang D, Chen F, Wei G, Jiang M, Dong M (2015) The mechanism of improved pullulan production by nitrogen limitation in batch culture of Aureobasidium pullulans. Carbohyd Polym 127:325–331
Wang M, Ding L, Gao L, Li Y, Shen Q, Guo S (2016) The interactions of aquaporins and mineral nutrients in higher plants. Int J Mol Sci 17

Wang X, Gao F, Bing J, Sun W, Feng X, Ma X, Zhou Y, Zhang G (2019) Overexpression of the Jojoba aquaporin gene, ScPIP1, enhances drought and salt tolerance in transgenic Arabidopsis. Int J Mol Sci

Wasternack C, Song S (2017) Jasmonates: Biosynthesis, metabolism, and signaling by proteins activating and repressing transcription. J Exp Bot 68:1303–1321

West G (2004) Cell cycle modulation in the response of the primary root of Arabidopsis to salt stress. Plant Physiol 135:1050–1058

White PJ, Broadley MR (2003) Calcium in plants. Ann Bot 92:487–511

Williamson LC, Ribrioux SPCP, Fitter AH, Leyser HMO (2001) Phosphate availability regulates root system architecture in Arabidopsis. Plant Physiol 126:875–882

Wong GR, Mazumdar P, Lau SE, Harikrishna JA (2018) Ectopic expression of a *Musa acuminata* root hair defective 3 (MaRHD3) in Arabidopsis enhances drought tolerance. J Plant Physiol 231:219–233

Wu X, Riaz M, Yan L, Du C, Liu Y, Jiang C (2017) Boron deficiency in trifoliate orange induces changes in pectin composition and architecture of components in root cell walls. Front Plant Sci 8:1–10

Xu G, Fan X, Miller AJ (2012) Plant nitrogen assimilation and use efficiency. Annu Rev Plant Biol 63:153–182

Xu P, Cai XT, Wang Y, Xing L, Chen Q, Bin Xiang C (2014) HDG11 upregulates cell-wall-loosening protein genes to promote root elongation in Arabidopsis. J Exp Bot 65:4285–4295

Yan J, He C, Wang J, Mao Z, Holaday SA, Allen RD, Zhang H (2004) Overexpression of the Arabidopsis 14-3-3 protein GF14λ in cotton leads to a 'stay-green' phenotype and improves stress tolerance under moderate drought conditions. Plant Cell Physiol 45:1007–1014

Ye H, Liu S, Tang B, Chen J, Xie Z, Nolan TM, Jiang H, Guo H, Lin HY, Li L et al (2017) RD26 mediates crosstalk between drought and brassinosteroid signalling pathways. Nat Commun 8:1–13

Yu LX, Ray JD, Otoole JC, Nguyen HT (1995) Use of wax-petrolatum layers for screening rice root penetration. Crop Sci 35:684–687

Yu L, Chen X, Wang Z, Wang S, Wang Y, Zhu Q, Li S, Xiang C (2013) Arabidopsis enhanced drought tolerance1/HOMEODOMAIN GLABROUS11 confers drought tolerance in transgenic rice without yield penalty. Plant Physiol 162:1378–1391

Yu P, Li X, White PJ, Li C (2015) A large and deep root system underlies high nitrogen-use efficiency in maize production. PLoS ONE 10:1–17

Yu LH, Wu SJ, Peng YS, Liu RN, Chen X, Zhao P, Xu P, Zhu JB, Jiao GL, Pei Y et al (2016) Arabidopsis EDT1/HDG11 improves drought and salt tolerance in cotton and poplar and increases cotton yield in the field. Plant Biotechnol J 14:72–84

Zhang H, Mao X, Wang C, Jing R (2010) Overexpression of a common wheat gene Tasnrk2.8 enhances tolerance to drought, salt and low temperature in Arabidopsis. PLoS ONE 5. https://doi.org/10.1371/journal.pone.0016041

Zhang L, Yan J, Vatamaniuk OK, Du X (2016) CsNIP2;1 is a plasma membrane transporter from cucumis sativus that facilitates urea uptake when expressed in saccharomyces cerevisiae and arabidopsis thaliana. Plant Cell Physiol 57:616–629

Zhang Y, Zhao H, Zhou S, He Y, Luo Q, Zhang F, Qiu D, Feng J, Wei Q, Chen L et al (2018) Expression of TaGF14b, a 14-3-3 adaptor protein gene from wheat, enhances drought and salt tolerance in transgenic tobacco. Planta 248:117–137

Zhang S, Feng M, Chen W, Zhou X, Lu J, Wang Y, Li Y, Jiang CZ, Gan SS, Ma N et al (2019) In rose, transcription factor PTM balances growth and drought survival via PIP2;1 aquaporin. Nat Plants 5:290–299

Zhao Y, Xing L, Xingang W, Hou Y-J, Gao J, Wang P, Duan C-G, Zhu X, Zhu J-K (2014) The ABA receptor PYL8 promotes lateral root growth by enhancing MYB77-dependent transcription of auxin-responsive genes. Sci Signal 7:1–7

Zheng HG, Babu RC, Pathan MS, Ali L, Huang N, Courtois B, Nguyen HT (2000) Quantitative trait loci for root-penetration ability and root thickness in rice: comparison of genetic backgrounds. Genome 43:53–61

Zhou L, Zhou J, Xiong Y, Liu C, Wang J, Wang G, Cai Y (2018) Overexpression of a maize plasma membrane intrinsic protein ZmPIP1;1 confers drought and salt tolerance in Arabidopsis. PLoS ONE 13. https://doi.org/10.1371/journal.pone.0198639

Zhu Z, Sun B, Xu X, Chen H, Zou L, Chen G, Cao B, Chen C, Lei J (2016) Overexpression of AtEDT1/HDG11 in Chinese kale (Brassica oleracea var. alboglabra) enhances drought and osmotic stress tolerance. Front Plant Sci 7

Plant Responses and Tolerance to Extreme Salinity: Learning from Halophyte Tolerance to Extreme Salinity

Waleed Fouad Abobatta

Abstract Salinity is one of the major problems facing agricultural production worldwide, particularly in arid and semiarid environments. Unfortunately, most major economic crops (glycophytes) cannot tolerate saline conditions, even at low concentrations (<40 mM NaCl). Halophytes are tolerant of high salinities, up to 200 mM NaCl. Therefore, there is increased interest in the production of new cultivars that have the potential to produce higher yields under saline conditions. There is also research interest in understanding the various strategies, including salt tolerance and salt avoidance, that halophytes use to adapt to saline conditions, besides conventional habits like accumulation of compatible solutes and polyamine production. Additionally, halophytes use a number of complicated mechanisms to grow and reproduce under such harsh conditions, such as regulating ion concentrations in cells, exuding excessive salt from plant tissues, and eliminating salts via glands and hairs to keep salt concentrations in leaves and shoots below a certain threshold. Some halophytes produce high leaf/stem succulence or accumulate excessive salt in old leaves. Other species use ion accumulation and control plant growth to cope with salinity. In this chapter we discuss the classification and distribution of halophytes and their potential role as resources for humankind (food, oil, forage, and biofuel). Glycophyte growth is affected by high salinity, which can lead to plant death. Therefore, glycophytes use various strategies to enhance growth under salinity stress, such as producing salt-tolerant genotypes and using different mechanisms—like excretion of salt ions and salt sequestration—to improve growth and productivity. Salinity stress reduces glycophyte growth due to its negative effects on the metabolism and phytohormone levels, such as cytokinin and gibberellic acid. In addition, high salt levels have an adverse effect on water relations and disturb the nutritional balance in plant cells.

W. F. Abobatta (✉)
Citrus Department, Horticulture Research Institute, Agriculture Research Center, Giza, Egypt
e-mail: wabobatta@yahoo.com

M. Hasanuzzaman and M. Tanveer (eds.), *Salt and Drought Stress Tolerance in Plants*, Signaling and Communication in Plants, https://doi.org/10.1007/978-3-030-40277-8_7

Abbreviations

GA	Gibberellic acid
ABA	Abscisic acid
IAA	Indole acetic acid
CK	Cytokinins
JA	Jasmonates
NPQ	Non-photochemical quenching coefficient
PV	Perennial vine
PSh	Perennial shrub
PSuSh	Perennial succulent shrub
PPP	Parasitic perennial plant
SuP	Submerged plant
Sh	Shrub
SSh	Subshrubs
F	Fern
V	Vine
T	Tree
RGR	Relative growth rate

1 Introduction

Plants are classified as halophytes or glycophytes based on their responses to conditions of salinity. This means of categorization is not absolute, there is variation in tolerance to salinity between the various species—from sensitive to moderately tolerant to highly tolerant—of major economic crops. Barley, cotton, sugar beet, and canola are the most tolerant; bread wheat is moderately tolerant; most legume species are sensitive; and strawberry plants highly sensitive (Kotuby-Amacher et al. 2000).

Salinity affects over 6% of the world's land, particularly in arid and semiarid regions. According to the FAO Land and Plant Nutrition Management Service (FAO 2005) there are 45 million ha of salt-affected land (19.5%) forming part of the 230 million ha of currently irrigated land, with about 32 million ha in the form of dryland. Considering all global cultivated land some 1,500 million ha are salt affected to different degrees (2.1%). In addition, globally there are about 800 million ha of cultivated land affected by soil salinity.

Generally, the common cations associated with salinity are Na^+, Ca^{2+}, and, Mg^{2+}, while common anions are Cl^-, $SO_4{}^{2-}$, and $HCO_3{}^-$ (Dudley 1994). However, Na^+ and Cl^- ions are considered the most important because Na^+ causes degradation of the physical structure of the soil and at high concentrations is toxic to plants, and Cl^- is also harmful to plants (Hasegawa et al. 2000; Tanveer et al. 2018).

Soil salinity occurs by natural factors or human-induced activities resulting in the accumulation of dissolved salts in soil water to such a high level that plant growth is inhibited. Clay soils become sodic as a result of salinity, due to leaching (water from rainfall or irrigation) of soluble salts into the subsoil, leaving sodium bound to the negative charges of the clay particles.

Different factors, both endogenous and exogenous, affect plant responses to salinity, nevertheless, phytohormones are considered the most important endogenous substances, being essential for modulating physiological responses that ultimately lead to adaptation to salinity stress.

Plant responses usually include fluctuations in the levels of various phytohormones, relating to changes in gene biosynthesis and the responses they regulate. It is a common assumption that glycophytes are sensitive to saline conditions while halophytes are resistant or tolerant. However, there are numerous species, traditionally considered as glycophytes, that might be tolerant to salt. In addition, halophytes can of course be sensitive to other abiotic stresses.

Halophytes do not just survive in high salinity levels, they grow and reproduce vegetatively and sexually. Generally there is distinct species zonation from lower salt tolerance (upper marsh species), such as *Elymus pycnanthus, Elymus repens, Festuca rubra, Agrostis stolonifera, Atriplex littoralis, Plantago lanceolata, Chenopodium album*—salt marsh plants obtained by van Diggelen (1988), to higher salt tolerance (lower marsh species), such as *Salicornia* spp., *Spartina anglica, Aster tripolium, Atriplex portulacoides, Limonium vulgare, Triglochin maritima,* and *Spergularia maritime* (Cooper 1982; Schat 1982; Rozema et al. 1985). Most plants, including conventional crops, cannot even withstand low (<40 mM NaCl) salt concentrations (Munns and Tester 2008).

Unfortunately, all the major crops are glycophytes—their average yields are reduced by more than 50% under saline conditions (Bray et al. 2000). Therefore, there is increasing attention directed toward producing new cultivars that have the potential for higher yields under abiotic stresses such as salinity, drought, and extreme temperatures.

In recent decades halophytes have received greater attention from plant scientists due to increased soil salinities in various regions of the world, particularly in arid and semiarid climates, triggered by low-quality irrigation water, high evaporation ratios, and rising sea levels associated with global warming, all of which reduce crop yields. Such a reduction in crop yield is considered a serious problem that humankind will need to face in the near future, exacerbated by a rapidly increasing world population—anticipated to be more than 9 billion by 2050 (United Nations Population Information Network).

There have been different investigations in recent decades focused on understanding the mechanisms of salt tolerance in halophytes and their ability to grow and reproductive under saline conditions, in order to enhance glycophyte tolerance and control ionic homeostasis and osmotic shock to reduce the adverse effects of salinity and thereby increase the total yield of economic glycophytes. It is also believed that biotechnology could play an important role via transgenics from halophytes to glycophytes, enhancing their tolerance to salinity.

2 Chapter Objectives

This chapter describes the classification and distribution of halophytes and considers different strategies and mechanisms in use for tolerating saline environments.

The chapter goes on to discuss glycophyte growth under salt stress and the potential tolerance strategies that can be used to cope with salinity and improve crop productivity, giving consideration to the various mechanisms that can help with adaptation to saline conditions. We discuss the role of plant hormones under salinity stress and the negative effects of salinity on glycophyte growth. At the end of the chapter we review the economic benefits of halophytes as a new source of profitable crops in areas influenced by saline conditions.

Throughout the chapter we mention more than 120 species of halophytes and discuss approximately 165 references.

3 Types of Salinity

Saline soil is defined as having a high concentration of soluble salts, high enough to affect plant growth, that is, having an electrical conductivity (ECe) of 4 dS/m or more (Scudiero et al. 2017).

Salt concentration in soil is measured in terms of ECe, representing the electrical conductivity of 'saturated paste extract.'

Salinity is considered a limiting factor for most plants, however, many crops are affected by soil with an ECe less than 4 dS/m, particularly in dry regions. In conditions where high evapotranspiration reduces soil moisture, actual salinities in soils that normally have an ECe of 4 dS/m may reach 8 dS/m.

A. Natural Salinity

Natural salinity occurs in arid and semiarid zones under rainfall of less than 300 mm, therefore, under these conditions improving the salt tolerance of crops is a useful approach.

B. Secondary or Human-induced Salinity

Human activities, such as excessive mineral fertilization, improper irrigation, and usage of poor quality irrigation water, may cause soil salinization and change the hydrological balance in the soil between precipitation (irrigation or rainfall) and water uptake by crops. The most common causes of such an imbalance are:

(i) Deforestation and the replacement of perennial vegetation with annual crops.
(ii) Irrigation schemes using poor quality water containing salts.
(iii) Insufficient drainage.
(iv) Excessive fertilization.

Due to human activities, particularly in arid and semiarid regions, the water balance is often changed because of the different water consumption of different annual crops, excessive irrigation, and poor drainage.

Excessive irrigation raises the water table and transfer salts from the subsoil to the upper layer—near to the root zone. With such a continued rising of the water table toward the soil surface, water evaporates, leaving salts behind. These form a salt scald that can penetrate back through the layers of the soil into watercourses.

Under irrigated conditions, salts increase in the soil due to irrigation water including soluble salts. Even good quality water, containing only 200–500 mg/l of soluble salt, can actually add about 0.5 t of salt to soil per 1000 m^3. Now, plants need 6,000–10,000 m^3 annually per hectare, so irrigation water adds noticeable amounts of salt—each hectare receives about 3–5 t of salt annually, which accumulates in the soil near the root zone. Consequently, such soils need annual leaching to remove significant quantities of salt from the root zone.

Stress associated with salinity is caused by various physiological and biochemical disorders that delay the germination of crops, and reduce plant growth and crop productivity (Ungar 1996). It is therefore imperative to reduce gaps in crop yield by increasing tolerance to salinity, thereby ensuring food security for our increasing population as well as benefiting poor farmers the world over.

In this context, crop stress tolerance is defined in terms of yield stability under abiotic stress conditions. However, yield losses due to abiotic stress vary depending on timing, intensity, and longevity of stress—tied to other abiotic stresses such as high light intensity and temperature.

Therefore, improving crop efficiency under conditions of salinity stress requires a complete understanding of the physiological and molecular basis of stress tolerance.

4 Halophytes

Halophytes have a remarkable ability to tolerate salinity conditions that would normally kill most other species. They can complete life cycles under salinity conditions of up to 200 mM NaCl or more (Flowers and Colmer 2008). Halophytes are native to saline habitats and are salt-resistant or salt-tolerant plants. It is important to note that saline soils generate an osmotic pressure in plant cells, thereby reducing the total yield of most crops (Breckle 1995). In addition, ion toxicity causes chlorosis and necrosis due to Na^+ accumulation that interferes with different physiological processes in plants.

Halophyte growth rates are stimulated at high salinity levels (150–300 mM NaCl), however, low salinity conditions inhibit their growth as in the cases of *Salicornioideas* and *Chenopodiaceae* (now included in *Amaranthaceae*) (Rozema and Schat 2013). Habitat-indifferent halophytes usually prefer to live in a salt-free soil but have the ability to cope with saline conditions (Cushman 2001). There is a significant difference between the growth responses of halophytes to increasing salinity—varying from impressive stimulation in the case of *Salicornia* spp. and inhibition in *Atriplex*.

It is generally accepted that K^+ ions are required by halophytes in a similar way to glycophytes, and that small quantities of K^+ in the cytoplasm of halophytes complete metabolic functions similar to those in glycophytes (Gorham 1995). There are clear differences between monocotyledonous and dicotyledonous halophytes both in their ability to use Na^+ and K^+ and their salt tolerance.

Dicotyledonous halophytes grow in high salinities, up to 15,000 ppm NaCl, however, higher salinity levels reduce growth. One example is *Salicornia* spp. (in the family Amaranthaceae) which is considered a genuine model for dicotyledonous halophyte plants, having a positive correlation between growth and salinity level up to 15,000 ppm NaCl. However, monocotyledonous halophytes grow slowly under salinity concentrations greater than 10,000 ppm NaCl. In addition, these plants are not stimulated at low to medium salinity levels. Salt tolerance in monocotyledonous halophytes is very important because cereals crops are monocotyledonous. Flowers and Yeo (1986) reported that there are differences between dicotyledonous and monocotyledonous halophytes in terms of their relative vacuolar volumes—being much bigger in dicotyledonous than monocotyledonous species.

4.1 Halophyte Distribution

For a long time, halophytes were defined as species that are adapted to permanent saline conditions. In the last decade, the term halophyte has gained a new definition, that is, as crops which can grow and reproduce in conditions of 200 mM NaCl (Flowers and Colmer 2008). Other research, Glenn et al. (1999), suggests a cutoff of 85 mM NaCl and suggests that there are as many as 6000 halophytes species.

Flowers (2014) identified more than 1500 species as salt tolerant and Saslis-Lagoudakis et al. (2014) identified about 1653 species as halophytes.

According to Flowers and Yeo (1986) there are two main groups of halophytes:

1. Chenopodiaceae, which contain over 380 halophytic species including (succulents, dicotyledonous herbs, and shrubs).
2. Poaceae, with more than 140 halophytic species—containing the major group of monocotyledonous halophytes.

Halophytes represent about 1% of world flora, existing in 37 of the 65 orders of plants. However, there are no complete halophyte families—some families, like Chenopodiaceae, contain high numbers of halophytes, in this case around 381 species.

At the genus level, there are many halophytic species, spread across some 500 genera with more than 71 families of flowering plants. Additionally, both halophytes and non-halophytes can be found in the same genus, such as *Glycine, Plantago*, and *Solanum*.

In this chapter, we consider around 120 genus and species (Table 1) from 35

Table 1 The major genus, species, and variety of halophytes considered in this chapter

Genus	Family	References
Suaeda maritime	Chenopodiaceae	Hajibagheri and Flowers (1989)
S. salsa	Chenopodiaceae	Zhao et al. (2002)
S. fruticosa	Chenopodiaceae	Hameed and Khan (2011)
S. monoica	Chenopodiaceae	Al-Said et al. (2017)
S. aegyptiaca	Chenopodiaceae	Guan et al. (2010)
S. vermiculata	Chenopodiaceae	Abd Elbar and Abd El-Maboud (2013)
S. pruinosa	Chenopodiaceae	Abd Elbar and Abd El-Maboud (2013)
Atriplex spongiosa	Chenopodiaceae	Storey et al. (1983)
Salicornia bigelovii	Chenopodiaceae	van Diggelen (1988)
A. nummularia	Chenopodiaceae	Mills et al. (1985)
A. farinose	Chenopodiaceae	Morsy (2008)
A. littoralis	Chenopodiaceae	Bylka (2004)
A. portulacoides	Chenopodiaceae	Redondo-Gomez et al. (2007)
A. spongiosa	Chenopodiaceae	Storey and Jones (1979)
S. bigelovii	Chenopodiaceae	Tajeddin and Behmadi (2019)
Anabasis setifera	Chenopodiaceae	Sukhorukov (2008)
A. aphylla	Chenopodiaceae	Mozaffarian (2001)
Salsola iberica	Chenopodiaceae	Schillinger (2007)
S. kali	Chenopodiaceae	Chinnock (2010)
S. tetrandra	Chenopodiaceae	Mahmoud et al. (2016)
Bassia dayphylla	Chenopodiaceae	Mo et al. (1997)
Halopeplis perfoliata	Chenopodiaceae	Ahmed and Lotfy (2015)
Camphorosma monspeliaca	Chenopodiaceae	Zhao et al. (2011)
Halostachys caspica	Chenopodiaceae	Hu et al. (2012)
Halimione	Chenopodiaceae	Kadereit et al. (2010)
Kochia scoparia	Chenopodiaceae	Dodd and Randall (2014)
Petrosimonia spp.	Chenopodiaceae	Grozeva et al. (2019)
Kalidium capsicum	Chenopodiaceae	Li and XiMing (2007)
Chenopodium quinoa	Chenopodiaceae	Mahmoud (2017)
C. album	Chenopodiaceae	Ohri (2015)
Kali komarovii	Amaranthaceae	Burundukovaa et al. (2017)
Arthrocnemum macrostachyum	Amaranthaceae	Ahmed and Lotfy (2015)
A. portulacoides	Amaranthaceae	Redondo-Gomez et al. (2007)
Halocnemum strobilaceum	Amaranthaceae	Amiri and Rasouli (2015)
S. tetrandra	Amaranthaceae	Khedr and Zahran (2002)

(continued)

Table 1 (continued)

Genus	Family	References
Haloxylon Persicum	Amaranthaceae	Alghanem (2018)
Puccinellia peisonis	Poaceae	Stelzer and Lauchli (2010)
Spartina townsendii	Poaceae	Koyro and Stelzer (1988)
S. anglica	Poaceae	Abideen et al. (2011)
Phragmites karka	Poaceae	Serag et al. (2002)
P. australis	Poaceae	Vaicekonyte et al. (2014)
Panicum turgidum	Poaceae	Hameed and Khan (2011)
Desmostachya bipinnata	Poaceae	Khan et al. (2009)
Distichlis spp.	Poaceae	Pearlstein et al. (2012)
Pennisetium spp.	Poaceae	van de Wouw et al. (1999)
Sporobolus ioclados	Poaceae	Gul and Khan (1998)
S. spicatus	Poaceae	Duvall et al. (2016)
Uniola paniculata	Poaceae	Franks et al. (2004)
Zizania latifolia	Poaceae	Sinha et al. (2016)
Aeluropus lagopoides	Poaceae	Ding et al. (2011)
A. sinenesis	Poaceae	Ding et al. (2011)
Imperata cylindrica	Poaceae	King and Grace (2000)
Achnatherum splendens	Poaceae	Li et al. (2019)
Halopyrum mucronatum	Poaceae	Siddiqui and Khan (2013)
Crypsis aculeate (L.)	Poaceae	Zhao et al. (2011)
Agrostis stolonifera	Poaceae	Kavut et al. (2014)
Elymus pycnanthus	Poaceae	Dizkirici et al. (2010)
Festuca rubra	Poaceae	Stukonis et al. (2015)
Phacelurus latifolius	Poaceae	Nam et al. (2018)
Bruguiera gynmorrhiza	Rhizophoraceae	Zhao et al. (2011)
Ceriops tagol	Rhizophoraceae	Liao et al. (2007)
Kandelia candel	Rhizophoraceae	Sheue et al. (2003)
Rhizophora apiculata	Rhizophoraceae	Gevana and Im (2016)
R. styosa	Rhizophoraceae	Gevana and Im (2016)
Juncus acutus	Juncaceae	Mohammed (1980)
J. rigidus	Juncuaceae	Ahmed and Lotfy (2015)
J. clayndrica	Juncaceae	Zahran (1993)
J. effuses	Juncaceae	Kaczmark-Derda et al. (2018)
Acacia tortilis	Fabaceae	Ahmed and Lotfy (2015)
Sesbania canabinia	Fabaceae	Rao and Gill (1995)
Alhagi sparsifolia	Fabaceae	Asghari et al. (2016)

(continued)

Table 1 (continued)

Genus	Family	References
A. marorum	Fabaceae	Asghari et al. (2016)
Zygophyllym album	Zygophyllaceae	Ghazanfar and Osborne (2015)
Z. coccineum	Zygophyllaceae	Ghazanfar and Osborne (2015)
Frankenia pulverulenta (L.)	Zygophyllaceae	Zhao et al. (2011)
Glycyrrhiza uralensis	Zygophyllaceae	Zhao et al. (2011)
Typha domingensis	Typhaceae	Khedr (1999)
T. elephantine	Typhaceae	Serag et al. (1999)
Tamarix chinensis	Tamaricaceae	Cui et al. (2010)
T. nilotica	Tamaricaceae	Abdelgawad (2017)
T. passerinoides	Tamaricaceae	Abdelgawad (2017)
Reaumria	Tamaricaceae	Zhang et al. (2016)
Aster tripolium	Asteraceae	Shennan et al. (1987)
Artemisia spp.	Asteraceae	Shukla et al. (2015)
Inula crithmoides	Asteraceae	Zurayk and Baalbaki (1996)
Limonium vulgare	Plumbaginaceae	van Diggelen (1988)
L. stocksii	Plumbaginaceae	Zia and Khan (2004)
L. axillare	Plumbaginaceae	Yasseen and Abu-Al-Basal (2008)
Spergularia marina	Caryophyllaceae	Lazof and Cheeseman (1988)
S. maritima	Caryophyllaceae	van Diggelen (1988)
Apocynum venetum	Apocynaceae	Thevs et al. (2012)
A. pictum	Apocynaceae	Thevs et al. (2012)
Thellungiella halophila	Brassicaceae	Wang et al. (2006)
T. salsuginea	Brassicaceae	Koch and German (2013)
Zostera marina	Zosteraceae	Koyro and Stelzer (1988)
Phyllospadix japonica	Zosteraceae	Tanaka et al. (2003)
Halodule spp.	Hydrocharitaceae	Osathanunkul et al. (2014)
H. spp.	*Hydrocharitaceae*	Osathanunkul et al. (2014)
Nitraria retusa	Nitrariaceae	Khedr and Zahran (2002)
Nitria tangutorum	Nitrariaceae	Zhao et al. (2011)
Leucophyllum pruinosum	Scrophulariaceae	Gándara and Sosa (2013)
L. pruinosum	Scrophulariaceae	Gándara and Sosa (2013)
Eleocharis uniglumis	Cyperaceae	Feizbakhsh and Naeemy (2011)
Triglochin maritima	Juncaginaceae	van Diggelen (1988)
Mesembryanthemum crystallinum	Aizoaceae	Bohnert and Cushman (2000)
Lumnitzera racemosa	Combretaceae	Ravikumar and Gnanadesigan (2011)

(continued)

Table 1 (continued)

Genus	Family	References
Acrostichum spp.	Pteridaceae	Castrejon-Varela et al. (2018)
Ruppia rostellata	Ruppiaceae	Davis and Tomlinson (1974)
Avicennia marina	Verbenaceae	Yasseen and Abu-Al-Basal (2008)
Plantago lanceolata	Plantaginaceae	Nichita et al. (2016)
Apocynum pictum	Apocynaceae	Thevs et al. (2012)
Acanthus spp.	Acanthaceae	Okoli et al. (2008)
Annona glabra	Annonaceae	Cochrane et al. (2008)
Asparagus persicus	Liliaceae	Zhao et al. (2011)
Elaeagnus omgustifolia	Elaeagnaceae	Tehranizadeh et al. (2016)
Ipomoea pestigridis (L.)	Convolvulaceae	Zhao et al. (2011)
Sonneratia alba	Lythraceae	Ragasa et al. (2015)
Xylocarpus granatum	Meliaceae	Zhao et al. (2011)
Cynanchum sibiricum	Asclepiadaceae	Podlipaev (1995)
Flagellaria indica	Flagellariaceae	Tomlinson and Posluszny (1977)
Orobanche amonena	Orobanchaceae	Athar et al. (2007)
Zannichellia palustris	Pteridaceae	Alwan (2006)
Pandanus tectorius	Pandanaceae	https://www.protreeworks.com.au/wp-content/uploads/download-ptectorius-pandanus-1.pdf

families (Chenopodiaceae with 28, followed by Poaceae with 23 species, Amaranthaceae with 7 species, Rhizophoraceae with 5 species, then Fabaceae, Juncaceae, Tamaricaceae, and Zygophyllaceae with 4 species each).

4.2 Classification of Halophytes

Different classifications of halophytes have been suggested based on various characteristics, such as the chemical composition of shoots (physiotypes; El-Khouly 2001), the ability to secrete ions (recreto-halophytes; Albert et al. 2000), the consequence of naturally saline habitats on physiological bases, in terms of ecological type, and life cycle (Breckle 2002). Also, some halophytes, like succulent species (*Suaeda*, *Halimione*, *Petrosimonia*) are considered extreme halophytes, adapted to high salinity conditions. Other species, such as *Atriplex*, *Bassia*, and *Camphorosma*, are not so closely tied to conditions of increased salinity, therefore, they can be classified as reversible halophytes.

There is always difficulty in terms of the distinct terminology associated with halophytes because their definition is still obscure.

Table 2 Halophyte growth forms

Growth form	Species	Family
Rhizomatous	*Juncus rigidus*	Juncaceae
	J. clayndrica	Juncaceae
	Typha domingensis	Typhaceae
	Cyperus laevigatus	Cyperaceae
Leafless succulent shrubs	*Suaeda aegyptiaca*	Chenopodiaceae
	Salsola tetrandra	Chenopodiaceae
	Arthrocnenuim macrostachyum	Amaranthaceae
Leafy non-succulent shrubs	*Tamarix nilotica*	Tamaricaceae
	Alhagi graecoriem	Fabaceae
Stoloniferous	*Phragmites australis*	Poaceae
	Aeluropus lagopoides	Poaceae
Non-succulent herbs	*Tripteris* amplexicaulis	Asteraceae
Succulent herbs	*Nitraria retusa,*	Nitrariaceae
Leafless trees	*T. aphylla*	Tamaricaceae
Succulent subshrubs	*Aptenia cordifolia*	*Aizoaceae*

4.3 Halophyte Growth Forms

There are many growth forms that can be distinguished in halophytic species, these reflect the high levels of stress associated with saline land (Table 2); also, some species modify their organs to store the water (El-Khouly and Khedr 2000), some produce rhizomatous growth forms, leafless succulent shrubs, leafy non-succulent herbs, and succulent herbs.

4.4 Halophyte Life Habits

Halophytes species are identified according to life forms (Table 3): annual plants (A) such as *Crypsis aculeate*, *Zygophyllum simplex*, and *C. quinoa*; annual herbs (AH) like *Acanthus* spp.; grasses (G) such as *E. pycnanthus* and *Astragalus orbiculantus*; perennials (P) such as *Suaeda fruticosa, Limonium vulgare*, *Inula crithmoides*, and *Zygophylium coccineum*; perennial vines (PV) like *Cynanchum sibiricum*; perennial shrubs (PSh) such as *A. portulacoides* and *Acacia tortilis*; perennial succulent shrubs

Table 3 Halophyte life habits

Life habits	Species	Family
Annual	*Crypsis aculeate*	Poaceae
	Ipomoea pestigridis	Convolvulaceae
	Apocynum venetum	Apocynaceae
	Frankenia pulverulenta	Zygophyllaceae
	Artemisia spp.	Asteraceae
	Chenopodium album	Chenopodiaceae
Annual herbs	*Acanthus* spp.	Acanthaceae
Grasses	*Elymus pycnanthus*	Poaceae
	Halophila spp.	Hydrocharitaceae
	Phacelurus latifolius	Poaceae
	Phyllospadix japonica	Zosteraceae
	C. quinoa	Chenopodiaceae
Perennials	*Limonium vulgare*	Plumbaginaceae
	Suaeda fruticosa	Chenopodiaceae
	Inula crithmoides	Asteraceae
	Zygophylium coccineum	Zygophyllaceae
Perennial vines	*Cynanchum sibiricum*	Asclepiadaceae
Perennial shrubs	*Atriplex portulacoides*	Amaranthaceae
	Acacia tortilis	Fabaceae
	Tamarix nilotica	Tamaricaceae
Perennial succulent shrubs	*Halopeplis perfoliata*	Chenopodiaceae
Perennial rhizomes	*Apocynum pictum*	Apocynaceae
Parasitic perennial plants	*Orobanche amonena*	Orobanchaceae
Submerged plants	*Zannichellia palustris*	Potamogetonaceae
Shrubs	*Lumnitzera racemosa*	Combretaceae
	Leucophyllum pruinosum	Scrophulariaceae

(continued)

Table 3 (continued)

Life habits	Species	Family
Subshrubs	*Halocnemum strobilaceum*	Amaranthaceae
	Kalidium capsicum	Chenopodiaceae
Ferns	*Acrostichum* spp.	Pteridaceae
Vines	*Flagellaria indica*	Flagellariaceae
Trees	*Rhizophora styosa*	Rhizophoraceae
	Xylocarpus granatum	Meliaceae

(PSuSh) like *Halopeplis pcrfoliata* from the Chenopodiaceae family; parasitic perennial plants (PPP) like *Orobanche amonena*; submerged plants (SuP) like *Zannichellia palustris*; shrubs (Sh) like *Lumnitzera racemose*; subshrubs (SSh) like *Halocnemum strobilaceum*; ferns (F) like *Acrostichum* spp.; vines (V) like *Flagellaria indica*; and trees (T) such as *Rhizophora styosa* and *Xylocarpus granatum*.

Halophyte species are a diverse group of plants, with varying degrees of actual salt tolerance (Le Houerou 1993), consisting of two groups that respond differently to increasing soil salinity:

(a) *Obligate halophytes*, also called euhalophytes, include varieties that need constantly high levels of salt for optimal growth rates (Flowers and Colmer 2015). They grow even better under high saline conditions (about 200–300 mM NaCl which is 35–55% seawater salinity) and therefore could serve as model organisms enabling the understanding of mechanisms associated with high salt tolerance found in Salicornioideae and some other genera in Amaranthaceae *and* Suaedoideae (Zhao et al. 2011), e.g., *H. strobilaceum, I. crithmoides, Arthrocemum macrostachyum,* and *Atriplex portulacoides. Juncus rigidus* is tolerant to aridity and soil salinity conditions (Zahran 1993).

(b) *Facultative halophytes* are plants with the ability to grow in salty soils. However, their growth is improved in soils that are devoid of salt or under low salt levels, e.g., *Chenopodium quinoa, Kali komarovii,* and *Aster tripolium* (Adolf et al. 2013; Burundukovaa et al. 2017; Barbour 1970; Grigore and Toma 2010).

Cushman (2001) used the presence of salt glands to differentiate between obligate, facultative, and habitat-indifferent halophytes. Both obligate and facultative halophytes depend on the availability of salt which is required for their growth and tolerance to sodium salts.

Grigore and Toma (2010) classified halophytes thus:

(a) *Extreme halophytes* are plants that are adapted to saline environments and grow exclusively in high-salinity conditions. Examples include some species in the Chenopodiaceae family*, S. Halimione,* and *Petrosimonia*, which are well-adapted extreme halophytes, moreover, the habitat of these halophytes may be irreversible or reversible.

(b) *Mesohalophytes* are species with intermediary anatomical adaptations between extreme halophytes and glycophytes.

However, halophytes such as *Atriplex*, *Bassia*, and *Camphorosma* are not strictly related to increased salinity and are therefore often classified as reversible halophytes.

Generally, different investigations have ascertained that water and different nutrients are important factors for halophyte growth, besides the availability of salts in natural saline conditions (Grigore et al. 2012).

4.5 Physiological Types of Halophytes

According to Breckle (1995) halophytes can be classified, based on salt uptake and storage, into three physiological types:

1. Salt-secreting halophytes (exorecretohalophytes and endorecretohalophytes).
    a. Exorecretohalophytes secrete salt from epidermal salt glands in the leaves and stems, including species of *Limonium*, *Tamarix*, *Spartina*, *Avicennia*, and *Frankenia*.
    b. Endorecretohalophytes release salt from epidermal bladders on the leaves, like species of *Atriplex* and *Chenopodium*.
2. Euhalophytes (salt-diluting halophytes). These plants have succulent leaves and stems (species with succulent leaves include *Suaeda, Salsola, Petoroimonia,* and *Reaumria*; species with succulent stems include *Halostachys, Halocnemum, Ralidium, Kalidium,* and *Salicornia*).
3. Pseudohalophytes (salt-excluding halophytes). These include species of *Phragmites, Artemisia,* and *Juncus.*

4.6 Ecological Types

Halophytes can be divided into three ecological types:

1. *Mesohalophytes*. These normally grow in salt meadows, salt shrublands, and other semiarid habitats with moderate soil water content. Mesohalophytes include species of *Atriplex, Limonium, Aeluropus, Apocynum,* and *Suaeda.*
2. *Xerohalophytes*. These grow mainly in salt deserts like the deserts of Xinjiang in China. Xerohalophytes includes species of *Haloxylon, Tamarix, Kalidium, Halocnemum, Halostachys, Suaeda, Glycyrrhiza, Zygophyyllum,* and *Anabasis aphylla.*
3. *Hydrohalophytes*. These are divided into two groups:

3.1. *Emergent halophytes* that grow in salt marshes and coastal marsh habitats like mangroves. Emergent halophytes include species of *Bruguiera, Kandelia, Rhizophora, Acanthus, Lumnitzera, Nypa, Sonneratia, Acrostichum, Triglochin,* and *Salicornia.*
3.2. *Submerged halophytes* grow beneath the sea surface and include *Ruppia rostellata, Zostera merina, Phyllospadix japonica, Cymodocae* spp., *Halodule* spp, and *Halophila* spp. (Ramadan 1998).

4.7 Salinity Tolerance in Halophytes

All halophytes display a common need to regulate cellular Na^+, Cl^-, and K^+ concentrations as they adjust to external water potential. Halophytes require a large amount of Na^+ and Cl^- in their vacuoles and produce a wide range of compatible solutes for their cytoplasm. Differences exist between various halophytes in their usage of Na^+ and K^+ to regulate cellular ion concentrations, with a clear difference in the balance of Na^+ and K^+ used by different halophytes, particularly in monocotyledonous halophytes which have high values of net SK–Na. Also, halophytes adjust to external water potential, with variation between species with salt glands and without, and between flood tolerant and flood sensitive plants.

4.8 Salt Tolerance Strategies in Halophytes

Halophytes utilize different strategies to withstand, and complete life cycles within, highly saline conditions in which most plants would die. These strategies depend on adaptive changes at the molecular, anatomical, and morphological levels.

Halophytes survive in a highly salty environment through a combination of strategies—some species even grow better in the presence of high NaCl concentrations, that is, up to 15,000 ppm. Halophytes from the Chenopodiaceae family have the ability to accumulate high levels of Na^+ and Cl^- in the vacuoles of their leaves and use these salts for osmotic adjustment to the low water potential in saline environments (Tanveer and Shah 2017). For example, quinoa (*C. quinoa*) can cope with salt levels as high as those found in sea water (Hariadi et al. 2011). Also, there are considerable differences in salt tolerance mechanisms between dicotyledonae, such as Salicornioideae, Chenopodioideae, and Suaedoideae, and monocotyledonae, like Poaceae.

In general, most halophytes use two main strategies to adapt to salinity conditions:

(a) Salt tolerance.
(b) Salt avoidance.

Besides that, halophytes use conventional habits to defend against and tolerate salinity stress, such as an accumulation of compatible solutes, such as proline, betaine, glycine, soluble sugars, and other inorganic ions, as well as polyamine production (Lokhande and Suprasanna 2012; Patel et al. 2016).

4.9 Physiology Basics of Salinity Tolerance in Halophytes

The basic physiological requirements for salinity tolerance include:

I. Growth.
II. Osmotic adjustment.
III. Ion compartmentation.
IV. Compatible solutes.

4.10 Halophyte Mechanisms for Salt Tolerance

Halophytes have evolved several complicated mechanisms to meet their requirements for survival under harsh conditions, including regulating ion concentrations in their cells or forcing excessive salt out of their tissues. Here are some such mechanisms:

(a) *Salt exclusion.* This prevents the entry of salts into the vascular system by exclusion of salt around the roots, e.g., *P. karka* and *Panicum turgidum.*
(b) *Salt elimination.* In some species, such as coastal grasses and mangroves, salt-secreting glands and hairs actively eliminate salts and aid the maintenance, over time, of relatively constant shoot ion concentrations, thus keeping the salt concentration in the leaves and shoots beneath a certain threshold (Shabala et al. 2014).
(c) *Salt succulence.* By increasing the storage volume of cells gradually with the uptake of salt, salt concentrations can be kept reasonably constant for extended periods (e.g., in *Suaeda* and *Arthrocnemum* species)—high leaf–stem succulence to dilute Na^+ concentrations in tissues was considered by Hameed and Khan (2011).
(d) *Salt redistribution.* Na^+ and Cl^- can be readily translocated in the phloem so that high concentrations arising in actively transpiring leaves can be redistributed throughout the plant (Slama et al. 2015).
(e) *Salt accumulation.* Excessive salt accumulates in old leaves which drop, thereby removing excess Na^+ tissue from a plant, e.g., *Atriplex* species.

However, other species use ion accumulation and control their growth to cope with salinity.

Nevertheless, it is important to remember that most halophytes do not utilize glands or external bladders to modulate their tissue ion concentrations.

Salinity tolerance in halophytes is dependent on molecular, physiological, and morphological bases.

At the molecular level halophytes impart salt tolerance by regulating stress-responsive genes through abscisic acid (ABA)-dependent or ABA-independent regulation mechanisms, generally, salt tolerance in a halophyte is a complex process that includes the interactions of numerous physiological responses directed by various genes and gene products.

Generally, defense mechanisms in halophytes include changes in ion homeostasis, the formation of osmoprotectants, activation of crosstalk genes, induction of antioxidants, and the development of salt glands or bladders (Himabindu et al. 2016; Cushman and Bohnert 2000).

Salt avoidance adaptation strategies include:

1. Secretion.
2. Shedding.
3. Succulence.

1. *Secretion* is a complex mechanism. Some halophytes are capable of excreting excess salts via salt hairs or salt glands—other halophytes can excrete excessive liquid salt that turns into visible crystals on leaf surfaces. Other species like *Mesembryanthemum crystallinum* accumulate excessive salt ions in bladder cells.
2. *Shedding* of old leaves to avoid toxicity happens in some halophytes under high salt concentrations.
3. *Succulence* is a strategy used by halophytes that are very tolerant to high-salinity conditions. This mechanism involves increasing the storage volume of the cell with increased uptake of salts, and provides the ability to accumulate salt ions in succulent leaves without secretion via salt glands, e.g., *Suaeda* genus (Rozema 1995; Aslam et al. 2011).

5 Glycophytes

According to Cheeseman (2015) a glycophyte is a species which has evolved by adaptation under natural selective pressures in ecosystems with low soil sodium levels and, at the same time, maintains low sodium levels in its aerial tissues—mainly its leaves.

The growth of glycophytes decreases when exposed to salinity stress. Sometimes this results in plant death due to adverse water relations, membrane permeability, and the effects of ions on the metabolism. Glycophytes use different strategies to improve their growth under salinity stress, like producing salt-tolerant genotypes in different crops. In addition, many other mechanisms have been used to adapt to saline environments, such as osmotic regulation, the production of reactive oxygen species, toxic ion uptake, ion homeostasis, CO_2 assimilation, and antioxidant defenses (Abobatta 2018; Ashraf 1994; Kingsbury and Epstein 1984; Shannon and Grieve 1999).

5.1 The Effect of Salinity on Glycophytes

Previous investigations (Carvajal et al. 1999; Yeo 1998; Grattana and Grieveb 1999) have reported the following main effects of salinity on plant growth:

(a) A reduction in the osmotic potential of the soil solution that decreases available water for the plant.
(b) A degradation in the physical structure of the soil, such as water permeability and soil aeration.
(c) An increase in the concentration of toxic ions, such as Na^+ and Cl^-, which inhibit plant metabolism and disturb plant nutrition.

5.2 Salinity Tolerance in Glycophytes

Under saline conditions plants need to maintain a high water content to grow and survive, therefore, plants continue to uptake water and exclude a large amount of salt, particularly Na^+ and Cl^- ions, in the soil solution to avoid increasing salt concentrations to levels dangerous to plant organs. Roots transport only 2% of salts to the shoots, excluding 98% of the salts in the soil solution. It is worth noting that there are differences between species with contrasting rates of Na^+ uptake—Munns (2005) reported that durum wheat and rice exclude about 94% of soil Na^+ from transpiration streams.

The three most widely grown economic crops in the world are wheat, rice, and maize. Wheat is most salt tolerant (Rozema et al. 2014), followed by maize, and then rice which is considered most sensitive.

1. Wheat can be grown in saline environments with up to 150 mM NaCl (15 dS/m). However, supplemental irrigation or rainfall are required during critical stages, particularly pollen meiosis and fertilization, to achieve appropriate yields.
2. Maize tolerates salinity less than wheat, but more than rice in terms of salt sensitivity (Fig. 1).
3. Rice is the most salt sensitive.

Variation in glycophyte salt tolerance can be classified as follows:

(a) Salt-tolerant crops include barley (*Hordeum vulgare*), cotton (*Gossypium hirsutum*), sugarbeet (*Beta vulgaris*), and wheat (*Triticum aestivum*) (Rozema et al. 2014).
(b) Moderately salt-tolerant crops can tolerate salinities from 5.0 to 6.8 dS/m and include soybean (*Glycine max*), cowbean (*Vigna unguiculata*), and millet (*Sorghum bicolor*).
(c) Moderately salt-sensitive crops include sugarcane (*Saccharum officinarum*), corn (*Zea mays*), broad beans (*Vicia faba*), Egyptian clover (*Trifolium alexandrium*), cucumber (*Cucumis sativus*), and lettuce (*Lactuca sativa*).

Fig. 1 Zea mays cultivation under salinity conditions in northern Egypt in the 'Sahl EL-Tina' area (2019)

(d) Salt-sensitive crops cannot tolerate salinities greater than 3 dS/m, including rice (*Oryza sativa*), potato (*Solanum tuberosum*), beans (*Phaseolus vulgaris*), carrots (*Daucus carota*), grapefruit (*Citrus paradise*), oranges (*Citrus sinensis*), and plums (*Prunus domestica*).

(e) Very sensitive crops include strawberry plants (*Ananas fregeria*). Such crops are unable to tolerate salinities of 1 dS/m.

Glycophytes can tolerate high salt concentrations because of the different mechanisms they have for withstanding adverse water relations induced by salinity or reducing the movement of toxic ions to the shoots (Munns and Tester 2008).

6 Salt Avoidance

Salt secretion is considered an avoidance strategy used to control and regulate salt content in plant organs, particularly in the leaves by specific glands scattered on their surfaces (Yeo 1998). By increasing secretion from salt glands the plant decreases its salt content in mesophyll cells in the leaves and accordingly improves photosynthetic activity (Atkinson et al. 1967).

7 Salt Exclusion

Under saline conditions, some glycophyte varieties, such as rice, white clover, and lucerne, accumulate Na^+ or Cl^- ions in their leaves; this characteristic has been used in breeding programs to improve productivity under conditions of stress.

There is a positive correlation in various glycophyte species between accumulating Na^+ or Cl^- ions in tissues and salinity tolerance as a genetic heritage, e.g., durum wheat (*T. durum*) has the ability to uptake Na^+, but cannot absorb Cl^-.

However, in other species the ability to tolerate salinity is correlated with Cl^- exclusion by accumulating Cl^- ions in the leaves and retaining Na^+ ions in the roots in exchange for K^+ ion uptake—the absorption of K^+ and the K^+/Na^+ balance is considered a result of genetic differences between species. There are various factors that affect plant tolerance including physiological, morphological, and molecular factors. In addition, the interaction of these factors also affects salinity tolerance as does the different responses to factors by genes.

Plants use different strategies to maximize growth and productivity under saline conditions. Different mechanisms are also being employed to improve salinity and increase plant growth under conditions of high salinity, such as:

a. *Excretion of salt ions*. Some plants like *Colocasia esculenta* have the ability to withstand salinity conditions through the exuding of salt on the surface of the plant (Chand and Pearson 1998).
b. *Production of salt-tolerant genotypes*. The availability of salt-tolerant cultivars requires efficient rapid multiplication techniques to enable effective distribution of planting material for farmers. Tissue culture is considered an effective technique that could be used for the rapid multiplication of plants (Barhoumi et al. 2007).
c. *Salt sequestration*.

8 The Role of Plant Hormones in Salinity Stress

It has been well reported that salinity stress decreases plant growth and reduces productivity by decreasing photosynthesis. This reduction in plant growth under saline environments is because of the action of specific ions on the metabolism. In conditions of adverse water, salinity reduces plant growth due to its negative effect on the metabolism (Rozema et al. 2014).

Plant hormones are active members of signal compounds involved in the induction of plant stress responses (Pedranzani et al. 2003). Salinity stress affects levels of phytohormones and reduces plant growth. Previous investigations have reported that cytokinin and gibberellic acid (GA) content decreased in plant tissues under saline conditions, while, ABA content increased.

Levels of different growth regulators, such as gibberellic acid (GA), ABA, and indole acetic acid (IAA), play an important role in the metabolism of plants under saline conditions and affect water relations and membrane permeability (Morgan 1990).

In general, hormone mobility in the xylem sap correlates with plant tissue response to salinity—an increased movement of hormones appear related by 'root-to-shoot communication' (Karmoker and Van Steveninck 1979). Salinity may influence hormone balance by affecting plant growth and development, therefore, a reduction in plant growth under conditions of stress could be an outcome of an altered hormonal balance (Davies et al. 1994).

8.1 Gibberellic Acid

It is well documented that GA3 accumulates rapidly in plants when exposed to both biotic and abiotic stress (Prakash and Prathapasenan 1990).

Plants use various types of phytohormones to alleviate the toxic effects of salinity, of these, GA3 has been reported to be helpful in enhancing the growth of different crops (such as wheat and rice) under salinity stress. In addition, GA3 application was found to improve the seed germination of wheat under saline conditions (Kumar and Singh 1996).

Nayyar et al. (1995) reported that wheat seeds treated with 20 mg/l GA3 had the highest percent germination, showing that GA3 application could improve salinity tolerance in crop plants grown under saline condition.

The Role of Gibberellic Acid

1. Enhancing the growth of various crops under salinity stress (Nayyar et al. 1995).
2. Improving wheat seed germination under saline conditions (Lehmann et al. 1995).
3. Exogenous treatment of GA3 increases the tolerance of crops to salinity (Kumar and Singh 1996).

8.2 Abscisic Acid

Salt stress stimulates the synthesis of ABA and is correlated to stomatal reactions. Different research has shown an increases in the endogenous ABA concentration of plant tissues under salt stress (He and Cramer 1996, in the leaf tissue of *Brassica*; He and Cramer 1996; Montero et al. 1998, in *Phaseolus* and *Z. mays*). It has also been noted that ABA concentration increases in the roots and maintains root growth (Cramer and Quarrie 2002; Lachno and Baker 1986).

Stress responses of root and shoot tissues also appear to be coordinated by increased amounts of hormones moving in the xylem sap by 'root-to-shoot' communication (Montero et al. 1997).

Jeschke et al. (1997) reported that inhibition of leaf growth and decreased leaf conductance correlated with increased ABA concentration in the xylem. It is clear that ABA acts as the root-to-shoot stress signal. They found increased ABA concentrations in the leaves of mature *Ricinus*, under conditions of 128 mM NaCl concentrations, representing significant evidence that ABA acts as the root-to-shoot stress signal. Also noted was that barley could tolerate salinity by increasing its ABA concentrations in the roots—inducing ion accumulation in the vacuoles as an adaptation to salt stress.

Montero et al. (1997) reported that salt-induced ABA mediated the inhibition of leaf expansion and limited the accumulation of Na^+ and Cl^- in the leaves. To date, the role of ABA as a signal of root stress is not completely understood.

Role of Abscisic Acid

1. Abscisic acid delays the deleterious effect of NaCl on plant growth (Munns and King 1988).
2. Improves ionic stress tolerance (Amzallag et al. 1990).
3. Limits the accumulation of Na^+ and Cl^- in the leaves.

8.3 *Indole Acetic Acid*

To date, little information is available about the relationship between salinity stress and hormone levels in plants. It is known that IAA is important in controlling plant growth by:

1. Controlling cell elongation.
2. Controlling vascular tissue development.
3. Controlling apical dominance.

Increased IAA content correlates with decreased growth under saline conditions and the quantity of IAA in plant tissues relative to those of ABA, though, little information seems to be available on the relationship between salinity stress and IAA levels in plants (Wang et al. 2001). The variations in IAA content under stress conditions appear to be similar to those of ABA, and increased levels of IAA are considered some kind of response to salinity stress and correlate with reduced growth (Ribaut and Pilet 1991).

Increased salinity levels appear to decrease IAA concentration in rice leaves (Prakash and Prathapasenan 1990; Nilsen and Orcutt 1996). Additionally, under saline conditions IAA concentration in tomato plants (*S. lycopersicum*) reduced to about 75% of regular levels (Nilsen and Orcutt 1996), therefore, an exogenous application of IAA could counter stress conditions for various plants.

Generally, further research needs to be conducted to understand the real mechanisms behind IAA in plant tissues under stress.

9 Negative Effects of Saline Conditions on Plant Growth

Increased salinity drastically reduces the growth of glycophytes—the number of leaves and roots produced by different varieties were adversely affected when grown in saline conditions. Very high salinity (5%) negatively affected leaf production and had adverse effects on rooting in taro (*C. Esculenta* L.) varieties (Vaurasi and Kant 2016).

Salt in the soil solution (osmotic stress) has negative effects on the growth of glycophytes, as follows:

1. Decreases leaf growth.
2. Accelerates the senescence of old leaves.
3. Inhibits root extending.
4. Reduces stomatal conductance.
5. Decreases photosynthesis.
6. Produces photoinhibition and oxidative stress.
7. Disturbs plant nutrition due to excessive sodium in the soil solution.

Average leaf death is a major factor for plant survival over time under saline conditions. Concurrently, the rate of new leaves produced depends on the water potential of the soil solution, therefore, if the rate of new leaves produced is greater than the rate old leaves die then there will be sufficient photosynthetic products for a plant to reproduce (i.e., flower and produce seeds).

However, if the rate of leaf death exceeds the rate of new leaves produced, then the plant may not survive to complete its life cycle.

Generally, salinity reduces growth rate, resulting in fewer and smaller leaves with reduced stature.

10 Influence of Saline Conditions on Growth Reduction

a. *Osmotic phase.* This is basically water stress and appears quickly because of salt outside the roots. During this phase there is little variation between different genotypes. Reduction in the growth of salt-sensitive plants is due to the osmotic effect of the salt outside the roots and is also linked to the in inability of plants to prevent salt from reaching toxic levels in transpiring leaves.
b. *Growth reduction phase.* During this phase, plants take time to develop, and continue to grow new leaves but at a slower rate than the rate of leaf death. This

results from internal injury due to a continued salt uptake which is crucial to plant survival (Atkinson et al. 1967). During this phase, the ability of species to cope with Na^+ or Cl^- ions in the soil determines their tolerance—species with a low Na^+ uptake rate continued to grow, however, plants with a high Na^+ uptake rate produce only slight biomass and eventually die.

c. *Yield reduction.* Consistent yield reduction demonstrates a clearer response to salinity levels than vegetative growth response.
   Plants exposed to salinity correlate with poor growth stages. Correlations show a linear reduction in plant growth when exposed to salinity at an early stage—reducing leaf area, new leaf numbers, total biomass, and consequently decreased plant productivity with increased salinity.
   Under such conditions the response of plants is two-phase, depending on life cycle duration. Annual species try to complete their life cycles and produce seeds while leaf area is sufficient to make essential photosynthates. However, perennial plants can start a dormancy period to survive such conditions.

11 Salt Tolerance and the Spatial Distribution of Halophytes and Glycophytes

In glycophytes the initial growth reduction is due to the osmotic effect of salt outside the roots, as well as an inability to avoid an accumulation of toxic levels of salt in transpiring leaves. However, halophytes can prevent salt from reaching toxic levels in transpiring leaves due to their ability to excrete excess salts in different forms. Halophytes also have ability to discriminate between Na^+ and K^+. However, under low-salinity conditions, the relative growth rate (RGR) of glycophytes is greater than halophytes as is their germination rate, thereby enabling an earlier start for vegetative growth due to various elements which provide a competitive advantage for glycophytes in non-saline environments, such as phenological, ecological, and genetic factors. These afford extended growing season and late senescence (Rozema and Schat 2013).

Halophytes tolerate being flooded with salty water for greater periods of time than glycophytes. In addition, halophytes are able to grow in soils that have low water rations compared with glycophytes (Munns 1993).

Flowers et al. (2010) reported that there are gradual changes in degrees of salt tolerance from salt-sensitive plants (glycophytes) to halophytes (high salt tolerance).

12 Economic Benefits of Halophytes

Under conditions of increased salinity and desertification of arable land, halophytes represent a new source of economic crops in areas with saline soils or rising water tables that are transferring salts to the surface.

Halophytes offer great economic potential as sources of food, forage, oil, fuel, and medicine:

1. Halophytes can be used in saline agriculture to provide food stuffs such as starch, sugar, protein, or even oil materials including *Salicornia* spp.—considered priceless leafy vegetables—and *C. quinoa* which is a grain crop that can be used as a replacement for wheat grain in bread; other examples include *Suaeda* spp., *Pandanus tectorius, Acacia* spp, *Glycyrrhiza uralensis, Asparagus persicus,* and *Aster tripolium* (Mahmoud 2017; Tackholm 1974; Pearlstein et al. 2012).
2. *Halophytic forage.* Halophytes can be used as animal fodder, e.g., *Salsola iberica, Aeluropus sinenesis, Phacelurus latifolius, Achnatherum splendens, Atriplex* spp., *Kochia* spp., *Halostachys caspica,* and *Alhagi sparsifolia.* Cultivation areas used for halophytic fodder currently stand at 450 million ha worldwide and produce about 4–5 billion tons annually (Attia-Ismail 2015), e.g., *Atriplex* spp. produces 12.6–20.9 t/ha of biomass containing 9.9–19.5% protein growing in full salinity seawater.
3. There are more than 100 species of halophytes used as medicinal crops, e.g., *Lycium chinense, Glycyrrhiza uralensis, Apocynum venetum, Nitria tangutorum, J. acutus, Annona glabra, Elaeagnus angustifolia, S. monoica,* and *S. kall* (Cochrane et al. 2008; Ahmed and Lotfy 2015; Tehranizadeh et al. 2016).
4. Halophytes are considered a source of a fiber, e.g., *Ashmatherum splendens, A. venetum, P. pictum, Tamarix chinensis, Sesbania cannabinia, P. australis, Typha elephantina,* and *J. effuses* (Zahran 1993; Thevs et al. 2012).
5. Halophytes provide tannin materials, e.g., *Kandelia candel, Ceriops tagol, Bruguiera gynmorrhiza, R. styosa, R. apiculata, Xylocarpus grantum,* and *Lumnitzera racemosa* (Gevana and Im 2016).
6. Halophytes are also used as windbreaks and for fixing sandy soils, e.g., *Tamarix* spp., *Alhagi sparsifolia,* and *Elaeagnus omgustifolia* (Tehranizadeh et al. 2016; Asghari et al. 2016).
7. Some halophytes can reduce air pollution, e.g., *Tamarix* spp. (an tolerate SO_2, Cl_2, and HF) and *P. Australia* (absorbs NH_3) (Abdelgawad 2017).
8. Some species can serve as nutrient "scrubbers in polluted aquatic systems," thereby playing a key role in nutrient cycling, e.g., *Typha Species (T. domingensis* and *T. elephantine*) (El-Ameir 2013; Serag et al. 1999).
9. Halophytes have been examined for their potential as bioenergy crops, e.g., *Desmostachya bipinnata, P. karka, H. mucronatum,* and *P. turgidum* (Sharma et al. 2016; Abideen et al. 2011).

Halophyte cultivation on salty soils and salt marshes could provide different economic benefits and indirect environmental effects.

13 Conclusion

Salinity represents the abiotic stress that most affects crop productivity, particularly in arid and semiarid climates. Unfortunately, most economic crops are glycophytes that cannot tolerate high salt levels, even low concentrations (<40 mM NaCl). Therefore, in this chapter we have considered halophyte growth under saline conditions and reviewed different strategies and mechanisms used to tolerate high salt levels, particularly salt tolerance and salt avoidance strategies—considered the main strategies used to adapt to saline conditions. In addition, there are various complicated mechanisms used by halophytes to grow and reproduce under high-salinity conditions, such as eliminating salts using glands and hairs to keep salt concentrations in the leaves and shoots below a certain threshold, exuding excessive salt from plant tissues, or regulating ion concentrations in cells.

Therefore, there is still much to learn from halophytes that might facilitate the enhancement of glycophyte crops to cope with abiotic stresses, particularly high salinity. Therefore, biotechnology could play a vital role in transferring different halophyte genes to glycophytes to increase their salt tolerance and improve productivity.

References

Abd Elbar OH, Abd El-Maboud MM (2013) Anatomical and physiological responses of three species of *Suaeda Forssk.* ex Scop. under different habitat conditions. J Appl Sci Res 9(8):5370–5379

Abdelgawad AAM (2017) *Tamarix nilotica* (Ehrenb) bunge: a review of phytochemistry and pharmacology. J Microb Biochem Technol 9(1):544–553. https://doi.org/10.4172/1948-5948.1000340

Abideen Z, Ansari R, Khan MA (2011) Halophytes: potential source of ligno-cellulosic biomass for ethanol production. Biomass Bioenergy 35:1818–1822

Abobatta WF (2018) Some physiological mechanisms of salt tolerance in the glycophytes plant: overview. Acta Sci Agric 2(12):154–156

Adolf VI, Jacobsen S-E, Shabala S (2013) Salt tolerance mechanisms in quinoa (*Chenopodium quinoa* Willd). Environ Exp Bot 92:43–54

Ahmed FA, Lotfy RA (2015) Phytochemical evaluation of some selected medicinal plants growing wildly in Southeastern Egypt. Middle East J Appl Sci 5(04):1239–1246

Albert R, Pfundner G, Hertenberger G, Kastenbauer T, Watzka M (2000) The physiotype approach to understanding halophytes and xerophytes. In: Breckle S-W, Schweizer B, Arndt U (eds) Ergebnisse weltweiter ökologischer Forschung. Verlag Günter Heimbach, Stuttgart, Germany, pp 69–87

Alghanem SM (2018) Ecological and botanical diversity in *Haloxylon Persicum* community at Al-Qassim Region in Kingdom of Saudi Arabia. Am J Environ Prot 6(2):43–49

Al-Said MS, Siddiqui NA, Mukhair MA, Parvez MK, Alam P, Ali M, Haque A (2017) A novel monocyclic triterpenoid and a norsesquaterpenol from the aerial parts of *Suaeda monoica Forssk.* ex J.F. Gmel with cell proliferative potential. Saudi Pharm J 25:1005–1010

Alwan ARA (2006) Past and present status of the aquatic plants of the marshlands of Iraq. Marsh Bull 2:160

Amiri B, Rasouli B (2015) Effects of Salinity on Ion Exchanges in *Halocnemum strobilaceum* and *Halostachys caspica*. J Range Sci 5(1):72

Amzallag GN, Lerner HR, Poljakoff-Mayber A (1990) Exogenous ABA as a modulator of response of sorghum to high salinity. J Exp Bot 41:1389–1394

Asghari MH, Fallah M, Moloudizargari M, Mehdikhani F, Sepehrnia P, Bigard Moradi B (2016) A systematic and mechanistic review on the phytopharmacological properties of Alhagi species. Anc Sci Life 36(2):65–71. https://doi.org/10.4103/asl.ASL_37_16

Ashraf M (1994) Breeding for salinity tolerance in plants. CRC Crit Rev Plant Sci 13:7–42

Aslam R, Bostan N, Amen N, Maria M, Safdar W (2011) A critical review on halophytes: salt tolerant plants. J Med Plant Res 5:7108–7118

Athar M, Chaudhary AH, Yousaf Z, Shabbir SM (2007) Taxonomic reflections on the parasitic angiosperms of Pakistan. Phytologia 89(3):339

Atkinson MR, Findlay GP, Hope AB, Pitman MG, Saddler HDW, West KR (1967) Salt regulation in the mangrove *Rhizophora mucronata* Lam. and *Aegiceras annulata* R. Br Aust J Biol Sci 20:589–599

Attia-Ismail SA (2015) Nutritional and feed value of halophytes and salt tolerant plants. In: El Shaer, Squires (eds) Halophytic and salt tolerant feedstuffs: impacts on nutrition, physiology and reproduction of livestock, 126, vol 106. CRC Press, New York

Barbour MG (1970) Is any angiosperm an obligate halophyte? Am Midl Nat 84:105–120

Barhoumi Z, Wahbi D, Abderrazak S, Wided C, Chedly A (2007) Salt impact on photosynthesis and leaf ultrastructure of Aeluropus littoralis. J Plant Physiol 164(7):842–850. https://doi.org/10.1016/j.jplph.2006.05.008

Bohnert HJ, Cushman JC (2000) The ice plant cometh: lessons in abiotic stress tolerance. J Plant Growth Regul 19:334–346

Bray EA, Bailey-Serres J, Weretilnyk E (2000) Responses to abiotic stresses. In: Gruissem W, Buchannan B, Jones R (eds) Biochemistry and molecular biology of plants. American Society of Plant Physiologists, Rockville, MD, pp 1158–1249

Breckle S-W (1995) How do halophytes overcome salinity? In: Khan MA, Ungar IA (eds) Biology of salt tolerant plants. University of Karachi, Department of Botany, Pakistan, pp 199–213

Breckle S-W (2002) Salinity, halophytes and salt affected natural ecosystems. In: Lauchli A, Luttge U (eds) Salinity: environment-plants-molecules. Kluwer Academic, New York, NY, pp 53–77

Burundukovaa OL, Shuyskaya EV, Rakhmankulova ZF, Burkovskayaa EV, Chubar EV, Gismatullina LG, Toderiche KN (2017) Kali komarovii (Amaranthaceae) is a xero-halophyte with facultative NADP-ME subtype of C4 photosynthesis. Flora 227:25–35. Contents lists available at Science Direct Journal homepage: www.elsevier.com/locate/flora

Bylka W (2004) A new acylated flavonol diglycosides from *Atriplex littoralis*. Acta Physiol Plant 24:393–398

Carvajal M, Martinez V, Alcaraz CF (1999) Physiological function of water channels as affected by salinity in roots of paprika pepper. Physiol Plant 105:95–101

Castrejon-Varela A, Perez-García B, Mendoza-Ruiz A, Espinosa-Matias S (2018) Gametophyte morphology of *Acrostichum aureum* and *A. danaeifolium (Pteridaceae)*. Rev Biol Trop (Int J Trop Biol ISSN-0034-7744) 66(1):178–188

Chand H, Pearson MN (1998) Rapid vegetative multiplication in Colocasia esculenta (L) Schott (taro). Plant Cell Tissue Organ Cult 55:223–226

Cheeseman JM (2015) The evolution of halophytes, glycophytes and crops, and its implications for food security under saline conditions. New Phytol 206:557–570. https://doi.org/10.1111/nph.13217

Chinnock RJ (2010) Some observations on *Salsola* L. (*Chenopodiaceae*) in Australia. J Adelaide Bot Gard 24:75–79

Cochrane CB, Nair PKR, Melnick SJ, Resesk AP, Ramachandran C (2008) Anticancer effects of *Annona glabra* plant extracts in human leukemia cell lines. Anticancer Res 28:965–972

Cooper A (1982) The effects of salinity and waterlogging on the growth and cation uptake of salt marsh plants. New Phytol 90:263–275

Cramer GR, Quarrie SA (2002) Abscsic acid is correlated with the leaf growth inhibition of four genotypes of maize differing in their response to salinity. Funct Plant Biol 29:111–115
Cui B, Yang Q, Zhang K, Zhao X, You Z (2010) Responses of saltcedar (*Tamarix chinensis*) to water table depth and soil salinity in the Yellow River Delta, China. Plant Ecol 209:279–290. https://doi.org/10.1007/s11258-010-9723-z
Cushman JC (2001) Osmoregulation in plants: implications for agriculture. Am Zool 414:758–769
Cushman JC, Bohnert HJ (2000) Genomic approaches to plant stress tolerance. Curr Opin Plant Biol 3:117–124. https://doi.org/10.1016/S1369-5266(99)00052-7
Davies WJ, Tardieu F, Trejo CL (1994) How do chemical signals work in plants that grow in drying soil. Plant Physiol 104:309–314
Davis JS, Tomlinson PB (1974) A new species of *Ruppia* in high salinity in Western Australia. J Arnold Arbor 55(1):59–66. https://www.jstor.org/stable/43781927?seq=1#page_scan_tab_contents
Ding W, Liu J, Wang Y, Wu D, Chang C, Wang R (2011) Salinity stress modulates habitat selection in the clonal plant *Aeluropus sinensis* subjected to crude oil deposition. J Torrey Bot Soc 138(3):262–271
Dizkirici A, Kaya Z, Cabi E, Dogan M (2010) Phylogenetic relationships of Elymus L. and related genera (Poaceae) based on the nuclear ribosomal internal transcribed spacer sequences. Turk J Bot 34:467–478
Dodd J, Randall RP (2014) Eradication of kochia (*Bassia scoparia* (L.) A.J. Scott, Chenopodiaceae) in Western Australia. In: Thirteenth Australian weeds conference, pp 300–303. https://doi.org/10.1007/s10592-006-9151-8
Dudley LM (1994) Salinity in the soil environment. In: Pessarakli M (ed) Handbook of plant and crop stress. Marcel Dekker, New York, pp 13–30
Duvall MR, Fisher AE, Columus JT, Ingram AL, Wysocki WP, Burke SV, Clark SLG, Kelchner SA (2016) Phylogenomics and plastome evolution of the chloridoid grasses (*Chloridiodeae: Poaceae*). Int J Plant Sci 177(3):235–246
El-Ameir YA (2013) Spatial distribution and nutritive value of two *Typha* species in Egypt. Egypt J Bot 53(1):91–113
El-Khouly AA (2001) Plant diversity in the dry land habitats of Siwa Oasis, in the Western desert of Egypt. J Environ Sci Mansoura Univ 22:125–143
El-Khouly AA, Khedr AA (2000) Species diversity and phenology in the wetland vegetation of Sewa Oasis, in the Western desert of Egypt. Desert Inst Bull Egypt 50:239–258
FAO (2005) Global network on integrated soil management for sustainable use of salt-affected soils. FAO Land and Plant Nutrition Management Service, Rome, Italy. http://www.fao.org/ag/AGL/agll/spush
Feizbakhsh A, Naeemy A (2011) Volatile constituents of essential oils of *Eleocharis pauciflora* (Light) link and *Eleocharis uniglums* (Link) *J. A. Schultes* growing wild in Iran. Bull Chem Soc Ethiop 25(3):461–464
Flowers TJ (2014) eHALOPH Halophytes Database. [WWW document]. http://www.sussex.ac.uk/affiliates/halophytes/. Accessed 9 Nov 2014
Flowers TJ, Colmer TD (2008) Salinity tolerance in halophytes. New Phytol 179:945–963. https://doi.org/10.1111/j.1469-8137.2008.02531.x
Flowers TJ, Colmer TD (2015) Plant salt tolerance: adaptations in halophytes. Ann Bot 115:327–331. https://doi.org/10.1093/aob/mcu267
Flowers TJ, Yeo AR (1986) Ion relations of plants under drought and salinity. Austr J Plant Physiol 13:75–91
Flowers TJ, Galal HK, Bromham L (2010) Evolution of halophytes: multiple origins of salt tolerance in land plants. Funct Plant Biol 37:604–612
Franks SJ, Richards CL, Gonzales E, Cousins JE, Hamrick JL (2004) Multi-scale genetic analysis of *Uniola paniculata* (*Poaceae*): a coastal species with a linear, fragmented distribution. Am J Bot 91(9):1345–1351

Gándara E, Sosa V (2013) Testing the monophyly and position of the North American shrubby desert genus *Leucophyllum (Scrophulariaceae: Leucophylleae)*. Bot J Linn Soc 171:508–518

Gevana DT, Im S (2016) Allometric models for Rhizophora Stylosa griff. In dense monoculture plantation in the Philippines. Malays For 79(1 & 2):39–53

Ghazanfar S A, Osborne J (2015) Typification of *Zygophyllum propinquum Decne. and Z. coccineum L. (Zygophyllaceae)* and a key to Tetraena in SW Asia. Kew Bull 70:38. https://doi.org/10.1007/s12225-015-9588-3

Glenn EP, Brown JJ, Blumwald E (1999) Salt tolerance and crop potential of halophytes. Crit Rev Plant Sci 18:227–255

Gorham J (1995) Mechanisms of salt tolerance of halophytes. In: Choukr-Allah R, Malcom CV, Hamdy A (ed) Halophytes and biosaline agriculture. Ciheam, Instituto Agronomico Mediterraneo, Bary, Italy, Marcel Dekker INC., pp 31–54

Grattana SR, Grieveb CM (1999) Salinity-mineral nutrient relations in horticultural crops. Sci Hortic 78:127–157

Grigore M-N, Toma C (2010) Structuri secretoare de sa˘ruri la halofite. O abordare integrativa˘. Edit. Acad. Rom., Bucures¸ti

Grigore M-N, Toma C, Zamfirache M-M, Boscaiu M, Olteanu Z, Cojocaru D (2012) Ecological anatomy in halophytes with C4 photosynthesis: discussing adaptative features in endangered ecosystems. Carp J Earth Environ Sci 7(2):13–21

Grozeva N, Todorova M, Pavlov D (2019) Karyological and morphological variation within *Petrosimonia brachiata* Bunge in Bulgaria. Botanica Serbica 43(1):13–21. https://doi.org/10.2298/BOTSERB1901013G

Gul B, Khan MA (1998) Population characteristics of a coastal halophyte *Arthrocnemum macrostachyum*. Pak J Bot 30:189

Hajibagheri MA, Flowers TJ (1989) X-ray microanalysis of ion distribution within root cortical cells of the halophyte *Suaeda maritima* (L.) *Dum*. Planta 177(1):131–134

Hameed A, Khan MA (2011) Halophytes: biology and economic potentials. Karachi Univ J Sci 39(1 & 2):40–44

Hariadi Y, Marandon K, Tian Y, Jacobsen S-E, Shabala S (2011) Ionic and osmotic relations in quinoa (Chenopodium quinoa Willd.) plant grown at various salinity levels. J Exp Bot 62(1):185–193

Hasegawa PM, Bressan RA, Zhu J-K, Bohnert HJ (2000) Plant cellular and molecular responses to high salinity. Ann Rev Plant Physiol Plant Mol Biol 51:463–499

He T, Cramer GR (1996) Abscisic acid concentrations are correlated with leaf area reductions in two salt-stressed rapid cycling *Brassica* species. Plant Soil 179:25–33

Himabindu Y, Chakradhar T, Reddy MC, Kanygin A, Redding KE, Chandrasekhar T (2016) Salt-tolerant genes from halophytes are potential key players of salt tolerance in glycophytes. Environ Exp Bot 124:39–63. https://doi.org/10.1016/j.envexpbot.2015.11.010

Hu YZ, Zeng YL, Guan B, Zhang FC (2012) Overexpression of a vacuolar H^+-pyrophosphatase and a B subunit of H^+-ATPase cloned from the halophyte *Halostachys caspica* improves salt tolerance in Arabidopsis thaliana. Plant Cell Tiss Organ Cult 108:63–71. https://doi.org/10.1007/s11240-011-0013-9

Jeschke WD, Peuke AD, Pate JS, Hartung W (1997) Transport, synthesis and catabolism of abscisic acid (ABA) in intact plants of castor bean (*Ricinus communis* L.) under phosphate deficiency and moderate salinity. J Exp Bot 48:1737–1747

Kaczmark-Derda W, Ostrem L, Myromslien M, Brandsaeter LO, Netland J (2018) Growth pattern of *Juncus effusus* and *Juncus conglomeratus* in response to cutting frequency. Weed Research published by Wiley on behalf of European Weed Research Society. Int J Weed Biol Ecol Veg Manag. https://doi.org/10.1111/wre.12338

Kadereit G, Mavrodiev EV, Zacharias EH, Sukhorukov AP (2010) Molecular phylogeny of *Atripliceae* (*Chenopodioideae, Chenopodiaceae*): implications for systematics, biogeography, flower and fruit evolution, and the origin of C4 photosynthesis. Am J Bot 97(10):1664–1687

Karmoker JL, Van Steveninck FM (1979) The effect of abscisic acid on the uptake and distribution of ions in intact seedlings of *Phaseolus vulgaris* cv. Redland Pioneer. Physiol Plant 45:453–459

Kavut YT, Avcioglu R, Demiroglu G, Geren H, Kir B (2014) Adaptability of newly improved Creeping Bentgrass (*Agrostis stolonifera* L.) genotypes in a Mediterranean environment. Turk J Field Crop 19(1):19–24
Khan HA, Ayub CM, Pervez MA, Bilal RM, Shahid MA, Ziaf K (2009) Effect of seed priming with NaCl on salinity tolerance of hot pepper (*Capsicum annuum* L.) at seedling stage. Soil Environ 28(1):81–87
Khedr AA (1999) Floristic composition and phytogeography in a Mediterranean deltaic lake (Lake Burollos), Egypt. Ecologia Mediterr 25:1–11
Khedr AA, Zahran MA (2002) The salt marsh visitation of lake Bardawil, North Sinai, an overview. In: Proceedings of international symposium on "The optimum resources utilization in salt-affected ecosystems in arid and semi-arid regions", pp 339–345, 8–10 April, 2002, Cairo, Egypt
King SE, Grace JB (2000) The effects of gap size and disturbance type on invasion of wet pine Savanna by *cogongrass, Imperata cylindrica (Poaceae)*. Am J Bot 87(9):1279–1286
Kingsbury RW, Epstein E (1984) Selection for salt resistance in spring wheat. Crop Sci 24:310–315
Koch MA, German DA (2013) Taxonomy and systematics are key to biological information: *Arabidopsis, Eutrema (Thellungiella), Noccaea* and *Schrenkiella (Brassicaceae)* as examples. Front Plant Sci 4:267. https://doi.org/10.3389/fpls.2013.00267, https://www.ncbi.nlm.nih.gov/pmc/articles/PMC3728732/
Kotuby-Amacher J, Koenig K, Kitchen B (2000) Salinity and plant tolerance. https://extension.usu.edu/files/publications/publication/AG-SO-03.pdf
Koyro H-W, Stelzer R (1988) Ion concentrations in the cytoplasm and vacuoles of rhizodermis cells from NaCl treated *Sorghum, Spartina* and *Puccinellia* plants. J Plant Physiol 133:441–446
Kumar B, Singh B (1996) Effect of plant hormones on growth and yield of wheat irrigated with saline water. Ann Agric Res 17:209–212
Lachno DR, Baker DA (1986) Stress induction of abscisic acid in maize roots. Physiol Plant 68:215–221
Lazof D, Cheeseman JM (1988) Sodium and potassium compartmentation and transport across the roots of intact *Spergularia marina*. Plant Physiol 88:1274–1278
Le Houerou HN (1993) Salt tolerant plants for the arid regions of the Mediterranean isoclimatic zone. In: Leith H, El-Masoom A (eds) Towards the rational use of high salinity tolerant plants, vol 1. Kluwer Academic Publications, Dordrecht, The Netherlands, p 403
Lehmann J, Atzorn R, Bruckner C, Reinbothe S, Leopold J, Wasternack C, Parthier B (1995) Accumulation of jasmonate, abscisic acid, specific transcripts and proteins in osmotically stressed barley leaf segments. Planta 197:156–162
Li L, XiMing Z (2007) Germination strategies of two halophytes in Salt Desert of northwestern China. Sci China Ser D 50(Suppl 1):115–121. https://doi.org/10.1007/s11430-007-5004-7
Li Z, Jia G, Ni X (2019) The complete chloroplast genome sequence of *Achnatherum splendens (Pooideae)*, a high-quality forage grass in Northern China. Mitochondrial DNA Part B 4(1):1841–1843. https://doi.org/10.1080/23802359.2019.1612720
Liao P-C, Havanond S, Huang S (2007) Phylogeography of *Ceriops tagal (Rhizophoraceae)* in Southeast Asia: the land barrier of the Malay Peninsula has caused population differentiation between the Indian Ocean and South China Sea. Conserv Genet 8:89–98
Lokhande VH, Suprasanna P (2012) Prospects of halophytes in understanding and managing abiotic stress tolerance. In: Ahmad P, Prasad MNV (eds) Environmental adaptations and stress tolerance of plants in the era of climate change. Springer, New York, NY, pp 29–56
Mahmoud AH (2017) Production of Quinoa (*Chenopodium quinoa*) in the marginal environments of South Mediterranean Region: Nile Delta, Egypt. Egypt J Soil Sci 57(3):329–337
Mahmoud AH, Soliman MS, Metwally NS, Farrag AH, Elsharabasy FS, Arafa S, Ibrahim AMM (2016) Tremendous effect of *Salsola tetrandra* and *Salsola baryosma* on a liver toxicity using paracetamol overdose. Der Pharma Chemica 8 (18):117–126. www.derpharmachemica.com
Mills DM, Robinson K, Hodges TK (1985) Sodium and potassium fluxes and compartmentation in roots of Atriplex and oat. Plant Physiol 78:500–509

Mo W, Natori T, Jiang S, Nishimura N, Omasa K (1997) Responses of photosynthesis and water use to drought in two desert annual *Agriophyllum squarrosum* and *Bassia dasyphylla*. J Arid Land Stud 7(2):185–195

Mohammed NE (1980) Studies in the genus *Juncus* in Egypt. MSc thesis, Fac. Sci., Cairo University, Egypt

Montero E, Cabot C, Barcelo J, Poschenrieder C (1997) Endogenous abscisic acid levels are linked to decreased growth of bush bean plants treated with NaCl. Physiol Plant 101:17–22

Montero E, Cabot C, Poschenrieder CH, Barcelo J (1998) Relative importance of osmotic-stress and ion-specific effects on ABA-mediated inhibition of leaf expansion growth in *Phaseolus vulgaris*. Plant Cell Environ 21:54–62

Morgan PW (1990) Effects of abiotic stresses on plant hormone systems. In: Alscher RG, Cumming JR (eds) Stress responses in plants: adaptation and acclimation mechanism. Wiley-Liss, New York

Morsy A (2008) Ecophysiological studies on *Atriplex farinosa Forssk* under different habitat conditions. Aust J Basic Appl Sci 2:272–281

Mozaffarian V (2001) Flora of Iran (Vol 38; Chenopodiaceae). Institute of Forests & Rangelands Publications, Iran, p 38

Munns R (1993) Physiological processes limiting plant growth in saline soils: some dogmas and hypotheses. Plant Cell Environ 16:15–24. https://doi.org/10.1111/j.1365-3040.1993.tb00840.x

Munns R (2005) Genes and salt tolerance: bringing them together. Plant Physiol 167:645–663

Munns R, King RW (1988) Abscisic acid is not the only stomatal inhibitor in the transpiration stream of wheat plants. Plant Physiol 88:703–708

Munns R, Tester M (2008) Mechanisms of salinity tolerance. Ann Rev Plant Biol 59:651–681

Nam BE, Hong MG, Park HJ, Kim JG (2018) Soil factors determining the distribution of *Phragmites australis* and *Phacelurus latifolius* in upper tidal zone. J Ecol Environ 42(25):2–6. https://doi.org/10.1186/s41610-018-0086-z

Nayyar H, Walia DP, Kaistha BL (1995) Performance of bread wheat (*Triticum aestivum* L.) seed primed with growth regulators and inorganic salts. Indian J Agric Sci 65:116–122

Nichita C, Georgeta Neagu G, Cucum A, Vulturescu V, Bertesteanu SVG (2016) Antioxidative properties of *Plantago lanceolata* L. extracts evaluated by chemiluminece method. AgroLife Sci J 5(2):95–102

Nilsen E, Orcutt DM (1996) The physiology of plants under stress—abiotic factors. Wiley, New York, pp 118–130

Ohri D (2015) The taxonomic riddle of *Chenopodium album* L. complex (*Amaranthaceae*). Nucleus 58(2):131–134. https://doi.org/10.1007/s13237-015-0143-2

Okoli CO, Akah PA, Onuoha NJ, Okoye TC, Nwoye AC, Nworu CS (2008) *Acanthus montanus*: an experimental evaluation of the antimicrobial, anti-inflammatory and immunological properties of a traditional remedy for furuncles. J Appl Bot Food Qual 85:159–167

Osathanunkul M, Suwannapoom S, Singtonat S, Poomipoo N, Jampeetong A, Madesis P (2014) Rapid analysis for the identification of the seagrass *Halophila ovalis (Hydrocharitaceae)*. Afr J Biotechnol 14(8):649–656. https://doi.org/10.5897/AJB204.13855

Patel MK, Mishra A, Jha B (2016) Untargeted metabolomics of halophytes. In: Kim S (ed) Marine Omics: principles and applications. CRC Press, Boca Raton, FL, pp 309–325. https://doi.org/10.1201/9781315372303-18

Pearlstein SL, Felger RS, Glenn EP, Harrington J, Al-Ghanem KA, Nelson SG (2012) Nipa (*Distichlis palmeri*): a perennial grain crop for saltwater irrigation. J Arid Environ 82:60–70

Pedranzani H, Racagni G, Alemano S, Miersch O, Ramirez I, Pena-Cortes H, Taleisnik E, Machado-Domenech E, Abdala G (2003) Salt tolerant tomato plants show increased levels of jasmonic acid. Plant Growth Regul 41:149–158

Podlipaev SA (1995) Plant Trypanosome (Mastigophora: Trypanosomatidae) from *Cynanchum acutum* Plants in Northern Israel. Phytoparasitica 23(4):351–353

Prakash L, Prathapasenan G (1990) NaCl and gibberellic acid induced changes in the content of auxin, the activity of cellulose and pectinlyase during leaf growth in rice (*Oryza sativa*). Ann Bot 365:251–257

Ragasa CY, Ebajo Jr VD, Reyes MM, Mandia EH, Brkljaca R, Urban S (2015) Triterpenes and Sterols from Sonneratia alba. Int J Curr Pharm Rev Res 6(6):256–261. www.ijcpr.com
Ramadan T (1998) Ecophysiology of salt excretion in the xerohalophyte reaumuria. New Phytol 139:273–281
Rao DLN, Gill HS (1995) Biomass and biofertilzer production by *Sesbania cannabina* in alkaline soil. Biores Technol 53:169–172
Ravikumar S, Gnanadesigan M (2011) Hepatoprotective and antioxidant activity of a mangrove plant *Lumnitzera racemosa*. Asian Pac J Trop Biomed 1(5):348–352
Redondo-Gomez S, Mateos-Naranjo E, Davy AJ, Fernandez-Munoz F, Castellanos EM, Luque T, Enrique Figueroa M (2007) Growth and photosynthetic responses to salinity of the salt-marsh shrub *Atriplex portulacoides*. Ann Bot 100:555–563
Ribaut JM, Pilet PE (1991) Effect of water stress on growth, osmotic potential and abscisic acid content of maize roots. Physiol Plant 81:156–162
Rozema J (1995) Biology of halophytes. In: Choukr-AllAh R, Malcolm CV, Hamdy A (eds) Halophytes and biosaline agriculture. Marcel Dekker Inc., New York, NY, pp 17–30
Rozema J, Schat H (2013) Salt tolerance of halophytes, research questions reviewed in the perspective of saline agriculture. Environ Exp Bot 82:83–95. https://doi.org/10.1016/j.envexpbot.2012.08.004
Rozema J, Bijwaard P, Prast G, Broekman R (1985) Ecophysiological adaptations of coastal halophytes from fore dunes and salt marshes. Vegetatio 62:499–521
Rozema J, Broekman R, Ji B, Bruning B, Katschnig D (2014) Saving fresh water by crop cultivation on salinizing soils, a survey. Report number KfC 146/2014. http://edepot.wur.nl/352728
Saslis-Lagoudakis CH, Moray C, Bromham L (2014) Evolution of salt tolerance in angiosperms: a phylogenetic approach. In: Rajakaruna N, Boyd RS, Harris TB (eds) Plant ecology and evolution in harsh environments. Nova Science Publishers, Hauppauge, NY, USA, pp 77–95
Schat H (1982) On the ecology of some Dutch dune slack plants. PhD thesis, Vrije Universiteit, Amsterdam, p 128
Schillinger WF (2007) Ecology and control of Russian thistle (*Salsola iberica*) after spring wheat harvest. Weed Sci 55:381–385
Scudiero E, Skaggs T, Corwin D (2017) The U.S. Salinity Laboratory (USDA-ARS) guidelines for assessing multi-scale soil salinity with proximal and remote sensing. Geophysical Research Abstracts 19, EGU2017-10576, 2017 EGU General Assembly 2017
Serag MS, Khedr AA, Zahran MA, Willis AJ (1999) Ecology of some aquatic plants in polluted water courses, Nile Delta, Egypt. J Union Arab Biol 9:85–89
Serag MS, Zahran MA, Khedr AA (2002) Ecology and economic potentialities of the dominant salt-tolerant reeds and rushes in the Nile Delta. In: Proceedings of international symposium on "The optimum resources utilization in salt-affected ecosystems in arid and semi-arid regions", pp 245–256, 8–10 April, 2002, Cairo, Egypt
Shabala S, Bose J, Hedrich R (2014) Salt bladders: do they matter? Trends Plant Sci 19:687–691. https://doi.org/10.1016/j.tplants.2014.09.001
Shannon MC, Grieve CM (1999) Tolerance of vegetable crops to salinity. Scientia Hortic 78:5–38
Sharma R, Wungrampha S, Singh V, Pareek A, Sharma MK (2016) Halophytes as bioenergy crops. Front Plant Sci 7:1372. https://doi.org/10.3389/fpls.2016.01372
Shennan C, Hunt R, Macrobbie EAC (1987) Salt tolerance in Aster tripolium L. I. The effect of salinity on growth. Plant Cell Environ 10(1):59–65
Sheue C-R, Liu H-Y, Yong JWH (2003) *Kandelia obovata (Rhizophoraceae)*, a new mangrove species from Eastern Asia. Taxon 52:287–294
Shukla V, Pala Z, Alok A, Desai N (2015) Screening of different *Artemisia* spp. from Western Ghats of Maharashtra for an Anti-Malarial Compound—Artemisinin. Am J Plant Sci 6:1619–1632. Published Online June 2015 in SciRes. http://www.scirp.org/journal/ajps, http://dx.doi.org/10.4236/ajps.2015.69162
Siddiqui ZS, Khan MA (2013) Some physiological attributes of dimorphic seeds of *Halopyrum Mucronatum* (L.) Stapf. Pak J Bot 45(6):1975–1979

Sinha K, Deka AC, Bharalee RB (2016) Morphogenesis of edible gall in *Zizania latifolia* (Griseb.) Turcz. ex Stapf due to Ustilago esculenta Henn. infection in India Swapan Afr J Microbiol Res 10(31):1215–1223. https://doi.org/10.5897/ajmr2016.8207
Slama I, Abdelly C, Bouchereau A, Flowers T, Savoure A (2015) Diversity, distribution and roles of osmoprotective compounds accumulated in halophytes under abiotic stress. Ann Bot 115:433–447. https://doi.org/10.1093/aob/mcu239
Stelzer R, Lauchli A (2010) Salt- and flooding tolerance of *Puccinellia peisonis* IV. Root respiration and the role of aerenchyma in providing atmospheric oxygen to the roots. Zeitschriftfü Pflanzenphysiologie 97(2):171–178. https://doi.org/10.1016/s0044-328x (80)80031-6
Storey R, Jones GW (1979) Responses of *Atriplex spongiosa* and *Suaeda monoica* to Salinity. Plant Physiol 63:156–162
Stukonis V, Juzenas S, Ceseviciene J, Norkeviciene E (2015) Assessment of morpho-anatomical traits of red fescue (*Festuca rubra* L.) germplasm differing in origin. Zemdirb-Agric 102(4):437–442. https://doi.org/10.13080/z-a.2015.102.056
Sukhorukov AP (2008) Fruit anatomy of the genus Anabasis (*Salsoloideae, Chenopodiaceae*). Aust Syst Bot 21:431–442
Tackholm V (1974) Students' flora of Egypt. Cairo University Press, Egypt
Tajeddin B, Behmadi H (2019) Respiration rate and some physicochemical characteristics of *Salicornia bigelovii* as a leafy green vegetable. J Food Biosci Technol 9(2):21–28
Tanaka N, Kuo J, Omori Y, Nakaoka M, Aioi K (2003) Phylogenetic relationships in the genera *Zostera and Heterozostera* (*Zosteraceae*) based on matK sequence data. J Plant Res 116:273–279
Tanveer M, Shah AN (2017) An insight into salt stress tolerance mechanisms of *Chenopodium album*. Environ Sci Poll Res 24:16531–16535
Tanveer M, Shahzad B, Sharma A, Biju S, Bhardwaj R (2018) 24-Epibrassinolide; an active brassinolide and its role in salt stress tolerance in plants: a review. Plant Physiol Biochem 130:69–79
Tehranizadeh ZA, Ali Baratian A, Hosseinzadeh H (2016) Russian olive (*Elaeagnus angustifolia*) as a herbal healer. BioImpacts 6(3):155–167
Thevs N, Zerbe S, Kyosev Y, Rozi A, Tang B, Abdusalih N, Novitskiy Z (2012) *Apocynum venetum L.* and *Apocynum pictum Schrenk (Apocynaceae)* as multi-functional and multi-service plant species in Central Asia: a review on biology, ecology, and utilization. J Appl Bot Food Qual 85:159–167
Tomlinson PB, Posluszny U (1977) Features of dichotomizing apices in *Flagellaria Indics* (Monocotyledones). Am J Bot 64(9):1057–1065
Ungar IA (1996) Effect of salinity on seed germination, growth and ion accumulation of *Atriplex patula (Chenopodiaceae)*. Am J Bot 83:604–607
United Nations Population Information Network available online at https://www.un.org/en/development/desa/population/
Vaicekonyte R, Kiviat E, Nsenga F, Ostfeld A (2014) An exploration of common reed (*Phragmites australis*) bioenergy potential in North America. Mires and Peat 13 (2013/14), Article 12:1–9. http://www.mires-and-peat.net
van de Wouw M, Hanson J, Luethi S (1999) Morphological and agronomic characterisation of a collection of napier grass (*Pennisetum purpureum*) and *P. purpureum* × *P. glaucum*. Trop Grassl 33:150–158
van Diggelen J (1988) A comparative study on the ecophysiology of salt marsh halophytes. PhD thesis, Vrije Universiteit, p 208
Vaurasi V, Kant R (2016) Effects of salinity and plant growth media on in vitro growth and development of taro (*Colocasia esculenta*) varieties. Acta Horticulturae et Regiotecturae 1:17–20. https://doi.org/10.1515/ahr-2016-0005
Wang Y, Mopper S, Hasentein KH (2001) Effects of salinity on endogenous ABA, IAA, JA, and SA in Iris hexagona. J Chem Ecol 27:327–342

Wang B, Davenport RJ, Volkov V, Amtmann A (2006) Low unidirectional sodium influx into root cells restricts net sodium accumulation in *Thellungiella halophila*, a salt-tolerant relative of *Arabidopsis thaliana*. J Exp Bot 57(5):1161–1170

Yasseen BT, Abu-Al-Basal MA (2008) Ecophysiology of *Limonium axillare* and *Avicennia marina* from the coastline of Arabian Gulf-Qatar. J Coast Conserv 12:35–42. https://doi.org/10.1007/s11852-008-0021-z

Yeo AR (1998) Molecular biology of salt tolerance in the context of whole plant physiology. J Exp Bot 49:915–929

Zahran MA (1993) Juncus and Kochi fiber-and fodder-producing halophytes under salinity and aridity stress. In: Pessarakli M (ed) Plant and crop stress. New York-Basal-Hong Kong, pp 505–528

Zhang H, Du C, Wang Y, Wang J, Zheng L, Wang Y (2016) The Reaumuria trigyna leucoanthocyanidin dioxygenase (RtLDOX) gene complements anthocyanidin synthesis and increases the salt tolerance potential of a transgenic *Arabidopsis* LDOX mutant. Plant Physiol Biochem 106:278–287. https://doi.org/10.1016/j.plaphy.2016.05.005

Zhao KF, Fan H, Jiang XY, Zhou S (2002) Critical day-length and photoinductive cycles for the induction of flowering in halophyte Suaeda salsa. Plant Sci 162:27–31

Zhao K, Song J, Zhao M, Feng G, Liu J (2011) Species, types, distribution, and economic potential of halophytes in China. Plant Soil 342:495–509. https://doi.org/10.1007/s11104-010-0470-7

Zia S, Khan MA (2004) Effect of light, salinity, and temperature on seed germination of *Limonium stocksii*. Can J Bot 82(2):151–157. https://doi.org/10.1139/B03-118

Zurayk RA, Baalbaki R (1996) *Inula crithmoides*: a candidate plant for saline agriculture. Arid Soil Res Rehabil 10(3):213–223

Programmed Cell Death and Drought Stress Signaling

Sadia Latif, Tariq Shah, Rizwana Nawaz, Fazal Munsif, Mudassir Ali, Muneeb ur Rehman and Hamad Khan

Abstract Programmed cell death (PCD) refers to the death of a cell that is genetically programmed. Alongside cell division and cell migration, PCD enables the organism to strictly control cell numbers and tissue size as well as to protect itself from unwanted cells that threaten homeostasis. To counteract the oxidative damage, plants have evolved efficient antioxidant defense mechanisms containing enzymatic and non-enzymatic reactive oxygen species (ROS) scavengers. The function of ROS also as the messenger molecules interact with several other signaling pathways including nitric oxide (NO), salicylic acid (SA), jasmonic acid (JA), and ethylene (ET). These interactions and ROS/hormonal balance determine whether the cell stays alive or enter the PCD pathway. Reactive oxygen species are also responsible for the transduction of stress signals and systemic acclimation to unfavorable environmental conditions. Although PCD causes crop yield losses, such selective death of cells under abiotic stress eventually provides survival benefits. Elucidation of the specific enzymes involved in PCD pathways is of a great interest from the context of applied agriculture and environmental protection, since it would enable the creation of plants that are more tolerant to stress. This chapter will provide an overview of PCD and its roles during development and in response to drought stress and physiological and molecular basis of manipulation of PCD pathways enhancing tolerance to drought stress in plant.

S. Latif · R. Nawaz
Department of Plant Sciences, Faculty of Biological Sciences, Quaid-e-Azam University, Islamabad, Islamabad, Pakistan

T. Shah (✉)
Department of Agronomy, Faculty of Crop Production Sciences, University of Agriculture Peshawar, Peshawar, Pakistan
e-mail: tariqshah@aup.edu.pk

F. Munsif · M. Ali · M. Rehman
Department of Agronomy, Faculty of Crop Production Sciences, University of Agriculture Peshawar, Amir Muhammad Khan Campus, Peshawar, Pakistan

H. Khan
Department of Agronomy, University of Agriculture Faisalabad, Faisalabad, Pakistan

M. Hasanuzzaman and M. Tanveer (eds.), *Salt and Drought Stress Tolerance in Plants*, Signaling and Communication in Plants, https://doi.org/10.1007/978-3-030-40277-8_8

1 Introduction

Programmed cell death (PCD) is a physiological and genetically controlled process that is evolutionarily conserved across kingdoms. PCD allows multicellular organisms to eliminate excessive or damaged cells, which arise during development and in response to abiotic and biotic stress (Williams and Dickman 2008; Fomicheva et al. 2012). Programmed cell death has been studied extensively in animals and the underlying mechanisms in plants are gradually being discovered.

The roles of PCD during the development of animals were thoroughly reviewed in Fuchs and Steller (2011), especially in the regulation of structure sculpting and driving morphogenesis, deletion of unwanted or redundant transient functional structures, control of cell numbers and elimination of unwanted and potentially dangerous cells. In plants, PCD is involved in many stages of development from the embryo to reproduction and aging such as embryogenesis, somatic embryogenesis (Giuliani et al. 2002; Suarez et al. 2004; Hill et al. 2013), sex determination in unisexual species (Dellaporta and Calderon-Urrea 1994; Beers 1997), seed development (Young and Gallie 2000), and senescence (Greenberg 1996; Yen and Yang 1998; Simeonova et al. 2000; Yoshida 2003). Programmed cell death also plays an important role in the elicitation of defence mechanisms. For example, the hypersensitive response, which occurs at the site of pathogen attack and involves programmed cell death of infected as well as uninfected adjacent bystander cells, is one of the strategies that plants employ to prevent pathogen invasion (Lam et al. 2001; Lam 2004).

Although programmed cell death plays important roles during development and in response to environmental stimuli, it may be beneficial or detrimental to the plant depending on the context (Table 1) (Williams and Dickman 2008). Being sessile, plants are particularly vulnerable to aberrant environmental conditions, including saline soils and water deficit. To mitigate osmotic stress, plants implement a range of strategies; however, if these mechanisms are unable to cope with the prolonged stress imposed, the plant will implement selective PCD as a last ditch effort to survive; sacrifice a few cells for the greater good of the organism as a whole (Hara-Nishimura et al. 2010; Greenberg 1996). Paradoxically, studies have shown that inhibition of PCD during stress promotes survival (Dickman et al. 2001; Awada et al. 2003; Shabala et al. 2007; Wang et al. 2009b; Li et al. 2010; Hoang et al. 2014, 2015). In the following sections, we will provide an overview of programmed cell death and the physiological and molecular basis of the enhancement of tolerance to osmotic stress associated with drought and salinity through the prevention of programmed cell death in plants.

2 Why Should a Cell Decide to Commit Suicide?

In order to maintain successful development and growth of complex multicellular organisms it is necessary for cells to undergo PCD process. PCD helps to regulate the rate of cell division and shaping of cells, tissues, and organs. It also controls

Table 1 PCD processes controlled by different proteins under different abiotic stresses

Gene	Function	Stress	Citation
AtbZIP60, TF	Transduces the UPR-response; important for salt stress tolerance	Salt, ER stress	Fujita et al. (2007)
AtBiP2, ER luminal binding protein 2	ER chaperone protein, negative regulator of PCD	Cd2 + stress	Xu et al. (2013)
ZmABP9, ABRE binding protein 9	Controls ABA signaling and ROS accumulation	Drought, salt, cold stress	Zhang et al. (2011)
AtbZIP28, TF	Transduces the UPR-response; part of the heat shock tolerance pathway	Heat, ER stress	Liu et al. (2007)
AtANAC013, TF	Conveys information about mitochondrial dysfunction	Mitochondrial stress	De Clercq et al. (2013)
AtSOLDAT, singlet oxygen-linked death activator	Suppresses 1O2-induced cell death	$^{1}O_2$ stress	Coll et al. (2009)
AtEXECUTER 1 and 2	Drives effects after 1O2 accumulation	$^{1}O_2$ stress	Lee et al. (2007)
AOX1a, alternative oxidase 1a	Mitigates mitochondrial ROS burst	Al stress	Liu et al. (2014)
VPE, vacuolar processing enzyme	A PCD-executing protease	Salt, osmotic, heat, ER-stress	Kim et al. (2014)
BCL-2, B-cell lymphoma 2	Suppressor of VPE	Salt, osmotic stress	Kim et al. (2014)
AtCYSa and AtCYSb, cystatins	Cysteine protease inhibitors	Salt, drought, cold	Zhang et al. (2008)
AtBOI, Botrytis susceptible1 interactor	Ubiquitinates BOS1, a transcription factor (TF) implicated in stress responses	Salt stress	Luo et al. (2010)
AtHSFA2, heat shock factor A2	Alleviates oxidative stress during heat shock	Heat stress	Zhang et al. (2009)
AtKOD, kiss of death	A peptide that induces PCD responses upstream of mitochondrial dysfunction	Heat, Oxidative stress	Blanvillain et al. (2011)

(continued)

Table 1 (continued)

Gene	Function	Stress	Citation
AtMC8, metacaspase 8	A part of the PCD pathway provoked by UV-radiation	UV-stress	He et al. (2008)
AtNTL-4, TF	Enhances ROS production during drought-induced senescence and heat stress	Drought stress	Lee et al. (2014)

cell populations and defense against invading microbes and drought stress. The cell commits suicide because PCD is needed to destroy cells that represent a threat to the integrity of organism. Examples are cell infected with virus and cell with DNA damage. In plants, the deletion of suspensor cells in embryo and stamen primordia cells in unisexual flowers are also due to PCD.

3 Different Types of PCD

There are two main types of PCD, which is called apoptosis or type I cell death and autophagic or type II cell death which is discussed briefly.

3.1 Apoptosis or Type I Cell Death

Apoptosis is the process of PCD that may occur in multicellular organisms (Green 2011). Biochemical events lead to characteristic cell changes (morphology) and death. These changes include blebbing, cell shrinkage, nuclear fragmentation, chromatin condensation, and chromosomal DNA fragmentation (Fig. 2). It is now thought that in a developmental context—cells are induced to positively commit suicide while in a homeostatic context—the absence of certain survival factors may provide the impetus for suicide. There appears to be some variation in the morphology and indeed the biochemistry of these suicide pathways; some treading the path of "apoptosis," others following a more generalized pathway to deletion, but both usually being genetically and synthetically motivated. There is some evidence that certain symptoms of "apoptosis" such as endonuclease activation can be spuriously induced without engaging a genetic cascade; however, presumably true apoptosis and programmed cell death must be genetically mediated. It is also becoming clear that mitosis and apoptosis are toggled or linked in some way and that the balance achieved depends on signals received from appropriate growth or survival factors (Bowen and Ivor 1993).

3.2 Autophagy or Type II Cell Death

Macroautophagy, often referred to as autophagy, is a catabolic process that results in the autophagosomic-lysosomal degradation of bulk cytoplasmic contents, abnormal protein aggregates, and excess or damaged organelles.

Autophagy is generally activated by conditions of nutrient deprivation, but has also been associated with physiological as well as pathological processes such as development, differentiation, stress, and infection.

3.3 Other Types

Besides the above two types of PCD, other pathways have been discovered (Kroemer and Martin 2005). Called "non-apoptotic programmed cell-death" (or "caspase-independent programmed cell-death" or "necroptosis"), these alternative routes to death are as efficient as apoptosis and can function as either backup mechanisms or the main type of PCD.

Necroptosis is a programmed form of necrosis or inflammatory cell death. Conventionally, necrosis is associated with unprogrammed cell death resulting from cellular damage or infiltration by pathogens, in contrast to orderly, programmed cell death via apoptosis.

Eryptosis is a form of suicidal erythrocyte death (Lang et al. 2006).

Aponecrosis is a hybrid of apoptosis and necrosis and refers to an incomplete apoptotic process that is completed by necrosis (Formigli et al. 2000).

NETosis is the process of cell-death generated by Neutrophil extracellular traps (NETs) (Fadini et al. 2015).

Paraptosis is another type of nonapoptotic cell death that is mediated by MAPK through the activation of IGF-1. It is characterized by the intracellular formation of vacuoles and swelling of mitochondria (Ross 2016).

Pyroptosis, an inflammatory type of cell death, is uniquely mediated by caspase 1, an enzyme not involved in apoptosis, in response to infection by certain microorganisms (Ross 2016).

Plant cells undergo particular processes of PCD similar to autophagic cell death. However, some common features of PCD are highly conserved in both plants and metazoa.

4 PCD Mechanism and Pathways Under Drought Stress in Plants

Programmed cell death particularly apoptosis, the physiological form of PCD, has been studied for more than 40 years and is known to occur in many species across all kingdoms. For example, human Bcl-2 can partially complement *Caenorhabditis elegans* Ced-9 mutants even though the two genes have limited sequence homology. The animal-derived anti-apoptotic genes Ced-9 and Bcl-2 confer tolerance to a wide range of biotic and abiotic stresses, upon overexpression in plants (Qiao et al. 2002; Chen and Dickman 2004; Shabala et al. 2007; Wang et al. 2009a; Paul et al. 2011).

Since the first evidence that a genetic program existed for physiological (programmed) cell death came from studying development in the nematode *C. elegans* (Kerr et al. 1972; Horvitz et al. 1982; Ellis and Horvitz 1986; Vaux et al. 1988), the study of pathways and regulation of programmed cell death has been carried out on several model systems including *C. elegans*, the vinegar fly *Drosophila melanogaster*, and the mouse. The conservation of the core apoptotic machinery has been found across vast evolutionary distances from worm to human; however, it is somewhat obscure in plants (Williams and Dickman 2008; Fuchs and Steller 2011).

As the core apoptotic machinery is conserved across kingdoms, details of a well-studied programmed cell death model would be helpful to establish an understanding of programmed cell death in plants. In the next section, we will review the literature of mammalian PCD pathways.

Programmed cell death is an integral physiological process in many aspects of eukaryotic development that involves selective elimination of damaged or unwanted cells (Fig. 1) (Fomicheva et al. 2012). In plants, PCD occurs during reproductive and vegetative stages of development, including sex determination, gamete development, embryogenesis, and leaf senescence. Programmed cell death can also be activated in response to abiotic and biotic stress factors. Hypersensitive response (HR), a rapid host cell death around the site of infection, is a common disease resistance strategy for plant under pathogen attack (Lam et al. 2001). Characteristics of PCD

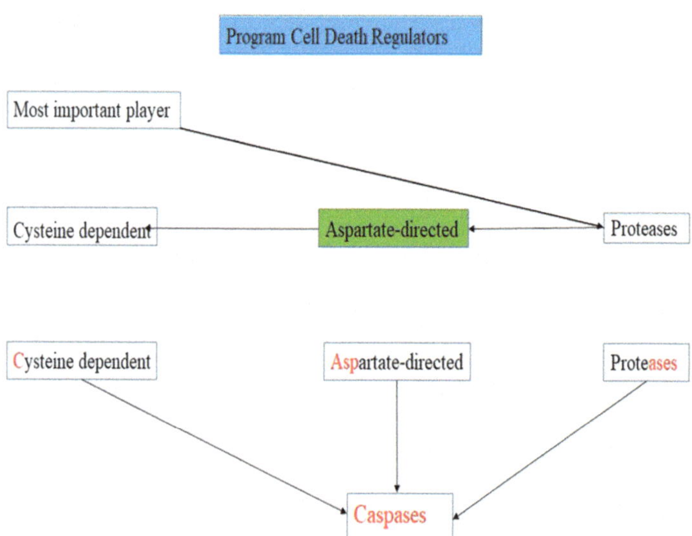

Fig. 1 Different regulators involved in program cell death

include chromatin aggregation, cytoplasmic and nuclear condensation, and formation of apoptotic bodies. During PCD chromatin is cleaved at the inter-nucleosomal sites into approximately 180 bp multimers of DNA fragments (Lam 2004). Abiotic stress factors such as drought can stimulate the production of ROS, which imposes oxidative damage to cellular components. (Dat et al. 2003). Due to additional roles of ROS in signaling, plant cells have developed sophisticated strategies of regulating intracellular ROS concentrations through avoidance and scavenging mechanisms (Mittler et al. 1999). PSII photo-inactivation under high lights through leaf movement and rolling helps plant cells to avoid excess ROS production while the use of scavenging enzymes such as superoxide dismutase (SOD), ascorbate peroxidase (APX), glutathione peroxidase, and catalase (CAT) and antioxidants such as ascorbate and glutathione detoxify excess ROS (Fomicheva et al. 2012; Tanveer and Shabala 2018). Excessive production of ROS beyond these control strategies results in oxidative stress causing cellular damage to proteins, nucleic acids and lipids, which eventually result in PCD (Mateo et al. 2004). The role of ROS in the activation of PCD was further confirmed in other stresses such as UV irradiation, ozone, aluminum toxicity, and heat shock (Mateo et al. 2004). Although PCD is induced by the effects of abiotic stresses, it has not been observed to occur in desiccation-tolerant (DT) plants subjected to extreme dehydration. Programmed cell death avoidance mechanisms in DT plants have not been previously reported. However, it can be hypothesized that whatever the survival mechanism, it must be geared toward suppression of PCD-related processes and promotion of pro-survival strategies of DT plants (Fig. 2).

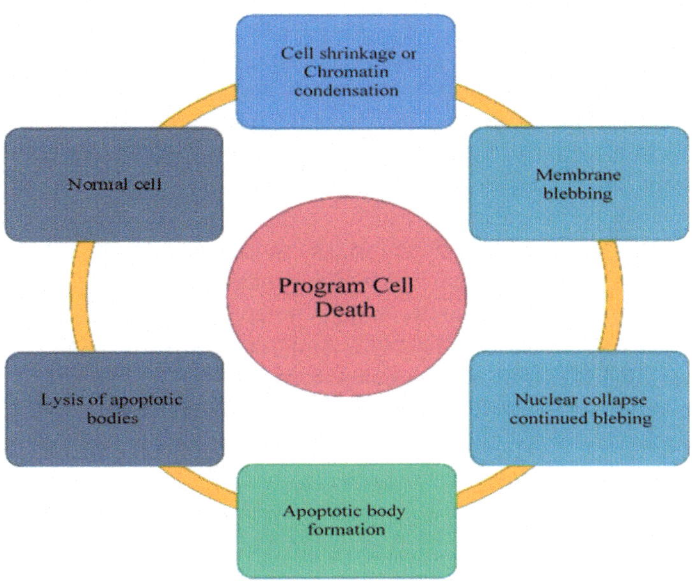

Fig. 2 Schematic diagram of program cell death mechanism

5 PCD: As a Host Defense Mechanism Against Drought Stress

Drought is one of the most unfavorable environmental factors that affect growth and development of plants and consequently limits plant productivity (Anjum et al. 2017a; Tanveer et al. 2018). Plants have developed specific acclimation and adaptation mechanisms to survive the soil water deficit (Anjum et al. 2017b). In response to drought, plants can exhibit either escape (ability to complete the life cycle before severe stress) or resistance mechanisms. Resistance mechanisms include drought avoidance and drought tolerance. The latter depends on the cell turgor maintenance by accumulating osmolytes and soluble sugars (Harb et al. 2010; Anjum et al. 2016). There are several examples of molecules that help to maintain an osmotic balance under dehydration conditions: sugars, polyols, and proline (Yoshiba et al. 1997). Proline is accumulated in the cytoplasm and chloroplast stroma while other solutes (sugars, organic acids, potassium) are cumulated in the vacuole. When the cellular water content decreases, they stabilize cellular structures through hydrophilic interactions and hydrogen bonding (Verslues and Sharma 2010). A similar role is fulfilled by late embryogenesis abundant (LEA) proteins—a family of unstructured proteins. LEA proteins accumulate in response to dehydration and ABA treatment. Because of their high hydrophilicity and solubility in water, it has been proposed that they play a role in protecting cytoplasmic structures during dehydration (Ramanjulu and Bartels 2002). The avoidance mechanism is possible by the maintenance of high water potential in plant tissue despite soil water deficit. It can be achieved by improved water uptake under stress, the ability to hold water as well as by the reduction of its loss through smaller leaf area and lesser stomatal and cuticular conductance. One of the first acclimation responses to drought is the decrease in leaf growth, which helps to maintain the cell turgor and reduces the transpiration area. In *Arabidopsis*, the size of leaf is regulated by both cell division and cell expansion. Under drought stress, *Arabidopsis* leaves have been demonstrated to compensate for the low rate of expansion by the extension of expansion duration (Aguirrezabal et al. 2006). The cell expansion is a process of cell wall loosening performed by enzymatic and non-enzymatic components, from which expansions are key proteins engaged in this process. The onset of cell expansion involves pumping of protons into the cell wall, which makes the surrounding more acidic. As a result, expansions become activated. They loosen connections between cellulose microfibrils, which lead to the cell wall relaxation, water uptake and consequently the cell expansion (Cosgrove 2005). It has been demonstrated that the mild osmotic stress causes the induction of expansin genes (Skirycz et al. 2010).

Photosynthesis is one of the major processes affected by water deficit since stomata closure causes reduced CO_2 diffusion to the chloroplast. As a result of the inhibition of photosynthesis and the predominance of photorespiration, ROS are generated (Noctor et al. 2002). It has been demonstrated that in drought-stressed plants, the ABA-controlled stomata closure is mediated by H_2O_2 (Pei et al. 2000). Under severe drought stress, some antioxidant enzymes have been shown to be highly induced

(Dat et al. 2000). However, studies on many drought-stressed crop species showed an inconsistency in their expression since in some cases they have been induced, but in other repressed, suggesting that different ROS balance may be required during different response phases (Cruz de Carvalho 2008).

During water deficit, ROS are responsible for the induction of leaf senescence, which is executed through the programmed cell death and play an important role in the plant survival. It contributes to the nutrients remobilization during stress and allows the rest of the plant to benefit from them and stay alive. Drought-induced PCD enables also the abscission of some leaves and thus the avoidance of further water loss through the transpiration. It occurs gradually and is manifested by specific biochemical and molecular changes such as chromatin condensation, thylakoid swelling, lipid peroxidation, degradation of chlorophyll (leaf yellowing), and proteins. Apart from ROS, cytokinins and ABA have been shown to be involved in the regulation of water-deficit-triggered senescence. Recent studies have shown that the water deficit triggers PCD not only in green tissues but also in plant root tips. Apical meristem cells of primary roots undergoing PCD, demonstrate increased size of vacuole, degradation of organelles, and the collapse of plasma membrane (Duan et al. 2010).

Early events in the perception of drought stress signals include the activation of transcription factors belonging to such classes as DREB/CBF (e.g., DREB1a, DREB2a), ABF (e.g., ABF2, ABF4), MYB (e.g., MYB2), MYC, NAC, and WRKY (Table 1). Many of them possess stress-responsive cis-regulatory elements in their promoter sequences like abscisic acid-responsive elements (ABRE) and drought-responsive elements (DRE) (Bartels and Sunkar 2005; Fujita et al. 2011; Rushton et al. 2012). The plasma membrane-associated NTL4 (NAC transcription factor) after drought or ABA treatment has been shown to be proteolytically activated and transported to the nucleus where it induces expression of NADPH oxidase involved in ROS generation (Lee and Park 2012). Moreover, the dehydration stimulates expression of BAX inhibitor-1 (AtBI-1). The atbi1 mutant has been shown to display more severe cell death, indicating that ER-located AtBI-1 modulates the water-deficit-induced PCD (Duan et al. 2010). Drought stress has been also proven to regulate the expression and activity of aquaporins—a family of channel proteins that facilitate the transport of water along transmembrane water potential gradients (Johansson et al. 2000).

Upon soil water deficit, the accumulation of ABA and the induction of ABA-associated signaling genes occur. ABA induces various second messengers such as cytosolic Ca^{2+}, ROS, and NO in guard cells. These signals evoke ion efflux through plasma membrane ion channels, resulting in the reduction of guard cell turgor pressure and stomata closure to reduce water loss through transpiration (Skirycz and Inzé 2010). Mutants with the perturbation of ABA synthesis or signaling display drought hypersensitivity, manifested in significant growth reduction, which suggests that ABA is needed for the proper response to drought. In *Arabidopsis*, stomata closure has been shown to be regulated by ABI1 and ABI3 (ABA-insensitive 1 and 3) belonging to a group of genes identified through mutant screens and being associated with ABA-mediated metabolic responses to stress. Under drought, ABI1 transcription is upregulated while ABI3 is usually downregulated. Recently, ABI3 has been hypothesized to be essential for the successful drought recovery (Pinheiro and Chaves 2011). The cell-surface ABA receptors have not been recognized yet. However, a recent study has proposed the flowering time control protein A (FCA) and the chloroplastic magnesium protoporphyrin-IX chelatase H subunit (CHLH) as candidates for ABA receptors, both of which have been shown to bind ABA in vitro (Wasilewska et al. 2008). Two genes have been found to play a crucial role in the prevention of stomata opening—GPA1 (G PROTEIN ALPHA SUBUNIT 1) and PLDα1 (PHOSPHOLIPASE D AL PHA 1) (Zhao et al. 2010). PLD-produced phosphatidic acid has been also shown to play an important role in the plant response to drought stress (Li et al. 2009).

Similarly to ABA, JA also triggers stomata closure and such response is conserved among various plant species (Xiong et al. 2002). At the early stage of moderate drought, plants accumulate high concentrations of ABA and induce ABA-responsive genes. At this stage, no significant differences in JA-responsive genes are observed. At later stage of drought stress, ABA level returns to normal, while JA synthesis and JA signaling genes are significantly downregulated. This suggests the negative correlation between ABA and JA pathways. The high concentration of JA is probably undesirable during drought stress, as it inhibits the cell expansion and results in stunted growth (Zhang and Turner 2008). Therefore, plants down-regulate JA synthesis and signaling pathways to minimize the inhibitory effect of JA on growth, establishing a new hormone homeostasis.

Downstream of early stress perception events, signaling molecules are activated. Such secondary messengers include Ca^{2+} ions and ROS. They induce further genes that are needed to establish a new cellular homeostasis leading to drought resistance and tolerance (Ramanjulu and Bartels 2002). Recent studies have strongly proven that drought response progresses through mitogen-activated protein kinase (MAPK) pathways (Mishra et al. 2006). In yeast and animals, MAPK-regulated pathways take part in the production of osmolytes and antioxidants. These MAPK pathways are activated by receptors/sensors such as protein tyrosine kinases, G-protein–coupled receptors, and histidine kinases. Among these, G-protein–associated receptors have been proposed to serve as one kind of membrane-bounded receptors for ABA. A family of histidine kinases (HK) have been also identified in plants (Xiong et al. 2002). An *Arabidopsis* AtHK1 has been suggested to be involved in the osmotic stress signal transduction (Urao et al. 1999). Other members of this family, ATHB7 and ATHB12 have been proposed to maintain the reduction of plant growth under drought, which is an acclimation response to survive prolonged drought stress (Harb et al. 2010). The elucidation of mechanisms controlling drought stress responses has enabled to engineer plants by the expression of specific stress-related genes. Although it was believed that the modulation of osmoregulatory genes would be the best strategy, attempts failed to result in any significant drought-stress tolerance improvement (Bhatnagar-Mathur et al. 2008). However, constitutive expression of some LEA proteins has conferred tolerance to soil water deficit in transgenic rice (Xu et al. 1996) and wheat (Xiong et al. 2002). Moreover, tomato plants overexpressing Arabidopsis CBF1 (DREB1B) have exhibited improved drought tolerance (Sivamani et al. 2000). Similar results have been obtained for transgenic *Arabidopsis* and rice plants overexpressing stress-responsive NAC genes (Nakashima et al. 2012).

6 Role of ROS in PCD Signaling Under Drought Stress

The first indications for an involvement of ROS in PCD were based mainly on spatiotemporal correlations between increased ROS levels and cell death. The initial experimental evidence that ROS also acted as a signal in plant PCD was obtained in cell suspensions by demonstrating that H_2O_2-induced cell death could be blocked by cycloheximide and protease inhibitors (Levine et al. 1994). Since then, there has been ever-growing support for a key role for ROS as triggers of PCD. Whereas initial results were generally obtained by exogenous application of oxidants, more recently, transgenic plants with perturbed levels of cellular antioxidants and mutants unable to stop the initiation or propagation of ROS-driven cell death have elegantly demonstrated the important role of ROS in the initiation phase of different plant PCD events (Lorrain et al. 2003).

The best-studied examples of ROS-derived PCD are those following the typical biphasic oxidative burst during both the HR and ozone stress. The ensuing cell death is characterized by discrete cellular lesions that are generally preceded by the appearance of several hallmarks of PCD such as chromatin condensation, DNA laddering, and cytochrome c release (Lam 2004). Moreover, these features can be halted by administrating either high concentrations of antioxidants or inhibitors of both translation and transcription and of known signal-transducing components, such as kinases or phytohormones. Accordingly, transgenic plants with low or high levels of several antioxidants (SOD, CAT, and APX) also exhibit an altered response to both pathogen- and ozone-driven PCD, again demonstrating the crucial importance of a tightly orchestrated redox balance (Van Camp et al. 1994; Örvar et al. 1997; Mittler et al. 1999). Further evidence for a ROS-dependent PCD pathway in plants has been provided by transgenic plants catalase-deficient transgenic (Cat1AS), the major H_2O_2-scavenging enzyme. In Cat1AS leaves of tobacco (*Nicotiana tabacum*), accumulation of photorespiratory H_2O_2 induced active cell death of leaf palisade parenchyma cells. In addition to the presence of several PCD hallmarks, this high light-driven cell death could be blocked by the infiltration of various antagonists of HR-like PCD (Dat et al. 2003). Interestingly, we have also been able to uncouple a necrosis-like response, provoked by continuous high light-dependent H_2O_2 accumulation, from PCD triggered by a transient, nonlethal, in planta ROS accumulation. The short H_2O_2 pulse sufficient to induce PCD in leaves had a lipoxygenase-dependent oxylipin signature similar to that induced by a pathogenic elicitor (cryptogein). In contrast, the continuous H_2O_2 accumulation generated by long-term high light exposure or H_2O_2 feeding led to necrosis and ROS-mediated lipid peroxidation (Montillet et al. 2005). The existence of a dose-dependent ROS threshold below, which PCD will be triggered and above which necrotic cell death will prevail might be the reason for the overwhelming effect that phytotoxic ROS levels have on PCD hallmarks.

Alternatively, similarly to what is reported in animals, high doses of oxidants might inhibit components of the PCD pathway (Kazzaz et al. 1996). Complementary to the data provided by transgenic plants, a number of *Arabidopsis* mutants have demonstrated the specificity in the signaling capacity of different ROS in initiating PCD. In two mutants of Arabidopsis, *lesion-stimulating disease1* (*lsd1*) and *radical-induced cell death1* (*rcd1*), elevated ROS levels are necessary and sufficient to induce spreading of cell death (Jabs et al. 1996; Mateo et al. 2004). The conditional *fluorescent* (*flu*) mutant of *Arabidopsis* that generates singlet oxygen upon a dark-to-light shift initiates a cell death response immediately after the release of singlet oxygen (op den Camp et al. 2003). Singlet oxygen is interpreted differently at the molecular level than H_2O_2 and superoxide, thus providing further evidence for a selective signaling effect of various ROS in plants (Danon et al. 2005). The ozone-sensitive mutant *vitamin c-1* (*vtc-1*), deficient in L-ascorbic acid, shows small patches of spontaneous lesions on its leaves (Pavet et al. 2005). These lesions were attributed to changes in redox homeostasis rather than to H_2O_2, as evidenced by lower ascorbate levels in *vtc* mutants than in the wild-type plants (Pavet et al. 2005). Finally, the localized expression pattern of the copper amine oxidase-encoding gene *ATAO1*, which produces H_2O_2 associated with PCD in lateral root cap cells and developing tracheary elements, also hints to H_2O_2 as a trigger for certain developmental cell death processes (Møller and McPherson 1998). In summary, various ROS species and cellular redox changes can trigger different signaling cascades leading to PCD. However, the understanding of how ROS signals are perceived and transduced is still in its infancy.

7 Molecular Aspects of PCD During ROS Signaling

One of the first genes identified in ROS-dependent cell death was isolated through a genetic screen. The *Arabidopsis lsd1* mutant is unable to cope with increased superoxide levels and develops spontaneous superoxide- and SA-dependent runaway lesions (Jabs et al. 1996). LSD1 encodes a zinc-finger protein that, together with two other zinc-finger proteins (LOL1 and LOL2), could act as a molecular rheostat, sensing changes in ROS homeostasis, thereby repressing a default death pathway or controlling an antideath mechanism through the regulation of cell death genes (Dietrich et al. 1997; Epple et al. 2003). The ozone-sensitive lesion mimic mutant, *rcd1*, also displays PCD lesions. RCD1 has been proposed to modulate the activity of target proteins through ADP ribosylation (Ahlfors et al. 2004; Overmyer et al. 2005). In a landmark paper, ROS-induced PCD has recently been shown not to be mediated by photooxidative damage but through a genetically determined PCD pathway (Wagner et al. 2004). In fact, inactivation of the *Arabidopsis Executer1* gene completely abolished singlet oxygen-induced cell death (Wagner et al. 2004). This hitherto unknown nucleus-encoded chloroplastic protein might perceive nonscavenged singlet oxygen species within the chloroplast. Whether *Executer1* is specifically involved in singlet oxygen perception or also in sensing other ROS species remains to be tested.

Other important stress signal transducers include mitogen-activated protein kinases (MAPKs) that act upstream of the oxidative burst during ozone treatment and the HR (Ren et al. 2002; Samuel and Ellis 2002). However, MAPKs might also work downstream in ROS-dependent cell death events. The primary ROS-activated tobacco MAPK is the SA-induced protein kinase, which is required during harpin-dependent PCD. A MAPKKK of alfalfa (*Medicago sativa*) activates cell death induced by H_2O_2 through a specific MAPK-scaffolding action (Nakagami et al. 2004). Overlapping features of animal and plant PCD have also inspired different functional approaches by introducing animal genes into plants to gather mechanistic insight into oxidative stress-dependent cell death events. No obvious homologs of the BCL-like animal cell death suppressors have been identified in plant genomes to date; nevertheless, protection against mitochondrial and chloroplast-derived ROS-dependent cell death is conferred by their overproduction in tobacco. Although functional plant homologs have been successful in protecting against ROS-mediated PCD in the transgenic lines, their existence remains unclear (Mitsuhara et al. 1999; Chen and Dickman 2004). However, recently an evolutionarily conserved Arabidopsis BCL2-associated athanogene protein has been shown to be induced by H_2O_2 and capable of provoking PCD in both yeast (*Saccharomyces cerevisiae*) and plants (Kang et al. 2006).

8 Conclusion and Perspectives

Plants have evolved various strategies to acclimate to different environmental stresses. The most fundamental strategy is the development of high plasticity of plant tissues. It has been demonstrated that programmed cell death plays an important role in this plasticity and subsequent adaptation to unfavorable conditions. There is a growing evidence that PCD is a crucial process in both morphogenetic changes execution and the following adaptation. Although each decade brings a vast number of researches, our understanding of plant PCD and its underlying mechanisms is still in its early stage. Two abiotic stress factors that significantly hinder world crop production are soil water deficit and salinization (Munns 2011). Researchers have shown that manipulation of PCD pathways can be applied to monocots and dicots for enhancing stress tolerance to a range of abiotic and biotic stresses. Currently, we are able to produce crops with enhanced drought and salinity tolerance that survive in the glasshouse; however, once applied in the field the tolerance fails due to combined stresses. Further insight into details of the PCD molecular machinery in plants is important, since it is an attractive target for improving stress tolerance and plant yield under adverse conditions. Essentially, it could lead to the generation of pathogen-resistant and stress-tolerant crops as well as fruit varieties with an extended shelf life.

References

Aguirrezabal L, Bouchier-Combaud S, Radziejwoski A, Dauzat M, Cookson SJ, Granier C (2006) Plasticity to soil water deficit in Arabidopsis thaliana: dissection of leaf development into underlying growth dynamic and cellular variables reveals invisible phenotypes. Plant Cell Environ 29:2216–2227

Ahlfors R, Lång S, Overmyer K, Jaspers P, Brosché M, Tauriainen A, Kollist H, Tuominen H, Belles-Boix E, Piippo M et al (2004) Arabidopsis RADICAL-INDUCED CELL DEATH1 belongs to the WWE protein-protein interaction domain protein family and modulates abscisic acid, ethylene, and methyl jasmonate responses. Plant Cell 16:1925–1937

Anjum SA, Tanveer M, Ashraf U, Hussain S, Shahzad B, Khan I, Wang L (2016) Effect of progressive drought stress on growth, leaf gas exchange, and antioxidant production in two maize cultivars. Environ Sci Pollu Res 23(17):17132–17141

Anjum SA, Ashraf U, Zohaib A, Tanveer M, Naeem M, Ali I, Nazir U (2017a) Growth and development responses of crop plants under drought stress: a review. Zemdirbyste 104(3):267–276

Anjum SA, Ashraf U, Tanveer M, Khan I, Hussain S, Shahzad B, Wang LC (2017b) Drought induced changes in growth, osmolyte accumulation and antioxidant metabolism of three maize hybrids. Front Plant Sci 8

Awada T, Dunigan DD, Dickman MB (2003) Animal anti-apoptotic genes ameliorate the loss of turgor in water-stressed transgenic tobacco. Can J Plant Sci 83:499–506

Bartels D, Sunkar R (2005) Drought and salt tolerance in plants. Crit Rev Plant Sci 24(1):23–58

Beers EP (1997) Programmed cell death during plant growth and development. Cell Death Differ 4:649–661

Bhatnagar-Mathur P, Vadez V, Sharma KK (2008) Transgenic approaches for abiotic stress tolerance in plants: retrospect and prospects. Plant Cell Rep 27(3):411–424

Blanvillain R, Young B, Cai YM, Hecht V, Varoquaux F, Delorme V et al (2011) The Arabidopsis peptide kiss of death is an inducer of programmed cell death. EMBO J 30:1173–1183

Bowen ID (1993) Cell Biol Int 17(4):365–380

Chen S, Dickman MB (2004) Bcl-2 family members localize to tobacco chloroplasts and inhibit programmed cell death induced by chloroplast-targeted herbicides. J Exp Bot 55:2617–2623

Coll NS, Danon A, Meurer J, Cho WK, Apel K (2009) Characterization of soldat8, a suppressor of singlet oxygen-induced cell death in *Arabidopsis* seedlings. Plant Cell Physiol 50:707–718

Cosgrove DJ (2005) Growth of the plant cell wall. Nat Rev Mol Cell Biol 6(11):850–861

Cruz de Carvalho MH (2008) Drought stress and reactive oxygen species. Plant Signal Behav 3(3):156–165

Danon A, Miersch O, Felix G, op den Camp RGL, Apel K (2005) Concurrent activation of cell death-regulating signaling pathways by singlet oxygen in *Arabidopsis thaliana*. Plant J 41:68–80

Dat J, Vandenabeele S, Vranová E, Van Montagu M, Inzé D, Van Breusegem F (2000) Dual action of the active oxygen species during plant stress responses. Cell Mol Life Sci 57(5):779–795

Dat JF, Pellinen R, Beeckman T, Van De Cotte B, Langebartels C, Kangasjärvi J, Inzé D, Van Breusegem F (2003) Changes in hydrogen peroxide homeostasis trigger an active cell death process in tobacco. Plant J 33:621–632

De Clercq I, Vermeirssen V, Van Aken O, Vandepoele K, Murcha MW, Law SR et al (2013) The membrane-bound NAC transcription factor ANAC013 functions in mitochondrial retrograde regulation of the oxidative stress response in *Arabidopsis*. Plant Cell 25:3472–3490

Dellaporta SL, Calderon-Urrea A (1994) The sex determination process in maize. Sci. 266:1501–1505

Dickman MB, Park YK, Oltersdorf T, Li W, Clemente T, French R (2001) Abrogation of disease development in plants expressing animal antiapoptotic genes. Proc Natl Acad Sci USA 98:6957–6962

Dietrich RA, Richberg MH, Schmidt R, Dean C, Dangl JL (1997) A novel zinc finger protein is encoded by the *Arabidopsis* LSD1 gene and functions as a negative regulator of plant cell death. Cell 88:685–694

Duan Y, Zhang W, Li B, Wang Y, Li K, Sodmergen et al (2010) An endoplasmic reticulum response pathway mediates programmed cell death of root tip induced by water stress in *Arabidopsis*. New Phytol 186(3):681–695

Ellis HM, Horvitz HR (1986) Genetic control of programmed cell death in the nematode C. elegans. Cell 44:817–829

Epple P, Mack AA, Morris VRF, Dangl JL (2003) Antagonistic control of oxidative stress-induced cell death in Arabidopsis by two related, plant-specific zinc finger proteins. Proc Natl Acad Sci USA 100:6831–6836

Fadini GP, Menegazzo L, Scattolini V, Gintoli M, Albiero M, Avogaro A (2015) A perspective on NETosis in diabetes and cardiometabolic disorders. Nutr Metab Cardiovasc Dis NMCD 26(1):1–8

Fomicheva AS, Tuzhikov AI, Beloshistov RE, Trusova SV, Galiullina RA, Mochalova LV, Chichkova NV, Vartapetian AB (2012) Programmed cell death in plants. Biochemistry (Mosc) 77:1452–1464

Formigli L, Papucci L, Tani A, Schiavone N, Tempestini A, Orlandini GE, Zecchi Orlandini S (2000) Aponecrosis: morphological and biochemical exploration of a syncretic process of cell death sharing apoptosis and necrosis. J Cellular Physio 182(1):41–49

Fuchs Y, Steller H (2011) Programmed cell death in animal development and disease. Cell 147:742–758

Fujita M, Mizukado S, Fujita Y, Ichikawa T, Nakazawa M, Seki M et al (2007) Identification of stress-tolerance-related transcription-factor genes via mini-scale Full-length cDNA Over-eXpressor (FOX) gene hunting system. Biochem Biophys Res Commun 364:250–257

Fujita Y, Fujita M, Shinozaki K, Yamaguchi-Shinozaki K (2011) ABA-mediated transcriptional regulation in response to osmotic stress in plants. J Plant Res 124(4):509–525

Giuliani C, Consonni G, Gavazzi G, Colombo M, Dolfini S (2002) Programmed cell death during embryogenesis in maize. Ann Bot 90:287–292

Green D (2011) Means to an end. Cold Spring Harbor Laboratory Press, New York. ISBN 978-0-87969-888-1

Greenberg JT (1996) Programmed cell death: a way of life for plants. Proc Natl Acad Sci USA 93:12094–12097

Hara-Nishimura I, Inoue K, Nishimura M (2010) A unique vacuolar processing enzyme responsible for conversion of several proprotein precursors into the mature forms. FEBS Lett 294:89–93

Harb A, Krishnan A, Ambavaram MMR, Pereira A (2010) Molecular and physiological analysis of drought stress in Arabidopsis reveals early responses leading to acclimation in plant growth. Plant Physiol 154(3):1254–1271

He R, Drury GE, Rotari VI, Gordon A, Willer M, Farzaneh T et al (2008) Metacaspase-8 modulates programmed cell death induced by ultraviolet light and H_2O_2 in *Arabidopsis*. J Biol Chem 283:774–783

Hill RD, Huang S, Stasolla C (2013) Hemoglobins, programmed cell death and somatic embryogenesis. Plant Sci 211:35–41

Hoang TML, Williams B, Khanna H, Dale J, Mundree SG (2014) Physiological basis of salt stress tolerance in rice expressing the antiapoptotic gene SfIAP. Funct Plant Biol 41:1168–1177

Hoang TML, Moghaddam L, Williams B, Khanna H, Dale J, Mundree SG (2015) Development of salinity tolerance in rice by constitutive-overexpression of genes involved in the regulation of programmed cell death. Front Plant Sci 6:175

Horvitz HR, Ellis HM, Sternberg PW (1982) Programmed cell death in nematode development. Neurosci Comment 1:56–65

Jabs T, Dietrich RA, Dangl JL (1996) Initiation of runaway cell death in an *Arabidopsis* mutant by extracellular superoxide. Science 273:1853–1856

Johansson I, Karlsson M, Johanson U, Larsson C, Kjellbom P (2000) The role of aquaporins in cellular and whole plant water balance. Biochim Biophys Acta 1465(1–2):324–342

Kang CH, Jung WY, Kang YH, Kim JY, Kim DG, Jeong JC, Baek DW, Jin JB, Lee JY, Kim MO et al (2006) AtBAG6, a novel calmodulin-binding protein, induces programmed cell death in yeast and plants. Cell Death Differ 13:84–95

Kazzaz JA, Xu J, Palaia TA, Mantell L, Fein AM, Horowitz S (1996) Cellular oxygen toxicity: oxidant injury without apoptosis. J Biol Chem 271:15182–15186

Kerr JF, Wyllie AH, Currie AR (1972) Apoptosis: a basic biological phenomenon with wide-ranging implications in tissue kinetics. Br J Cancer 26:239–257

Kim Y, Wang M, Bai Y, Zeng Z, Guo F, Han N et al (2014) Bcl-2 suppresses activation of VPEs by inhibiting cytosolic Ca2+ level with elevated K+ efflux in NaCl-induced PCD in rice. Plant Physiol Biochem 80:168–175

Kroemer G, Martin SJ (2005) Caspase-independent cell death. Nat Med 11(7):725–730

Lam E (2004) Controlled cell death, plant survival and development. Nat Rev Mol Cell Biol 5:305–315

Lam E, Kato N, Lawton M (2001) Programmed cell death, mitochondria and the plant hypersensitive response. Nature 411:848–853

Lang F, Lang KS, Lang PA, Huber SM, Wieder T (2006) Mechanisms and significance of eryptosis. Antioxid Redox Signal 8(7–8):1183–1192

Lee S, Park CM (2012) Regulation of reactive oxygen species generation under drought conditions in *Arabidopsis*. Plant Signal Behav 7(6):599–601

Lee KP, Kim C, Landgraf F, Apel K (2007) EXECUTER1- and EXECUTER2-dependent transfer of stress-related signals from the plastid to the nucleus of *Arabidopsis thaliana*. Proc Natl Acad Sci USA 104:10270–10275

Lee S, Seo PJ, Lee HJ, Park CM (2012) A NAC transcription factor NTL4 promotes reactive oxygen species production during drought-induced leaf senescence in *Arabidopsis*. Plant J 70(5):831–844

Levine A, Tenhaken R, Dixon R, Lamb C (1994) H_2O_2 from the oxidative burst orchestrates the plant hypersensitive disease resistance response. Cell 79:583–593

Li M, Hong Y, Wang X (2009) Phospholipase D- and phosphatidic acid-mediated signaling in plants. Biochim Biophys Acta 1791(9):927–935

Li W, Kabbage M, Dickman MB (2010) Transgenic expression of an insect inhibitor of apoptosis gene, SfIAP, confers abiotic and biotic stress tolerance and delays tomato fruit ripening. Physiol Mol Plant Pathol 74:363–375

Liu JX, Srivastava R, Che P, Howell SH (2007) An endoplasmic reticulum stress response in *Arabidopsis* is mediated by proteolytic processing and nuclear relocation of a membrane-associated transcription factor, bZIP28. Plant Cell 19:4111–4119

Liu J, Li Z, Wang Y, Xing D (2014) Overexpression of ALTERNATIVE OXIDASE1a alleviates mitochondria-dependent programmed cell death induced by aluminium phytotoxicity in *Arabidopsis*. J Exp Bot 65:4465–4478

Lorrain S, Vailleau F, Balagué C, Roby D (2003) Lesion mimic mutants: keys for deciphering cell death and defense pathways in plants? Trends Plant Sci 8:263–271

Luo H, Laluk K, Lai Z, Veronese P, Song F, Mengiste T (2010) The *Arabidopsis* botrytis susceptible1 interactor defines a subclass of RING E3 ligases that regulate pathogen and stress responses. Plant Physiol 154:1766–1782

Mateo A, Mühlenbock P, Rustérucci C, Chi-Chen Chang C, Miszalski Z, Karpinska B, Parker JE, Mullineaux PM, Karpinski S (2004) LESION SIMULATING DISEASE 1 is required for acclimation to conditions that promote excess excitation energy. Plant Physiol 136:2818–2830

Mishra NS, Tuteja R, Tuteja N (2006) Signaling through MAP kinase networks in plants. Arch Biochem Biophys 452(1):55–68

Mitsuhara I, Malik KA, Miura M, Ohashi Y (1999) Animal cell-death suppressors Bcl-xL and Ced-9 inhibit cell death in tobacco cells. Curr Biol 9:775–778

Mittler R, Herr EH, Orvar BL, Van Camp W, Willekens H, Inzé D, Ellis BE (1999) Transgenic tobacco plants with reduced capability to detoxify reactive oxygen intermediates are hyperresponsive to pathogen infection. Proc Nat Acad Sci 96(24):14165–14170

Møller SG, McPherson MJ (1998) Developmental expression and biochemical analysis of the Arabidopsis atao1 gene encoding an H2O2-generating diamine oxidase. Plant J 13:781–791

Montillet J-L, Chamnongpol S, Rustérucci C, Dat J, Van de Cotte B, Agnel J-P, Battesti C, Inzé D, Van Breusegem F, Triantaphylidès C (2005) Fatty acid hydroperoxides and H2O2 in the execution of hypersensitive cell death in tobacco leaves. Plant Physiol 138:1516–1526

Munns R (2011) Plant adaptations to salt and water stress: differences and commonalities. In: Ismail T (ed) Advances in botanical research. Academic Press, Waltham, pp 1–32

Nakagami H, Kiegerl S, Hirt H (2004) OMTK1, a novel MAPKKK, channels oxidative stress signaling through direct MAPK interaction. J Biol Chem 279:26959–26966

Nakashima K, Takasaki H, Mizoi J, Shinozaki K, Yamaguchi-Shinozaki K (2012) NAC transcription factors in plant abiotic stress responses. Biochim Biophys Acta 1819(2):97–103

Noctor G, Veljovic-Jovanovic S, Driscoll S, Novitskaya L, Foyer CH (2002) Drought and oxidative load in the leaves of C3 plants: a predominant role for photorespiration? Ann Bot 89(7):841–850

op den Camp RGL, Przybyla D, Ochsenbein C, Laloi C, Kim C, Danon A, Wagner D, Hideg E, Göbel C, Feussner I et al (2003) Rapid induction of distinct stress responses after the release of singlet oxygen in Arabidopsis. Plant Cell 15:2320–2332

Örvar BL, McPherson J, Ellis BE (1997) Pre-activating wounding response in tobacco prior to high-level ozone exposure prevents necrotic injury. Plant J 11:203–212

Overmyer K, Brosché M, Pellinen R, Kuittinen T, Tuominen H, Ahlfors R, Keinänen M, Saarma M, Scheel D, Kangasjärvi J (2005) Ozone-induced programmed cell death in the Arabidopsis radical-induced cell death1 mutant. Plant Physiol 137:1092–1104

Paul J-Y, Becker DK, Dickman MB, Harding RM, Khanna HK, Dale JL (2011) Apoptosis-related genes confer resistance to Fusarium wilt in transgenic 'Lady Finger' bananas. Plant Biotechnol J. 9:1141–1148

Pavet V, Olmos E, Kiddle G, Mowla S, Kumar S, Antoniw J, Alvarez ME, Foyer CH (2005) Ascorbic acid deficiency activates cell death and disease resistance responses in Arabidopsis. Plant Physiol 139:1291–1303

Pei ZM, Murata Y, Benning G, Thomine S, Klüsener B, Allen GJ et al (2000) Calcium channels activated by hydrogen peroxide mediate abscisic acid signalling in guard cells. Nature 406(6797):731–734

Pinheiro C, Chaves MM (2011) Photosynthesis and drought: can we make metabolic connections from available data? J Exp Bot 62(3):869–882

Qiao J, Mitsuhara I, Yazaki Y, Sakano K, Gotoh Y, Miura M, Ohashi Y (2002) Enhanced resistance to salt, cold and wound stresses by overproduction of animal cell death suppressors Bcl-xl and Ced-9 in tobacco cells—their possible contribution through improved function of organella. Plant Cell Physiol 43:992–1005

Ramanjulu S, Bartels D (2002) Drought- and desiccation-induced modulation of gene expression in plants. Plant Cell Environ 25(2):141–151

Ren D, Yang H, Zhang S (2002) Cell death mediated by MAPK is associated with hydrogen peroxide production in Arabidopsis. J Biol Chem 277:559–565

Ross M (2016) Histology: a text and atlas (7th ed.), p 94. ISBN 978-1451187427

Rushton DL, Tripathi P, Rabara RC, Lin J, Ringler P, Boken AK et al (2012) WRKY transcription factors: key components in abscisic acid signalling. Plant Biotechnol J 10(1):2–11

Samuel MA, Ellis BE (2002) Double jeopardy: both overexpression and suppression of a redox-activated plant-mitogen-activated protein kinase render tobacco plants ozone sensitive. Plant Cell 14:2059–2069

Shabala S, Cuin TA, Prismall L, Nemchinov LG (2007) Expression of animal CED-9 anti-apoptotic gene in tobacco modifies plasma membrane ion fluxes in response to salinity and oxidative stress. Planta 227:189–197

Simeonova E, Sikora A, Charzyńska M, Mostowska A (2000) Aspects of programmed cell death during leaf senescence of mono- and dicotyledonous plants. Protoplasma 214:93–101

Sivamani E, Bahieldin1 A, Wraith JM, Al-Niemi T, Dyer WE, Ho THD et al (2000) Improved biomass productivity and water use efficiency under water deficit conditions in transgenic wheat constitutively expressing the barley HVA1 gene. Plant Sci 155(1):1–9

Skirycz A, Inzé D (2010) More from less: plant growth under limited water. Curr Opin Biotechnol 21(2):197–203

Skirycz A, De Bodt S, Obata T, De Clercq I, Claeys H, De Rycke R et al (2010) Developmental stage specificity and the role of mitochondrial metabolism in the response of Arabidopsis leaves to prolonged mild osmotic stress. Plant Physiol 152(1):226–244

Suarez MF, Filonova LH, Smertenko A, Savenkov EI, Clapham DH, Von Arnold S, Zhivotovsky B, Bozhkov PV (2004) Metacaspase-dependent programmed cell death is essential for plant embryogenesis. Curr Biol 14:R339–R340

Tanveer M, Shabala S (2018) Targeting redox regulatory mechanisms for salinity stress tolerance in crops. In: Salinity responses and tolerance in plants, vol 1. Springer, Cham, pp 213–234

Tanveer M, Shahzad B, Sharma A, Khan EA (2018) 24-Epibrassinolide application in plants: An implication for improving drought stress tolerance in plants. Plant Physiol Biochem

Urao T, Yakubov B, Satoh R, Yamaguchi-Shinozaki K, Seki M, Hirayama T et al (1999) A transmembrane hybrid-type histidine kinase in Arabidopsis functions as an osmo-sensor. Plant Cell 11(9):1743–1754

Van Camp W, Willekens H, Bowler C, Van Montagu M, Inzé D, Reupold-Popp P, Sandermann H Jr, Langebartels C (1994) Elevated levels of superoxide dismutase protect transgenic plants against ozone damage. Biotechnology 12:165–168

Vaux DL, Cory S, Adams JM (1988) Bcl-2 gene promotes haemopoietic cell survival and cooperates with c-myc to immortalize pre-B cells. Nature 335:440–442

Verslues PE, Sharma S (2010) Proline metabolism and its implications for plant-environment interaction. Arabidopsis Book

Wagner D, Przybyla D, op den Camp R, Kim C, Landgraf F, Lee KP, Würsch M, Laloi C, Nater M, Hideg E et al (2004) The genetic basis of singlet oxygen-induced stress responses of Arabidopsis thaliana. Science 306:1183–1185

Wang W, Pan J, Zheng K, Chen H, Shao H, Guo Y, Hongwu B, Han N, Wang J, Zhu M (2009a) Ced-9 inhibits Al-induced programmed cell death and promotes Al tolerance in tobacco. Biochem Biophys Res Commun 383:141–145

Wang Z, Song J, Zhang Y, Yang B, Chen S (2009b) Expression of baculovirus anti-apoptotic p35 gene in tobacco enhances tolerance to abiotic stress. Biotechnol Lett 31:585–589

Wasilewska A, Vlad F, Sirichandra C, Redko Y, Jammes F, Valon C et al (2008) An update on abscisic acid signaling in plants and more. Mol Plant 1(2):198–217

Williams B, Dickman M (2008) Plant programmed cell death: can't live with it; can't live without it. Mol Plant Pathol 9:531–544

Xiong L, Schumaker KS, Zhu J-K (2002) Cell signaling during cold, drought, and salt stress. Plant Cell 14(Suppl):S165–183

Xu D, Duan X, Wang B, Hong B, Ho T, Wu R (1996) Expression of a late embryogenesis abundant protein gene, HVA1, from barley confers tolerance to water deficit and salt stress in transgenic rice. Plant Physiol 110(1):249–257

Xu H, Xu W, Xi H, Ma W, He Z, Ma M (2013) The ER luminal binding protein (BiP) alleviates Cd2+-induced programmed cell death through endoplasmic reticulum stress-cell death signaling pathway in tobacco cells. J Plant Physiol 170:1434–1441

Yen C-H, Yang C-H (1998) Evidence for programmed cell death during leaf senescence in plants. Plant Cell Physiol 39:922–927

Yoshiba Y, Kiyosue T, Nakashima K, Yamaguchi-Shinozaki K, Shinozaki K (1997) Regulation of levels of proline as an osmolyte in plants under water stress. Plant Cell Physiol 38(10):1095–1102

Yoshida S (2003) Molecular regulation of leaf senescence. Curr Opin Plant Biol 6:79–84

Young TE, Gallie DR (2000) Programmed cell death during endosperm development. Plant Mol Biol 44:283–301

Zhang Y, Turner JG (2008) Wound-induced endogenous jasmonates stunt plant growth by inhibiting mitosis. PLoS ONE 3(11):e3699

Zhang X, Liu S, Takano T (2008) Two cysteine proteinase inhibitors from *Arabidopsis thaliana*, AtCYSa and AtCYSb, increasing the salt, drought, oxidation, and cold tolerance. Plant Mol Biol 68131–143

Zhang L, Li Y, Xing D, Gao C (2009) Characterization of mitochondrial dynamics and subcellular localization of ROS reveal that HsfA2 alleviates oxidative damage caused by heat stress in *Arabidopsis*. J Exp Bot 60:2073–2091

Zhang X, Wang L, Meng H, Wen H, Fan Y, Zhao J (2011) Maize ABP9 enhances tolerance to multiple stresses in transgenic *Arabidopsis* by modulating ABA signaling and cellular levels of reactive oxygen species. Plant Mol Biol 75:365–378

Zhao Z, Stanley BA, Zhang W, Assmann SM (2010) ABA-regulated G protein signaling in Arabidopsis guard cells: a proteomic perspective. J Proteome Res 9(4):1637–1647

Overview of Signal Transduction in Plants Under Salt and Drought Stresses

Aida Shomali and Sasan Aliniaeifard

Abstract Drought and salt stresses are among abiotic stresses that hinder plants from reaching their maximum productivity and thus menacing food security by reducing crop yield worldwide. The mechanisms that stressors affect the plants are very complicated. To cope with environmental stresses, evolutionary processes have been developed through natural selection, which enables plants to integrate molecular and cellular mechanisms to develop adaptive responses. Mechanism by which plant perceive stress signals and further transmission of perceived signals to cell machinery has long been of concern and is still an ongoing point of discussion. In the last decade, introduction of molecular techniques has been used to unravel the mechanism of perception and transmission of stress signals along with plant's adaptive responses. Understanding the physiological and molecular mechanisms of abiotic stress response and tolerance mechanisms may advance our knowledge about candidate genes involving in genetic or epigenetic aspects of stress response, which are essential for improving stress tolerance in plants. In this chapter, the most recent literature on the mechanisms involving in salt and drought stress signal transduction including their signaling pathways and signaling components are used to best elaborate underlying stress response mechanisms involved in salt and drought stresses.

1 Introduction

Plants are frequently exposed to primary abiotic stresses such as low temperature, salt, drought, heat, hypoxia, air pollutants, and heavy metal toxicity, which result in the occurrence of oxidative stress as the secondary stress in plants. The performance and adaption of modern plants to their environments is bestowed by numerous evolutionary processes in response to various types of stresses, especially salt and drought stresses. These abiotic stresses are the major threats for agricultural sustainability

A. Shomali · S. Aliniaeifard (✉)
Photosynthesis Laboratory, Department of Horticulture, College of Aburaihan, University of Tehran, Tehran, Iran
e-mail: aliniaeifard@ut.ac.ir

M. Hasanuzzaman and M. Tanveer (eds.), *Salt and Drought Stress Tolerance in Plants*, Signaling and Communication in Plants, https://doi.org/10.1007/978-3-030-40277-8_9

and food security because they hinder plants to exhibit their full genetic potentials (Zhu 2002; Mahajan and Tuteja 2005). More than half of yield loss in agriculture is due to abiotic stresses (Bray 2000). The climate change in recent decades also exacerbated the determinal effects of abiotic stresses on the crops (Fedoroff et al. 2010).

Abiotic stresses apply adverse effects on plant from both general and specific aspects. Minimizing stomatal and cuticular water loss and disturbance in plant water status are among drought stress responses. Through imbalance in plant water relations and stomatal closure, consequently, drought reduces the rate of photosynthesis. Salinity not only imposes physiological drought to the plants via osmotic stress but also causes ion toxicity in different plant compartments (Zhu 2002). Therefore, mechanism of osmotic adjustment that leads to osmotic hemostasis is an important part of stress tolerance mechanism during salinity. Regulation of gene expression through major or minor changes in transcript level of genes is also among stress responses (Shinozaki and Yamaguchi-Shinozaki 2000). Considering the variety of general and specific stress responses and the fact that some of the stressors may bombard plants in corporation, it can be expected that several hierarchy pathways are developed by plants that act in association to mitigate biotic and abiotic stresses (Chinnusamy et al. 2004; Mahajan and Tuteja 2005). The signal transduction in response to salt and drought stresses involves ionic and osmotic hemostasis signaling pathway, detoxification response pathway, growth regulation response pathway (Zhu 2002), and gas exchange response pathways (Aliniaeifard and van Meeteren 2013; Aliniaeifard et al. 2016a, b; Aliniaeifard and Van Meeteren 2018; Kalhor et al. 2018).

The mechanisms underlie stress signal perception and further transduction of signals to cell machinery, and mechanisms that lead to the generation of adaptive responses are of particular importance. Perception of stress signals links with the genetic responses through signal transduction pathways. The activation of signal transduction cascade in plant cells leads to coordination of physiological and biochemical responses that lead to environmental adaptions (Hasegawa et al. 2000; Zhu 2002; Golldack et al. 2011).

For drought and salt stresses, it is of particular importance to differentiate between the primary and secondary signals that are the result of water deficiency or excess salt. Disturbance in water relations and hyperosmotic stress are the primary signals of drought, which is an important problem for plant cells. In salt stress, the effect of both osmotic and ion toxicity on plant cells is considered as the primary signal. The secondary signals of drought and salt are unique and overlapping that result in complicated effects on plant cells including oxidative stress; damage to cellular components such as membrane lipids, proteins, and nucleic acids; and metabolic dysfunction. In both salt and drought stresses, accumulation of abscisic acid (ABA) as a consequence of turgor-sensing mechanism or hyperosmotic signals is an important phenomenon which is the prerequisite for generation of many adaptive responses (Aliniaeifard and van Meeteren 2013, 2014; Aliniaeifard et al. 2014; Zhu 2016). Also both salt and drought stresses impose osmotic stress at the cellular levels.

2 Osmotic Stress Signaling

2.1 *Protein Kinase Signaling*

Perception is the first step of stress signal transduction. Stress signals are sensed by receptors, which are plasma membrane proteins that bind to and interact with the extracellular molecules. The second messengers are generated once the stress signal is transduced to downstream signaling pathway. Second messengers such as ROS and inositol phosphates regulate the elevation of intracellular Ca^{2+} levels. Elevated Ca^{2+} is sensed by Ca^{2+} sensors, which results in confrontational changes in sensors then, a phosphorylation cascade is initiated which eventually results in regulation of stress-responsive genes. As the stress-responsive genes expressed, TFs that regulate the expression of downstream stress-regulated genes are activated (Xiong et al. 2002; Mahajan and Tuteja 2005). Stress-inducible genes may generate two groups of products, the first group includes protein kinases, phosphatases, cis-regulatory elements, and TFs that regulate the intracellular signaling and stress-inducible gene expression. The second group consists of water channel proteins and late embryogenesis abundant (LEA) proteins that directly act in stress tolerance mechanisms (Hirayama and Shinozaki 2010) (Fig. 1).

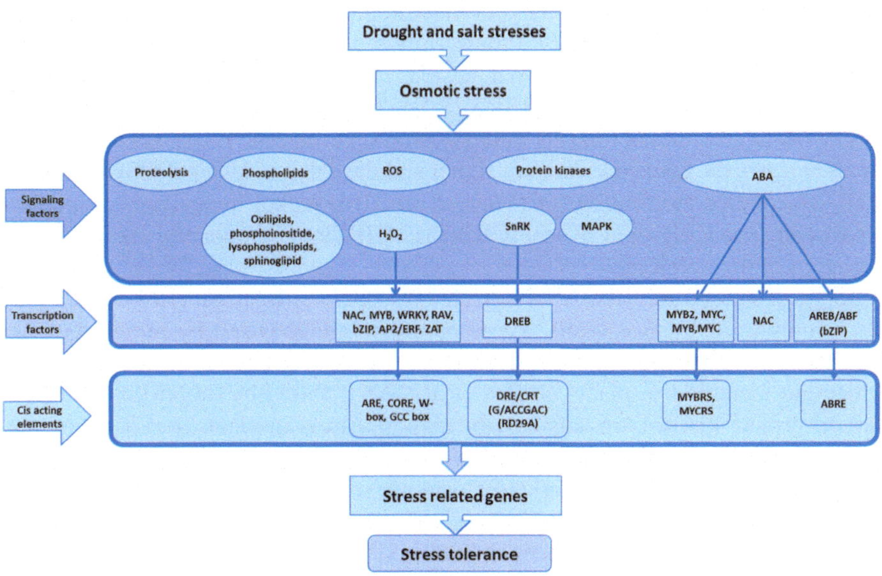

Fig. 1 Simplified model of signaling pathway following drought and salt-induced osmotic stress. Signaling factors including phospholipid, ROS, proteolysis activities, protein kinases, and ABA cause activation/deactivation of transcription factors and consequently influence expression of Cis-acting elements which further results in expression of stress-related genes

Other factors like protein modifiers such as enzymes for glycosylation, methylation, myristoylation, and ubiquitination; scaffold proteins; and adaptors are among accessory molecules that may not directly participate in stress signaling but may function in modification, delivery, or assembly of various signaling components (Xiong and Zhu 2001).

Salt, drought, and osmotic stress treatments as well as ABA activate the SnRK2 family of protein kinases (Boudsocq et al. 2004). It was revealed that under osmotic stress, Arabidopsis mutant plants defected in all 10 SnRK2s showed very sensitive phenotypes that hindered plant growth (Fujii et al. 2011). However, it is yet to find out how osmotic stress activates the SnRK2, while the mechanism of activation of SnRK2s by ABA has been clarified (Zhu 2016). Evidence showed that decuple *snrk2* mutant plants showed disrupted regulation of osmotic stress-related genes and impaired accumulation of ABA, proline, and the second messenger IP3 (Zhu 2016). However, osmotic stress-induced accumulation of ROS is not affected in the mutant plants, which led the researchers to postulate that activation of SnRK2 is upstream of ABA accumulation, which results in osmotic adjustment and other adaptive responses to osmotic stress (Zhu 2016). However, many others have reported that SnRK2 is acting downstream of ABA in signaling pathways of stomata guard cells and functions as positive regulator of ABA signaling cascade (Kim et al. 2010; Aliniaeifard and van Meeteren 2013). Arabidopsis SRK2C/SnRK2.8 is activated by osmotic stress known as a positive regulator of drought tolerance. Overexpression of *SRK2C* has shown to result in hyper-induction of some drought stress-responsive genes, such as the *DREB1A/CBF3* transcription factor. Moreover, based on reverse genetic studies, it was postulated that, besides involving in ABA signaling, subclass II SnRK2s, SRK2C/SnRK2.8, and SRK2F/SnRK2.7 are candidated as the major factors involved in osmotic-stress signaling (Kuromori et al. 2015). The fact that osmotic stress elevates the cytosolic calcium signal and the calcium channel OSCA1 is generally considered as osmosensor (Yuan et al. 2014) makes the calcium-responsive kinases such as CPKs and SCaBP/CBL-PKS/CIPKs to be potential candidates for upstream of the SnRK2s.

In yeast and in animals, MAPK pathways operate the production of compatible osmolytes and antioxidants. These pathways are activated by protein tyrosine kinases, G-protein–coupled receptors, as well as two-component histidine kinases which are identified in plants (Xiong et al. 2002). Based on in-gel kinase assays, several protein kinases are activated by osmotic stress in plants, but it is not clarified if osmotic stress is the activator of the pathway or other factors are involved (Zhu 2002). It was revealed that under osmotic stress, the transcript levels of some protein kinases including a two-component histidine kinase, MAPKKK, MAPKK, and MAPK increase (Mizoguchi et al. 2000). SIMK is a MAP kinase in alfalfa cells shown to be activated under moderate hyperosmotic stress but became inactive in severe hyperosmotic stress (>750 mM NaCl) which make the researchers to postulate that the type of activated kinases corresponds the level of stress (Munnik et al. 1999). A tobacco MAP kinase SIPK that is induced by salicylic acid is also reported to be activated under hyperosmotic stress (Mikolajczyk et al. 2000). A remarkable finding in this regard is that Arabidopsis 40-kD protein kinase activation was not affected by the *sos3* mutation, therefore, it is activated independent of the SOS pathway, which may be an indication for another osmotic stress signaling pathway in plants (Zhu 2002).

2.2 Phospholipid Signaling

Stress signals can be also generated by phospholipids that are important parts of the plasma membrane and located close to the membrane receptors, which make them potential candidates as signaling molecules (Munnik 2001). Generally, signaling lipid molecules are low abundant which is retrieve with their fast turnover (Hou et al. 2016). It is known that various lipid signals including phosphatidic acid, phosphoinositides, sphingolipids, lysophospholipids, oxylipins, N acylethanolamines, and many others are generated in response to osmotic stresses (Hou et al. 2016). These signaling molecules are phospholipids with inositol and Phosphatidylinositol (PI). In addition, PI monophosphate and PI bisphosphate are isomers of the phosphorylated molecules that are also involved in signaling pathways. Phosphatidylinositol 4,5 bisphosphate [PI (4,5) P2] split by Phospholipase C (PLC) results in production of inositol 1,4,5-triphosphate (IP3) and diacylglycerol (DAG) which act as second messengers in stress signaling (Jenks and Haseqawa 2007). Exogenous ABA results in increased level of IP3 in Arabidopsis (Xiong et al. 2001) and protoplast of *vicia faba* guard cells (Lee et al. 1996).

It was also demonstrated that Arabidopsis genes that encode the synthesis of [PI(4,5) P2] have been upregulated in response to water deficit (Jenks and Haseqawa 2007). The phospholipase D (PLD), that contains a large family of enzymes, also acts as a signal transduction molecule (Bargmann et al. 2009). This molecule plays its role in stress signaling by the removal of the terminal phosphodiester bond of phospholipids through hydrolysis reaction, which results in the production of phosphatidic acid (PA) (Testerink and Munnik 2005, 2011). ABA also found to accelerate the reaction and thus signal transduction (Munnik et al. 2000). Phosphatidic acid is transiently generated in plant cells and involved in the osmotic stress response caused by dehydration (Hong et al. 2008a) and salt (Munnik et al. 2000; Yu et al. 2010).

It has been shown that PLD produces PA in the presence of ABA and promotes stomatal closure through inhibition of negative role of ABI1 on the SnRK2 activity (Zhang et al. 2004; Aliniaeifard and van Meeteren 2013). PA also facilitates the translocation of an Arabidopsis MYB transcription factor to the nucleus (Yao et al. 2013). The fact that PA acts as a negative regulator upon binding to the constitutive triple response kinase (CTR1) in Arabidopsis, while is a positive regulator when binds to a sphingosine kinase (Testerink et al. 2007; Guo et al. 2011) suggests that PA–protein interactions can lead to both activation or repression of the protein function. Based on the unique properties of PA, an 'electrostatic/hydrogen-bond switch' model is introduced to explain the molecular basis of PA binding. This model is introduced to fill the gap for lack of information on PA-binding motif that even may not exist (Kooijman and Testerink 2010). PA was also demonstrated to protect the cell by maintaining potassium homeostasis under salt stress (McLoughlin et al. 2012). Under salinity, a PA species couples the MAPK cascades with the SOS1 through taking part in accumulation, binding, and activating the MPK6, which promotes the phosphorylation of the plasma membrane Na^+/H^+ antiporter SOS1 (Yu et al. 2010).

2.3 Proteolysis

Proteolysis controls the magnitude of signaling factors at the protein level, which makes it to turn up as a major signal transduction regulatory mechanism beside post-translational modifications and second messengers. Ubiquitination of target proteins by E3 ubiquitin ligases renders the proteolysis occur selectively. In plants, several types of E3 ligases including RING E3 ligases, such as KEEP ON GOING (KEG), RHA2a/b, SDIR1, AIP2, or components of CUL4-based E3 ligases, DWA1 and DWA2 exist. These E3 ligases are functioning in ABA signaling. Furthermore, some other modifications rather than ubiquitination (e.g. SUMOylation) are involved in ABA and osmotic-stress signaling (Kuromori et al. 2015).

2.4 ABA Signaling

Activation of stress-responsive genes can also mediate the biosynthesis of hormones such as ABA, which can propagate the initial signal or generate a second round of signaling pathways. It is also likely that ABA acts as a completely different signaling component (Xiong and Zhu 2001). Production of ABA is a general phenomenon during drought stress events. It plays role in plant osmotic stress tolerance by regulating plant water status through regulating the swelling/shrinking of guard cells and induction of genes that encode enzymes and proteins involved in cellular dehydration tolerance (Luan 2002; Zhu 2002). ABA-dependent and ABA-independent signal transduction pathways have been suggested in plant response to drought, osmotic stress, and high VPD conditions (Yoshida et al. 2006; Planchet et al. 2010; Riera et al. 2005; Aliniaeifard and van Meeteren 2013). This emerged from the finding of those researchers who reported that exogenous ABA induces a number of genes that are related to respond to dehydration, while others demonstrated that some genes that are induced by dehydration do not respond to exogenous ABA (Zhu 2002; Qiu et al. 2004; Lopez-Perez et al. 2009). Numerous genes that respond to dehydration have been identified in Arabidopsis that are categorized as early response to dehydration genes and responsive to dehydration genes. For gene expression in response to water stress, at least four independent regulatory systems exist. Among them, two are ABA-dependent and the rest are ABA-independent (Shinozaki and Yamaguchi-Shinozaki 2000; Valliyodan and Nguyen 2006).

2.5 ABA-Dependent Signaling Pathway

The crucial role of ABA in osmotic stress-regulated gene expression was revealed by genetic analysis of the ABA-deficient mutant of Arabidopsis, *los5/aba3* and *los6/aba1*. It was demonstrated that in *los5* mutant, the expression of stress-responsive genes such as *RD29A*, *RD22*, *COR15A*, *COR47*, and *P5CS* was reduced or blocked completely (Xiong and Zhu 2001).

In Arabidopsis, expression levels of several ABA synthesis genes are upregulated by drought and salt stress but not affected by cold stress. These genes are including the zeaxanthin epoxidase (*ZEP*; also *LOS6* or *ABA1*), a 9-cis-epoxycarotenoid dioxygenase gene (*NCED3*), the aldehyde oxidase gene (*AAO3*), and the molybdenum cofactor sulfurase gene (*MCSU*; also known as *LOS* or *ABA3*) (Cheng et al. 2002; Zhu 2002; Valliyodan and Nguyen 2006).

Based on cis-element analysis, the promotor region of ABA-dependent drought-related genes contains a conserved ABA-responsive element (ABRE) with ACGT cores (Uno et al. 2000).

The basic-domain leucine zipper (bZIP), including ABRE-binding proteins (AREB) or ABRE-binding factors, predominantly regulated the expression of these genes (ABFs) (Uno et al. 2000).

Based on the requirement for synthesis of new proteins, the ABA-dependent pathway may have two different routes (Shinozaki and Yamaguchi-Shinozaki 1997; Bray 2002): When protein synthesis is not required, an ABA-responsive element ABRE exists in the promoter domain in all ABA-responsive genes. ABRE can lead to ABA-induced gene expression by bounding with its corresponding bzip family of transcription factors such as EmBP-1. Through overexpression studies, it was revealed that among the AREB/ABF subfamily of bZIP TFs in the Arabidopsis genome, *AREB1/ABF2*, *AREB2/ABF4,* and *ABF3* are positive regulators of the ABA signaling under drought stress (Kang 2002; Kim et al. 2004; Furihata et al. 2006). These TFs cooperate to regulate ABRE-dependent gene expression in ABA signaling during drought stress response (Liu et al. 2018). In Arabidopsis, dehydration-responsive rd29B gene, which possesses two ABREs essential for its expression, is mediated by ABA. ABA activates protein kinases needed for ABA-mediated expression of *rd29B* by two bzip TFs, AREB1 and AREB2 (Uno et al. 2000).

At the second route, synthesis of de novo protein is required for the ABA-induced gene expression. ABA-responsive elements combine with the MYC family transcription factors because these genes have no ABREs. In this route, the activation of the genes for MYC family TFs and the synthesis of the TFs have to take place before the activation of any ABA-initiated genes (Shinozaki and Yamaguchi-Shinozaki 1997; Leung and Giraudat 1998; Bray 2002).

2.6 ABA-Independent Signaling Pathway

Many drought-induced genes do not respond to ABA treatment (Shinozaki and Yamaguchi-Shinozaki 2000). These genes are regulated via the drought responsible element (DRE) and C-repeat (CRT) cis-acting elements, in combination with DRE-binding protein (DREB) or transcription factors of C-repeat-binding factor (CBF). Their APETALA2 (AP2) DNA binding domain, such as DREB2, has an important function in ABA-independent gene expression in response to drought stress (Shinozaki and Yamaguchi-Shinozaki 2000; Tran et al. 2004; Sakuma et al. 2006). It was also revealed that transcription factors such as MYB/MYC and WRKY involve in ABA-independent signaling and regulate drought stress response in plants (Abe et al. 1997).

U-box E3 ubiquitin ligases such as AtPUB18 and AtPUB19 have negative roles over ABA signaling and their activity depends on SnRK2, while AtPUB22 and AtPUB23, independent from SnRK2 and ABA, act as negative regulators of drought responses (Seo et al. 2012).

2.7 Shared Components of Both ABA-Dependent and ABA-Independent Pathways

Some components of drought and osmotic signaling pathways are involved in both ABA-dependent and ABA-independent pathways (Chandler and Robertson 1994; Ingram and Bartels 1996). NAC TF triggers both ABA-dependent and ABA-independent signaling pathways under drought stress (Hu et al. 2006). NAC TF that is encoded by the gene *ANAC096* in the ABA-independent signaling pathway triggers response to drought stress by physical interaction with the ABA-dependent transcription factors ABF2 and ABF4 (Xu et al. 2013).

ABA-inducible *RD29* (Responsive to Desiccation) genes, *RD29A* and *RD29B*, are actively involved in plant response to abiotic stresses such as drought and salinity. *RD29B* is an element of ABA-dependent signaling pathway, while *RD29A* is involved in both ABA-dependent and ABA-independent signaling pathways (Yamaguchi-Shinozaki et al. 1995; Narusaka et al. 2003; Kasuga et al. 2004; Hua et al. 2006; Ma et al. 2010). Involvement of *RD29A* in response to drought (Yamaguchi-Shinozaki et al. 1995; Narusaka et al. 2003; Kasuga et al. 2004; Hua et al. 2006; Ma et al. 2010) and salinity (Narusaka et al. 2003; Hua et al. 2006; Qiu et al. 2012) stresses has been frequently reported. The promoter of *RD29A* gene has both ABA-dependent and ABA-independent regulatory cis-elements. The DRE/C-repeat element in the ABA-independent cis-element is needed for water stress-induced gene expression, the process that is independent of ABA (Ingram and Bartels 1996; Leung and Giraudat 1998).

It was demonstrated that the relationship between ABA-dependent and ABA-independent signaling pathways is very complicated (Nakashima et al. 2009,

2014; Roychoudhury et al. 2013). Through crosstalk of ABA-dependent and ABA-independent pathways under drought, the expression of AREB/ABFs and DREB are regulated by participation of core components in ABA signaling and the three subclass III SNF1-related protein kinases 2 (SnRK2s). Moreover, it was revealed that the interaction of ABA-dependent proteins AREB1/ABF2, AREB2/ABF4, and ABF3 with the ABA-independent proteins DREB2A, DREB1A, and DREB2C results in the regulation of the drought stress response (Lee et al. 2010; Kim et al. 2011).

A recent study on *Arabidopsis thaliana* that conducted for network analysis of ABA-dependent and ABA-independent drought responsive genes demonstrated a systematic view of the interplay between ABA-dependent and ABA-independent pathways in response to drought (Liu et al. 2018). However, there are still lots of open questions in this regard.

2.8 Putative Sensors of Salt/Drought Stresses

The fact that plants show specific alteration in gene expression, metabolism, and physiological functions under stress conditions indicates that plant cells are capable of sensing various environmental stimuli. The identification of responsible genes in stress sensing and response faced difficulties due to functional redundancy of genes encoding sensor proteins (Zhu 2016). A sensor in plant may be of essential function, to the extent that loss-of-function mutants may be lethal. As well, dysfunction in one gene may not result in phenotype changes in response to stress. These nature of sensor proteins make their identification difficult, so that, experimental evidence are lacked even for widely accepted stress sensors in microorganisms or mammalian systems (Zhu 2016).

2.9 Sensing Salt/Ionic Stress

Since high concentrations of salt in plant cells results in ionic stress, sensing salt signals is essential for cellular ionic homeostasis under stress conditions. Therefore, the existence of some mechanism in plant cells to sense the changes in ion concentration is likely. It is believed that increase in the levels of Na^+ is sensed by plants via a putative sensor or receptor in the cytosol. However, any sensor or receptor for Na^+ has not yet been identified. Therefore, it's still an open question that how Na^+ is sensed by plants (Zhu 2016; Yang and Guo 2018a).

As far as the elevation in cytosolic Ca^{2+} levels occurs within seconds of exposure to NaCl or mannitol, it is expected that hyperosmotic sensors of plants be coupled with Ca^{2+} channels (Knight et al. 1997; Deinlein et al. 2014). This finding has led the researchers to hypothesize that hyperosmotic stress may be sensed by a mechanically gated Ca^{2+} channel (Kurusu et al. 2013). In response to ionic stress, the protein kinases, including calcium-dependent protein kinases (CDPKs), calcineurin B-like proteins (CBLs), and CBL-interacting protein kinases (CIPKs) which are the downstream of Ca^{2+}, operate the transduction of the hyperosmotic signal to downstream gene transcription and protein activity (Harmon et al. 2000; Weinl and Kudla 2009). Moreover, the TFs including calmodulin-binding transcription activators (CAMTAs) (Pandey et al. 2013), GT element-binding-like proteins (GTLs) (Weng et al. 2012), and MYBs (Yoo et al. 2005) are reported to be directly activated by Ca^{2+}/calmodulin. It is probable that beside Ca^{2+} dependent pathways, Ca^{2+} independent pathway also exists.

Reactive oxygen species (ROS) are both second messenger and the hallmark of abiotic stresses. In other words, it functions in the plant stress response as a signal transduction molecule despite it is a toxic by-product of stress metabolism (Miller et al. 2010). ROS are among second messengers that link to Ca^{2+} signaling in response to salt or hyperosmotic stresses (Jiang et al. 2013). It is likely that ROS acts upstream of the Ca^{2+} signaling. In *Arabidopsis thaliana*, response to high extracellular NaCl depends on ROS-activated Ca^{2+} influx across the plasma membrane with mediation of AtANN1 (Laohavisit et al. 2013).

A ROS-dependent signaling pathway has been suggested during stress response in Arabidopsis plants (Boursiac et al. 2008). In stress condition, ROS is sensed by its sensors, which result in activation of signal transduction pathways that subsequently transports the signal to the nucleus via pathways including redox reactions and Mitogen-Activated Proteins Kinase (MAPK). As a result, stress-regulated cis-acting elements (ARE, CORE, W-box, GCC box, as-1 like) and stress-responsive transcription factors (NAC, MYB, WRKY, RAV, bZIP, AP2/ERF, ZAT) mediate alternation in gene expression pattern which results in prevention of oxidative damage (Singh et al. 2019). ROSs have been also shown to take part in outward-rectifying K^+ channels in Arabidopsis roots (Demidchik et al. 2010). Sever salt stress results in generation of damaging ROS that activates K^+ efflux channels and speeds up cellular K^+ leak whereas moderate decrease in the cytosolic K^+ generates a low level of ROS needed for signaling, therefore, it was postulated that the programmed cell death (PCD) is the result of acute ROS toxicity due to fast and significant K^+ loss (Wu et al. 2018).

Besides ROS, cyclic nucleotides such as cGMP are among the components in NaCl sensing. It has been reported that at the onset of salt stress, a rapid transient in both cGMP and ROS occurred (Kiegle et al. 2000; Donaldson et al. 2004). Increase in cellular cGMP is an indication of salt and osmotic stresses (Donaldson et al. 2004) which further inhibits Na^+ influx (Maathuis and Sanders 2001; Essah et al. 2003; Rubio et al. 2003) and promotes K^+ uptake (Maathuis 2006; 2014; Isner and Maathuis 2018). In this regard, a crosstalk between Ca^{2+} and cGMP signaling has been suggested (Donaldson et al. 2004). The role of cGMP in plant signal transduction, stress response, and ion homeostasis has been also revealed. It was suggested that cGMP exerts its role through regulating cation fluxes by cyclic nucleotide gated channels and modulating stomatal movements. However, there are still lots of open questions in this regard that new technologies should be utilized to unravel the mechanisms by which cGMP functions in stress response (Isner and Maathuis 2018).

An interesting finding in regard to sensing stress is the role of cutin in osmotic stress responses. It was shown that cutin biosynthesis mutants interfere with downstream abscisic acid (ABA) production and many other osmotic-induced responses (Wang et al. 2011).

Salt Overly Sensitive (SOS) is a key regulator of Na^+ homeostasis. Discovery of *SOS* genes results in the description of an ionic stress signaling pathway (Zhu 2002). The *sos1* mutants, led to the discovery of the SOS (salt overly sensitive) pathway (Zhu 2000). The SOS components constitute a typical signaling system that maintains ionic homeostasis in Arabidopsis under salt-stress conditions (Yang and Guo 2018b). Since the response of *sos* mutants are not altered in response to general osmotic stress, it has been postulated that SOS genes specifically function in alleviating the ionic effects of salt stress (Zhu 2001).

SOS1 locates on the plasma membrane and acts as an exchanger of Na^+/H^+ (Qiu et al. 2002; Shi et al. 2000). SOS2 has a regulatory domain (C-terminal) and a catalytic domain (N-terminal) and acts as a Ser/Thr kinase (Liu et al. 2000). SOS3 is a Ca^{2+}-binding protein that perceives calcium signals induced by salt stress (Ishitani et al. 2000). The phosphorylation of the SOS1 antiporter and activation of its transport activity are regulated by SOS2–SOS3 complex during salt stress (Batelli et al. 2007; Quintero et al. 2002).

Generally under salt stress, cytosolic Ca^{2+} signal occurs in response to excess intracellular or extracellular Na^+ (Zhu 2002). The Ca^{2+} signal is decoded by the EF-hand Ca^{2+} binding proteins, SOS3 and SCaBP8/CBL10, and translated to SOS2, a serine/threonine protein kinase (Liu and Zhu 1998; Guo et al. 2001; Lin et al. 2009; Quan et al. 2007; Zhu 2016).

2.10 *SOS3*

SOS3 also known as CBL4 (calcineurin B-like protein) (Mahajan and Tuteja 2005) is a Ca^{2+} sensor which plays pivotal roles in the salt stress-induced Ca^{2+} signal transduction and thus, salt stress tolerance in Arabidopsis. A Ca^{2+} binding protein with an N-myristoylation motif and three Ca^{2+}-binding EF-hands is encoded by *SOS3* (Liu and Zhu 1998). Interestingly, significant similarities have been identified in the amino acid sequence of SOS3 and the regulatory subunit of animal neuronal Ca^{2+} sensors and yeast calcineurin (Ishitani et al. 2000).

The functional role of myristoylation in protein–protein interaction or for protein-membrane attachment has been reported. AtCBL4/SOS3 harbors conserved myristoylation motifs. In addition, based on in vitro myristoylation assays, this protein can be myristoylated which is considered essential for salt tolerance (Ishitani et al. 2000).

Studying the *sos3-1* (a loss-of-function mutant) revealed that it was hypersensitive to salt due to reduced Ca^{2+} binding capacity of SOS3. Supplementation of growth medium with high level of Ca^{2+} resulted in partial reduction in adverse effects of mutation (Ishitani et al. 2000; Liu and Zhu 1998). Calcium sensors vary with respect to their affinity with Ca^{2+}. The level of affinity can be considered as a criteria for distinguishing the variations in Ca^{2+} signals (Chinnusamy et al. 2004). The affinity of SOS3 with Ca^{2+} is lower in comparison to the affinity of caltractin and calmodulin to Ca^{2+} (Ishitani et al. 2000).

2.11 SOS2

SOS2 gene was isolated through positional cloning (Liu and Zhu 1998). It has catalytic and regulatory domains; both are essential for *SOS2* function (Liu and Zhu 1998). In Arabidopsis, *SOS2* encodes a 446-amino acid serine/threonine protein kinase that contains N-terminal catalytic domain. SOS2 is known as the central signaling component. The catalytic domain of SOS2 is similar to the yeast sucrose non-fermenting 1 (SNF1) and mammalian AMP-activated kinase (AMPK) (Zhu 2016). In Arabidopsis, these proteins are known as SNF1-related kinases (SnRKs) which have 3 members in subfamily1 (SnRK1s), 10 members in subfamily2 (SnRK2s), and 25 members in subfamily3 (SnRK3s) (Hrabak et al. 2003). SOS2 also encodes a novel C-terminal regulatory domain (Liu and Zhu 1998). It was shown that this pathway plays important role in sodium and potassium ion homeostasis and salt tolerance, since phosphorylation by protein kinases is a requisite for regulation of signaling cascades (Liu and Zhu 1998).

An increase in intracellular Ca^{2+} levels occurs in response to salt stress which is perceived by Ca^{2+} binding protein SOS3. After binding Ca^{2+}, conformational changes in SOS3 take place in calcium-dependent manner and then SOS3 physically interacts with SOS2. Studying the *sos2sos3* double mutant revealed that these genes act in the same pathway and supports the idea of interaction between SOS2 and SOS3 (Halfter et al. 2000). The yeast two-hybrid analysis and the in vitro binding assay revealed the necessity and efficiency of 21-amino acid motif in the C-terminal regulatory region (FISL motif) of SOS2 for interaction with SOS3. Moreover, it was demonstrated that N-terminal kinase domain and C-terminal regulatory domain interact with each other which suggests the role of regulatory domain in inhibition of kinase activity through blocking the access of substrate to the catalytic site (Guo et al. 2001).

2.12 SOS1

The *SOS1* locus was isolated through positional cloning. Based on computational analysis, the molecular mass of SOS1 is 127 kDa and encodes a polypeptide of 1146 amino acid residues. Based on hydrophobic plot analysis, the N-terminal region of SOS1 is highly hydrophobic and has 12 predicted transmembrane domains. In this regard, substantial similarity between the transmembrane regions of SOS1 and Na^+/H^+ antiporters of animals and microbial origins has been reported on the basis of blast analysis. On the other hand, the C-terminal region of SOS1 is highly hydrophilic. Moreover, based on GenBank database, there are no similarities between C-terminal domain of SOS1 and other identified antiporters (Mahajan and Tuteja 2005).

Molecular genetic studies clarified a relationship between function and molecular structure of the antiporter. In vivo and in vitro studies on the oligomerization state of SOS1 in addition to electron microscopy and single-particle reconstruction of negatively stained full-length and active SOS1 demonstrated that this protein is a homodimer that consists of a membrane domain similar to other antiporters of the family and an elongated, large, and structured cytosolic domain that both contribute to the dimerization of the antiporter. The link between regulation and transport activity of the antiporter is the result of the close contacts between the transmembrane and the cytosolic domains (Núñez-Ramírez et al. 2012).

The SOS1 is known as the largest Na^+/H^+ antiporter sequence due to its long C-terminal hydrophilic tail. It was revealed by phylogenic analysis that SOS1 is related with plasma membrane Na^+/H^+ antiporters such as SOD2 (superoxide dismutase), NHA1, NhaA, and NhaP.

Studies on *Schizosaccharomyces pombe* and *Saccharomyces cerevisae* demonstrated that SOD2 and NHA1 function as plasma membrane Na^+/H^+ antiporters and take part in efflux of excess Na^+ out of the cytosol to the extracellular space. Moreover, the calcineurin pathway plays a major role in Na^+ stress signaling and tolerance in *Saccharomyces cerevisae* (Thewes 2014). The fact that SOS1 is more distantly related to organellar Na^+/H^+ antiporters such as AtNHX1 or NHE6 in comparison with plasma membrane Na^+/H^+ antiporters makes the researchers postulate that SOS1 may function as plasma membrane Na^+/H^+ antiporter to efflux excess ions out of the cytosol (Shi et al. 2000).

Under salt stress condition, the transcription level of *SOS1* was significantly upregulated while ABA was not significantly affecting the *SOS1* gene regulation (Shi et al. 2000). In *sos3* plant, NaCl stress resulted in the upregulation of *SOS1* transcript in either root or shoot and in *sos2* mutant, upregulation of *SOS1* was observed only in roots which suggest that *SOS3*/*SOS2* pathway takes part in mediating of *SOS1* pathway (Shi et al. 2000). The fact that *SOS1* is expressed in root epidermal cells and xylem parenchyma cells, suggests that activated SOS1 can extrude Na^+ into the soil solution and load Na^+ into the xylem for long-distance transport to leaves facilitated by transpirational stream (Shi et al. 2002; Zhu 2016). The long-distance role of *SOS1* is not yet fully understood because *SOS1* is expressed in the leaves xylem parenchyma and its function is not clarified so far. It is suggested that a casparin strip-like structure exists in the leaves xylem parenchyma cells which hinders the direct entrance of xylem stream to the apoplastic space of mesophyll cells. Therefore, it is postulated that SOS1 may take part in extruding Na^+ from the xylem parenchyma cells into the apoplastic space of mesophyll cells. This model predicts that SOS1 may preferentially localize on the side of the cell subjected to the mesophyll apoplastic space (Zhu 2016).

2.13 Sensing Osmotic/Drought Stress

For hyperosmotic stress, OSCA1 (reduced hyperosmolality-induced calcium increase 1) is identified as a putative sensor in Arabidopsis. Under the osmotic stress treatments such as sorbitol and mannitol, reduced calcium spike is identified in *osca1* loss-of-function mutants of Arabidopsis in comparison to wild type plants (Yuan et al. 2014). Since increase in cytosolic free Ca^{2+} in plants is considered as a stress response, reduced calcium spikes in *osca1* demonstrates inability of this mutant plant to combat stress (Yildizli et al. 2018). A plasma membrane protein that forms hyperosmolality-gated calcium permeable channels is encoded by *OSCA1*. The physiological significance of *OSCA1* under the stress conditions is unclear since *osca1* is not characterized with any drought or salt stress phenotypes and the mechanism by which osmotic stress is sensed is not clarified yet (Yuan et al. 2014).

A family of MscS-like proteins and one Piezo homolog are identified in plants which are known as mechano-sensitive channels (Arnadottir and Chalfie 2010; Hedrich 2012). MSL8, a MscS-like protein in Arabidopsis, senses and responds to changes in membrane tension to survive from hypo-osmotic shock during hydration and germination (Hamilton et al. 2015). Therefore, this protein can be a potential candidate for osmotic sensor protein.

2.14 Salt Stress and Ion Signaling

A complex signaling network during salt stress regulates stress-related gene expression which results in the regulation of physiological, developmental, and biochemical adaptions. Some stress adaption mechanisms and factors that are involved in stress tolerance have been revealed now. However, the involving genetic network has not been unraveled (Deinlein et al. 2014). Major signal transduction pathways have been categorized into three pathways, including (1) calcium-dependent signaling pathway that involves dehydration-responsive element/C-repeat (DRE/CRT) or late embryogenesis abundant (LEA) like proteins, (2) mitogen-activated protein kinases (MAPK) cascades renders the production of compatible osmolytes, antioxidants and regulate cell cycle, and (3) salt overly sensitive (SOS) pathway involved in regulating ion homeostasis (Zhu 2001; Kaleem et al. 2018).

Signal perception is at the onset of signal transduction pathway followed by generation of second messengers such as Ca^{2+}, ROSs, and inositol phosphates, which are acting in further modulation of the intracellular calcium level. In the next step, Ca^{2+} sensors perceive elevation in cytosolic Ca^{2+}, and then interact with their interacting partners that leads to initiation of a phosphorylation cascade, as a consequence, alteration in stress-responsive genes or their TFs takes place. The expression of stress-responsive gene enables plant to survive stress condition. The biosynthesis of plant hormones such as ABA, salicylic acid, and ethylene is also the result of alteration in gene expression. The signaling pathways can be more complicated than a single pathway since a second round of signaling may be initiated by amplification of initial signals which may follow the same signaling pattern or a different component of signaling may be applied (Huang et al. 2012; Shao et al. 2007; Mahajan and Tuteja 2005).

Salt stress causes accumulation of Na^+ and deficiency of essential ions such as K^+ in the cytoplasm resulting in disruption of plant ion hemostasis (Binzel and Ratajczak 2002). To prioritize the import of K^+ over Na^+ into plant cells, ion transporters limit Na^+ entry into and exit out of plants, which results in further regulation of Na^+ compartmentation in the vacuole (Huang et al. 2012). To maintain favorable ion ratios in the cytoplasm during salt stress, the salt overly sensitive (SOS) pathway plays a crucial role. Under salt stress, the Ca^{2+} signal is decoded by the EF-hand Ca^{2+} binding proteins, SOS3, and SCaBP8/CBL10 (a SOS3 paralog), then, translated to SOS2, a serine/threonine protein kinase (Liu and Zhu 1998; Guo et al. 2001; Quan et al. 2007; Zhu 2016). SOS3 interact with and activate SOS2 kinase activity in the plasma membrane in roots and SCaBP8 operates the same process in shoots where it is mainly expressed (Halfter et al. 2000; Quan et al. 2007; Lin et al. 2009; Zhu 2016). Then, SOS1 a PM Na^+/H^+ antiporter is phosphorylated and activated as a consequence of SOS2 activation (Lin et al. 2009; Qiu et al. 2002; Shi et al. 2000). Further, SOS1 promotes the transport of Na^+ from the cytoplasm to the apoplast (Shi et al. 2000; Zhu 2002).

An intermolecular interaction between cytosolic and transmembrane domains of SOS1 results in auto-inhibition of SOS1 activity (Núñez-Ramírez et al. 2012). SOS2 relieves auto-inhibition of SOS1 via phosphorylating the C-terminal auto-inhibitory domain of serine 1,044 (Quintero et al. 2011; Jarvis et al. 2014). SOS2 and SOS3 cooperate to regulate SOS1 expression which involves in the transport of Na^+ from the cytoplasm to the apoplast (Shi et al. 2000; Zhu 2002).

A remarkable finding in regard to the SCaBP8 protein is that it occurs in both soluble and membrane fractions whereas the phosphorylated form of SCaBP8 particularly appears in membrane fraction (Quan et al. 2007). It has been shown that for SCaBP8 targets to the plasma membrane, phosphorylation is not a precondition since the truncated, cytosolic mutant SCaBP8DN protein is demonstrated to be phosphorylated in vitro by SOS2 but without in vivo phosphorylation. Moreover, mutation of *Ser-237* did not couple with the targeting of SCaBP8 to the plasma membrane. It is turns out that the interaction between SCaBP8 and SOS2 which can theoretically occur in the cytoplasm is the following event of additional inputs, suggesting that the phosphorylation of SCaBP8 and stabilization of the ternary complex are induced by the interaction of the SCaBP8-SOS2 complex with its downstream target (Lin et al. 2009).

Arabidopsis loss-of-function mutants of the Ca^{2+} sensor *CBL10* showed a severe salt-sensitive phenotype, which indicates the role of Ca^{2+} signaling in the extrusion of NaCl from the cell and the intracellular sequestration of the toxic NaCl (Köster et al. 2019). Another SOS1 regulating mechanism in response to sever salt stress is related to the nuclear Ca^{2+} signaling (Guan et al. 2013). Signals of nuclear Ca^{2+} activates RSA1, a Ca^{2+} binding protein that complexes with RITF1 (RSA1 interacting transcription factor) further, the RSA1-RITF1 complex binds the SOS1 promoter to propagate its transcription (Guan et al. 2013).

Another issue that rises in regard to Na^+ sequestration is the finding that Arabidopsis Vacular-ATPase at the tonoplast activates for efficient nutrient storage but not for Na^+ accumulation (Krebs et al. 2010; Schumacher and Krebs 2010). It was demonstrated that lack of tonoplast-localized V-ATPase subunits VHA-A2 and VHA-A3 in double mutants did not affect salt tolerance and accumulation while defect in V-ATPase activity in the trans-Golgi network/early endosome reduced VHA-A1 accumulation and increased salt sensitivity. Based on previous studies, a role for the endosomal/pre-vacuole compartment for developing salt tolerance and Na^+ sequestration is suggested in plants (Hernandez et al. 2009; Krebs et al. 2010). However, the way endosomal compartment functions in salt tolerance has to be elucidated. It should be clarified if the endosomal system represents a primary target of Na^+ toxicity or endosomal vesicles take up Na^+ and discharge it into the vacuole. In this regard, it was demonstrated that double mutants of the endosomal/Golgi/trans-Golgi-localized proteins NHX5 and NHX6 showed severe salt sensitivity and the vacuolar trafficking of proteins were defected in these mutants (Bassil et al. 2011a, b) which may be an indication of the role of endosomal vesicles in taking up and discharge of Na^+ into vacuole (Köster et al. 2019).

3 Ion Channel Regulation Under Salt Stress

3.1 *NHX and SOS*

Under salt stress, low cytoplasmic Na^+ concentrations in plant cells are maintained via the tonoplast-localized Na^+/H^+ exchanger 1 (NHX1) (Blumwald and Poole 1985) and plasma membrane-localized SOS1 (NHX7) Na^+/H^+ antiporters (Qiu et al. 2002; Yamaguchi et al. 2013). Most of the NHXs function in Na^+ detoxification by transferring Na^+ to the vacuole, while the SOS signaling pathway acts in export of Na^+ out of the cell. Previous researches revealed that increased salinity tolerance in plant is attained by constitutive overexpression of *AtNHX1* and its orthologs in Arabidopsis and other plant species, such as tomato (*Solanum lycopersicum*) and rice (*Oriza sativa*) (Apse et al. 1999; Zhang and Blumwald 2001).

NHX-type proteins are also shown to take part in compartmentalization of K^+ into vacuoles and homeostasis of cellular pH (Barragán et al. 2012) and also in turgor adjustment and stomatal regulation (Barragan et al. 2012; Andres et al. 2014). *AtNHX1* overexpressed plants have increased vacuolar K^+ and increased transport of K^+ from root to shoot. Furthermore, reduced Na^+ stress due to enhanced intra-cellular K^+/Na^+ ratios have been observed in tomato when *AtNHX1* was overexpressed (Bassil et al. 2011a; Barragán et al. 2012).

Under salt stress, the activity of NHX, a vacuolar Na^+/H^+ exchanger, as well as H^+-ATPase are regulated positively by the SOS3 and SOS2 complex (Batelli et al.

2007; Qiu et al. 2004). Moreover, the inhibition of PKS5 kinase activity results in activation of J3 (DnaJ homolog3) which consequences in activation of PM H^+-ATPase by which a proton gradient is generated across the plasma membrane which is the driving force for the activation of the Na^+/H^+ antiporter (Yang and Guo 2018a). Another event is the release of SOS2 kinase activity as a result of degradation of the 14-3-3 and GI proteins through the 26S proteasomal pathway (Kim et al. 2013; Tan et al. 2016). However, the involvement of the salt-induced increase in ABA levels in release of ABI2-dependent SOS2 inhibition under salt-stress conditions remained controversial (Yang and Guo 2018b).

When there is no salt stress, SCaBP (SOS3-like Ca^{2+}-binding protein) and PKS5 (SOS2-like protein5)/PKS24 (SOS2-like protein24) suppress the plasma membrane H^+-ATPase activity. The kinase activity of SOS2 is inhibited by 14-3-3, GI (GIGANTEA), and ABI2 (ABA INSENSITIVE2) (Kim et al. 2013; Tan et al. 2016).

For generation of stress response, where Ca^{2+} plays the role of second messenger, especially in regulation of ion transporter activities, the CBL-CIPK module is essential for various abiotic stress signaling pathways. Notably, a large number of possible SCaBP/CBL-PKS/CIPK combinations is discovered in plants, suggesting that the Ca^{2+}-SOS3-SOS2 signaling principle is widely used in plants (Zhu 2016). On the basis of previous researches, it is widely accepted that Ca^{2+}-SCaBP/CBL-PKS/CIPK signaling pathways take part in plant adaption to stress condition (Yu et al. 2014; Yang and Guo 2018b).

Based on the most recent reports on salt stress signaling, the elevation in Ca^{2+} is perceived by SCaBP8 and SOS3, then interacts with SOS2 and regulates activation of SOS1 and NHX (a possible vacuolar Na^+/H^+ exchanger) activity. SOS1 Na^+/H^+ antiporter activation is regulated when PLD1-derived PA interacts with MPK6. Vacuolar Na^+/H^+ exchanger activity is derived by a proton gradient across the vacuolar membrane which is the result of activation of Vacuolar $H/^+$-ATPase by SOS2 (Yang and Guo 2018a). The H^+ gradient established by vacuolar H^+-pyrophosphatase and H^+-ATPase drives the transport of Na^+ from the cytoplasm to the vacuole (Brini and Masmoudi 2012; Drozdowicz and Rea 2001). However, it is yet to find out which NHX is responsible for this cellular activity (Bassil et al. 2011b; Barragán et al. 2012).

3.2 AKT

AKT1 is known as the most investigated K^+ channel in Arabidopsis (Ren et al. 2013). This channel regulates the K^+/Na^+ ratio in plant cells under salt stress (Qi and Spalding 2004; Ma et al. 2009; Ardie et al. 2010; Assaha et al. 2017). In many studies, it was reported that CBL-CIPK signaling pathways regulate the activities of some plants ion channels or transporters such as AKT1 and SOS1 (Ren et al. 2013). In this regard, the protein kinase CBL1/9-CIPK23 that phosphorylates and activates the AKT1 channel identified as the key regulatory factor of AKT1-mediated K^+ uptake under low-K^+ stress (Li et al. 2006; Xu et al. 2006). On the other hand, the CBL4-CIPK24 (SOS3-SOS2) complex is found to regulate the SOS1-mediated salt tolerance (Liu and Zhu 1998; Qiu et al. 2002). To enhance K^+ uptake under the low-K^+ conditions, phosphorylation and activation of the potassium channel AKT1 is regulated by CBL1/CBL9-CIPK23 (Xu et al. 2006). Moreover, in Arabidopsis, CBL10 is found to directly interact with the K^+ channel AKT1 in a CIPK-independent manner in order to maintain K^+ homeostasis (Ren et al. 2013). Besides, AKT1-mediated K^+ uptake in Arabidopsis roots under K^+ limiting conditions is negatively regulated by Potassium channel α-subunit AtKC1 (Geiger et al. 2009; Wang et al. 2010).

It was also demonstrated that CBL10 modulates AKT1-mediated inward K^+ currents. Altogether, adjustment and modulation of AKT1 by various regulators is a key phenomenon leading to plant adaption to various stimuli such as low-K^+ stress (Ren et al. 2013).

For maintaining the K^+ permeability that is an important factor for cell hemostasis and plant growth, the presence of Na^+, the SOS1 Na^+-H^+ antiporter is necessary. The interactions between homeostatic mechanisms of K^+ and Na^+ are due to their physicochemical similarities. Aspects of K^+ nutrition in the Arabidopsis *sos* mutants, and salt sensitivity in the K^+ transport mutants *akt1* (Arabidopsis K^+ transporter) and skor (shaker-like K^+ outward-rectifying channel) were investigated in regard of the physiological interactions between K^+ and Na^+. Arabidopsis mutants defected in K^+ uptake (*akt1*) or distribution (skor), and mutant sensitive to salt (sos1, 2, and 3) have been studied for unraveling the mechanism of protecting effects of K^+ against Na^+ stress. It was revealed that SOS1 plays a protective role in the K^+ permeability of the membrane and the AKT1 K^+ channel protects the cell from inhibiting effects of Na^+. Therefore, researchers concluded that AKT1 is a target of salt stress, which explains the mitigating effect of K^+ on Na^+ toxicity. Interestingly, the efficiency of Arabidopsis roots to uptake K^+ is not influenced by sos1 and its regulators in the absence of Na^+ (Qi and Spalding 2004). It is worth to mention that the effect of Na^+ on K^+ transport is reported to be more important at the uptake stage than at the xylem loading stage, since the *akt1* seedlings that was defected in K^+ uptake was salt sensitive during early seedling development, but seedlings of *skor* that was defected in K^+ distribution were normal (Qi and Spalding 2004).

3.3 HKT

HKT is another transporter that plays an important role in long-distance Na^+ transport and root-to-shoot Na^+ partitioning (Mäser et al. 2002a; Deinlein et al. 2014). Two distinct subgroups of HKT transporter revealed through sequence and transport analyses include the class I that mediates Na^+-selective transport (Uozumi et al. 2000; Mäser et al. 2002b) and class II which is Na^+–K^+ co-transport (Rubio et al. 1995).

Analysis of the Na^+ and K^+ conductance of AtHKT1;1 in root stelar cells, revealed that Na^+ currents have been observed in AtHKT1 clearly, while AtHKT1;1-dependent K^+ currents were not observed suggesting this transporter is Na^+ selective (Xue et al. 2011). Furthermore, this transporter has shown no strong voltage dependence. The Na^+ selectivity of AtHKT1 has been also reported in *Xenopus oocytes* and yeast (Uozumi et al. 2000; Mäser et al. 2002b).

Despite it is believed that response to ABA leads to salt tolerance, unexpectedly, ABA is reported to enhance salt sensitivity via AtHKT1;1 regulation. It was revealed that the two TFs, ABI4 and AtZIP24 act as negative regulators of *AtHKT1;1* expression. Decreased Na^+ accumulation and increased expression levels of *AtHKT1;1* gene have been observed in the loss-of-function mutant of *ABI4*, an ABA-responsive transcription factor. In contrast, decreased salt tolerance due to suppressed *AtHKT1;1* expression is observed in *ABI4*-overexpressor plants (Shkolnik-Inbar et al. 2013). It was demonstrated that AtHKT1;1 in Arabidopsis and OsHKT1;5 in rice reduce Na^+ accumulation in the xylem sap and increase K^+ concentrations in xylem sap and in leaves (Horie et al. 2005; Ren et al. 2005).

Moreover, a soil microbe, *Bacillus subtilis*, has been shown to help plant to cope with salinity stress (Lastochkina et al. 2017). Inoculation with this microbe induces the expression of *AtHKT1;1* in shoots and the repression of the expression in roots of Arabidopsis which results in Na^+-tolerance (Zhang et al. 2008).

3.4 MAPK

Another involving factor in salt stress response is Mitogen-activated protein kinase (MAPK) cascades, which transduce signals from the extracellular matrix to the nucleus in signal transduction of salt-exposed plants. Three sequentially phosphorylating and activating elements form the (MAPK) cascades including a MAP kinase kinase kinase (MAPKKK/MEKK), a MAP kinase kinase (MAPKK/MKK), and a MAP kinase (MAPK/MPK) (Mishra et al. 2006). Arabidopsis genome encodes 60 MAPKKKs, 10 MAPKKs, and 20 MAPKs (Nakagami et al. 2005).

In Arabidopsis, MEKK1-MKK2-MPK4/MPK6 cascade involves in salt-stress signaling (Teige et al. 2004). *AtMPK6* involves in salt stress responses via phosphorylating and activating the SOS1 Na^+/H^+ antiporter activity (Yu et al. 2010). In alfalfa (*Medicago sativa*), the MKK-MK cascade is activated in response to salt stress (Kiegerl et al. 2000). In rice (*Oryza sativa*), the MKK1-MPK4 cascade regulates the expression of transcription factor of genes that involve in salt-stress signaling (Wang et al. 2014). It was revealed that in Arabidopsis, at least three MAP kinases were enzymatically activated by salt and other environmental signals (Ichimura et al. 2000).

It is probable that induction of SOS1 activity by salt stress may also depend on phosphorylation in a phospholipase D (PLD) signaling pathway-dependent manner beside the SOS3–SOS2 complex (Yu et al. 2010). In response to High Na^+ concentrations, an increase in enzyme activity of PLDα1 occurs that leads to fast accumulation of PA, a lipid second messenger. Accumulation of PA activates MPK6 which in turn directly phosphorylates SOS1 in loss-of-function mutants of PLDα1 and MPK6 leading to accumulation of more Na^+ in their shoots under salt stress (Yu et al. 2010; Isayenkov and Maathuis 2019).

ROS accumulation as an indicator of salt stress (Hong et al. 2008b) takes part in activation of downstream MAPK cascades (Miller et al. 2010; Maathuis et al. 2014) via transcriptional regulation. In this regard, it was shown that ROS generated by plasma membrane localized NADPH oxidase that facilitates the stabilization of *AtSOS1* transcripts (Chung et al. 2008). Furthermore, ROS-sensitive transcription factor ERF1 (ethylene response factor) in rice can bind to multiple promoters such as MAPKs, which leads plants to revive under salt stress conditions (Schmidt et al. 2013).

4 Concluding Remark

Salinity and drought stresses are among the major environmental threats that hinder plants to exhibit their highest genetic potentials. Plants have adopted various adaption mechanisms to avoid or tolerate drought and salt stresses, which are discussed in many research papers thus far. However, there are still lots of open questions in this regard that should be answered to shed light on the complex network of drought and salt stress signaling pathways in plants. Identifying the mechanism for stress sensing, signal transduction, and adaptive responses and finding out whether the mechanism is specific or general in plant species are challenging issues in the field of plant salinity and drought stress signal transduction.

References

Abe H, Yamaguchi-Shinozaki K, Urao T, Iwasaki T, Hosokawa D, Shinozaki K (1997) Role ofArabidopsis MYC and MYB homologs in drought- and abscisic acid-regulated gene expression. Plant Cell 9(10):1859–1868

Aliniaeifard S, van Meeteren U (2013) Can prolonged exposure to low VPD disturb the ABA signalling in stomatal guard cells? J Exp Bot 64(12):3551–3566

Aliniaeifard S, van Meeteren U (2014) Natural variation in stomatal response to closing stimuli among *Arabidopsis thaliana* accessions after exposure to low VPD as a tool to recognize the mechanism of disturbed stomatal functioning. J Exp Bot 65(22):6529–6542

Aliniaeifard S, Van Meeteren U (2018) Natural genetic variation in stomatal response can help to increase acclimation of plants to dried environments. In: International symposium on the role of plant genetic resources in reclaiming lands and environment deteriorated by human 1190, pp 71–76

Aliniaeifard S, Malcolm Matamoros P, van Meeteren U (2014) Stomatal malfunctioning under low vapor pressure deficit (VPD) conditions: induced by alterations in stomatal morphology and leaf anatomy or in the ABA signaling? Physiol Plant 152(4):688–699

Aliniaeifard S, Hajilou J, Tabatabaei S-J (2016a) Photosynthetic and growth responses of olive to proline and salicylic acid under salinity condition. Not Bot Horti Agrobot Cluj Napoca 44(2):579–585

Aliniaeifard S, Hajilou J, Tabatabaei S-J, Sifi-Kalhor M (2016b) Effects of ascorbic acid and reduced glutathione on the alleviation of salinity stress in olive plants. Int J Fruit Sci 16(4):395–409

Andres Z, Perez-Hormaeche J, Leidi E-O, Schlucking K, Steinhorst L, McLachlan D-H, Schumacher K, Hetherington A-M, Kudla J, Cubero B, Pardo JM (2014) Control of vacuolar dynamics and regulation of stomatal aperture by tonoplast potassium uptake. PNAS 111:E1806–E1814

Apse M-P, Aharon GS, Snedden WA, Blumwald E (1999) Salt tolerance conferred by overexpression of a vacuolar Na+/H+ antiport in Arabidopsis. Science 285(5431):1256–1258

Ardie SW, Liu S, Takano T (2010) Expression of the AKT1-type K^+ channel gene from Puccinellia tenuiflora, PutAKT1, enhances salt tolerance in Arabidopsis. Plant Cell Rep 29(8):865–874

Arnadottir J, Chalfie M (2010) Eukaryotic mechanosensitive channels. Annu Rev Biophys 39:111–137

Assaha DV, Ueda A, Saneoka H, Al-Yahyai R, Yaish M-W (2017) The role of Na^+ and K^+ transporters in salt stress adaptation in glycophytes. Front Physiol 8:509

Bargmann B-O-R, Laxalt A-M, Riet B-t, van Schooten B, Merquiol E, Testerink C, Munnik T (2009) Multiple PLDs required for high salinity and water deficit tolerance in plants. Plant Cell Physiol 50:78–89

Barragán V, Leidi E-O, Andrés Z, Rubio L, De Luca A, Fernández J-A, Cubero B, Pardo J-M (2012) Ion Exchangers NHX1 and NHX2 mediate active potassium uptake into vacuoles to regulate cell turgor and stomatal function in Arabidopsis. Plant Cell 111

Bassil E, Ohto M-A, Esumi T, Tajima H, Zhu Z, Cagnac O, Belmonte M, Peleg Z, Yamaguchi T, Blumwald E (2011a) The Arabidopsis intracellular Na^+/H^+ antiporters NHX5 and NHX6 are endosome associated and necessary for plant growth and development. Plant Cell 110

Bassil E, Tajima H, Liang Y-C, Ohto M-A, Ushijima K, Nakano R, Esumi T, Coku A, Belmonte M, Blumwald E (2011b) The Arabidopsis Na^+/H^+ antiporters NHX1 and NHX2 control vacuolar Ph and K^+ homeostasis to regulate growth, flower development, and reproduction. Plant Cell 111

Batelli G, Verslues PE, Agius F, Qiu Q, Fujii H, Pan S, Schumaker K-S, Grillo S, Zhu C (2007) SOS2 promotes salt tolerance in part by interacting with the vacuolar H^+-ATPase and upregulating its transport activity. Mol Cell Biol 27(22):7781–7790

Binzel M, Ratajczak R (2002) Function of membrane transport systems under salinity: tonoplast. In: Salinity: environment-plants-molecules. Springer, pp 423–449

Blumwald E, Poole R (1985) Na^+/H^+ antiport in isolated tonoplast vesicles from storage tissue of *Beta vulgaris*. Plant Physiol 78(1):163–167

Boudsocq M, Barbier-Brygoo H, Laurie're C (2004) Identification of nine sucrose nonfermenting 1 related protein kinases 2 activated by hyperosmotic and saline stresses in *Arabidopsis thaliana*. J Biol Chem 279:41758–41766

Boursiac Y, Boudet J, Postaire O, Luu D-T, Tournaire-Roux C, Maurel C (2008) Stimulus-induced downregulation of root water transport involves reactive oxygen species-activated cell signalling and plasma membrane intrinsic protein internalization. Plant J 56(2):207–218

Bray E-A (2000) Response to abiotic stress. Biochem Mol Biol Plants 1158–1203

Bray E-A (2002) Abscisic acid regulation of gene expression during waterdeficit stress in the era of the Arobidopsis genome. Plant Cell Environ 25(2):153–161

Brini F, Masmoudi K (2012) Ion transporters and abiotic stress tolerance in plants. ISRN molecular biology

Chandler PM, Robertson M (1994) Gene expression regulated by abscisic acid and its relation to stress tolerance. Annu Rev Plant Biol 45(1):113–141

Cheng WH, Endo A, Zhou L et al (2002) A unique short-chain Dehydrogenase/reductase in Arabidopsis glucose signaling and abscisic acid biosynthesis and functions. Plant Cell 14(11):2723–2743

Chinnusamy V, Schumaker K, Zhu J (2004) Molecular genetic perspectives on cross-talk and specificity in abiotic stress signalling in plants. J Exp Bot 55(395):225–236

Chung Y-M, Lee S-B, Kim H-J, Park S-H, Kim J-J, Chung J-S, DoYoo Y (2008) Replicative senescence induced by Romo1-derived reactive oxygen species. J Biol Chem 283(48):33763–33771

Deinlein U, Stephan A-B, Horie T, Luo W, Xu G, Schroeder J (2014) Plant salt-tolerance mechanisms. Trends Plant Sci 19(6):371–379

Demidchik V, Cuin T-A, Svistunenko D (2010) Arabidopsis root K^+ efflux conductance activated by hydroxyl radicals: single-channel properties, genetic basis and involvement in stress-induced cell death. J Cell Sci 064352

Donaldson L, Ludidi N, Knight M-R, Gehring C, Denby K (2004) Salt and osmotic stress cause rapid increases in *Arabidopsis thaliana* cGMP levels. FEBS Lett 569(1–3):317–320

Drozdowicz Y-M, Rea P (2001) Vacuolar H^+ pyrophosphatases: from the evolutionary backwaters into the mainstream. Trends Plant Sci 6(5):206–211

Essah P-A, Davenport R, Tester M (2003) Sodium influx and accumulation in Arabidopsis. Plant Physiol 133(1):307–318

Fedoroff N-V, Battisti D-S, Beachy R-N, Cooper P-J, Fischhoff D-A, Hodges C, Knauf V-C, Lobell D, Mazur B-J, Molden D (2010) Radically rethinking agriculture for the 21st century. Science 327(5967):833–834

Fujii H, Verslues P-E, Zhu J-K (2011) Arabidopsis decuple mutant reveals the importance of SnRK2 kinases in osmotic stress responses in vivo. PNAS 108(4):1717–1722

Fujita Y, Nakashima K, Yoshida T, Katagiri T, Kidokoro S, Kanamori N, Umezawa T, Fujita M, Maruyama K, Ishiyama K, Kobayashi M (2009) Three SnRK2 protein kinases are the main positive regulators of abscisic acid signaling in response to water stress in Arabidopsis. Plant Cell Physiol 50(12):2123–2132

Furihata T, Maruyama K, Fujita Y, Umezawa T, Yoshida R, Shinozaki K, Yamaguchi-Shinozaki K (2006) Abscisic acid-dependent multisite phosphorylation regulates the activity of a transcription activator AREB1. PNAS 103(6):1988–1993

Geiger D, Becker D, Vosloh D, Gambale F, Palme K, Rehers M, Anschuetz U, Dreyer I, Kudla J, Hedrich R (2009) Heteromeric AtKC1/AKT1 channels in Arabidopsis roots facilitate growth under K^+ limiting conditions. J Biol Chem M109

Golldack D, Lüking I, Yang O (2011) Plant tolerance to drought and salinity: stress regulating transcription factors and their functional significance in the cellular transcriptional network. Plant Cell Rep 30(8):1383–1391

Guan Q, Wu J, Yue X, Zhang Y, Zhu J (2013) A nuclear calcium sensing pathway is critical for gene regulation and salt stress tolerance in Arabidopsis. PLoS Genet 9(8):e1003755

Guo Y, Halfter U, Ishitani M, Zhu J (2001) Molecular characterization of functional domains in the protein kinase SOS2 that is required for plant salt tolerance. Plant Cell 13(6):1383–1400

Guo L, Mishra G, Taylor K, Wang X (2011) Phosphatidic acid binds and stimulates Arabidopsis sphingosine kinases. J Biol Chem 286:13336–13345

Halfter U, Ishitani M, Zhu J (2000) The Arabidopsis SOS2 protein kinase physically interacts with and is activated by the calcium-binding protein SOS3. PNAS 97(7):3735–3740

Hamilton E-S, Jensen G-S, Maksaev G, Katims A, Sherp A-M, Haswell E-S (2015) Mechanosensitive channel MSL8 regulates osmotic forces during pollen hydration and germination. Science 350(6259):438–441

Harmon A-C, Gribskov M, Harper J (2000) CDPKs—a kinase for every Ca^{2+} signal? 5:154–159

Hasegawa P-M, Bressan R-A, Zhu J-K, Bohnert H (2000) Plant cellular and molecular responses to high salinity. Annu Rev Plant Biol 51(1):463–499

Hedrich R (2012) Ion channels in plants. Physiol Rev 92(4):1777–1811

Hernandez A, Jiang X, Cubero B, Nieto P-M, Bressan R-A, Hasegawa P-M, Pardo J-M (2009) Mutants of the Arabidopsis thaliana Cation/H^+ Antiporter AtNHX1 conferring increased salt tolerance in yeast. J Biol Chem 284:14276–14285

Hirayama T, Shinozaki K (2010) Research on plant abiotic stress responses in the post-genome era: past, present and future. Plant J 61(6):1041–1052

Hong Y, Zheng S, Wang X (2008a) Dual functions of phospholipase Dα1 in plant response to drought. Mol Plant 1(2):262–269
Hong J-K, Yun B-W, Kang J-G, Raja M-U, Kwon E, Sorhagen K (2008b) Nitric oxide function and signaling in plant disease resistance. J Exp Bot 59(2):147–154
Horie T, Motoda J, Kubo M, Yang H, Yoda K, Horie R, Chan W-Y, Leung H-Y, Hattori K, Konomi M (2005) Enhanced salt tolerance mediated by AtHKT1 transporter-induced Na^+ unloading from xylem vessels to xylem parenchyma cells. Plant J 44(6):928–938
Hou Q, Ufer G, Bartels D (2016) Lipid signalling in plant responses to abiotic stress. Plant Cell Environ 39(5):1029–1048
Hrabak E-M, Chan C-W, Gribskov M, Harper J-F, Choi J-H, Halford N, Kudla J, Luan S, Nimmo H-G, Sussman M (2003) The Arabidopsis CDPK-SnRK superfamily of protein kinases. Plant Physiol 132(2):666–680
Hu H, Dai M, Yao J, Xiao B, Li X, Zhang Q, Xiong L (2006) Overexpressing a NAM, ATAF, and CUC (NAC) transcription factor enhances drought resistance and salt tolerance in rice. PNAS 103(35):12987–12992
Hua Z-M, Yang X, Fromm M-E (2006) Activation of the NaCl- and drought-induced RD29A and RD29B promoters by constitutively active Arabidopsis MAPKK or MAPK proteins. Plant Cell Environ 29:1761–1770
Huang G-T, Ma S-L, Bai L-P, Zhang L, Ma H, Jia P, Liu J, Zhong M, Guo Z (2012) Signal transduction during cold, salt, and drought stresses in plants. Mol Biol Rep 39(2):969–987
Ichimura K, Mizoguchi T, Yoshida R, Yuasa T, Shinozaki K (2000) Various abiotic stresses rapidly activate Arabidopsis MAP kinases ATMPK4 and ATMPK6. Plant J 24(5):655–665
Ingram J, Bartels D (1996) The molecular basis of dehydration tolerance in plants. Annu Rev Plant Biol 47(1):377–403
Isayenkov S, Maathuis F-J-M (2019) Plant salinity stress. Many unanswered questions remain. Front Plant Sci 10
Ishitani M, Liu J, Halfter U, Kim C-S, Shi W, Zhu J (2000) SOS3 function in plant salt tolerance requires N-myristoylation and calcium binding. Plant Cell 12(9):1667–1677
Isner J-C, Maathuis F-J (2018) cGMP signalling in plants: from enigma to main stream. Funct Plant Biol 45(2):93–101
Jarvis D-E, Ryu C-H, Beilstein M-A, Schumaker K (2014) Distinct roles for SOS1 in the convergent evolution of salt tolerance in *Eutrema salsugineum* and *Schrenkiella parvula*. Mol Biol Evol 31(8):2094–2107
Jenks M-A, Haseqawa P-M (2007) Plant abiotic stress. Blackwell Publishing, Oxford, UK
Jiang Z, Zhu S, Ye R, Xue Y, Chen A, An L, Pei Z (2013) Relationship between NaCl- and H_2O_2-induced cytosolic Ca^{2+} increases in response to stress in Arabidopsis. PLoS ONE 8(10):e76130
Kaleem F, Shabir G, Aslam K, Rasul S, Manzoor H, Shah S-M, Khan A (2018) An overview of the genetics of plant response to salt stress: present status and the way forward. Appl Biochem Biotechnol 1–29
Kalhor M, Aliniaeifard S, Seif M, Asayesh E, Bernard F, Hassani B, Li T (2018) Enhanced salt tolerance and photosynthetic performance: implication of ɤ-amino butyric acid application in salt-exposed lettuce (*Lactuca sativa* L.) plants. Plant Physiol Biochem 130:157–172
Kang JY (2002) Arabidopsis basic leucine zipper proteins that mediate stress-responsive abscisic acid aignaling. Plant Cell 14(2):343–357
Kasuga M, Miura S, Shinozaki K, Yamaguchi-Shinozaki K (2004) A combination of the Arabidopsis DREB1A gene and stress-inducible rd29A promoter improved drought- and low-temperature stress tolerance in tobacco by gene transfer. Plant Cell Physiol 45(3):346–350
Kiegerl S, Cardinale F, Siligan C, Gross A, Baudouin E, Liwosz A, Eklöf S, Till S, Bögre L, Hirt H (2000) SIMKK, a mitogen-activated protein kinase (MAPK) kinase, is a specific activator of the salt stress–induced MAPK, SIMK. Plant Cell 12(11):2247–2258
Kiegle E, Moore CA, Haseloff J, Tester M-A, Knight M-R (2000) Cell-type-specific calcium responses to drought, salt and cold in the Arabidopsis root. Plant J 23(2):267–278

Kim S, Kang JY, Cho DI, Park JH, Kim SY (2004) ABF2, an ABRE-binding bZIP factor, is an essential component of glucose signaling and its overexpression affects multiple stress tolerance. Plant J 40(1):75–87

Kim T-H, Böhmer M, Hu H, Nishimura N, Schroeder J-I (2010) Guard cell signal transduction network: advances in understanding abscisic acid, CO_2, and Ca^{2+} signaling. Annu Rev Plant Biol 61(561):591

Kim J-S, Mizoi J, Yoshida T, Fujita Y, Nakajima J, Ohori T, Todaka D, Nakashima K, Hirayama T, Shinozaki K (2011) An ABRE promoter sequence is involved in osmotic stress-responsive expression of the DREB2A gene, which encodes a transcription factor regulating drought inducible genes in Arabidopsis. Plant Cell Physiol 52(12):2136–2146

Kim W-Y, Ali Z, Park H-J, Park S-J, Cha J-Y, Perez-Hormaeche J, Quintero F-J, Shin G, Kim M-R, Qiang Z (2013) Release of SOS2 kinase from sequestration with GIGANTEA determines salt tolerance in Arabidopsis. Nat Commun 4:1352

Knight H, Trewavas A-J, Knight M (1997) Calcium signalling in *Arabidopsis thaliana* responding to drought and salinity. Plant J 12(5):1067–1078

Kooijman E, Testerink C (2010) Phosphatidic acid: an electrostatic/hydrogenbond switch? In: Munnik T (ed) Lipid signaling in plants. Springer, Berlin Heidelberg, pp 203–222

Köster P, Wallrad L, Edel K-H, Faisal M, Alatar A-A, Kudla J (2019) The battle of two ions: Ca^{2+} signalling against Na^+ stress. Plant Biol 21:39–48

Krebs M, Beyhl D, Gorlich E, Al-Rasheid K-A-S, Marten I, Stierhof Y-D, Hedrich R, Schumacher K (2010) Arabidopsis V-ATPase activity at the tonoplast is required for efficient nutrient storage but not for sodium accumulation. PNAS 107(7):3251–3256

Kuromori T, Mizoi J, Umezawa T, Yamaguchi-Shinozaki K, Shinozaki K (2015) Stress signaling networks: drought stress. Mol Biol 1–23

Kurusu T, Kuchitsu K, Nakano M, Nakayama Y, Iida H (2013) Plant mechanosensing and Ca^{2+} transport. Trends Plant Sci 18(4):227–233

Laohavisit A, Richards S-L, Shabala L, Colaço R-D-D-R, Swarbreck S-M (2013) Salinity-induced calcium signaling and root adaptation in Arabidopsis require the calcium regulatory protein Annexin1. Plant Physiol 113

Lastochkina O, Pusenkova L, Yuldashev R, Babaev M, Garipova S, Blagova D-y, Khairullin R, Aliniaeifard S (2017) Effects of *Bacillus subtilis* on some physiological and biochemical parameters of *Triticum aestivum* L. (wheat) under salinity. Plant Physiol Biochem 121:80–88

Lee Y, Choi Y-B, Suh J, Lee J, Assmann S-M (1996) Abscisic acid–induced phosphoinositide turnover in guard cell protoplasts of *Vicia faba*. Plant Physiol 110(3):987–996

Lee S-J, Kang J-Y, Park H-J, Kim M-D, Bae M-S, Choi H-I Kim S-Y (2010) DREB2C interacts with ABF2, a bZIP proteinregulating abscisic acid-responsive gene expression, and its overexpression affects abscisic acid sensitivity. Plant Physiol 110

Leung J, Giraudat J (1998) Abscisic acid signal transduction. Annu Rev Plant Biol 49(1):199–222

Li L, Kim B-G, Cheong Y-H, Pandey G-K, Luan S (2006) A Ca^{2+} signaling pathway regulates a K^+ channel for low-K response in Arabidopsis. PNAS 103(33):12625–12630

Lin H, Yang Y, Quan R, Mendoza I, Wu Y, Du W, Zhao S, Schumaker K-S, Pardo J-M, Guo Y (2009) Phosphorylation of SOS3-LIKE CALCIUM BINDING PROTEIN8 by SOS2 protein kinase stabilizes their protein complex and regulates salt tolerance in Arabidopsis. Plant Cell 21(5):1607–1619

Liu J, Zhu J-K (1998) A calcium sensor homolog required for plant salt tolerance. Science 280(5371):1943–1945

Liu J, Ishitani M, Halfter U, Kim C-S, Zhu J (2000) The *Arabidopsis thaliana* SOS2 gene encodes a protein kinase that is required for salt tolerance. PNAS 97(7):3730–3734

Liu S, Lv Z, Liu Y, Li L, Zhang L (2018) Network analysis of ABA-dependent and ABA independent drought responsive genes in *Arabidopsis thaliana*.. Genetics Mol Biol 41(3):624–637

Lopez-Perez L, Martınez-Ballesta Mdel C, Maurel C, Carvajal M (2009) Changes in plasma membrane lipids, aquaporins and proton pump of broccoli roots, as an adaptation mechanism to salinity. Phytochemistry 70(4):492–500

Luan S (2002) Signalling drought in guard cells. Plant Cell Environ 25(2):229–237
Ma Y, Szostkiewicz I, Korte A, Moes D, Yang Y, Christmann A, Grill E (2009) Regulators of PP2C phosphatase activity function as abscisic acid sensors. Science 324:1064–1068
Ma C, Hong B, Wang T, Yang Y-J, Tong Z, Zuo Z-R, Yamaguchi-Shinozaki K, Gao J-P (2010) DREB1A regulon expression in rd29A:DREB1A transgenic chrysanthemum under low temperature or dehydration stress. J Hortic Sci Biotech 85(6):503–510
Maathuis F (2006) The role of monovalent cation transporters in plant responses to salinity. J Exp Bot 57(5):1137–1147
Maathuis F-J, Sanders D (2001) Sodium uptake in Arabidopsis roots is regulated by cyclic nucleotides. Plant Physiol 127(4):1617–1625
Maathuis F-J-M, Ahmad I, Patishtan J (2014) Regulation of Na^+ fluxes in plants. Front Plant Sci 5:467
Mahajan S, Tuteja N (2005) Cold, salinity and drought stresses: an overview. Science 324(5930):1064–1068
Mäser P, Eckelman B, Vaidyanathan R, HorieT F-J, Kubo M, Yamagami M, Yamaguchi K, Nishimura M, Uozumi N (2002a) Altered shoot/root Na^+ distribution and bifurcating salt sensitivity in Arabidopsis by genetic disruption of the Na^+ transporter AtHKT1. FEBS Lett 531(2):157–161
Mäser P, Hosoo Y, Goshima S, Horie T, Eckelman B, Yamada K, Yoshida K, Bakker EP, Shinmyo A, Oiki S (2002b) Glycine residues in potassium channel-like selectivity filters determine potassium selectivity in four-loop-per-subunit HKT transporters from plants. PNAS 99(9):6428–6433
McLoughlin F, Galvan-Ampudia CS, Julkowska MM, Caarls L, van der Does D, Laurière C, Testerink C (2012) The Snf1-related protein kinases SnRK2.4 and SnRK2.10 are involved in maintenance of root system architecture during salt stress. Plant J 72(3):436–449
Mikolajczyk M, Olubunmi SA, Muszynska G, Klessig DF, Dobrowolska G (2000) Osmotic stress induces rapid activation of a salicylic acid–induced protein kinase and a homolog of protein kinase ASK1 in tobacco cells. Plant Cell 12(1):165–178
Miller G, Suzuki N, Ciftci-Yilmaz S, Mittler R (2010) Reactive oxygen species homeostasis and signalling during drought and salinity stresses. Plant Cell Environ 33(4):453–467
Mishra N-S, Tuteja R, Tuteja N (2006) Signaling through MAP kinase networks in plants. Arch Biochem Biophys 452(1):55–68
Mizoguchi T, Ichimura K, Yoshida R, Shinozaki K (2000) MAP kinase cascades in Arabidopsis: their roles in stress and hormone responses. In: MAP kinases in plant signal transduction. Springer, Berlin, Heidelberg, pp 29–38
Munnik T (2001) Phosphatidic acid: an emerging plant lipid second messenger. Trends Plant Sci 6(5):227–233
Munnik T, Ligterink W, Meskiene I, Calderini O, Beyerly J et al (1999) Distinct osmosensing protein kinase pathways are involved in signalling moderate and severe hyperosmotic stress. Plant J 20(4):381–388
Munnik T, Meijer H-J-G, Ter Riet B, Hirt H, Frank W, Bartels D, Musgrave A (2000) Hyperosmotic stress stimulates phospholipase D activity and elevates the levels of phosphatidic acid and diacylglycerol pyrophosphate. Plant J 22(2):147–154
Nakagami H, Pitzschke A, Hirt H (2005) Emerging MAP kinase pathways in plant stress signalling. Trends Plant Sci 10(7):339–346
Nakashima K, Ito Y, Yamaguchi-Shinozaki K (2009) Transcriptional regulatory networks in response to abiotic stresses in Arabidopsis and grasses. Plant Physiol 149(1):88–95
Nakashima K, Yamaguchi-Shinozaki K, Shinozaki K (2014) The transcriptional regulatory network in the drought response and its crosstalk in abiotic stress responses including drought, cold, and heat. Front Plant Sci 5:170
Narusaka Y, Nakashima K, Shinwari Z-K, Sakuma Y, Furihata T, Abe H, Narusaka M, Shinozaki K, Yamaguchi-Shinozaki K (2003) Interaction between two cis-acting elements, ABRE and DRE, in ABA-dependent expression of Arabidopsis rd29A gene in response to dehydration and high salinity stresses. Plant J 34(2):137–148

Núñez-Ramírez R, Sánchez-Barrena MJ, Villalta I, Vega J-F, Pardo J-M, Quintero F-J, Martinez-Salazar J, Albert A (2012) Structural insights on the plant salt-overly-sensitive 1 (SOS1) Na^+/H^+ antiporter. J Mol Biol 424(5):283–294

Pandey N, Ranjan A, Pant P, Tripathi R-K, Ateek F, Pandey H-P, Patre U-V, Sawant S (2013) CAMTA 1 regulates drought responses in *Arabidopsis thaliana*. BMC Genomics 14(1):216

Planchet E, Rannou O, Ricoult C, Boutet-Mercey S, Maia-Grondard A, Limami A-M (2010) Nitrogen metabolism responses to water deficit act through both abscisic acid (ABA)-dependent and independent pathways in *Medicago truncatula* during post-germination. J Exp Bot 62(2):605–615

Qi Z, Spalding E (2004) Protection of plasma membrane K^+ transport by the salt overly sensitive1 Na^+-H^+ antiporter during salinity stress. Plant Physiol 136(1):2548–2555

Qiu Q-S, Guo Y, Dietrich M-A, Schumaker K-S, Zhu J (2002) Regulation of SOS1, a plasma membrane Na^+/H^+ exchanger in *Arabidopsis thaliana*, by SOS2 and SOS3. PNAS 99(12):8436–8441

Qiu QS, Guo Y, Quintero FJ, Pardo JM, Schumaker KS, Zhu JK (2004) Regulation of vacuolar Na^+/H^+ exchange in *Arabidopsis thaliana* by the salt-overly-sensitive (SOS) pathway. J Biol Chem 279(1):207–215

Qiu W, Liu M, Qiao G, Jiang J, Xie L, Zhuo R (2012) An isopentyl transferase gene driven by the stress-inducible rd29A promoter improves salinity stress tolerance in transgenic tobacco. Plant Mol Biol Rep 30(3):519–528

Quan R, Lin H, Mendoza I, Zhang Y, Cao W, Yang Y, Shang M, Chen S, Pardo J-M, Guo Y (2007) SCABP8/CBL10, a putative calcium sensor, interacts with the protein kinase SOS2 to protect Arabidopsis shoots from salt stress. Plant Cell 19(4):1415–1431

Quintero F-J, Ohta M, Shi H, Zhu J-K, Pardo J (2002) Reconstitution in yeast of the Arabidopsis SOS signaling pathway for Na^+ homeostasis. PNAS 108(6):2611–2616

Quintero F-J, Martinez-Atienza J, Villalta I, Jiang X, Kim W-Y, Ali Z, Fujii H, Mendoza I, Yun D-J, Zhu J (2011) Activation of the plasma membrane Na/H antiporter Salt-Overly-Sensitive1 (SOS1) by phosphorylation of an auto-inhibitory C-terminal domain. PNAS 108(6):2611–2616

Ren Z-H, Gao J-P, Li L-G, Cai X-L, Huang W, Chao D-Y, Zhu M-Z, Wang Z-Y, Luan S, Lin H (2005) A rice quantitative trait locus for salt tolerance encodes a sodium transporter. Nat Genet 37(10):1141

Ren X-L, Qi G-N, Feng H-Q, Zhao S, Zhao S-S, Wang Y, Wu W (2013) Calcineurin B-like protein CBL 10 directly interacts with AKT 1 and modulates K^+ homeostasis in Arabidopsis. Plant J 74(2):258–266

Riera M, Valon C, Fenzi F, Giraudat J, Leung J (2005) The genetics of adaptive responses to drought stress: abscisic acid-dependent and abscisic acid-independent signalling components. Physiol Plant 123(2):111–119

Roychoudhury A, Paul S, Basu S (2013) Cross-talk between abscisic acid-dependent and abscisic acid-independent pathways during abiotic stress. Plant Cell Rep 32(7):985–1006

Rubio F, Gassmann W, Schroeder J (1995) Sodium-driven potassium uptake by the plant potassium transporter HKT1 and mutations conferring salt tolerance. Science 270(5242):1660–1663

Rubio F, Flores P, Navarro J-M, Martínez V (2003) Effects of Ca^{2+}, K^+ and cGMP on Na^+ uptake in pepper plants. Plant Sci 165(5):1043–1049

Sakuma Y, Maruyama K, Osakabe Y, Qin F, Seki M, Shinozaki K, Yamaguchi-Shinozaki K (2006) Functional analysis of an Arabidopsis transcription factor, DREB2A, involved in drought responsive gene expression. Plant Cell 18(5):1292–1309

Schmidt R, Mieulet D, Hubberten H-M, Obata T, Hoefgen R, Fernie A-R (2013) Salt-responsive ERF1 regulates reactive oxygen species-dependent signaling during the initial response to salt stress in rice. Plant Cell 113

Schumacher K, Krebs M (2010) The V-ATPase: small cargo, large effects. Curr Opin Plant Biol 13(6):724–730

Seo D-H, Ryu M-Y, Jammes F, Hwang J-H, Turek M, Kang B-G, Kwak J-M, Kim W-T (2012) Roles of four Arabidopsis U-box E3 ubiquitin ligases in negative regulation of ABA-mediated drought stress responses. Plant Physiol 112

Shao H-B, Guo Q-J, Chu L-Y, Zhao X-N, Su Z-L, Hu Y-C, Cheng J (2007) Understanding molecular mechanism of higher plant plasticity under abiotic stress. Colloids Surf B Biointerfaces 54(1):37 45

Shi H, Ishitani M, Kim C, Zhu J (2000) The *Arabidopsis thaliana* salt tolerance gene SOS1 encodes a putative Na^+/H^+ antiporter. PNAS 97(12):6896–6901

Shi H, Quintero F-J, Pardo J-M, Zhu J (2002) The putative plasma membrane Na^+/H^+ antiporter SOS1 controls long-distance Na^+ transport in plants. Plant Cell 14(2):465–477

Shinozaki K, Yamaguchi-Shinozaki K (1997) Gene expression and signal transduction in water-stress response. Plant Physiol 115(2):327

Shinozaki K, Yamaguchi-Shinozaki K (2000) Molecular responses to dehydration and low temperature: differences and cross-talk between two stress signaling pathways. Curr Opin Plant Biol 3(3):217–223

Shkolnik-Inbar D, Adler G, Bar-Zvi D (2013) ABI4 downregulates expression of the sodium transporter HKT1; 1 in Arabidopsis roots and affects salt tolerance. Plant J 73(6):993–1005

Singh A, Kumar A, Yadav S, Singh I (2019) Reactive oxygen species-mediated signaling during abiotic stress. Plant Gene 100173

Tan T, Cai J, Zhan E, Yang Y, Zhao J, Guo Y, Zhou H (2016) Stability and localization of 14-3-3 proteins are involved in salt tolerance in Arabidopsis. Plant Mol Biol 92(3):391–400

Teige M, Scheikl E, Eulgem T, Dóczi R, Ichimura K, Shinozaki K, Dangl J-L, Hirt H (2004) The MKK2 pathway mediates cold and salt stress signaling in Arabidopsis. Mol Cell 15(1):141–152

Testerink C, Munnik T (2005) Phosphatidic acid: a multifunctional stress signaling lipid in plants. Trends Plant Sci 10(8):368–375

Testerink C, Munnik T (2011) Molecular, cellular, and physiological responses to phosphatidic acid formation in plants. J Exp Bot 62(7):2349–2361

Testerink C, Larsen P-B, van der Does D, van Himbergen J-A, Munnik T (2007) Phosphatidic acid binds to and inhibits the activity of Arabidopsis CTR1. J Exp Bot 58(14):3905–3914

Thewes S (2014) Calcineurin-Crz1 signaling in lower eukaryotes. Eukaryot Cell 13(6):694–705

Tran L-S, Nakashima K, Sakuma Y, Simpson SD, Fujita Y, Maruyama K, Fujita M, Seiki M, Shinozaki K, Yamaguchi-Shinozaki K (2004) Isolation and functional analysis of Arabidopsis stress inducible NAC transcription factors that bind to a drought-responsive cis-element in the early responsive to dehydration stress 1 promoter. Plant Cell 16(9):2481–2498

Uno Y, Furihata T, Abe H, Yoshida R, Shinozaki K, Yamaguchi-Shinozaki K (2000) Arabidopsis basic leucine zipper transcription factors involved in an abscisic acid-dependent signal transduction pathway under drought and high-salinity conditions. PNAS 97(21):11632–21163

Uozumi N, Kim EJ, Rubio F, Yamaguchi T. Muto S, Tsuboi A, Bakker E-P, Nakamura T, Schroeder J (2000) The Arabidopsis HKT1 gene homolog mediates inward Na^+ currents in *Xenopus laevis* oocytes and Na^+ uptake in *Saccharomyces cerevisiae*. Plant Physiol 122(4):1249–1260

Valliyodan B, Nguyen HT (2006) Understanding regulatory networks and engineering for enhanced drought tolerance in plants. Curr Opin Plant Biol 9(2):189–195

Wang Y, He L, Li H-D, Xu J, Wu W (2010) Potassium channel α-subunit AtKC1 negatively regulates AKT1-mediated K^+ uptake in Arabidopsis roots under low-K^+ stress. Cell Res 20(7):826

Wang Z-Y, Xiong L, Li W, Zhu J-K, Zhu J (2011) The plant cuticle is required for osmotic stress regulation of abscisic acid biosynthesis and osmotic stress tolerance in Arabidopsis. Plant Cell 110

Wang F, Jing W, Zhang W (2014) The mitogen-activated protein kinase cascade MKK1–MPK4 mediates salt signaling in rice. Plant Sci 227:181–189

Weinl S, Kudla J (2009) The CBL–CIPK Ca^{2+}-decoding signaling network: function and perspectives. New Phytol 184(3):517–528

Weng H, Yoo CY, Gosney M-J, Hasegawa P-M, Mickelbart M (2012) Poplar GTL1 is a Ca^{2+}/calmodulin binding transcription factor that functions in plant water use efficiency and drought tolerance. PLoS ONE 7(3):e32925

Wu H, Zhang X, Giraldo J-P, Shabala S (2018) It is not all about sodium: revealing tissue specificity and signalling roles of potassium in plant responses to salt stress. Plant Soil 431(1–2):1–17

Xiong L, Zhu JK (2001) Abiotic stress signal transduction in plants: molecular and genetic perspectives. Physiol Plant 112(2):152–166
Xiong L, Lee BH, Ishitani M, Lee H, Zhang C, Zhu JK (2001) FIERY1 encoding an inositol polyphosphate 1-phosphatase is a negative regulator of abscisic acid and stress signaling in Arabidopsis. Genes Dev 15(15):1971–1984
Xiong L, Schumaker K-S, Zhu J-K (2002) Cell signaling during cold, drought, and salt stress. Plant Cell 14(suppl 1):S165–S183
Xu J, Li H-D, Chen L-Q, Wang Y, Liu L-L, He L, Wu W (2006) A protein kinase, interacting with two calcineurin B-like proteins, regulates K^+ transporter AKT1 in Arabidopsis. Cell 125(7):1347–1360
Xu Z-Y, Kim S-Y, Hyeon Do Y, Kim D-H, Dong T, Park Y, Jin J-B, Joo S-H, Kim S-K, Hong J-C (2013) The Arabidopsis NAC transcription factor ANAC096 cooperates with bZIPtype transcription factors in dehydration and osmotic stress responses. Plant Cell 113
Xue S, Yao X, Luo W, Jha D, Tester M, Horie T, Schroeder J (2011) AtHKT1; 1 mediates nernstian sodium channel transport properties in Arabidopsis root stelar cells. PLoS ONE 6(9):e24725
Yamaguchi T, Hamamoto S, Uozumi N (2013) Sodium transport system in plant cells. Front Plant Sci 4:410
Yamaguchi-Shinozaki K, Urao T, Shinozaki K (1995) Regulation of genes that are induced by drought stress in *Arabidopsis thaliana*. J Plant Res 108(1):127–136
Yang Y, Guo Y (2018a) Unraveling salt stress signaling in plants. J Integr Plant Biol 60(9):796–804
Yang Y, Guo Y (2018b) Elucidating the molecular mechanisms mediating plant salt-stress responses. New Phytol 217(2):523–539
Yao H, Wang G, Guo L, Wang X (2013) Phosphatidic acid interacts with a MYB transcription factor and regulates its nuclear localization and function in Arabidopsis. Plant Cell 113
Yildizli A, Çevik S, Ünyayar S (2018) Effects of exogenous myo-inositol on leaf water status and oxidative stress of Capsicum annuum under drought stress. Acta Physiol Plant 40(6):122
Yoo J-H, Park C-Y, Kim J-C, Do Heo W, Cheong M-S, Park H-C, Kim M-C, Moon B-C, Choi M-S, Kang Y (2005) Direct interaction of a divergent CaM isoform and the transcription factor, MYB2, enhances salt tolerance in Arabidopsis. J Biol Chem 280(5):3697–3706
Yoshida R, Umezawa T, Mizoguchi T, Takahashi S, Takahashi F, Shinozaki K (2006) The regulatory domain of SRK2E/OST1/SnRK2. 6 interacts with ABI1 and integrates abscisic acid (ABA) and osmotic stress signals controlling stomatal closure in Arabidopsis. J Biol Chem 281(8):5310–5318
Yu L, Nie J, Cao C, Jin Y, Yan M, Wang F, Liu J, Xiao Y, Liang Y, Zhang W (2010) Phosphatidic acid mediates salt stress response by regulation of MPK6 in *Arabidopsis thaliana*. New Phytol 188(3):762–773
Yu Q, An L, Li W (2014) The CBL–CIPK network mediates different signaling pathways in plants. Plant Cell Rep 33(2):203–214
Yuan F, Yang H, Xue Y, Kong D, Ye R, Li C, Zhang J, Theprungsirikul L, Shrift T, Krichilsky B (2014) OSCA1 mediates osmotic-stress-evoked Ca^{2+} increases vital for osmosensing in Arabidopsis. Nature 514(7522):367
Zhang H-X, Blumwald E (2001) Transgenic salt-tolerant tomato plants accumulate salt in foliage but not in fruit. Nat Biotechnol 19(8):765
Zhang W, Qin C, Zhao J, Wang X (2004) Phospholipase Dα1-derived phosphatidic acid interacts with ABI1 phosphatase 2C and regulates abscisic acid signaling. PNAS 101(25):9508–9513
Zhang H, Kim M-S, Sun Y, Dowd S-E, Shi H, Paré P-W (2008) Soil bacteria confer plant salt tolerance by tissue-specific regulation of the sodium transporter HKT1. MPMI 21(6):737–744
Zhu J (2000) Genetic analysis of plant salt tolerance using Arabidopsis. Plant Physiol 124(3):941–948
Zhu J (2001) Cell signaling under salt, water and cold stresses. Curr Opin Plant Biol 4(5):401–406
Zhu J (2002) Salt and drought stress signal transduction in plants. Annu Rev Plant Biol 53(1):247–273
Zhu J (2016) Abiotic stress signaling and responses in plants. Cell 167(2):313–324

Calcium Signaling in Plants Under Drought

Sasan Aliniaeifard, Aida Shomali, Maryam Seifikalhor and Oksana Lastochkina

Abstract Environmental stresses are among the most important threats for food security since it adversely affects agricultural productivity. Among them, drought stress and water scarcity stand as major problems, especially in developing countries. Thus, understanding molecular mechanisms involving drought stress response is of essential importance for coping with its detrimental effects. Calcium is an essential macro-element, taking part in the regulation of many aspects of plant growth and development along with playing the role of ubiquitous secondary messenger, thus taking part in the generation of adaptive responses toward various developmental and environmental stimuli. Therefore, understanding the calcium signaling and calcium-dependent processes is crucial for improving plant productivity under environmental stresses. Formation of calcium signature through subsequent oscillations is at the onset of calcium signaling which further activates calcium-binding proteins which are acting as signal decoders, makes it possible to perceive the nature of stimuli. Further, protein kinases and transcription factors are activated in a stimuli-specific manner to create a stress-specific response. Understanding of calcium signaling processes and modulations is a promising step for unraveling the molecular mechanism of drought stress tolerance and engineering stress-tolerant crops. In this chapter, calcium-signaling components involving in drought stress signal perception and transition through calcium-signaling pathway and final formation of stress response have been discussed in detail.

S. Aliniaeifard (✉) · A. Shomali
Photosynthesis Laboratory, Department of Horticulture, College of Aburaihan, University of Tehran, Tehran, Iran
e-mail: aliniaeifard@ut.ac.ir

S. Aliniaeifard · M. Seifikalhor
Department of Plant Biology, College of Science, University of Tehran, Tehran, Iran

O. Lastochkina
Ufa Federal Research Centre, Bashkir Research Institute of Agriculture, Russian Academy of Sciences, Ufa, Russia

Ufa Federal Research Centre, Institute of Biochemistry and Genetics, Russian Academy of Sciences, Ufa, Russia

M. Hasanuzzaman and M. Tanveer (eds.), *Salt and Drought Stress Tolerance in Plants*, Signaling and Communication in Plants, https://doi.org/10.1007/978-3-030-40277-8_10

1 Introduction

Calcium (Ca^{2+}), an intracellular secondary messenger, known as a versatile ion takes part in modulating the numerous cellular processes (Berridge et al. 2000; White and Broadley 2003; Dodd et al. 2010) and plays a pivotal role in plant growth and development. Due to its crucial roles in various signaling pathways, it has been addressed as "signaling component of the age" (Scrase-Field and Knight 2003) and "convergence point of signaling pathway" (Tuteja and Mahajan 2007).

The presence of various intracellular Ca^{2+} stores like vacuoles, endoplasmic reticulum, mitochondria, and cell wall enables the cells to dynamic release the of the Ca^{2+} in and balance the internal Ca^{2+} concentration (Mahajan and Tuteja 2005; Mahajan et al. 2006). Interestingly, doubled-membrane organelles like mitochondria, chloroplasts, and nuclei are specialized to generate Ca^{2+} signal per se and keep a profound effect on the integrated cell function (Xiong et al. 2006). Ca^{2+} as a central component of the complicated signaling pathways modulating the cell physiology and gene expression in response to variable conditions (Dodd et al. 2010). In fact, numerous environmental stimuli such as biotic and abiotic stresses as well as extracellular signals remodel the cellular Ca^{2+} levels (reviewed by Mahajan and Tuteja 2005).

It is widely accepted that almost all kinds of stresses alter gene expression in plants (Thomashow 1999; Shinozaki and Yamaguchi-Shinozaki 2007; Hirayama and Shinozaki 2010). These gene expressions regulated by transcription factors (TFs) results in reprogramming of the transcriptome profile, which is apparently a significant consequence of stress signaling and adaptation (Brivanlou and Darnell 2002). The fact that Ca^{2+} functions as a messenger in the plants lead the researchers to introduce Ca^{2+} as a regulator of gene expression (Poovaiah et al. 1987). Because signal-induced changes in gene expression is followed by a rapid signal-specific shift in cellular Ca^{2+}, it is presumed that gene expression adjustment in the cell is mediated by Ca^{2+} signaling (Hu et al. 2004; Kaplan et al. 2006; McAinsh and Pittman 2009).

During the normal development and in hypersensitive responses to pathogens, sustained high cytosolic Ca^{2+} represented an apoptosis event (Levine et al. 1996). This indicates that although an increase in cytosolic Ca^{2+} is necessary for signal transduction to trigger an appropriate response in a diverse range of biotic and abiotic stresses, a prolonged increase in cytosolic Ca^{2+} has detrimental effects. To generate an appropriate adaptive response, the cytosolic Ca^{2+} perturbations have to be either transient or of low amplitude. Once the cytosolic Ca^{2+} elevates transiently, it can be run as single (spike), double (biphasic), or multiple (oscillations) frequencies (reviewed by Evans et al. 2001; Hetherington and Brownlee 2004; Tuteja and Mahajan 2007). In addition, cellular location, extent of propagation, and amplitude during propagation are factors that vary among generated Ca^{2+} perturbation (reviewed by Tuteja and Mahajan 2007).

Spatial and temporal dynamics of stimulus-induced changes in cytosolic Ca^{2+} has been referred to as a "calcium signature" (McAinsh and Hetherington 1998; Ng and McAinsh 2003; reviewed by Lecourieux et al. 2006; Oldroyd and Downie

2008). Ca^{2+} signature is unique to a particular stimulus in regard to subcellular location and/or the kinetics of magnitude of the cytosolic Ca^{2+} perturbation which makes it possible to functions as a specific physiological response (Rudd and Franklin-Tong 2001; reviewed by White and Broadley 2003). The apoplast, across the plasma membrane, or the intracellular organelles comprise the main routes for Ca^{2+}. Cellular type and localization and the abundance of Ca^{2+} permeable channels determine the spatial characteristics of cytosolic Ca^{2+} perturbations.

Calcium ion signatures have shown to be dependent on cell types (Moore et al. 2002) and the nature of the stimuli. Various stimuli can induce specific Ca^{2+} signals bearing various amplitude, duration, and frequency, in the cytoplasm or organelles of the cell (Stael et al. 2012; Nomura and Shiina 2014). These cell-stimuli-specific signatures evolved from multiple pathways to decode the natural signals and convert the stimuli message in spatiotemporal dynamics (Bothwell and Ng 2005) mediated by Ca^{2+} influx channels in the plasma membrane and endomembrane. Further, Ca^{2+} efflux transporters restate the cytosolic Ca^{2+} to stationary values by rapid removal of Ca^{2+} from the cytosol (reviewed by McAinsh and Pittman 2009).

Calcium ion-dependent protein kinase induced by drought stress in *Arabidopsis* (Urao et al. 1994), expression of putative Ca-ATPase under saline condition in tobacco (Perez-Prat et al. 1992) and expression of Ca^{2+}-binding protein in maize and sorghum under polyethylene glycol treatment (Pestacz and Erdei 1996) have been addressed as the first evidences of Ca^{2+} calcium participation involvement in stress response. Later, many other reports confirmed the crucial role of Ca^{2+} in plant response to different stresses including biotic elicitors (Knight et al. 1991; Kosuta et al. 2008), salt, and drought stresses (Ranf et al. 2008; Seifikalhor et al. 2019), PGRs (McAinsh et al. 1990; Allen et al. 2001) ABA treatment (Knight et al. 1996, 1997), cold stress (Knight et al. 1996), light stress (Shacklock et al. 1992), osmotic stress (Knight et al. 1997), oxidative stress, and gaseous pollutants (Evans et al. 2005). It was also demonstrated that Ca^{2+} influx from the plasma membrane and intracellular Ca^{2+} oscillation is a downstream event in ABA-induced stomatal closure (Kim et al. 2010; Takahashi et al. 2007; Aliniaeifard and van Meeteren 2013).

In general, during stress conditions, a signaling network is set up in plant cell starting with stress perception at the cell membrane level and stretched to the cellular response. In this regard, Tuteja and Mahajan (2007) have described a generic signal transduction pathway defined in four biological steps:

1. Perception of the signal through membrane receptors.
2. Generation of secondary messengers.
3. Formation of a protein phosphorylation/dephosphorylation cascade to modulate a specific set of stress-regulated genes that may also target transcription factors.
4. Induction of stress tolerance and plant adaption, which may concomitant with phenotypic responses.

To create an appropriate response or signal transduction against abiotic stresses, proper spatial and temporal coordination of all signaling molecules are required. It has been widely accepted that Ca^{2+} is increased during drought stress, enhances drought tolerance of plants, and implicates in the transduction of drought and salt stress-induced signals (Yildizli et al. 2018; Seifikalhor et al. 2019). However, the question of how stimulus-specific response raises from the complex network of Ca^{2+} signaling pathways in plant cells is still open to be answered. Generally, the specificity of Ca^{2+} signaling may result from the contribution of several mechanisms to form a specific signal which is suggested to occur in the following five steps (Sanders et al. 2002):

1. Additional signaling events that occur accompanied by changes in cytosolic Ca^{2+}.
2. Appropriate signaling machinery takes part in the transduction of a given signal refers to as “physiological address” (McAinsh and Hetherington 1998).
3. A defined response must arise in cells that express different Ca^{2+} sensors (Young et al. 2006).
4. The compartmentalization of signaling pathways and/or the selective activation of discrete response elements takes place in response to localized elevations in Ca^{2+}.
5. Decoding of signaling information of cytosolic Ca^{2+} oscillations and transients (Evans et al. 2001).

Plant cells have developed a complicated mechanism to sense the simple signals like alteration in the concentration of intracellular Ca^{2+}, thereby, the Ca^{2+} signals fashions (amplitude and duration) are precisely regulated (Schwaller 2012). To regulate the Ca^{2+} oscillation signal and transduction pathway, calcium transporter proteins including Ca^{2+}-permeable channels, divalent H^+ antiporters, and Ca^{2+}-P-type ATPases, are required (McAinsh and Pittman 2009). In addition to ascertaining intracellular Ca^{2+} signaling, the Ca^{2+}-signaling components are also taking part in the regulation of their own expression (Schwaller 2012). In response to environmental stimuli, Ca^{2+} channels and transporter activities must be programmed to generate an appropriate Ca^{2+} signature to influence cytosolic Ca^{2+} concentration. The cytosolic Ca^{2+} concentration is about 100 nM at cell rest, while, it reaches the levels higher than 1 μM for signaling events (White and Broadley 2003; Dodd et al. 2010).

Since the Ca^{2+} comprises the low diffusion through the cytoplasm while its buffering is high (0.1–1 mM), Ca^{2+} channels open to produce a local increase in cytosolic Ca^{2+} followed by a rapid dissipation in consequence of channel closure. Accordingly, the subcellular localization of Ca^{2+} channels is determinative for the operation of different cellular processes (Trewavas 1999; Tuteja and Mahajan 2007). Trewavas (1999) indicated that for coordination of cellular responses, cytosolic Ca^{2+} “waves” are produced within the cytoplasm when Ca^{2+} passes through particular Ca^{2+} channels recurrently. Suggesting that, increase in Ca^{2+} concentration might result in generation of soluble secondary messengers, such as IP3 (inositol triphosphate) or cADPR (cyclic adenosine diphosphate ribose), that diffuse through the cytoplasm by which a relay of spatially separated Ca^{2+} channels will be activated; channels close to the plasma membrane and channels that are adjacent to the vacuole, respectively (Allen et al. 1999). Ca^{2+} waves are responsible for the tandem opening of Ca^{2+} channels at the plasma membrane that activated by hyperpolarization and Ca^{2+}-activated channels in the tonoplast (Schroeder et al. 2001).

In stress conditions, different waveforms are generated in cells with elevated Ca^{2+} levels. This kind of "high calcium cellular waves" has been detected in cotyledons exposed to cold stress (Knight et al. 1993) and in root tissue stimulated by mechanical injury or salinity (Fasano et al. 2002; Moore et al. 2002; Seifikalhor et al. 2019). These waves are triggered by electrical potentials, osmotic perturbations, or chemical signals as well.

2 Ca^{2+} Roles in Plant

Ca^{2+}-as a bivalent cation is an essential element that plays a substantial role in all organisms. In the context of Calcium importance Trewavas (1999) stated: "Every reader knows that he or she started life as a single cell. Less familiar is the debt we owe to calcium in our earliest seconds. Penetration of the egg by the paternal sperm initiated an epigenetic calcium wave that moved quickly as a hollow band across the cytoplasm. In the wake of this calcium wave, processes were activated that led to cell division, differentiation, growth, and our eventual appearance as mature adults. As it is said in France, "Le calcium, c'est la vie." A calcium wave marked the onset of our existence, and will quite probably mark our demise: Irreversible failure of calcium-wave generation in the heart is the most common cause of death. Therefore, calcium waves are a life-and-death issue."

Calcium is an essential element that plays crucial structural and signaling functions in vascular plants. It is the counter-cation for inorganic and organic anions in the vacuole, as well as a secondary messenger in multiple intracellular signaling events that alleviate the adverse effect of stress by triggering stress responses. As calcium pectate, it is required for structural roles in cell wall and membranes where it accumulates and binds the cells together. One of the most well-known roles of Ca^{2+} in the plant is in membrane stability and cell integrity. Ca^{2+} takes part in cell membrane structure by stabilizing the lipid bilayers of cell membranes via bridging phosphate and carboxylate groups of phospholipids and proteins at the external site of the plasma membrane (Epstein 1972; Hepler and Winship 2010). The growth and elongation of pollen tube also require Ca^{2+}. Ca^{2+} ion helps the formation of microtubules, the prerequisite for the anaphasic movement of chromosomes during cell division via mitosis. Therefore, Ca^{2+} functions in the growth and development of plant, especially in the root and shoot tips (Marschner 1986; Sanders et al. 2002).

Plant root uptakes Ca^{2+} from the rhizosphere and delivers it to the shoot via xylem. Both symplastic and apoplastic pathways are appropriate passages for Ca^{2+} through the root system. Ca^{2+} shows a high affinity to the array of Ca^{2+}-binding proteins, this makes the concentration of cytosolic Ca^{2+} far higher than the concentration of free Ca^{2+} which is generally maintained at 200 nM (Bush 1995). The relative contribution of the apoplastic and symplastic pathways must be balanced in a way that root cells permitted to use the cytosolic Ca^{2+} concentration for signaling, controlling the rate of Ca^{2+} delivery to the xylem, and preventing the accumulation of toxic cations in the shoot. However, the balance between these pathways is unknown (White 2001).

In this regard, Tuteja and Mahajan, 2007 stated the fact that the concentration of free Ca^{2+} within the cell is low and it is prone to rapid chelation makes the apoplastic pathway predominant rather than symplastic pathway (Tuteja and Mahajan 2007).

The ability of plants to uptake Ca^{2+} from the soil and their requirement for Ca^{2+} varies in different plant species and plant organs. It is widely accepted that for optimal growth monocots require less Ca^{2+} than the dicots. Generally, Ca^{2+} content varies between 0.1 and >5.0% dry weight indicating the contrasting Ca^{2+} requirement of various plant species in various growth stages and physiological conditions (Loneragan and Snowball 1969; Marschner 1986; Fageria 2016). Ca^{2+} competes with other cations both for entrance sites and for the uptake from the soil and enters plant cells through Ca^{2+}-permeable ion channels (I_{Ca}) located in their plasma membranes (White 2001). The presence of other cations profoundly influence Ca^{2+} uptake as high levels of K and Mg reduces Ca^{2+} uptake, while the presence of high levels of Ca^{2+} is known to ameliorate uptake of toxic cations (Al^{3+} and Na^{+}) from the soil (Burstrom 1968). In normal conditions, a high level of cytosolic Ca^{2+} is cytotoxic, therefore, Ca^{2+}-ATPases and H/Ca^{2+}-antiports direct the ion traffic to maintain the concentration of Ca^{2+} at sub-micromolar concentrations in unstimulated cells (Sze et al. 2000; Hirschi 2001).

3 Ca^{2+} Uptake and Transport in Plant Under Drought Stress

Ca^{2+} is taken up from the rhizosphere solution through the root system especially root tip and the regions initiating lateral roots and transported to the above-ground parts via the xylem (White 2001). The distribution of Ca^{2+} within the plant occurs mainly in the xylem vessels (Clarkson 1984). The Ca^{2+} transport to the xylem of growing plants is about 40 nmol Ca^{2+} h^{-1} g^{-1} root fresh weight (White 1998). The Casparian strip is the main barrier for Ca^{2+} loading into the root (White 2001). Therefore, Ca^{2+} loading into the xylem is mainly restricted to the root tip and to regions initiating lateral roots. Lack or interrupted Casparian strip in those regions permit the Ca^{2+} fluxes to the apoplast and thereafter to the xylem. Movement of Ca^{2+} toward leaves and fruits can be performed in both apoplastic and symplastic pathways (White and Broadley 2003). Ca^{2+} moves in xylem vessel in chelate form and its movement is considerably dependent on the transpiration rate. In fact, low Ca^{2+} content in plant tissues is a consequence of low transpiration rate in those tissues. Therefore, environmental cues that influence the transpiration rate would also alter Ca^{2+} uptake and movement in the plant. For instance, reducing transpiration as a result of drought exposure will decrease Ca^{2+} uptake and movement to the shoot (del Carmen Martínez-Ballesta et al. 2011). Confirming this, it has been shown that the genotype of one species having more stomatal opening and more transpiration rate is capable of higher accumulation of Ca^{2+} in their tissues and as a result, is more drought tolerant (de Freitas and Mitcham 2012; Kaczmarek et al. 2017). Lower transpiration rate of fruits in comparison with leaves is the main reason for lower Ca^{2+} concentration and the occurrence of Ca^{2+}-related disorders in fruit tissue (de Freitas and Mitcham 2012; de Freitas et al. 2011).

4 Calcium Channels

Based on voltage dependence, Ca^{2+} channels are classified into three categories including hyperpolarization-activated (HACC), depolarization-activated (DACC), and voltage-independent (VICC) cation channels (Sanders et al. 2002; White and Broadley 2003; Marschner 2012). HACC opening is stimulated by ABA that play a central role in stomatal closure during water deficit conditions. This channel promotes the entrance of Ca^{2+} that depolarizes the plasma membrane and contributes to Ca^{2+}-dependent events (Murata et al. 2001; Kohler and Blatt 2002). DACCs are located in the plasma membrane and are permeable to both mono and bivalent cations and participate in calcium and other cations uptake. This channel has been shown to trigger plant adaption to low temperature (White and Broadley 2003; White 2009). VICCs are encoded by genes from the cyclic nucleotide-gated channel (CNGC) and glutamate-like receptor families (GLR). Studying the Arabidopsis protein channels revealed that CNGC and GLR gene families are more involved in processes associated with Ca^{2+} signaling than with Ca^{2+} storage (Conn and Gilliham 2010), therefore, CNGC and GLR are considered as the main channel in Ca^{2+} signature.

VICCs involve in maintaining of the cytosolic Ca^{2+} hemostasis in resting cells by retaining the permanent Ca^{2+} efflux balance through Ca^{2+} ATPase and H^+/Ca^{2+} antiporters (Marchner 2012). They are permeable to both mono- and bivalent cations. In the plasma membrane, many different VICCs are presented and characterized by their differences in cation selectivity (White and Broadley 2003; Demidchik et al. 2002). Furthermore, several membrane channels are involved in generating a Ca^{2+} influx from Ca^{2+} reservoir compartment (apoplast and organelles) to cytosol in order to create Ca^{2+} signatures (Dodd et al. 2010).

Guard cell plasma membrane I_{Ca} channels which play a role in Ca^{2+} influx from extracellular space (Hamilton et al. 2000), release Ca^{2+} from intracellular stores (Blatt 2000) are also involved in Ca^{2+} trafficking in guard cells of stomata.

5 Calcium Signal Transduction

Prototypical animal Ca^{2+} channels have been well characterized in the genomes of the evolutionary basal lineage of Chlorophyta algae; whereas, it was missed in land plants (Wheeler and Brownlee 2008). This can imply that the plant-specific Ca^{2+} signaling mechanism developed in compensation for the loss of these Ca^{2+} channels. Tuning of the "Ca^{2+} toolkit" on the level of the Ca^{2+} binding proteins, which decode the information presented by Ca^{2+} signatures and modulate the Ca^{2+}

signal transduction to downstream responses are also involved in this evolutionary trajectory (Batistič and Kudla 2012). A more recent notion in this regard specifically considered the Ca^{2+}-sensing protein CBL/CIPKs states that during land plant evolution CBL/CIPKs diversified gradually and eventually formed a complicated network involved in various aspects of stress signaling events (Beckmann et al. 2016; Kudla et al. 2018).

Despite the presence of many signaling molecules acting as a secondary messenger in plant cells, only Ca^{2+} has been shown to be involved in the regulation of gene expression of regulatory proteins that result in stress tolerance responses by altering the metabolism followed by phenotypic responses. Based on the number of genes and the type of gene and the regulation mode, the response to elevated Ca^{2+} levels varies from growth inhibition to cell death (Tuteja and Mahajan 2007). Large numbers of Ca^{2+}-responsive genes (230) are identified by analysis of transcriptome changes containing known early stress-responsive genes as well as genes with unknown function. Among Ca^{2+}-responsive genes, 162 genes are up-regulated and 68 are down-regulated under stress conditions (Kaplan et al. 2006).

During evolution adaption emerges upon plant response to stresses; this indicates that plant functions as a dynamic and endogenously integrated organism in response to changes in its surrounding environment. Under drought condition a wide variety of plant responses such as alterations in gene expression, accumulation of metabolites, e.g., osmotically active compounds and PGRs (especially ABA), and the synthesis of specific proteins (Ramachandra-Reddy et al. 2004) are promoted in plants; all are regulated through certain signaling pathways. Signal perception is at the onset of the signal transduction pathway. Once the signal perception takes place, the generation of second messengers occurs. Second messengers can alter intracellular Ca^{2+} levels; consequently, a protein phosphorylation cascade initiated which targets proteins that are directly involved in cellular protection or TFs controlling stress-regulated genes. Therefore, in response to stress conditions, large numbers of genes are expressed to produce proteins involved in stress tolerance pathways (Shinozaki and Yamaguchi-Shinozaki 1997). These genes may influence regulatory molecules, e.g., PGRs especially ABA. Further, these regulatory molecules take part in another signaling event with different components that may cross-talk with the above generic pathway. The proper spatial and temporal coordination of all signaling molecules should be provided for signal transduction. Therefore, coordination of certain molecules (protein modifier, scaffolds, or adapters that are known as signaling partners) in modification, delivery, or assembly of signaling components is necessary despite they do not directly relay the signal (Xiong and Zhu 2001).

In plants, studying of Ca^{2+} signaling in stomatal guard cells in legumes provides compelling evidence that signaling information can be encoded in the spatiotemporal dynamics of plant Ca^{2+} signatures. Therefore, an overview of the background of Ca^{2+} signaling in stomatal aperture facilitates addressing its role in drought stress.

Changes in guard cell cytosolic Ca^{2+} in response to a wide range of stimuli that modulate stomatal aperture led the researchers to postulate that cytosolic Ca^{2+} is at the center of the guard cell signaling network (Hetherington and Woodward 2003; Israelsson et al. 2006; Li et al. 2006; Aliniaeifard and van Meeteren 2013). The mechanism of encryption of signaling information in guard cells in response to Ca^{2+} oscillation attracted the attention of several researchers (Allen et al. 2000, 2001; Mori et al. 2006; Young et al. 2006). Reverse genetic studies on the model plant Arabidopsis provided ample evidence for the contribution of Ca^{2+}-dependent signaling pathway in ABA-induced stomatal closure (Roelfsema and Hedrich 2010; Steinhorst and Kudla 2013; Murata et al. 2015; Edel and Kudla 2016). Ca^{2+}-Dependent Protein Kinases (CDPKs) are activated in response to ABA-induced intracellular Ca^{2+} transients (Mori et al. 2006). In *Arabidopsis thaliana* 8 of the 34 detected CDPKs contribute to ABA signaling and reactive oxygen species (ROS) homeostasis in guard cells (Boudsocq and Sheen 2013; Zou et al. 2015; Simeunovic et al. 2016). CDPK3, 21 and 23 that belong to subgroup II of CDPKs are of particular importance in the regulation of SLAC1 (Slow Anion Channel-Associated 1) and KAT (potassium channel protein) channel activities in guard cell signaling (Cheng et al. 2002; Brandt et al. 2015). Mori et al. (2006) have shown that the *Arabidopsis* CDPK double mutant, *cpk3-1cpk6-1* are characterized by impaired ABA- and Ca^{2+}-induced stomatal closure. However, long-term Ca^{2+}-programmed stomatal closure was not affected by mutation; suggesting that, these are functionally separated responses (Mori et al. 2006). Studying of the *Vicia faba* guard cells showed that cytosolic ABA application maximizes the activation of anion channel within 1.5–2.5 min (Levchenko et al. 2005). Previously the activation and autophosphorylation of a Ca^{2+}-independent kinase in the same time span have been reported (Li and Assmann 1996; Li et al. 2000; Takahashi et al. 2007). However, electrophysiological studies of *V. faba* guard cells have shown that ABA-induced stomatal closure does not require Ca^{2+} elevation (Levchenko et al. 2005; Hubbard et al. 2011). Moreover, applying live-cell Ca^{2+} imaging technique in Arabidopsis revealed that only less than 50% of guard cells were characterized with elevated cytosolic calcium which is affected by the degree of humidity (Hubbard et al. 2011).

Localized increase and oscillations of cytosolic Ca^{2+} ($[Ca^{2+}]_{cyt}$) in guard cells follow a distinct pattern (McAinsh et al. 1995). The pattern of external Ca^{2+}- or ABA-induced oscillations in guard cells and the strength of the external Ca^{2+}-have been shown to induce stomatal closure (McAinsh et al. 1995; Staxén et al. 1999) which provides the possibility of encoding information about the nature of stimuli (McAinsh and Hetherington 1998; Ng and McAinsh 2003; McAinsh et al. 1995). Patch-clamp studies have revealed that both rapid- and slow-type of plasma membrane anion channels are activated in a Ca^{2+}-dependent manner (Hedrich et al. 1990). In *Arabidopsis*, steady-state stomatal closure that was programmed by Ca^{2+} oscillation have been observed in response to stimuli including external Ca^{2+}, hydrogen peroxide, cold stress, and ABA (Allen et al. 2000, 2001). However, reduced vacuolar H^+-ATPase activity in the *det3* mutant (Arabidopsis *de-etiolated 3*) was characterized with oscillations in guard cell cytosolic Ca^{2+} and steady-state stomatal closure only in response to cold and ABA (Allen et al. 2000). The growth controlled by ABA mutant (*gca2*) of *Arabidopsis* (cv. Landsberg erecta), which causes a strong high CO_2 insensitivity have impaired stomatal closure and cytosolic Ca^{2+} transient rate in response to CO_2 treatment and elevated CO_2 levels, respectively, led the researcher to propose a new model of cellular Ca^{2+}-signaling specificity in eukaryotes. Based on the proposed model, responses to cytosolic Ca^{2+}-elevations in a single cell may be mediated by priming and de-priming of Ca^{2+}-sensors by physiological stimuli (Young et al. 2006).

Inconformity between recent findings indicates additional complexities in the guard cell signaling network regarding the existence and importance of a Ca^{2+}-independent mechanism for stomatal closure. To deal with these complexities a recent study presented a more general principle to ensure specificity and robustness of stomatal Ca^{2+}-signaling on a cellular, genetic, and biochemical level. It was demonstrated that calcium-dependent protein kinase quadruple mutant of *Arabidopsis* has impaired ABA signal transduction in stomatal guard cells. According to the fact that protein phosphatase 2Cs prevent nonspecific Ca^{2+}-signaling, it was unexpectedly revealed that Ca^{2+}-dependent and Ca^{2+}-independent ABA-signaling branches are interdependent; as the triple *snrk2.2/snrk2.3/ost1* mutant showed impaired Ca^{2+} activation of S-type anion channels of guard cells. In addition, the in vivo simultaneous phosphorylation at S59 and S120 phosphorylation sites in SLAC1 are required intact ABA-induced stomatal closure (Brandt et al. 2015). SLAC1 channel activity is controlled by cooperative action of OST1 and CDPK3, 21 and 23 in counterbalance with competitive dephosphorylation by PP2Cs (Brandt et al. 2015), because the three CDPKs preferentially phosphorylate a residue different from that targeted by OST1 with respect to SLAC1 (Geiger et al. 2010; Brandt et al. 2012).

Notably, Ca^{2+}-activation of CDPKs and ROS production by NADPH oxidases are interdependent processes. These all represent a mechanism of guard cell regulation that remained to be investigated. Moreover, in experimental studies, epidermal peels of unstressed plants or heterologous systems have been utilized thus far which hinders the researchers to investigate the shared features of SnRK2-independent ABA/Ca^{2+} signaling in stomatal closure under drought stress in the scale of whole plants (Geiger et al. 2010; Brandt et al. 2012, 2015).

5.1 Calcium Signal Transduction Under Drought Stress

Under drought conditions, stomata closure is the main cause of limited gas exchange between the plant and the surrounding environment. Importance of stomata (aspects related to fine-tuned functioning and morphology) and natural variations among populations has been highlighted as important keys to improve plant acclimation to drought stress (Aliniaeifard and van Meeteren 2014; Aliniaeifard and Van Meeteren 2018; Daszkowska-Golec and Szarejko 2013). Abscisic acid is the main phytohormone involved in stomatal closure responses to water deficit conditions (low humidity in both above- and below-ground environments) (Aliniaeifard 2014; Aliniaeifard et al. 2014; Aliniaeifard and van Meeteren 2013; Endo et al. 2008). Drought (in both above- and below-ground environments) induces production of ABA and inhibits its catabolism in vascular tissues (Endo et al. 2008; Okamoto et al. 2009) as well as in guard cells (Aliniaeifard and van Meeteren 2013; Bauer et al. 2012; Okamoto et al. 2009). Stomatal response to ABA depends on the level of

$[Ca^{2+}]_{cyt}$ and its oscillation (Allen et al. 2000; Kim et al. 2010; Neill et al. 2008; Wang et al. 2011). As indicated before, induction of SLAC1 and inactivation of potassium channel (e.g., inward-rectifying K^+ channel KAT1) can be Ca^{2+}-dependent and/or Ca^{2+}-independent (Geiger et al. 2009, 2010; Joshi-Saha et al. 2011; Levchenko et al. 2005; Li and Assmann 1996; Marten et al. 2007; Siegel et al. 2009; Sutter et al. 2007). Among different components of ABA signaling pathway, type 2C protein phosphatases (PP2Cs) ABI1 (ABA-insensitive 1) and the SNF1-related protein kinase (SnRK2) open stomata 1 (OST1) are proposed to be Ca^{2+}-independent proteins. OST1 activates anion channels in an ABA-sensitive, while Ca^{2+}-independent manner (Li and Assmann 1996). Ca^{2+}-sensing receptors (CAS) are localized in chloroplasts its function is crucial for the regulation of stomatal responses (Weinl et al. 2008). At the onset of water deficit condition, increase in the extracellular Ca^{2+} concentration ($[Ca^{2+}]_o$) due to high transpiration rate will result in an increase in $[Ca^{2+}]_{cyt}$ with the mediation of CAS (Han et al. 2003).

CDPKs (Calcium-dependent protein kinase) are the main components in Ca^{2+}-dependent ABA responses (Zhu et al. 2007), which activate the SLAC1 channels (Geiger et al. 2010; Mori et al. 2006), while inhibiting KAT1 (Li et al. 1998). Under drought and high VPD conditions, guard cell plasma membrane ICa channels will be activated by ABA, which lead to Ca^{2+} influx from extracellular space (Hamilton et al. 2000), and also Ca^{2+} from intracellular stores (Blatt 2000). As a result of $[Ca^{2+}]_{cyt}$ elevation, CDPK activates SLAC1 leading to stomatal closure. CAS induces reactive oxygen (H_2O_2) and nitrogen [nitric oxide (NO)] species in the guard cells in response to an increase in $[Ca^{2+}]_o$ (Wang et al. 2011). Both H_2O_2 and NO elevates $[Ca^{2+}]_{cyt}$, and activates ICa and SLAC1 channels (Pei et al. 2000), and then stomatal closure (Chen et al. 2004; Li et al. 2009). A simplified model for signaling pathway involved in stomatal closure due to water deficit conditions is presented in Fig. 1.

6 Calcium-Sensing Proteins

There are many abiotic stress signals that require multiple sensors. Therefore, the presence of only one sensor for perceiving stress and controlling all subsequent signaling is unlikely, however, a single sensor might only initiate the onset of signaling cascade that is initiated by stimuli like stress signal. Therefore, various sensors are expected for stress signal transduction (Xiong et al. 2002). Studying Ca^{2+} signaling in *Arabidopsis thaliana* exposed to drought and salinity demonstrated that Ca^{2+} influx occurred in response to both drought and salinity. Notably, similar magnitude and duration of Ca^{2+} efflux observed for both salinity and drought stresses, leading the researchers to postulate that rather than Ca^{2+} different factor is involved in the discrimination between drought and salinity signals in Arabidopsis. Various signaling factors for drought tolerance identified including CDPK, GSK3/Shaggy MAPKKK, and SnRK2 which belong to protein kinases and other signaling factors are CBL1, 14-3-3 Protein, CC-NBS-LRR, Farnesyltransferase (reviewed by Umezawa et al. 2006). It has been demonstrated that there is a common event referred to as "calcium

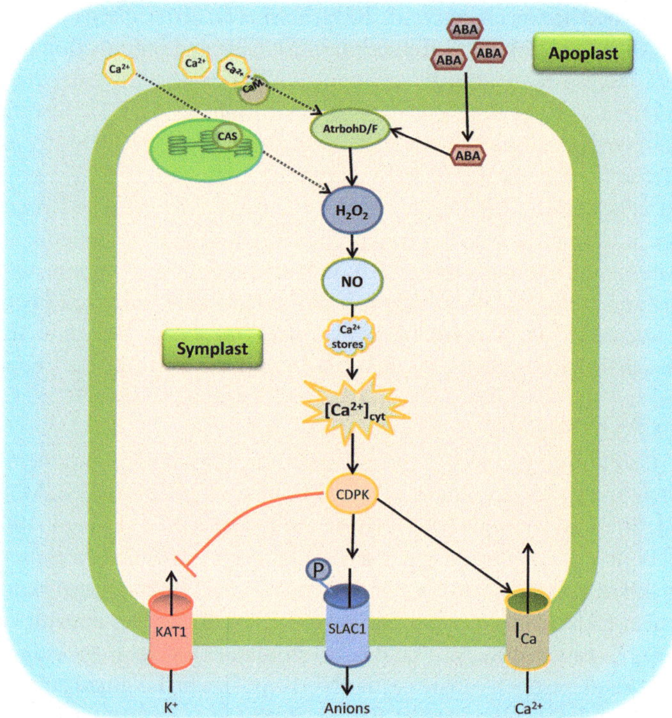

Fig. 1 Simplified model of calcium signaling under water deficit conditions. Increase in the level of extracellular calcium as well as apoplastic ABA results in elevation of $[Ca^{2+}]_{cyt}$ which leads to activation of calcium and anion and inhibition of potassium channels through calcium-dependent protein kinases (CDPK)

signature" for various types of stress transient Ca^{2+} influx into the cell cytoplasm (reviewed by Sanders et al. 1999; Knight et al. 1993; Wang and Nick 2001; Sangwan et al. 2001; Rudd and Franklin-Tong 2001). Furthermore, $[Ca^{2+}]_{cyt}$ elevation is required for the expression of certain stress-induced genes in plants (Knight et al. 1996). Then, activation of stress gene promoters leading to the $[Ca^{2+}]_{cyt}$ elevation is sufficient for their activation (Sheen 1996). It is generally accepted that Ca^{2+} is a transmitter of stress signal downstream in the pathway by interacting with protein sensors. In other words, Ca^{2+} sensor proteins synchronize alteration in metabolism, gene expression, and turnover of proteins by affecting the activity of downstream effectors (La Verde et al. 2018).

There are two groups of Ca^{2+}-binding proteins including membrane-intrinsic Ca^{2+} binding protein and Ca^{2+}-modulated proteins. Ca^{2+} binding proteins could be subdivided into two categories: The former modulates Ca^{2+} concentration in the environment by transporting it across cell membranes and the latter includes proteins that both contribute to the control of Ca^{2+} concentration and Ca^{2+} signals decoding, acting as Ca^{2+} sensors (Carafoli et al. 2001).

The spatial and temporal changes in cytosolic Ca^{2+} concentration form signals that are perceived and transduce through proteins termed as "Ca^{2+} sensors", thereby, the conformation or catalytic activity of Ca^{2+} sensors changes once Ca^{2+} is bound. Among various identified signaling factors potential Ca^{2+} sensors are CDPKs, calcineurin B-like protein (CBL), calmodulin (CAM), and calmodulin-like protein (CML) and the Ca^{2+} and calmodulin-dependent protein kinase (CCaMK) (Harmon et al. 2000; Luan et al. 2002; Sanders et al. 2002; Cheng et al. 2002; Wang et al. 2015; Zhou et al. 2015a, b). Except for CDPKs, that contain a kinase domain, the other Ca^{2+} sensors had no apparent enzymatic domain, conveys their interacting role in transmitting cellular signals to their downstream targets. Having no enzymatic activity (except CDPK) these small Ca^{2+} sensors interact with their target proteins and regulate the function of their target proteins (Zielinski 1998; Luan et al. 2002; Hashimoto and Kudla 2011). The high affinity of these proteins with Ca^{2+} together with their subcellular location within the cell determines their prominent function. Binding of Ca^{2+} to related sensors results in a conformational change that renders it to associate with downstream target proteins or directly stimulate the kinase activity (Harmon et al. 2000). Both the diversity of Ca^{2+} sensors and their downstream targets bring forward the induction of the specific responses (Hashimoto and Kudla 2011).

In a general conceptual characterization, Ca^{2+}-sensing proteins have been classified into two functional groups: (1) sensor relays and (2) sensor responders (Sanders et al. 2002). Among Ca^{2+} sensors, CDPKs classified as sensor responders due to their Ca^{2+} sensing (EF-hand motifs) and responding (protein kinase activity) functions. CDPKs translate the information encoded in the Ca^{2+} signatures into phosphorylation events of specific target proteins. In comparison with CDPK, CaM/CML family members have no enzymatic function thus, functions as sensor relay proteins. These Ca^{2+} sensors take part in signaling events by altering downstream target activities via protein–protein interactions in a Ca^{2+}-dependent manner. CBL proteins also categorized as sensor relay proteins with representing no enzymatic activity. However, it is considered as bimolecular sensor responders when interacting with a family of protein kinases designated as CBL-interacting protein kinases (CIPKs) (Hashimoto and Kudla 2011).

6.1 CBL

CBL are structurally very similar to calcineurin B (CNB) and neuronal calcium sensors (NCS) of animals (Liu and Zhu 1998; Kudla et al. 1999). Up to now, the CIPK gene family has been identified in Arabidopsis (Kolukisaoglu et al. 2004; Yu et al. 2007), poplar (*Populus trichocarpa*) (Yu et al. 2007), maize (*Zea mays*) (Chen et al. 2011), rice (*Oryza sativa*) (Kanwar et al. 2014), canola (*Brassica napus*) (Zhang et al. 2014), and Cassava (Hu et al. 2015). To date, at least 10 CBLs and 25 CIPKs in *Arabidopsis* and 10 CBLs and 31 CIPKs in rice have been identified. These proteins integrate to form a network-like signaling system for coupling of specific and synergistic stimulus–response (Luan et al. 2002; Batistic and Kudla

2004; Kolukisaoglu et al. 2004; Piao et al. 2010). In the signaling pathways, they act as a threshold factor in linking the Ca^{2+} signature to the downstream components (Sanders et al. 2002). Binding to Ca^{2+} ions enables CBLs to interact with CBL-interacting protein kinase (CIPK) and form Ca^{2+}-CBL-CIPK network that enables it to regulate related physiological processes (Sanyal et al. 2015).

CIPKs have a serine/threonine-protein kinase domain in the N-terminus and a self-inhibitory NAF domain in C-terminus (Albrecht et al. 2001). The activation of CIPKs is mediated when the NAF domain interacts with CBLs (Kleist et al. 2014). Therefore, CBLs modulate the activity of CBL-interacting protein kinases (CIPKs), known as sucrose non-fermenting 1 related kinase 3 (SnRK3) (Shi et al. 1999; Luan et al. 2002). The CBL-CIPK modules function downstream of Ca^{2+} flow in guard cells (Cheong et al. 2007). The CBL and CIPK proteins form Ca^{2+} decoding signaling network and play an important role in plant responses to abiotic stresses (Zhu et al. 2016).

Reportedly, CIPK gene family could be divided into an intron-rich clade and an intron-poor clade, which subgroups in Arabidopsis, maize, rice, and canola CIPK genes were induced by drought (Chen et al. 2011; Ye et al. 2013). Recently, in an attempt for increasing drought tolerance in soybean, 52 CIPK family members have been identified suggesting that intron-poor CIPK genes might be originated in seed plants. Using quantitative real-time polymerase chain reaction (qRT-PCR) confirmed that 18 genes were drought-inducible (Zhu et al. 2016).

It has been shown that overexpression of CBL1, CBL9, CIPK1, and CIPK6 increased drought tolerance in *Arabidopsis* (D'Angelo et al. 2006; Pandey et al. 2004; Tsou et al. 2012), conversely, the *cbl1*, *cbl9,* and *cipk3* mutants are shown to be more sensitive to drought stress (Cheong et al. 2003; Pandey et al. 2008). It was demonstrated that the CBL1 gene is induced by drought, cold, and wounding (Kudla et al. 1999). Cheong et al. 2003 studied the functions of a CBL type Ca^{2+} sensor—CBL1 in stress gene expression and stress tolerance in plants. Their research revealed that CBL1 functions as a critical Ca^{2+} sensor in abiotic stress responses. Interestingly, investigating the role of CBL in both cold and drought stress showed that, the same Ca^{2+} sensor can act as both positive and/or negative regulators in different signaling pathways. Overexpression of CBL1 exhibited elevated drought and salt tolerance and up-regulated the expression of stress-responsive genes. Moreover, it was revealed that the CBL1b gene is expressed during all developmental stages of plants and in all plant organs. Overexpression of CBL1 in transgenic plants activated the expression of some late stress-responsive genes under "normal" conditions and their tolerance to salt and drought was enhanced however reduced tolerance to freezing was reported as well. Suggesting that, the quantity of CBL1 is a rate-limiting factor in Ca^{2+}-mediated stress signaling with respect to the level of expression of downstream target genes (Cheong et al. 2003, 2010). In another study, the function of CBL1 was investigated by analyzing T-DNA-induced knock-out mutant and plants that overexpressed CBL1. Plants lacking CBL1 have shown impaired response to drought and salt stress though their sensitivity to ABA was not affected by the mutation. On the other hand, overexpression of CBL1 enhanced stress tolerance via the induction of stress-regulated genes. These findings suggest that CBL1 may form an integrative node in abiotic stress signaling and take part in the regulation of early stress-related TFs of CBF/DREB type (Albrecht et al. 2003). Overexpression of *CBL5* has shown to result in enhanced tolerance to high salt or drought stress. Furthermore, these transgenic plants were more resistant to high salt or hyperosmotic stress during the early developmental stage; nevertheless, their response to ABA didn't change. In addition, the expression of stress gene markers, such as *RD29A*, *RD29B*, and *Kin1* was modified by overexpression of *CBL5*, indicating that, *CBL5* may play a functional role in the regulation of salt or drought stress responses in plants (Cheong et al. 2010). In transgenic tobacco, ZmCIPK8 involved in drought stress tolerance by regulating the expression of related stress genes (Tai et al. 2015). Different studies in rice revealed that OsCIPK23 acts as a positive regulator under drought stress as *oscipk23* was sensitive to drought (Yang et al. 2008). Moreover, overexpression of the BrCIPK1 in rice plant resulted in enhanced proline concentration which subsequently led to osmoregulation and further drought stress tolerance (Abdula et al. 2015). Studying the grapevine under drought stress, clarified that, a CBL Ca^{2+} sensor 1–protein kinase, CIPK23 network is expressed in grape berries and activates grapevine Shaker inward K^+ channel which triggers in drought stress response (Cuéllar et al. 2010). Another type of CIPK2 was detected in Arabidopsis under drought and osmotic stress treatments that took part in osmotic stress tolerance when it was overexpressed (Li et al. 2012). Quantitative RT-PCR studies in Arabidopsis revealed a multifold increase in CIPK21 transcript in seedlings treated with NaCl or mannitol in comparison with control. Moreover, treatment with polyethylene glycol, abscisic acid (ABA), cold, and drought stress-induced CIPK21 transcript, which suggests the pivotal role of CIPK21 in salinity and osmotic stress tolerance (Pandey et al. 2015). CBL-CIPK complex also triggers stress response indirectly by regulating ABA signaling. In this regard, CBL9 has shown to function as a negative regulator in the ABA-signaling pathway (Nakashima and Yamaguchi-Shinozaki 2013; Pandey et al. 2004). The role of CIPK3 as a negative regulator of ABA signaling has been also revealed by studying seed germination of *cipk3* mutants of Arabidopsis (Kim et al. 2003) and increased germination rate of overexpressed *AtCBL9/AtCIPK3* Arabidopsis exposed to ABA (Pandey et al. 2008). CIPK6 can also involve in the regulation of the ABA-signaling pathway as it was found that *cipk6* mutant can accumulate more ABA under ABA treatment (Chen et al. 2013a).

In a recent study, it was shown that TaCIPK27A confers drought tolerance and sensitivity to exogenous ABA in transgenic Arabidopsis. It was found that TaCIPK27 is up-regulated by multiple abiotic stresses. TaCIPK27 interacts with AtCBL1, AtCBL3, AtCBL4, AtCBL5, and AtCBL9. In Arabidopsis, ectopic overexpression of *TaCIPK27* enhanced drought tolerance which was concomitant with the regulation of some drought-related genes and ABA concentration (Wang et al. 2018). In another study, the participation of TaCIPK23 in the regulation of several drought- and ABA-responsive genes has been demonstrated. Biochemical and reverse genetic studies confirmed the contribution of TaCIPK23 to plant ABA signaling and drought tolerance. Exogenous ABA induced the expression of TaCIPK23 and delayed seed germination and small stomatal aperture under exogenous ABA treatment observed.

Accordingly, under drought stress, the lower stomatal conductance in TaCIPK23-overexpression lines resulted in the reduced water loss and consequent drought tolerance. Based on these results, researchers concluded that TaCIPK23 positively modulates plant drought tolerance through ABA-dependent and -independent pathways (Cui et al. 2018).

Interaction network and co-expression analyses of cassava CIPKs, besides transcriptome analysis of different cassava accessions in response to drought indicated that the pathways controlled by CIPK networks may contribute to responses to drought stress in different tissue or different accessions of cassava (Hu et al. 2015). Overexpression of a cotton CIPK gene, *GhCIPK6*, in Arabidopsis was induced by drought, and enhanced plant tolerance to drought (He et al. 2013). It was demonstrated that the expression pattern of several *ZmCIPKs* under drought stress involved in ABA and H_2O_2 signaling (Tai et al. 2013) and in tobacco overexpression of *ZmCIPK8* enhanced drought tolerance by inducing the expression of the *NAC*, *CBF*, and *Rd29A* genes (Tai et al. 2016). It was also revealed that, although CBL1 and CBL9 are structurally and evolutionary homolog, they function differently. Both *cbl1* and *cbl9* mutants are highly sensitive to abiotic stress, but *cbl1* did not respond to stresses under ABA treatment, which led the researcher to postulate that CBL1 may trigger abiotic stress via ABA-independent pathway (Albrecht et al. 2003; Cheong et al. 2003). Reportedly, the diverse CBL–CIPK Ca^{2+} signaling pathway plays its role in stress tolerance by regulation of ions transport in plant under stress conditions. The CBL–CIPK pathways have been shown to operate the transport of sodium, potassium, nitrate, and phosphorous in plant (reviewed by Thoday-Kennedy et al. 2015).

Up to now, the majority of research focused on identifying the interactions between CBLs and CIPKs and the location of their interaction in Arabidopsis and the downstream targets of CIPKs signaling have been identified poorly while it is essential to be identified for unraveling the mechanism controlling plant ontogeny under abiotic stresses and the genes expressed and protein synthesized during this process. Studies on rice and Arabidopsis revealed that in each species CBL families follow species-specific evolutionary pathways with specific gene duplication events, despite the equal number of CBL genes encoded in the rice and Arabidopsis genome. Thus, precise gene-for-gene predictions of CBL function from one genus to the other may not be reliable (Batistic and Kudla 2004).

6.2 CAM

Calmodulin (CAM), one of the best distinguished Ca^{2+} sensors is a small acidic protein consist of four EF-hands. The binding of Ca^{2+} to CaM induces a conformational change in this protein. These changes enable the protein to contribute to hydrophobic and electrostatic interactions with its target proteins by exposing hydrophobic surfaces surrounded by negative charges (Snedden and Fromm 1998; Hoeflich and Ikura 2002). Up to now, more than 50 different types of CaM-binding proteins were identified. Ca^{2+}/CaM complex modulates the activities of numerous target proteins which yields in regulation of a variety of cellular responses including regulation of metabolism, cytoskeleton function, phytohormone signaling, ion transport, protein folding, protein phosphorylation and dephosphorylation, phospholipid metabolism, and transcriptional regulation (Snedden and Fromm 2001; Yang and Poovaiah 2003; Bouché et al. 2005).

The presence of CaM in the nucleus beside the identification of several nuclear molecules as CaM-binding proteins indicates the crucial role of CaM in decoding of Ca^{2+} signaling, in both of the cytosol and nucleus (Dauwalder et al. 1986; Schuurink et al. 1996). Mechanism of Ca^{2+} responses in nuclei is yet to be understood; however, the independent responses of cytosolic Ca^{2+} and nucleic Ca^{2+} have been revealed (Pauly et al. 2000). The identification of various TFs regulated directly or indirectly by CaM strongly supports the crucial role of CaM as a nuclear Ca^{2+} signaling decoder, Therefore, a role for CaM in regulation of gene expression that regulate cellular responses to developmental and environmental stimuli is highly likely (Snedden and Fromm 2001; Yang and Poovaiah 2003; Bouché et al. 2005). The possible role of nucleus-translocated CaM53 protein in the Ca^{2+}-mediated sugar sensing pathway in plant cells (Rodríguez-Concepción et al. 1999) accompanied with identification of numerous nuclear CaM-binding proteins like pea nuclear apyrase that binds to CaM in a Ca^{2+}-dependent manner which is activated by Ca^{2+}/CaM-binding proteins (Hsieh et al. 2000) and the potato CaM-binding protein (PCBP) (Reddy et al. 2002) are among evidence indicating significant role of CaM as a nuclear Ca^{2+} signaling decoder.

It is widely accepted that the Ca^{2+}/CaM complex mediates plant response to biotic and abiotic stimuli (Snedden and Fromm 1998; Bender and Snedden 2013; Virdi et al. 2015). There are two defined pathways through which Ca^{2+}/CaM complex functions either directly or indirectly to trigger a stress response. Direct pathway is binding to target proteins that are a coordinator of a variety of cellular functions and the indirect pathway is regulating the expression of genes encoding downstream effectors which result in cellular responses.

A novel CaM-binding transcription factor family that contains a CG-1 DNA-binding domain distinguished in plant proteins during the isolation of a partial cDNA clone encoding a sequence-specific DNA-binding protein from parsley (da Costa e Silva 1994). Later, homologs of CG-1 identified in various plant species including Arabidopsis, rice, tobacco, and rapeseed (Reddy et al. 2000; Yang and Poovaiah 2000; Bouché et al. 2002; Yang and Poovaiah 2002; Choi et al. 2005a, b). Several studies suggest that the CG-1 proteins act as transcriptional activators of downstream gene expression (Mitsuda et al. 2003; Choi et al. 2005a). Two different kinds of CaM-binding motifs of CG-1 proteins exist; a Ca^{2+}-dependent CaM-binding domain and a Ca^{2+}-independent CaM-binding domain, termed the IQ motif. The function of IQ motif is not clearly understood (Choi et al. 2005b).

Calmodulin-binding transcription activator (CAMTA) TF triggers the expression of several downstream genes involved in stress response (Janiak et al. 2015). Altered expression of CAMTA gene family in response to drought stress has been reported. Enhanced drought susceptibility has been observed in *Arabidopsis thaliana camta* mutant in comparison to the wild-type plants (Pandey et al. 2013). The possible role of CAMTA family in the regulation of drought stress indicated by the involvement of members of CAMTA family in regulation of different drought-responsive genes like ethylene-responsive element-binding factor 13 (ERF13), C-repeat/DRE binding factor 2 (CBF2), and WRKY33 (Pandey et al. 2013). In a recent study, the expression pattern of MuCAMTA (Musa acuminate Calmodulin-binding transcription activator) genes was identified by qRT-PCR which up-regulated (40-fold) under drought stress; suggesting that MuCAMTA1 plays an important role in drought stress tolerance in banana plants (Meer et al. 2019).

A plant-specific family of TFs that consists of CBP60 proteins bind to CaM (Bouché et al. 2005). One of the members of this family, interact with CaM in a Ca^{2+}-dependent manner (Wang et al. 2009). Overexpression of CBP60g in Arabidopsis resulted in increased sensitivity to ABA as well as increased tolerance to biotic and drought stress (Wan et al. 2012). A family of trihelix TFs or GT factors, identified in Arabidopsis and rice (Riechmann et al. 2000; Wang et al. 2014). One of the members of this family, GTL1 (GT-2 LIKE 1), negatively regulates water use efficiency, inasmuch as loss of function mutation in this gene in Arabidopsis showed enhanced tolerance to water stress (Yoo et al. 2010). On the other hand, overexpression of the same gene homolog in poplar (*Populus tremula* × *Populus alba*) downgraded the drought tolerance. Interestingly, the Ca^{2+}-dependent interaction of PtaGTL1 with CaM through the C-terminus amino acid residues 528–551 and 555–575 have been reported (Weng et al. 2012).

Beside direct regulation of TFs including DNA-binding or transcription activities, Ca^{2+}/CaM complex also regulates gene expression indirectly by either modulating posttranslational modification of TFs (Snedden and Fromm 2001) or by its function through a CaM-binding protein kinase and a CaM-binding protein phosphatase (Liu et al. 2007, 2008). The indirect regulation of gene expression by modulating the transcription elongation reaction has been also reported (Nelissen et al. 2003).

6.3 CML

CMLs are another type of Ca^{2+} sensor protein, which abundantly present in plants. In Arabidopsis, various CMLs exist with various lengths ranging from 83 to 330 amino acids. At the primary sequence level, CLMs shows between 16 and 75% amino acid identity with CaMs. Contrary to CaMs that contain 4 EF-hands, CMLs contain one to six EF-hands and not all motifs are functional. This variation in the number of EF-hands is hypothesized to be the reason for different responses to Ca^{2+} signals (Bender and Snedden 2013). Studying biochemical and structural properties of CMLs, revealed that CMLs generally, but not always function as Ca^{2+} sensors (Song et al. 2004; Bender and Snedden 2013).

A research on expression patterns of Arabidopsis *CML37*, *CML38,* and *CML39* genes under drought stress treatment demonstrated the increased expression of all three CMLs within 8 h in the shoot apex, hypocotyl and root (primarily in the vasculature) compared to the controls, suggesting that, these CMLs likely play important roles as Ca^{2+}-sensors and take part in stress response pathways (Vanderbeld and Snedden 2007). Genome-wide analyses of Ca^{2+} sensors indicated the possible involvement of Ca^{2+} sensors in storage root deterioration in Cassava. Interaction network and co-expression analyses identified some gene pairs uniformly up-regulated at 48 h/0 h after harvest. Based on the transcriptomic data, some Ca^{2+} sensor genes including MeCML-4, -10, -16, -17, -20, -22, -25, -26, and MeCBL2 induced after drought treatment and during storage root deterioration which can be possibly nominated for genetic improvement of cassava resistance to drought and storage root deterioration (Hu et al. 2018). In a recent study, TaCML20 overexpressing transgenic lines in wheat have shown to accumulate higher water-soluble carbohydrate concentrations in the shoots and a significant increase in transcript level of sucrose (sucrose-1-fructosyltransferase (1-SST) in the internodes compared with the control plants. Moreover, a role for TaCML20 in drought stress suggested, as a significant increase of TaCML20 transcripts observed in the shoots of wild-type plants under water deficit condition (Kalaipandian et al. 2018). Microarray and quantitative real-time RT-PCR analyses have revealed that expression of *OsMSR2*, a novel calmodulin-like protein gene isolated from rice Pei'ai 64S (*Oryza sativa* L.) was strongly up-regulated by various ranges of stresses in different tissues at different developmental stages of rice. In Arabidopsis, the expression of *OsMSR2* concomitant with altered expression of stress/ABA-responsive genes resulted in enhanced tolerance to high salt and drought stresses (Xu et al. 2011). Moreover, The ERECTA-overexpressing Arabidopsis exhibited reduced stomatal density (Masle et al. 2005). Interestingly, the transcript level of ERECTA in *OsMSR2*-transgenic plants is significantly higher than that in the wild type. Possibly, increased expression of *ERECTA* by expression of *OsMSR2* may take part in the reduced stomatal density. As far as stomatal density is a determinative factor in water and CO_2 exchange, it is possibly responsible for lower respiration rate in the transgenic plant (Hetherington and Woodward 2003; Xu et al. 2011). *Atcml9* null mutants showed enhanced tolerance to drought and salt stress concomitant with a hypersensitive response to ABA during early growth stages. Moreover, alterations in the expression of several stress and ABA-responsive and defense-related genes have been reported (Ranty et al. 2016).

In this regard, Scholz et al. (2015) reported antagonist involvement of AtCML37 and AtCML42 in drought stress. It was revealed that *CML37* knock-out line, *cml37*, showed high susceptibility to drought stress, in contrast, *CML42* knockout line, *cml42*, showed no difference in drought stress response compared to wild type (WT) plants. In accordance, a significant reduction of ABA under drought stress observed in *cml37* plants, whereas in *cml42* plants, an increase of ABA was detected which led the researchers to speculate that, CML37 is a positive regulator of ABA, while CML42 is a negative regulator of ABA accumulation, which is inducible by drought stress. Therefore, both CML37 and CML42 are antagonistically involved in drought stress response (Scholz et al. 2015). The antagonist regulation of CML37 and CML42 has been demonstrated in herbivory defense response as well (Vadassery et al. 2012; Scholz et al. 2014). Besides, *cml9* knockout mutants also showed enhanced tolerance to drought and salt stresses in Arabidopsis suggesting that CML9 negatively regulated drought tolerance responses (Magnan et al. 2008).

6.4 CDPK

The CDPKs are encoded by a multigene family and have been identified in a wide range of vascular and nonvascular plants as well as in green algae and certain protozoa (Harmon et al. 2001). CDPKs have been characterized in various plant species, such as Arabidopsis (Cheng et al. 2002), rice (Wan et al. 2007), cotton (Huang et al. 2008), and wheat (Li et al. 2008). The expression of CDPKs occurs both ubiquitously or in a tissue-specific manner in response to environmental stimuli (Hrabak et al. 2003). CDPKs function diversely by occurring in various subcellular locations including cytosol, nucleus, the plasma membrane, endoplasmic reticulum, peroxisomes, mitochondrial outer membrane, and oil bodies (Harper et al. 2004). Containing EF-hands, CDPKs have the feature of Ca^{2+}-dependence. Degeneration and loss of some EF-motifs alter their Ca^{2+} dependence or sensitivities. The ability of CDPKs in Ca^{2+} perception and loading might be affected by specific cellular circumstances and the surrounding regulators. Therefore, the interactive protein effectors and other surrounding regulators are determinative in Ca^{2+} loading of CDPKs (Shi et al. 2018).

It is widely revealed that CDPKs take part in plant responses to various abiotic stresses, including cold, salt, drought, and wounding (Cheng et al. 2002; Ludwig et al. 2004; Klimecka and Muszynska 2007; DeFalco et al. 2010; Seifikalhor et al. 2019). It has been shown that in normal conditions activation of some CDPK members is sufficient to activate a stress gene promoter in maize protoplasts, suggesting that these CDPK members are positive regulators of stress gene expression (Sheen 1996). The expression of *CPK13* in Arabidopsis guard cells and the inhibition of light-induced stomatal opening via *CPK13* overexpression have been also reported (Ronzier et al. 2014). Transgenic *Arabidopsis* plants overexpressing *PeCPK10* a CDPK gene cloned from *Populus euphratica* has shown stronger ABA-induced promotion of stomatal closure in response to drought stress (Chen et al. 2013b). Another study indicated that *AtCPK6* functions redundantly as a positive regulator of salt/drought stress in Arabidopsis as *AtCPK6* overexpressing plants showed altered regulation of several stress-regulated genes and enhanced tolerance to salt/drought stresses (Xu et al. 2010). Modulation of ABA- and Ca^{2+}-regulated stomatal movements by CPK10 (At1g18890) and involvement of a heat shock protein, HSP1 (At4g14830), as a CPK10-interacting protein has been indicated in plant drought stress responses which led scientists to conclude that for plant response to drought stress, especially for regulation of stomatal movements, the interaction between CPK10 and HSP1 is of particular importance (Zou et al. 2010). CPK8 have been also found to function in ABA-mediated stomatal closure in response to drought stress via CAT3 (CATALASE 3) regulation. It has been verified that CPK8 is able to interact and phosphorylate CAT3 (catalyzes the dismutation of H_2O_2 to water and oxygen). In addition, both *cpk8* and *cat3* mutant plants were more sensitive to drought stress than wild-type plants, and inhibition of K^+ flow by ABA and Ca^{2+} was reduced in guard cells of mutant plants which suggested the possible role of CPK8 functions in ABA-mediated stomatal movement in response to drought stress through the regulation of CAT3 (Zou et al. 2015). Overexpression of ginger *CDPK1* gene (*ZoCDPK1*) in tobacco resulted in high percentage of seed germination, higher relative water content, induction of expression of stress-responsive genes, higher leaf chlorophyll content, increased photosynthetic efficiency, and other photosynthetic parameters and 50% more growth than the wild-type plants, indicating increased tolerance to salinity and drought stress. This confirms the involvement of *ZoCDPK1* in positive regulation of salinity and drought stress response signaling pathways in ginger (Vivek et al. 2013). Increased tolerance to drought stress was also observed when a rice CDPK, *OsCPK13*, was overexpressed (Saijo et al. 2000). Phosphorylation of drought-responsive zinc-finger domain protein (Di19) by AtCPK4/1, proposing a role for CDPKs in the regulation of abiotic stress responses (Milla et al. 2006). In Arabidopsis, five of the seven AtDi19 family members are known to function in ABA-independent drought and salinity tolerance pathways, these AtDi19 members are phosphorylated by AtCPK3 and AtCPK11 (Milla et al. 2006). In other reports, both ABA-dependent and -independent drought-response pathways have been assigned to AtCPK11 and AtCPK3 (Mori et al. 2006; Zhu et al. 2007). It turns out from the evidence that AtCPK3 participates in drought stress responses in both ABA-dependent and -independent manners. At the onset of functional analysis of AtCDPKs, it was revealed that transcription of AtCPK10 and AtCPK11 was induced by drought and high salt stresses in *A. thaliana* (Urao et al. 1994). In *Mesembryanthemum crystallinum* exposure to dehydration stress and salinity increased transcription of McCPK1 was observed (Chehab et al. 2004). AtCPK32 is an ABA-signaling component that regulates the ABA-responsive gene expression via ABF4 (leucine zipper class transcription factors) (Choi et al. 2005a). Two Arabidopsis guard cell expressed *CDPK* genes, *CPK3* and *CPK6*, have been investigated. Among single and double mutants, *cpk3 cpk6* showed impaired ABA and Ca^{2+} activation of slow-type anion channels and ABA activation of plasma membrane Ca^{2+}-permeable channels which resulted in impaired stomatal closure; therefore, CPK3 and CPK6 identified as important regulators of guard cell ion channels and ABA-regulated stomatal signaling (Mori et al. 2006). It was demonstrated that ABA, PEG, and NaCl treatments induced the transcription of a rice *CDPK* gene, *OsCPK9*. Overexpression of *OsCPK9* (*OsCPK9*-OX) and *OsCPK9* RNA interference (*OsCPK9*-RNAi) analyses showed that *OsCPK9* induced drought stress tolerance by enhancement of stomatal closure and improvement of osmotic adjustment (Wei et al. 2014).

In contrast with mentioned evidences, it was reported that the deficiency of CPK21 and CPK23 enhanced the tolerance to hyperosmotic stress and drought/salt stresses, while the overexpressing lines showed reverse phenotypes proposing that CPK21 and CPK23 act as negative regulators in *Arabidopsis* responses to corresponding stresses (Franz et al. 2011; Latz et al. 2013). Based on biochemical analysis CPK21 and CPK23 are able to phosphorylate the guard cell anion channel SLAC1 in a Ca^{2+}-sensitive and Ca^{2+}-insensitive manner. Geiger et al. (2010) provided evidence of a new Ca^{2+}-dependent ABA-signaling pathway in guard cells, composed of CPK21 (strongly dependent)/CPK23 (weakly dependent), SLAC1 and ABI1. Two distinct *cpk33* mutants displayed a significantly smaller stomatal aperture in comparison to wild-type plants and showed drought tolerance and an increased activity of the slow anion channel under both drought stress and ABA treatment, whereas, *CPK33* overexpressing lines showed impaired stomatal closure, increased water loss, and less tolerance to drought stress (Li et al. 2016). In this regard, it was shown that overexpression of *AtCPK23* increased stomatal apertures. On the other hand, stomatal apertures reduced when *AtCPK23* expression was disrupted, suggesting that, modified *Arabidopsis* response to drought stress caused by the alteration of stomatal apertures due to alteration in *AtCPK23* expression (Ma and Wu 2007).

7 Involvement of Calcium in Other Signaling Pathways

7.1 ABA-Signaling Pathways

Ca^{2+} plays an important role in plant cell signaling and known as an important secondary messenger involved in ABA signal transduction (reviewed by Finkelstein and Rock 2002). The sensitivity of stomatal closing to increased intracellular Ca^{2+} in response to stimuli especially ABA termed as "Ca^{2+}-sensitivity priming" (Hubbard et al. 2011). Three regulatory steps bring about ABA signaling including: (1) perception by receptors, (2) protein kinases and phosphatases mediators, (3) targets such as transcription factors. The Ca^{2+} signals and ABA signaling can be integrated at all three steps (Kudla et al. 2018). It was revealed that a family of small proteins harboring a Ca^{2+}-binding C2 domain, namely CAR proteins (C2-domain ABA-related proteins) controls the establishment of ABA receptors in the membrane (Rodriguez et al. 2014).

Activation of onion channels occurs in both Ca^{2+}-independent and a Ca^{2+}-dependent manner (Levchenko et al. 2005; Stange et al. 2010). In both signaling pathways, the same ABA receptors and PP2C phosphatases such as ABI1, ABI2, and HAB1 are affected (Ma et al. 2009; Melcher et al. 2010; Hua et al. 2012; Edel and Kudla 2016). Recent experiments found that Ca^{2+}-sensing proteins such as CDPKs and CBL/CIPKs potentially link the Ca^{2+}-signaling with ABA responses (Edel and Kudla 2016). In addition, ABA sensitivity and consequent ABA response are affected by altered expression of CDPKs and CBLs/CIPKs (Choi et al. 2005a, b; Zhu et al. 2007; Lynch et al. 2012). Strong ABA-insensitive phenotypes in for both stomatal closure and inhibition of stomatal opening have been observed in double mutants of Arabidopsis *cpk4-1 cpk11-1* and *cpk4-1 cpk11-2*. These mutants are characterized by more water loss in comparison to single mutants. Moreover, in vitro experiment demonstrated that immunoprecipitated natural proteins of both CPK4 and CPK11 phosphorylate the ABA-responsive ABF1 and ABF4 TFs. Notably, ABA treatment significantly stimulated the phosphorylation of these TFs. These observations revealed evidence of participation of Ca^{2+}-dependent kinases in stomatal closure in an ABA-dependent manner via regulating ABA-responsive TFs (Zhu et al. 2007). In maize (*Zea mays*) leaf protoplasts, stress- and ABA-inducible promoter activated as a result of constitutive ectopic expression of two *A. thaliana* CDPKs, CPK10/CDPK1, and CPK30/CDPK1a, clarified the linkage between CDPKs and ABA signaling (Sheen 1996). Moreover, it was shown that *ZmCPK4*—a maize CDPK gene, positively regulated ABA signaling and enhanced drought stress tolerance in transgenic Arabidopsis. Researchers suggest that *ZmCPK4* may play its role in alleviating drought stress via ABA-mediated regulation of stomatal closure (Jiang et al. 2013).

The Ca^{2+}-regulated protein kinases share similarities with SnRK2s in their PP2C interaction and target regulation (Cutler et al. 2010; Raghavendra et al. 2010). ABA response regulatory SnRK2s such as OST1 directly phosphorylates and activates the ABA-responsive TFs by binding to PP2Cs (Kobayashi et al. 2005; Ng et al. 2011; Xie et al. 2012). The impairment of ABA signal transduction in stomata of CDPK quadruple mutant plants clarified the interdependence of CPKs and SnRK2s for a functional ABA-signaling pathway (Brandt et al. 2015). It was revealed that despite SnRK2s, CDPKs and CIPKs specifically address distinct targets, they have overlapping substrate spectrum (Steinhorst and Kudla 2014; Edel and Kudla 2016).

It has been suggested that certain types of Ca^{2+}-sensing proteins negatively regulate ABA signal transduction. The CBL-interacting protein kinase CIPK11/PKS5 negatively impacts the ABA responses by phosphorylation of the bZIP transcription factor ABI5 (Zhou et al. 2015a, b). CIPK15 interacts with two calcium-modulated protein phosphatases 2C, ABI1 and ABI2, known as negative regulators of ABA signaling (Gosti et al. 1999; Merlot et al. 2001; Guo et al. 2002). An AP2 transcription factor (At ERF7kinase) as a substrate of CIPK15, negatively regulates the ABA responses, therefore, CIPK15 may regulate ABA signaling directly by phosphorylating a TF and modulating gene expression (Song et al. 2005). CIPK3 and CBL9 are also characterized as negative regulators of ABA signaling (Kim et al. 2003; Pandey et al. 2004). It was also reported that disruption of CIPK3 function does not influence gene induction by drought and hyperosmotic stress. This evidence suggests that CIPK3 regulates specific pathways that lead to the stress gene expression. It was also stated that CIPK3 functions as a cross-talk "node" mediating the interaction between ABA and abiotic stress signal transduction pathways (Kim et al. 2003).

Despite the numerous reports on involvement of Ca^{2+}- and ABA in stress response, evidence for the role of Ca^{2+}-regulated kinases in activation of ABA-related TFs is still missing, however, an alternative scenario is CDPKs and CIPKs perceive the cues from environmental stimuli such as drought, salinity or flooding via endogenous Ca^{2+}-release and function in controlling the machinery of ABA signal transduction; therefore, these protein kinases turn out to be the prime candidates for acclimation responses in stress condition (Kudla et al. 2018; Seifikalhor et al. 2019).

7.2 ROS Signaling Pathway

ROSs play an essential role as signaling molecules in the regulation of numerous biological processes such as growth, development, hormonal signaling, polar growth, autophagy, programmed cell death, and responses to biotic and abiotic stimuli in plants (reviewed by Demidchik 2015; Foyer and Noctor 2016).

In the plasma membrane of Arabidopsis guard cells, activation of hyperpolarization-dependent I_{Ca} channels by ROS induces an increase in $[Ca^{2+}]_{cyt}$ (Pei et al. 2000). Other cell types also indicated ROS-induced increase in $[Ca^{2+}]_{cyt}$ indicating that the activation of Ca^{2+} channels could be a key step in many ROS-mediated processes (Mori and Schroeder 2004) (Fig. 1).

Reactive oxygen species and Ca^{2+} signals have been shown to regulate the same physiological reactions in the same time span; notably under biological conditions, H_2O_2 as relatively long-lived ROS does not exist for longer than one second. Apparently, ROS and Ca^{2+} stimulate each other in a way that ROS target the same or nearest cells that are effective autocrine/paracrine stimulators of Ca^{2+} signaling (Halliwell 2006). Increase in $[Ca^{2+}]_{cyt}$ results in an elevated ROS production and vice versa (Foreman et al. 2003). This self-amplification extends the duration and amplitude of weak signals; thereby, appropriate signaling causes dramatic ROS-Ca^{2+} alterations (Demidchik and Maathuis 2007; Demidchik et al. 2009).

Reactive oxygen species are generated by multiple biological systems including electron-transporting chains (ETC) of chloroplasts and mitochondria, small ETC of peroxisomes, class III cell wall peroxidases, oxalate oxidase, quinine reductase, and polyamine oxidase. These systems produce ROS as by-products of their constitutive physiological functions. NADPH oxidases contribute to extracellular ROS synthesis for mediation of signaling and regulation processes (Bose et al. 2014; Demidchik 2015; Foyer and Noctor 2016). In addition to direct ROS synthesis, primary stress stimuli induce NADPH oxidases leading to ROS production during signaling events. Superoxide generated by NADPH oxidases is at the onset of ROS generation and other signaling events and scavenging superoxide is a leading mechanism for terminating ROS and Ca^{2+} signals (Foyer and Noctor 2016). The joint function of Ca^{2+} and NADPH oxidase is due to a lack of NADPH oxidase function at the absence of Ca^{2+}. This Ca^{2+}-depended function is due to the presence of two Ca^{2+}-binding helix–loop–helix structural domains (EF-hands) in the NADPH oxidase structure. The conformational change and intramolecular interaction of the N-terminal Ca^{2+}-binding domain with the C-terminal superdomain in consequence of binding to Ca^{2+}, enables electron transfer (Bànfi et al. 2004). Reportedly, binding of CaM to the NADPH binding domain or phosphorylation of serine/tryptophan residues in the FAD-binding domain by protein kinase C enhances the capability of binding to Ca^{2+} (Jagnandan et al. 2007; Kobayashi et al. 2007; Tirone et al. 2010). Biding capacity to Ca^{2+} increases after CaM attachment to the NADPH binding domain or phosphorylation of serine/tryptophan residues in the FAD-binding domain by protein kinase C (Jagnandan et al. 2007; Kobayashi et al. 2007; Tirone et al. 2010).

A parallel increase in NADPH oxidase activity and CDPK expression have been reported (Xing et al. 2001). It was also reported that an increase in NADPH oxidase activity via small G proteins (Rac/Rop GTPases) occurs in a Ca^{2+}-dependent manner (Baxter-Burrell et al. 2002). It was hypothesized that the Ca^{2+}-activated NADPH oxidases work cooperatively with the ROS-activated I_{Ca} channel to generate and amplify stress-induced Ca^{2+} and ROS signals which are known as the "ROS-Ca^{2+} hub" (Demidchik and Maathuis 2007; Demidchik 2015; Shabala et al. 2015). It was reported that overexpression of *OsACA6* (Ca^{2+}ATPase gene from *Oryza sativa*) is triggered during salinity and drought stresses. Besides several physiological aspects, overexpressing lines also showed higher leaf chlorophyll and reduced accumulation of ROS compared to the wild-type. Transcription profiles of transgenic plants declared altered expression of genes encoding many TFs, stress-related proteins and signaling components. This suggests that Ca^{2+}-ATPases have diverse roles as regulators of stress signaling pathways (Huda et al. 2013).

Transcriptomic analysis of H_2O_2-regulated genes in *Arabidopsis* resulted in the identification of 175 H_2O_2 responsive genes; among them 113 up-regulated and 62 down-regulated by H_2O_2. Fourteen ESTs with the known function were selected from up-regulated genes for expression studies by RNA blots. Rapid dehydration of plants induced the expression of a couple of the selected genes; it was speculated that H_2O_2 is the mediator of observed effects, in a way that, these genes encoding CaM, a Ca^{2+}-binding protein, the DREB2A transcription factor, the Arabidopsis MAP kinase ATMPK3, and a zinc-finger protein, suggesting that H_2O_2 plays a pivotal role in plant drought stress response regulation by modulating the H_2O_2 signaling, MAPK cascades, and gene expression (Desikan et al. 2001).

8 Signal Transduction for Rhizobium-Induced Nodule Formation

Establishing of legume-rhizobium endosymbiosis requires an exchange of molecular signals between two partners (plant and bacterium). Bacterial signals (Nod factor) are perceived by host plants receptors and trigger a plethora of symbiotic responses, ranging from nuclear Ca^{2+} oscillation to symbiotic gene expression in host plants (Oldroyd 2013). Activation of the Nod factor receptors induces intracellular signaling, which is associated with oscillations of the Ca^{2+} concentration in the perinuclear region. These oscillations may occur through localized Ca^{2+} influx from outside of the cell or the intracellular release of Ca^{2+}. Bacterial induced Ca^{2+} flux has been reported in diverse legume plants indicating the common role for Ca^{2+} in endosymbiosis establishment. The Nod factor induced Ca^{2+} oscillation signal is—most probably—decoded by a nuclear-localized calcium and CaM-dependent kinase (CCaMK). This protein triggers a transcriptional network and phosphorylates the LjCYCLOPS as a target protein and makes it an active transcription factor. LjCYCLOPS binds to the promoter of the *ERN1* and *NIN* gene eventually leads to the root nodule initiation.

The essential role for ROS in the generation of the Ca^{2+} gradient in root-hair growing has been reported (Foreman et al. 2003). It has been revealed that ROS activates the Ca^{2+}-dependent protein kinase (CPK5) phosphorylation which is required for systemic defense responses including drought stress (Dubiella et al. 2013). This will let us speculate that ROS-mediated responses in the cell can play a key role in the signal transmission of distinct stimuli. Recent evidence also represent tolerance to drought stress by rhizobium infection in legume plants. Delayed leaf senescence and induced growth rate in soybean plants inoculated with rhizobium exposed to drought stress imply that rhizobium endosymbiosis commit a function to relief drought stress (Staudinger et al. 2016) though the central question that remains is which regulatory mechanism/s occurred so that upon the activation of nodule symbiosis drought tolerance is triggered.

9 Concluding Remark

Various aspects of Ca^{2+} signaling have attracted the attention of numerous researchers so far. The growing trend of research attempts in this issue highlights the Ca^{2+} signaling as a distinct realm of research. Despite the presence of myriad numbers of reports on Ca^{2+}-signal transduction under drought stress, there are not ample evidence on the cooperative mechanism of Ca^{2+}-dependent and -independent signaling network and the underlying molecular principles in regulation of stress response. The circumstance of the specificity of Ca^{2+} signals within the same cell is yet to be investigated. Certain Ca^{2+}-sensing proteins have shown to up-regulate drought tolerance signaling events, whereas negative regulation of drought stress is also attributed to these proteins. Unraveling the specific mechanism underlying the Ca^{2+} signaling events may simplify the complex network addressing the role of Ca^{2+} sensing proteins. And pave the way to provide drought-resistant plants in the coming future.

References

Abdula SE, Lee H-J, Ryu H, Kang KK, Nou I, Sorrells ME, Cho YG (2015) Overexpression of *BrCIPK1* gene enhances abiotic stress tolerance by increasing proline biosynthesis in Rice. Plant Mol Biol Rep 34:501–511

Albrecht V, Ritz O, Linder S, Harter K, Kudla J (2001) The NAF domain defines a novel protein protein interaction module conserved in Ca^{2+} regulated kinases. EMBO J 20(5):1051–1063

Albrecht V, Weinl S, Blazevic D, D'angelo C, Batistic O, Kolukisaoglu Ü, Bock R, Schulz B, Harter K, Kudla J (2003) The calcium sensor CBL1 integrates plant responses to abiotic stresses. Plant J 36(4):457–470

Aliniaeifard S (2014) Signal transduction pathways in guard cells after prolonged exposure to low vapour pressure deficits. Wageningen University (PhD thesis)

Aliniaeifard S, van Meeteren U (2013) Can prolonged exposure to low VPD disturb the ABA signaling in stomatal guard cells? J Exp Bot 64(12):3551–3566

Aliniaeifard S, Van Meeteren U (2018) Natural genetic variation in stomatal response can help to increase acclimation of plants to dried environments. Acta Hortic 1190:71–76

Aliniaeifard S, van Meeteren U (2014) Natural variation in stomatal response to closing stimuli among *Arabidopsis thaliana* accessions after exposure to low VPD as a tool to recognize the mechanism of disturbed stomatal functioning. J Exp Bot 65(22):6529–6542

Aliniaeifard S, Malcolm Matamoros P, van Meeteren U (2014) Stomatal malfunctioning under low vapor pressure deficit (VPD) conditions: induced by alterations in stomatal morphology and leaf anatomy or in the ABA signaling? Physiol Plant 152:688–699

Allen GJ, Kwak JM, Chu SP, Llopis J, Tsien RY, Harper JF, Schroeder JI (1999) Cameleon calcium indicator reports cytoplasmic calcium dynamics in Arabidopsis guard cells. Plant J 19(6):735–747

Allen GJ, Chu SP, Schumacher K, Shimazaki CT, Vafeados D, Kemper A, Hawke S, Tallman G, Tsien RY, Harper JF, Chory J (2000) Alteration of stimulus-specific guard cell calcium oscillations and stomatal closing in Arabidopsis det3 mutant. Science 289(5488):2338–2342

Allen GJ, Chu SP, Harrington CL, Schumacher K, Hoffman T, Tang YY, Grill E, Schroeder JI (2001) A defined range of guard cell calcium oscillation parameters encodes stomatal movements. Nature 411(6841):1053

Bànfi B, Tirone F, Durussel I, Knisz J, Moskwa P, Molnàr GZ, Krause KH, Cox JA (2004) Mechanism of Ca^{2+} activation of the NADPH oxidase 5 (NOX5). J Biol Chem 279:18583–18591

Barry H (2006) Oxidative stress and neurodegeneration: where are we now? J Neurochem 97(6):1634–1658

Batistic O, Kudla J (2004) Integration and channeling of calcium signaling through the CBL calcium sensor/CIPK protein kinase network. Planta 219(6):915–924

Batistič O, Kudla J (2012) Analysis of calcium signaling pathways in plants. Biochim Biophys Acta 1820(8):1283–1293

Bauer H, Ache P, Lautner S, Fromm J, Hartung W, Al-Rasheid Khaled AS, Sonnewald S, Sonnewald U, Kneitz S, Lachmann N, Mendel Ralf R, Bittner F, Hetherington Alistair M, Hedrich R (2012) The stomatal response to reduced relative humidity requires guard cell-autonomous ABA synthesis. Curr Biol 23(1):53–57

Baxter-Burrell A, Yang Z, Springer PS, Bailey-Serres J (2002) RopGAP4-dependent Rop GTPase rheostat control of Arabidopsis oxygen deprivation tolerance. Science 296:2026–2028

Beckmann L, Edel KH, Batistič O, Kudla J (2016) A calcium sensor–protein kinase signaling module diversified in plants and is retained in all lineages of Bikonta species. Sci Rep 6:31645

Bender KW, Snedden WA (2013) Calmodulin-related proteins step out from the shadow of their namesake. Plant Physiol 113

Berridge MJ, Lipp P, Bootman MD (2000) The versatility and universality of calcium signalling. Nat Rev Mol 1(1):11

Blatt MR (2000) Ca^{2+} signalling and control of guard-cell volume in stomatal movements. Curr Opin Pant Biol 3(3):196–204

Bose J, Rodrigo-Moreno A, Shabala S (2014) ROS homeostasis in halophytes in the context of salinity stress tolerance. J Exp Bot 65(5):1241–1257

Bothwell JHF, Ng CKY (2005) The evolution of Ca^{2+} signalling in photosynthetic eukaryotes. New Phytol 166(1):21–38

Bouché N, Scharlat A, Snedden W, Bouchez D, Fromm H (2002) A novel family of calmodulin binding transcription activators in multicellular organisms. J Biol Chem 277(24):21851–21861

Bouché N, Yellin A, Snedden WA, Fromm H (2005) Plant specific calmodulin-binding proteins. Annu Rev Plant Biol 56:435–466

Boudsocq M, Sheen J (2013) CDPKs in immune and stress signaling. Trends Plant Sci 18(1):30–40

Brandt B, Brodsky DE, Xue S, Negi J, Iba K, Kangasjärvi J, Ghassemian M, Stephan AB, Hu H, Schroeder JI (2012) Reconstitution of abscisic acid activation of SLAC1 anion channel by CPK6 and OST1 kinases and branched ABI1 PP2C phosphatase action. PNAS 109(26):10593–10598

Brandt B, Munemasa S, Wang C, Nguyen D, Yong T, Yang PG, Poretsky E, Belknap TF, Waadt R, Alemán F, Schroeder JI (2015) Calcium specificity signaling mechanisms in abscisic acid signal transduction in Arabidopsis guard cells. Elife 4:e03599

Brivanlou AH, Darnell JE (2002) Signal transduction and the control of gene expression. Science 295(5556):813–818

Burstrom HG (1968) Calcium and plant growth. Biol Rev 43(3):287–316

Bush DS (1995) Calcium regulation in plant cells and its role in signaling. Annu Rev Plant Biol 46(1):95–122

Carafoli E, Santella L, Branca D, Brini M (2001) Generation, control, and processing of cellular calcium signals. Crit Rev Biochem Mol Biol 36(2):107–260

Chehab EW, Patharkar OR, Hegeman AD, Taybi T, Cushman JC (2004) Autophosphorylation and subcellular localization dynamics of a salt and water deficit-induced calcium-dependent protein kinase from ice plant. Plant Physiol 135(3):1430–1446

Cheng SH, Willmann MR, Chen HC, Sheen J (2002) Calcium signaling through protein kinases. The Arabidopsis calcium dependent protein kinase gene family. Plant Physiol 129(2):469–485

Chen YL, Huang R, Xiao YM, Lu P, Chen J, Wang XC (2004) Extracellular calmodulin-induced stomatal closure is mediated by heterotrimeric G protein and H_2O_2. Plant Physiol 136(4):4096–4103

Chen X, Gu Z, Xin D, Hao L, Liu C, Huang J, Ma B, Zhang H (2011) Identification and characterization of putative CIPK genes in maize. J Genet Genom 38(2):77–87

Chen J, Xue B, Xia X, Yin W (2013b) A novel calcium-dependent protein kinase gene from *Populus euphratica*, confers both drought and cold stress tolerance. Biochem Biophys Res Commun 441(3):630–636

Chen L, Wang QQ, Zhou L, Ren F, Li DD, Li XB (2013a) Arabidopsis CBL-interacting protein kinase (CIPK6) is involved in plant response to salt/osmotic stress and ABA. Mol Biol Rep 40(8):4759–4767

Cheong YH, Kim KN, Pandey GK, Gupta R, Grant JJ, Luan S (2003) CBL1, a calcium sensor that differentially regulates salt, drought, and cold responses in Arabidopsis. Plant Cell 15(8):18331845

Cheong YH, Pandey GK, Grant JJ, Batistic O, Li L, Kim BG, Lee SC, Kudla J, Luan S (2007) Two calcineurin B-like calcium sensors, interacting with protein kinase CIPK23, regulate leaf transpiration and root potassium uptake in Arabidopsis. Plant J 52(2):223–239

Cheong YH, Sung SJ, Kim BG, Pandey GK, Cho JS, Kim KN, Luan S (2010) Constitutive overexpression of the calcium sensor CBL5 confers osmotic or drought stress tolerance in Arabidopsis. Mol Cells 29(2):159–165

Choi H, Park HJ, Park JH, Kim S, Im M-Y, Seo H-H, Kim Y-W, Hwang I, Kim SY (2005a) Arabidopsis calcium dependent protein kinase AtCPK32 interacts with ABF4, a transcriptional regulator of abscisic acid responsive gene expression, and modulates its activity. Plant Physiol 139(4):1750–1761

Choi MS, Kim MC, Yoo JH, Moon BC, Koo SC, Park BO, Lee JH, Koo YD, Han HJ, Lee SY, Chung WS (2005b) Isolation of a calmodulin binding transcription factor from rice (*Oryza sativa* L.). J Biol Chem 280(49):40820–40831

Clarkson DT (1984) Calcium transport between tissues and its distribution in the plant. Plant Cell Environ 7(6):449–456

Conn S, Gilliham M (2010) Comparative physiology of elemental distributions in plants. Ann Bot 105(7):1081–1102

Cuéllar T, Pascaud F, Verdeil JL, Torregrosa L, Adam-Blondon AF, Thibaud JB, Sentenac H, Gaillard I (2010) A grapevine Shaker inward K^+ channel activated by the calcineurin B-like calcium sensor 1–protein kinase CIPK23 network is expressed in grape berries under drought stress conditions. Plant J 61(1):58–69

Cui XY, Du YT, Fu JD, Yu TF, Wang CT, Chen M, Chen J, Ma YZ, Xu ZS (2018) Wheat CBL interacting protein kinase 23 positively regulates drought stress and ABA responses. BMC Plant Biol 18(1):93

Cutler SR, Rodriguez PL, Finkelstein RR, Abram SR (2010) Abscisic acid: emergence of a core signaling network. Annu Rev Plant Biol 61:651–679

D'angelo C, Weinl S, Batistic O, Pandey GK, Cheong YH, Schültke S, Albrecht V, Ehlert B, Schulz B, Harter K, Luan S (2006) Alternative complex formation of the Ca^{2+}-regulated protein kinase CIPK1 controls abscisic acid-dependent and independent stress responses in Arabidopsis. Plant J 48(6):857–872
Daszkowska-Golec A, Szarejko I (2013) Open or close the gate–stomata action under the control of phytohormones in drought stress conditions. Front Plant Sci 4:138
Dauwalder M, Roux SJ, Hardison L (1986) Distribution of calmodulin in pea seedlings: immune cytochemical localization in plumules and root apices. Planta 168(4):461–470
DeFalco TA, Bende KW, Snedden WA (2010) Breaking the code: Ca^{2+} sensors in plant signalling. Biochem J 425(1):27–40
De Freitas ST, Mitcham EI (2012) 3 factors involved in fruit calcium deficiency disorders. Hortic Rev 40(1):107–146
De Freitas ST, Shackel KA, Mitcham EJ (2011) Abscisic acid triggers whole-plant and fruit-specific mechanisms to increase fruit calcium uptake and prevent blossom end rot development in tomato fruit. J Exp Bot 62(8):2645–2656
del Carmen Martínez-Ballesta M, del Carmen Rodríguez-Hernández M, Alcaraz-López C, Mota-Cadenas C, Muries B, Carvajal M (2011) Plant hydraulic conductivity: The aquaporins contribution. Hydraulic Conductivity: Issues, Determination and Applications, p 103
Demidchik V (2015) Mechanisms of oxidative stress in plants: from classical chemistry to cell biology. EEB 109:212–228
Demidchik V, Maathuis FJM (2007) Physiological roles of nonselective cation channels in plants: from salt stress to signalling and development. New Phytol 175(3):387–404
Demidchik V, Davenport RJ, Tester M (2002) Nonselective cation channels in plants. Annu Rev Plant Biol 53(1):67–107
Demidchik V, Shang Z, Shin R, Thompson E, Rubio L, Chivasa S, Slabas AR, Glover BJ, Schachtman DP, Shabala SN, Davies JM (2009) Plant extracellular ATP signaling by plasma membrane NADPH oxidase and Ca^{2+} channels. Plant J 58(6):903–913
Desikan R, Soheila AH, Hancock JT, Neill SJ (2001) Regulation of the Arabidopsis transcriptome by oxidative stress. Plant Physiol 127(1):159–172
Dodd AN, Kudla J, Sanders D (2010) The language of calcium signaling. Annu Rev Plant Biol 61(593):620
Dubiella U, Seybold H, Durian G, Komander E, Lassig R, Witte C-P, Schulze WX, Romeis T (2013) Calcium dependent protein kinase/NADPH oxidase activation circuit is required for rapid defense signal propagation. PNAS 110(21):8744–8749
E Silva ODC (1994) CG-1, a parsley light-induced DNA-binding protein. Plant Mol Biol 25(5):921–924
Edel KH, Kudla J (2015) Increasing complexity and versatility: how the calcium signaling toolkit was shaped during plant land colonization. Cell Calcium 57(3):231–246
Edel KH, Kudla J (2016) Integration of calcium and ABA signaling. Curr Opin Plant Biol 33:83–91
Endo A, Sawada Y, Takahashi H, Okamoto M, Ikegami K, Koiwai H, Seo M, Toyomasu T, Mitsuhashi W, Shinozaki K, Nakazono M (2008) Drought induction of Arabidopsis 9-cis-epoxycarotenoid dioxygenase occurs in vascular parenchyma cells. Plant Physiol 147(4):1984–1993
Epstein E (1972) Mineral nutrition of plants: principles and perspectives. Wiley, New York
Evans NH, McAinsh MR, Hetherington AM (2001) Calcium oscillations in higher plants. Curr Opin Plant Biol 4(5):415–420
Evans NH, McAinsh MR, Hetherington AM, Knight MR (2005) ROS perception in *Arabidopsis thaliana*: The ozone induced calcium response. Plant J 41(4):615–626
Fageria NK (2016) The use of nutrients in crop plants. CRC Press
Fasano JM, Mass GD, Gilroy S (2002) Ionic signaling in plant responses to gravity and touch. J Plant Growth Regul 21(2):71–88

Finkelstein RR, Rock C (2002) Abscisic acid biosynthesis and signaling. In: Somerville CR, Meyerowitz EM (eds) The Arabidopsis book. American Society of Plant Biologists, Rockville, MD, pp 1–48
Foreman J, Demidchik V, Bothwell JH, Mylona P, Miedema H, Torre MA, Linstead P, Costa S, Brownlee C, Jones JD, Davies JM (2003) Reactive oxygen species produced by NADPH oxidase regulate plant cell growth. Nature 422(6930):442
Foyer CH, Noctor G (2016) Stress-triggered redox signalling: what's in prospect? Plant Cell Environ 39(5):951–964
Franz S, Ehlert B, Liese A, Kurth J, Cazalé AC, Romeis T (2011) Calcium-dependent protein kinase CPK21 functions in abiotic stress response in *Arabidopsis thaliana*. Mol Plant 4(1):83–96
Geiger D, Scherzer S, Mumm P, Stange A, Marten I, Bauer H, Ache P, Matschi S, Liese A, Al-Rasheid KAS, Romeis T, Hedrich R (2009) Activity of guard cell anion channel SLAC1 is controlled by drought-stress signaling kinase-phosphatase pair. PNAS 106(50):21425–21430
Geiger D, Scherzer S, Mumm P, Marten IAPPL, Ache P, Matschi S, Liese A, Wellmann C, Al-Rasheid KAS, Grill E, Romeis T (2010) Guard cell anion channel SLAC1 is regulated by CDPK protein kinases with distinct Ca^{2+} affinities. PNAS 107(17):8023–8028
Gosti F, Beaudoin N, Serizet C, Webb AA, Vartanian N, Giraudat J (1999) ABI1 protein phosphatase 2C is a negative regulator of abscisic acid signaling. Plant Cell 11(10):1897–1909
Guo Y, Xiong L, Song CP, Gon D, Halfter U, Zhu JK (2002) A calcium sensor and its interacting protein kinase are global regulators of abscisic acid signaling in Arabidopsis. Dev Cell 3(2):233–244
Hamilton DWA, Hills A, Kohler B, Blatt MR (2000) Ca^{2+} channels at the plasma membrane of stomatal guard cells are activated by hyperpolarization and abscisic acid. PNAS 97(9):4967–4972
Han S, Tang R, Anderson LK, Woerner TE, Pei Z-M (2003) A cell surface receptor mediates extracellular Ca^{2+} sensing in guard cells. Nature 425(6954):196
Harmon AC, Gribskov M, Harper JF (2000) CDPKs—a kinase for every Ca^{2+} signal? Trends Plant Sci 5(4):154–159
Harmon AC, Gribskov M, Gubrium E, Harper JF (2001) The CDPK superfamily of protein kinases. New Phytol 151(1):175–183
Harper JF, Breton G, Harmon A (2004) Decoding Ca^{2+} signals through plant protein kinases. Annu Rev Plant Biol 55:263–288
Hashimoto K, Kudla J (2011) Calcium decoding mechanisms in plants. Biochimie 93(12):2054–2059
He L, Yang X, Wang L, Zhu L, Zhou T, Deng J, Zhang X (2013) Molecular cloning and functional characterization of a novel cotton CBL-interacting protein kinase gene (GhCIPK6) reveals its involvement in multiple abiotic stress tolerance in transgenic plants. Biochem Biophys Res Commun 435(2):209–215
Hedrich R, Busch H, Raschke K (1990) Ca^{2+} and nucleotide dependent regulation of voltage dependent anion channels in the plasma membrane of guard cells. EMBO J 9(12):3889–3892
Hepler PK, Winship LJ (2010) Calcium at the cell wall-cytoplast interface. J Integr Plant Biol 52(2):147–160
Hetherington AM, Brownlee C (2004) The generation of Ca^{2+} signals in plants. Annu Rev Plant Biol 55:401–427
Hetherington AM, Woodward FI (2003) The role of stomata in sensing and driving environmental change. Nature 424(6951):901
Hirayama T, Shinozaki K (2010) Research on plant abiotic stress responses in the post-genome era: past, present and future. Plant J 61(6):1041–1052
Hirschi K (2001) Vacuolar H^+/Ca^{2+} transport: who's directing the traffic? Trends Plant Sci 6(3):100–104
Hoeflich KP, Ikura M (2002) Calmodulin in action: diversity in target recognition and activation mechanisms. Cell 108(6):739–742

Hrabak EM, Chan CW, Gribskov M, Harper JF, Choi JH, Halford N, Kudla J, Luan S, Nimmo HG, Sussman MR, Thomas M (2003) The Arabidopsis CDPK-SnRK superfamily of protein kinases. Plant Physiol 132(2):666–680

Hsieh HL, Song CJ, Roux SJ (2000) Regulation of a recombinant pea nuclear apyrase by calmodulin and casein kinase II. BBA-Gene Struct Expr 1494(3):248–255

Hu XY, Neill SJ, Cai WM, Tang ZC (2004) Induction of defence gene expression by oligogalacturonic acid requires increases in both cytosolic calcium and hydrogen peroxide in *Arabidopsis thaliana*. Cell Res 14(3):234

Hu W, Xia Z, Yan Y, Ding Z, Tie W, Wang L, Zou M, Wei Y, Lu C, Hou X, Wang W (2015) Genome wide gene phylogeny of CIPK family in cassava and expression analysis of partial drought induced genes. Front Plant Sci 6:914

Hu W, Yan Y, Tie W, Ding Z, Wu C, Ding X, Wang W, Xia Z, Guo J, Peng M (2018) Genome wide analyses of calcium sensors reveal their involvement in drought stress response and storage roots deterioration after harvest in cassava. Genes 9(4):221

Huang QS, Wang HY, Gao P, Wan, GY, Xia GX (2008) Cloning and characterization of a calcium dependent protein kinase gene associated with cotton fiber development. Plant Cell Rep 27(12):1869

Hua D, Wang, He J, Liao H, Duan Y, Zhu Z, Guo Y, Chen Z, Gong Z (2012) A plasma membrane receptor kinase, GHR1, mediates abscisic acid-and hydrogen peroxide-regulated stomatal movement in Arabidopsis. Plant Cell 112

Hubbard KE, Siegel RS, Valerio G, Brandt B, Schroeder JI (2011) Abscisic acid and CO_2 signalling via calcium sensitivity priming in guard cells, new CDPK mutant phenotypes and a method for improved resolution of stomatal stimulus–response analyses. Ann Bot 109(1):5–17

Huda KM, Banu MSA, Garg B, Tula S, Tuteja R, Tuteja N (2013) OsACA 6, a P-type IIB Ca^{2+} ATPase promotes salinity and drought stress tolerance in tobacco by ROS scavenging and enhancing the expression of stress-responsive genes. Plant J 76(6):997–1015

Israelsson M, Siegel RS, Young J, Hashimoto M, Iba K, Schroeder JI (2006) Guard cell ABA and CO_2 signaling network updates and Ca^{2+} sensor priming hypothesis. Curr Opin Plant Biol 9(6):654–663

Jagnandan D, Church JE, Banfi B, Stuehr DJ, Marrero MB, Fulton DJ (2007) Novel mechanism of activation of NADPH oxidase 5 calcium sensitization via phosphorylation. J Biol Chem 282(9):6494–6507

Janiak A, Kwaśniewski M, Szarejko I (2015) Gene expression regulation in roots under drought. J Exp Bot 67(4):1003–1014

Jiang S, Zhang D, Wang L, Pan Liu Y, Kong X, Zhou Y, Li D (2013) A maize calcium dependent protein kinase gene, ZmCPK4, positively regulated abscisic acid signaling and enhanced drought stress tolerance in transgenic Arabidopsis. Plant Physiol Biochem 71:112–120

Joshi-Saha A, Valon C, Leung J (2011) Abscisic acid signal off the STARTing block. Mol Plant 4(4):562–580

Kaczmarek M, Fedorowicz-Strońska O, Głowacka K, Waśkiewicz A, Sadowski J (2017) $CaCl_2$ treatment improves drought stress tolerance in barley (*Hordeum vulgare* L.). Acta Physiol Plant 39(1):41

Kalaipandian S, Xue GP, Rae AL, Glassop D, Bonnett GD, McIntyre LC (2018) Overexpression of TaCML20, a calmodulin-like gene, enhances water soluble carbohydrate accumulation and yield in wheat. Physiol Plant 165(4):790–799

Kanwar P, Sanyal SK, Tokas I, Yadav AK, Pandey A, Kapoor S, Pandey GK (2014) Comprehensive structural, interaction and expression analysis of CBL and CIPK complement during abiotic stresses and development in rice. Cell Calcium 56(2):81–95

Kaplan B, Davydov O, Knight H, Galon Y, Knight MR, Fluhr R, Fromm H (2006) Rapid transcriptome changes induced by cytosolic Ca^{2+} transients reveal ABRE-related sequences as Ca^{2+} responsive cis elements in Arabidopsis. Plant Cell 18(10):2733–2748

Kim KN, Cheong H, Grant JJ, Pandey GK, Luan S (2003) CIPK3, a calcium sensor associated protein kinase that regulates abscisic acid and cold signal transduction in Arabidopsis. Plant Cell 15(2):411–423

Kim TH, Böhmer M, Hu H, Nishimura N, Schroeder JI (2010) Guard cell signal transduction network: advances in understanding abscisic acid, CO_2, and Ca^{2+} signaling. Annu Rev Plant Biol 61(561):591

Kleist TJ, Spencley AL, Luan S (2014) Comparative phylogenomics of the CBL-CIPK calcium decoding network in the moss Physcomitrella, Arabidopsis, and other green lineages. Front Plant Sci 5:187

Klimecka M, Muszynska G (2007) Structure and functions of plant calcium-dependent protein kinases. Acta Biochim (Pol-English Edition) 54(2):219

Knight MR, Read ND, Campbell AK, Trewavas AJ (1993) Imaging calcium dynamics in living plants using semi-synthetic recombinant aequorins. J Cell Biol 121(1):83–90

Knight H, Trewavas AJ, Knight MR (1996) Cold calcium signaling in Arabidopsis involves two cellular pools and a change in calcium signature after acclimation. Plant Cell 8(3):489–503

Knight H, Trewavas AJ, Knight MR (1997) Calcium signalling in *Arabidopsis thaliana* responding to drought and salinity. Plant J 12(5):1067–1078

Kobayashi Y, Murata M, Minami H, Yamamoto S, Kagaya Y, Hobo T, Yamamoto A, Hattori T (2005) Abscisic acid-activated SNRK2 protein kinases function in the gene-regulation pathway of ABA signal transduction by phosphorylating ABA response element-binding factors. Plant J 44(6):939–949

Kobayashi M, Ohura I, Kawakita K, Yokota N, Fujiwara M, Shimamoto K, Doke N, Yoshioka H (2007) Calcium-dependent protein kinases regulate the production of reactive oxygen species by potato NADPH oxidase. Plant Cell 19(3):1065–1080

Kohler B, Blatt MR (2002) Protein phosphorylation activates the guard cell Ca^{2+} channel and is a prerequisite for gating by abscisic acid. Plant J 32(2):185–194

Kolukisaoglu Ü, Weinl S, Blazevic D, Batistic O, Kudla J (2004) Calcium sensors and their interacting protein kinases: genomics of the Arabidopsis and rice CBL-CIPK signaling networks. Plant Physiol 134(1):43–58

Kosuta S, Hazledine S, Sun J, Miwa H, Morris R, Downie JA, Oldroyd GE (2008) Differential and chaotic calcium signatures in the symbiosis signaling pathway of legumes. PNAS 105(28):9823–9828

Kudla J, Xu Q, Harter K, Gruissem W, Luan S (1999) Genes for calcineurin B-like proteins I Arabidopsis are differentially regulated by stress signals. PNAS 96(8):4718–4723

Kudla J, Becker D, Grill E, Hedrich R, Hippler M, Kummer U, Parniske M, Romeis T, Schumacher K (2018) Advances and current challenges in calcium signaling. New Phytol 218(2):414–431

La Verde V, Dominici P, Astegno A (2018) Towards understanding plant calcium signaling through calmodulin-like proteins: a biochemical and structural perspective. Int J Mol Sci 19(5):1331

Latz A, Mehlmer N, Zapf S, Mueller TD, Wurzinge B, Pfister B, Csaszar E, Hedrich R, Teige M, Becker D (2013) Salt stress triggers phosphorylation of the Arabidopsis vacuolar K^+ channel TPK1 by calcium-dependent protein kinases (CDPKs). Mol Plant 6(4):1274–1289

Lecourieux D, Ranjeva R, Pugin A (2006) Calcium in plant defence-signalling pathways. New Phytol 171(2):249–269

Levchenko V, Konrad KR, Dietrich P, Roelfsema MRG, Hedrich R (2005) Cytosolic abscisic acid activates guard cell anion channels without preceding Ca^{2+} signals. PNAS 102(11):4203–4208

Levine A, Pennell RI, Alvarez M, Palmer R, Lamb C (1996) Calcium-mediated apoptosis in a plant hypersensitive disease resistance response. Curr Biol 6(4):427–437

Li J, Assmann SM (1996) An abscisic acid-activated and calcium independent protein kinase from guard cells of faba bean. Plant Cell 8(12):2359–2368

Li J, Lee YRJ, Assmann SM (1998) Guard cells possess a calcium-dependent protein kinase that phosphorylates the KAT1 potassium channel. Plant Physiol 116(2):785–795

Li J, Wang XQ, Watson MB, Assmann SM (2000) Regulation of abscisic acid-induced stomatal closure and anion channels by guard cell AAPK kinase. Science 287(5451):300–303

Li S, Assmann SM, Albert R (2006) Predicting essential components of signal transduction networks: a dynamic model of guard cell abscisic acid signaling. PLoS Biol 4(10):e312

Li AL, Zhu YF, Tan XM, Wang X, Wei B, Gu HZ, Zhang ZL, Chen XB, Zhao GY, Kong XY, Jia JZ (2008) Evolutionary and functional study of the CDPK gene family in wheat (*Triticum aestivum* L.). Plant Mol Biol 66(4):429–443

Li J-H, Liu Y-Q, Lu P, Lin H-F, Bai Y, Wang X-C, Chen Y-L (2009) A signaling pathway linking nitric oxide production to heterotrimeric G protein and hydrogen peroxide regulates extracellular calmodulin induction of stomatal closure in Arabidopsis. Plant Physiol 150(1):114–124

Li R, Zhang J, Wu G, Wang H, Chen Y, Wei J (2012) HbCIPK2, a novel CBL interacting protein kinase from halophyte *Hordeum brevisubulatum*, confers salt and osmotic stress tolerance. Plant Cell Environ 35(9):1582–1600

Li CL, Wang M, Wu X, Chen DH, Lv HJ, Shen JL, Qiao Z, Zhang W (2016) THI1, a thiamine thiazole synthase, interacts with Ca^{2+}-dependent protein kinase CPK33 and modulates the S-type anion channels and stomatal closure in Arabidopsis. Plant Physiol 170(2):1090–1104

Liu J, Zhu JK (1998) A calcium sensor homolog required for plant salt tolerance. Science 280(5371):1943–1945

Liu HT, Li GL, Chang H, Sun DY, Zhou RG, Li B (2007) Calmodulin-binding protein phosphatase PP7 is involved in thermotolerance in Arabidopsis. Plant Cell Environ 30(2):156–164

Liu HT, Gao F, Li GL, Han JL, Liu DL, Sun DY, Zhou RG (2008) The calmodulin-binding protein kinase 3 is part of heat shock signal transduction in *Arabidopsis thaliana*. Plant J 55(5):760–773

Loneragan JF, Snowball K (1969) Calcium requirements of plants. Aust J Agric Res 20(3):465–478

Luan S, Kudla J, Rodriguez-Concepcion M, Yalovsky S, Gruissem W (2002) Calmodulins and alcineurin B-like proteins: calcium sensors for specific signal response coupling in plants. Plant Cell 14(Suppl 1):S389–S400

Ludwig AA, Romeis T, Jones JD (2004) CDPK-mediated signalling pathways: specificity and cross talk. J Exp Bot 55(395):181–188

Lynch T, Erickson BJ, Finkelstein RR (2012) Direct interactions of ABA-insensitive (ABI)-clade protein phosphatase (PP) 2Cs with calcium-dependent protein kinases and ABA response element binding bZIPs may contribute to turning off ABA response. Plant Mol Biol 80(6):647–658

Ma SY, Wu WH (2007) AtCPK23 functions in Arabidopsis responses to drought and salt stresses. Plant Mol Biol 65(4):511–518

Ma Y, Szostkiewicz I, Korte A, Moe D, Yang Y, Christmann A, Grill E (2009) Regulators of PP2C phosphatase activity function as abscisic acid sensors. Science 324(5930):1064–1068

Magnan F, Ranty B, Charpenteau M, Sotta B, Galaud JP, Aldon D (2008) Mutations in AtCML9, a calmodulin-like protein from Arabidopsis thaliana, alter plant responses to abiotic stress and abscisic acid. Plant J 56(4):575–589

Mahajan S, Tuteja N (2005) Cold, salinity and drought stresses: an overview. Arch Biochem Biophys 444(2):139–158

Mahajan SHILPI, Sopory SK, Tuteja NARENDRA (2006) CBL-CIPK paradigm: role in calcium and stress signaling in plants. Proc Indian Natl Sci Acad B Biol Sci 72(2):63

Marschner H (1986) Mineral nutrition of higher plants. Academic Press, London, pp 252–254

Marschner P (2012) Rhizosphere biology. In: Marschner's mineral nutrition of higher plants, 3rd edn, pp 369–388

Marschner H (2012) In: Marschner P (ed) Marschner's mineral nutrition of higher plants, vol 89. Academic press, London

Marten H, Konrad KR, Dietrich P, Roelfsema MRG, Hedrich R (2007) Ca^{2+}-dependent and –independent abscisic acid activation of plasma membrane anion channels in guard cells of *Nicotiana tabacum*. Plant Physiol 143(1):28–37

Masle J, Gilmore SR, Farquhar GD (2005) The ERECTA gene regulates plant transpiration efficiency in Arabidopsis. Nature 436(7052):866

McAinsh MR, Hetherington AM (1998) Encoding specificity in Ca^{2+} signalling systems. Trends Plant Sci 3(1):32–36

McAinsh MR, Brownlee C, Hetherington AM (1990) Abscisic acid-induced elevation of guard cell cytosolic Ca^{2+} precedes stomatal closure. Nature 343(6254):186
McAinsh MR, Webb AAR, Taylor JE, Hetherington AM (1995) Stimulus-induced oscillations in guard cell cytosolic-free calcium. Plant Cell 7(8):1207–1219
McAinsh MR, Pittman JK (2009) Shaping the calcium signature. New Phytol 181(2):275–294
Meer L, Mumtaz S, Labbo AM, Kha MJ, Sadiq I (2019) Genome-wide identification and expression analysis of calmodulin-binding transcription activator genes in banana under drought stress. Sci Hort 244:10–14
Melcher K, Xu Y, Ng LM, Zhou XE, Soon FF, Chinnusamy V, Suino-Powell KM, Kovach A, Tham FS, Cutler SR, Li J (2010) Identification and mechanism of ABA receptor antagonism. Nat Struct Mol Biol 17(9):1102
Merlot S, Gosti F, Guerrier D, Vavasseur A, Giraudat J (2001) The ABI1 and ABI2 protein phosphatases 2C act in a negative feedback regulatory loop of the abscisic acid signalling pathway. Plant J 25(3):295–303
Milla MAR, Townsend J, Chang F, Cushman JC (2006) The Arabidopsis AtDi19 gene family encodes a novel type of Cys2/His2 zinc-finger protein implicated in ABA-independent dehydration, high salinity stress and light signaling pathways. Plant Mol Biol 61(1–2):13–30
Mitsuda N, Isono T, Sato MH (2003) Arabidopsis CAMTA family proteins enhance V-PPase expression in pollen. Plant Cell Physiol 44(10):975–981
Moore CA, Bowen HC, Scrase-Field S, Knight MR, White PJ (2002) The deposition of suberin lamellae determines the magnitude of cytosolic Ca^{2+} elevations in root endodermal cells subjected to cooling. Plant J 30(4):457–465
Mori IC, Schroeder JI (2004) Reactive oxygen species activation of plant Ca^{2+} channels. A signaling mechanism in polar growth, hormone transduction, stress signaling, and hypothetically mechanotransduction. Plant Physiol 135(2):702–708
Mori IC, Murata Y, Yang Y, Munemasa S, Wang YF, Andreoli S, Tiriac H, Alonso JM, Harper JF, Ecker JR, Kwak JM (2006) CDPKs CPK6 and CPK3 function in ABA regulation of guard cell S type anion-and Ca^{2+}-permeable channels and stomatal closure. PLoS Biol 4(10):e327
Murata Y, Pei ZM, Mori IC, Schroeder J (2001) Abscisic acid activation of plasma membrane Ca^{2+} channels in guard cells requires cytosolic NAD (P) H and is differentially disrupted upstream and downstream of reactive oxygen species production in abi1-1 and abi2-1 protein phosphatase 2C mutants. Plant Cell 13(11):2513–2523
Murata Y, Mori IC, Munemasa S (2015) Diverse stomatal signaling and the signal integration mechanism. Annu Rev Plant Biol 66:369–392
Nakashima K, Yamaguchi-Shinozaki K (2013) ABA signaling in stress-response and seed development. Plant Cell Rep 32(7):959–970
Neill S, Barros R, Bright J, Desikan R, Hancock J, Harrison J, Morris P, Ribeiro D, Wilson I (2008) Nitric oxide, stomatal closure, and abiotic stress. J Exp Bot 59(2):165–176
Nelissen H, Clarke JH, De Block M, De Block S, Vanderhaeghen R, Zielinski RE, Dyer T, Lust S, Inze´ D, Van Lijsebettens M (2003) DRL1, a homolog of the yeast TOT4/KTI12 protein, has a function in meristem activity and organ growth in plants. Plant Cell 15(3):639–654
Ng CKY, McAinsh MR (2003) Encoding specificity in plant calcium signalling: hot-spotting the ups and downs and waves. Ann Bot 92(4):477–485
Ng LM, Soon FF, Zhou XE, West GM, Kovach A, Suino-Powell M, Chalmers MJ, Li J, Yong EL, Zhu JK, Griffin PR (2011) Structural basis for basal activity and autoactivation of abscisic acid (ABA) signaling SnRK2 kinases. PNAS 108(52):21259–21264
Nomura H, Shiina T (2014) Calcium signaling in plant endosymbiotic organelles: mechanism and role in physiology. Mol Plant 7(7):1094–1104
Okamoto M, Tanaka Y, Abrams SR, Kamiya Y, Seki M, Nambara E (2009) High humidity induces abscisic acid 8′ hydroxylase in stomata and vasculature to regulate local and systemic abscisic acid responses in Arabidopsis. Plant Physiol 149(2):825–834
Oldroyd GE (2013) Speak, friend, and enter: signalling systems that promote beneficial symbiotic associations in plants. Nat Rev Microbiol 11(4):252

Oldroyd GE, Downie JA (2008) Coordinating nodule morphogenesis with rhizobial infection in legumes. Annu Rev Plant Biol 59:519–546

Pandey GK, Cheong Y, Kim KN, Grant JJ, Li L, Hung W, D'Angel C, Weinl S, Kudla J, Luan S (2004) The calcium sensor calcineurin B-like 9 modulates abscisic acid sensitivity and biosynthesis in Arabidopsis. Plant Cell 16(7):1912–1924

Pandey GK, Grant JJ, Cheong YH, Kim BG, Luan S (2008) Calcineurin-B-like protein CBL9 interacts with target kinase CIPK3 in the regulation of ABA response in seed germination. Mol 1(2):238–248

Pandey N, Ranjan A, Pant P, Tripathi RK, Ateek F, Pandey HP, Patre U, Sawant SV (2013) CAMTA 1 regulates drought responses in *Arabidopsis thaliana*. BMC Genom 14(1):216

Pandey GK, Kanwar P, Singh A, Steinhorst L, Pandey A, Yadav AK, Tokas I, Sanyal S, Kim BG, Lee SC, Cheong YH (2015) CBL-interacting protein kinase, CIPK21, regulates osmotic and salt stress responses in Arabidopsis. Plant Physiol 00623

Pauly N, Knight MR, Thuleau P, van der Luit AH, Morea M, Trewavas AJ, Ranjeva R, Mazars C (2000) Control of free calcium in plant cell nuclei. Nature 405(6788):754

Pei ZM, Murata Y, Benning G, Thomine S, Klusener B, Allen GJ, Grill E, Schroeder JI (2000) Calcium channels activated by hydrogen peroxide mediate abscisic acid signalling in guard cells. Nature 406(6797):731

Perez-Prat E, Narasimhan ML, Binzel ML, Botella A, Chen Z, Valpuesta V, Bressan RA, Hasegawa PM (1992) Induction of a putative calcium ATPase mRNA in sodium chloride adapted cells. Plant Physiol 100(3):1471–1478

Pestacz A, Erdei L (1996) Calcium-dependent protein kinase in maize and sorghum induced by polyethylene glycol. Physiol Plant 97(2):360–364

Piao HL, Xuan Y, Park SH, Je BI, Park SJ, Park SH, Kim CM, Huang J, Wang GK, Kim MJ, Kang SM (2010) OsCIPK31, a CBL-interacting protein kinase is involved in germination and seedling growth under abiotic stress conditions in rice plants. Mol Cell 30(1):19–27

Poovaiah BW, Reddy ASN, Leopold AC (1987) Calcium messenger system in plants. Crit Rev Plant Sci 6(1):47–103

Raghavendra AS, Gonugunta VK, Christmann A, Grill E (2010) ABA perception and signalling. Trends Plant Sci 15(7):395–401

Ramachandra-Reddy A, Chaitanya KV, Vivekanandan M (2004) Drought-induced responses of photosynthesis and antioxidant metabolism in higher plants. J Plant Physiol 161(11):1189–1202

Ranf S, Wunnenberg P, Lee J, Becker D, Dunkel M, Hedrich R, Scheel D, Dietrich P (2008) Loss of the vacuolar cation channel, AtTPC1, does not impair Ca^{2+} signals induced by abiotic and biotic stresses. Plant J 53(2):287–299

Ranty B, Aldon D, Cotelle V, Galaud JP, Thuleau P, Mazars C (2016) Calcium sensors as key hubs in plant responses to biotic and abiotic stresses. Front Plant Sci 7:327

Reddy AS, Reddy VS, Golovkin M (2000) A calmodulin binding protein from Arabidopsis is induced by ethylene and contains a DNA-binding motif. Biochem Biophys Res Commun 279(3):762–769

Reddy AS, Day IS, Narasimhulu SB, Safadi F, Reddy VS, Golovkin M, Harnly MJ (2002) Isolation and characterization of a novel calmodulin-binding protein from potato. J Biol Chem 277(6):4206–4214

Riechmann JL, Heard J, Martin G, Reuber L, Jiang CZ, Keddie J, Adam L, Pineda O, Ratcliffe OJ, Samaha RR, Creelman R (2000) Arabidopsis transcription factors: genome-wide comparative analysis among eukaryotes. Science 290(5499):2105–2110

Rodríguez-Concepción M, Yalovsky S, Zik M, Fromm H, Gruissem W (1999) The prenylation status of a novel plant calmodulin directs plasma membrane or nuclear localization of the protein. EMBO J 18(7):1996–2007

Rodriguez L, Gonzalez-Guzman M, Diaz M, Rodrigues A, Izquierdo-Garcia AC, Peirats-Llobet M, Fernandez MA, Antoni R, Fernandez D, Marquez JA, Mulet JM (2014) C2-domain abscisic acid related proteins mediate the interaction of PYR/PYL/RCAR abscisic acid receptors with the plasma membrane and regulate abscisic acid sensitivity in Arabidopsis. Plant Cell 114

Roelfsema MR, Hedrich R (2010) Making sense out of Ca^{2+} signals: their role in regulating stomatal movements. Plant Cell Environ 33(3):305–321

Ronzier E, Corratgé-Faillie C, Sanchez F, Prado K, Brière C, Leonhardt N, Thibaud JB, Xiong TC (2014) CPK13, a non-canonical CPK, specifically inhibits KAT2 and KAT1 Shaker channels and reduces stomatal opening. Plant Physiol 114

Rudd JJ, Franklin-Tong VE (2001) Unravelling response-specificity in Ca^{2+} signalling pathways in plant cells. New Phytol 151(1):7–33

Saijo Y, Hata S, Kyozuka J, Shimamoto K, Izui K (2000) Over-expression of a single Ca^{2+} dependent protein kinase confers both cold and salt/drought tolerance on rice plants. Plant J 23(3):319–327

Sanders D, Brownlee C, Harper JF (1999) Communicating with calcium. Plant Cell 11(4):691–706

Sanders D, Pelloux J, Brownlee C, Harper JF (2002) Calcium at the crossroads of signaling. Plant Cell 14(Suppl 1):S401–S417

Sangwan V, Foulds I, Singh J, Dhindsa RJ (2001) Cold activation of *Brassica napus* BN115 promoter is mediated by structural changes in membranes and cytoskeleton, and requires Ca^{2+} influx. Plant J 27(1):1–12

Sanyal SK, Pandey A, Pandey GK (2015) The CBL-CIPK signaling module in plants: a mechanistic perspective. Physiol Physiol Plant 155(2):89–108

Scholz SS, Vadassery J, Heyer M, Reichelt M, Bender KW, Snedden WA, Boland W, Mithöfer A (2014) Mutation of the Arabidopsis calmodulin-like protein CML37 deregulates the jasmonate pathway and enhances susceptibility to herbivory. Mol Plant 7(12):1712–1726

Scholz SS, Reichelt M, Vadassery J, Mithöfer A (2015) Calmodulin-like protein CML37 is a positive regulator of ABA during drought stress in Arabidopsis. Plant Signal Behav 10(6):e1011951

Schroeder JI, Allen GJ, Hugouvieux V, Kwak JM, Waner D (2001) Guard cell signal transduction. Annu Rev Plant Biol 52(1):627–658

Schuurink RC, Chan V, Jones RL (1996) Modulation of calmodulin mRNA and protein levels in barley aleurone. Plant Physiol 111(2):371–380

Schwaller B (2012) The regulation of a cell's Ca^{2+} signaling toolkit: the Ca^{2+} homeostasome. In: Calcium signaling. Springer, Dordrecht, pp 1–25

Scrase-Field SA, Knight MR (2003) Calcium: just a chemical switch? Curr Opin Plant Biol 6(5):500–506

Seifikalhor M, Aliniaeifard S, Shomali A, Azad N, Hassani B, Lastochkina O, Li T (2019) Calcium signaling and salt tolerance are diversely entwined in plants. Plant Signal Behav 14:1665455

Shabala S, Wu H, Bose J (2015) Salt stress sensing and early signalling events in plant roots: current knowledge and hypothesis. Plant Sci 241:109–119

Shacklock PS, Read ND, Trewavas AJ (1992) Cytosolic free calcium mediates red light-induced photomorphogenesis. Nature 358(6389):753

Sheen J (1996) Ca^{2+}-dependent protein kinases and stress signal transduction in plants. Science 274(5294):1900–1902

Shi J, Kim KN, Ritz O, Albrecht V, Gupta R, Harter K, Luan S, Kudla J (1999) Novel protein kinases associated with calcineurin B-like calcium sensors in Arabidopsis. Plant Cell 11(12):393–2405

Shi S, Li S, Asim M, Mao J, Xu D, Ullah Z, Liu G, Wang Q, Liu H (2018) The Arabidopsis calcium dependent protein kinases (CDPKs) and their roles in plant growth regulation and abiotic stress responses. Int J Mol Sci 19(7):1900

Shinozaki K, Yamaguchi-Shinozaki K (1997) Gene expression and signal transduction in water-stress response. Plant Physiol 115(2):327

Shinozaki K, Yamaguchi-Shinozaki K (2007) Gene networks involved in drought stress response and tolerance. J Exp Bot 58(2):221–227

Siegel RS, Xue S, Murata Y, Yang Y, Nishimura N, Wang A, Schroeder JI (2009) Calcium elevation dependent and attenuated resting calcium-dependent abscisic acid induction of stomatal closure and abscisic acid-induced enhancement of calcium sensitivities of S-type anion and inward rectifying K^+ channels in Arabidopsis guard cells. Plant J 59(2):207–220

Simeunovic A, Mair A, Wurzinger B, Teige M (2016) Know where your clients are: subcellular localization and targets of calcium-dependent protein kinases. J Exp Bot 67(13):3855–3872

Snedden WA, Fromm H (1998) Calmodulin, calmodulin related proteins and plant responses to the environment. Trends Plant Sci 3(8):299–304
Snedden WA, Fromm H (2001) Calmodulin as a versatile calcium signal transducer in plants. New Phytol 151(1):35–66
Song J, Zhao Q, Thao S, Frederick RO, Markley JL (2004) Letter to the editor: solution structure of a calmodulin-like calcium-binding domain from *Arabidopsis thaliana*. J Biomol NMR 30(4):451–456
Song CP, Agarwal M, Ohta M, Guo Y, Halfter U, Wang P, Zhu JK (2005) Role of an Arabidopsis AP2/EREBP-type transcriptional repressor in abscisic acid and drought stress responses. Plant Cell 17(8):2384–2396
Stael S, Wurzinger B, Mair A, Mehlmer N, Vothknecht UC, Teige M (2012) Plant organellar calcium signalling: an emerging field. J Exp Bot 63(4):1525–1542
Stange A, Hedrich R, Roelfsema MRG (2010) Ca2 + -dependent activation of guard cell anion channels, triggered by hyperpolarization, is promoted by prolonged depolarization. Plant J 62(2):265–276
Staudinger C, Mehmeti-Tershani V, Gil-Quintana E, Gonzalez EM, Hofhansl F, Bachmann G, Wienkoop S (2016) Evidence for a rhizobia-induced drought stress response strategy in *Medicago truncatula*. J Proteomics 136:202–213
Staxén I, Pical C, Montgomery LT, Gray JE, Hetherington AM, McAinsh MR (1999) Abscisic acid induces oscillations in guard-cell cytosolic free calcium that involve phosphoinositide-specific phospholipase C. PNAS 96(4):1779–1784
Steinhorst L, Kudla J (2013) Calcium and reactive oxygen species rule the waves of signaling. Plant Physiol 113
Steinhorst L, Kudla J (2014) Signaling in cells and organisms—calcium holds the line. Curr Opin Plant Biol 22:14–21
Sutter JU, Sieben C, Hartel A, Eisenach C, Thiel G, Blatt MR (2007) Abscisic acid triggers the endocytosis of the Arabidopsis KAT1 K^+ channel and its recycling to the plasma membrane. Curr Biol 17(16):1396–1402
Sze H, Liang F, Hwang I, Curran AC, Harper JF (2000) Diversity and regulation of plant Ca^{2+} pumps: insights from expression in yeast. Annu Rev Plant Biol 51(1):433–462
Tai F, Wang Q, Yuan Z, Yuan Z, Li H, Wang W (2013) Characterization of five CIPK genes expressions in maize under water stress. Acta Physiol Plant 35(5):1555–1564
Tai F, Yuan Z, Li S, Wang Q, Liu F, Wang W (2015) ZmCIPK8, a CBL-interacting protein kinase, regulates maize response to drought stress. PCTOC 124(3):459–469
Tai F, Yuan Z, Li S, Wang Q, Liu F, Wang W (2016) ZmCIPK8, a CBL-interacting protein kinase, regulates maize response to drought stress. PCTOC 124(3):459–469
Takahashi Y, Kinoshita T, Shimazaki K (2007) Protein phosphorylation and binding of a 14-3-3 protein in Vicia guard cells in response to ABA. Plant Cell Physiol 48(8):1182–1191
Thoday-Kennedy EL, Jacobs AK, Roy SJ (2015) The role of the CBL–CIPK calcium signaling network in regulating ion transport in response to abiotic stress. Plant Growth Regul 76(1):3–12
Thomashow MF (1999) Plant cold acclimation: freezing tolerance genes and regulatory mechanisms. Annu Rev Plant Biol 50(1):571–599
Tirone F, Radu L, Craescu CT, Cox JA (2010) Identification of the binding site for the regulatory calcium-binding domain in the catalytic domain of NOX5. Biochemistry 49(4):761–771
Trewavas A (1999) Le calcium, c'est la vie: calcium makes waves. Plant Physiol 120(1):1–6
Tsou PL, Lee SY, Allen NS, Winter-Sederoff H, Robertson D (2012) An ER-targeted calcium binding peptide confers salt and drought tolerance mediated by CIPK6 in Arabidopsis. Planta 235(3):539–552
Tuteja N, Mahajan S (2007) Calcium signaling network in plants: an overview. Plant Signal Behav 2(2):79–85
Umezawa T, Fujita M, Fujita Y, Yamaguchi-Shinozaki K, Shinozaki K (2006) Engineering drought tolerance in plants: discovering and tailoring genes to unlock the future. Curr Opin Biotech 17(2):113–122

Urao T, Katagiri T, Mizoguchi T, Yamaguchi-Shinozaki K, Hayashida N, Shinozaki K (1994) Two genes that encode Ca^{2+}-dependent protein kinases are induced by drought and high-salt stresses *Arabidopsis thaliana*. MGG 244(4):331–340
Vadassery J, Reichelt M, Hause B, Gershenzon J, Boland W, Mith€ofer A (2012) CML42-mediated calcium signaling coordinates responses to Spodoptera herbivory and abiotic stresses in Arabidopsis. Plant Physiol 112
Vanderbeld B, Snedden WA (2007) Developmental and stimulus-induced expression patterns of Arabidopsis calmodulin-like genes CML37, CML38 and CML39. Plant Mol Biol 64(6):683–697
Virdi AS, Singh S, Singh P (2015) Abiotic stress responses in plants: roles of calmodulin-regulated proteins. Front Plant Sci 6:809
Vivek PJ, Tuteja N, Soniya EV (2013) CDPK1 from ginger promotes salinity and drought stress tolerance without yield penalty by improving growth and photosynthesis in *Nicotiana tabacum*. PLoS One 8(10):e76392
Wan B, Lin Y, Mou T (2007) Expression of rice Ca^{2+}-dependent protein kinases (CDPKs) genes under different environmental stresses. FEBS Lett 581(6):1179–1189
Wan D, Li R, Zou B, Zhang X, Cong J, Wang R, Xia Y, Li G (2012) Calmodulin-binding protein CBP60g is a positive regulator of both disease resistance and drought tolerance in Arabidopsis. Plant Cell Rep 31(7):1269–1281
Wang QY, Nick P (2001) Cold acclimation can induce microtubular cold stability in a manner distinct from abscisic acid. Plant Cell Physiol 42(9):999–1005
Wang L, Tsuda K, Sato M, Cohen JD, Katagiri F, Glazebrook J (2009) Arabidopsis CaM binding protein CBP60g contributes to MAMP-induced SA accumulation and is involved in disease resistance against Pseudomonas syringae. PLoS Pathogens 5(2):e1000301
Wang W-H, Yi X-Q, Han A-D, Liu T-W, Chen J, Wu F-H, Dong X-J, He J-X, Pei Z-M, Zheng H-L (2011) Calcium sensing receptor regulates stomatal closure through hydrogen peroxide and nitric oxide in response to extracellular calcium in Arabidopsis. J Exp Bot 63(1):177–190
Wang JP, Munyampundu JP, Xu YP, Cai Z (2015) Phylogeny of plant calcium and calmodulin dependent protein kinases (CCaMKs) and functional analyses of tomato CCaMK in disease resistance. Front Plant Sci 6:1075
Wang Y, Li T, John SJ, Chen M, Chang J, Yang G, He G (2018) A CBL-interacting protein kinase TaCIPK27 confers drought tolerance and exogenous ABA sensitivity in transgenic Arabidopsis. Plant Physiol Biochem 123:103–113
Wei S, Hu W, Deng X, Zhang Y, Liu X, Zhao X, Luo Q, Jin Z, Li Y, Zhou S, Sun T (2014) A rice calcium-dependent protein kinase OsCPK9 positively regulates drought stress tolerance and spikelet fertility. BMC Plant Biol 14(1):133
Weinl S, Held K, Schlücking K, Steinhorst L, Kuhlgert S, Hippler M, Kudla J (2008) A plastid protein crucial for Ca^{2+}-regulated stomatal responses. New Phytol 179(3):675–686
Weng H, Yoo CY, Gosney MJ, Hasegawa PM, Mickelbart MV (2012) Poplar GTL1 is a Ca^{2+}/calmodulin binding transcription factor that functions in plant water use efficiency and drought tolerance. PLoS One 7(3):e32925
Wheeler GL, Brownlee C (2008) Ca^{2+} signalling in plants and green algae–changing channels. Trends Plant Sci 13(9):506–514
White PJ (1998) Calcium channels in the plasma membrane of root cells. Ann Bot 81(2):173–183
White PJ (2001) The pathways of calcium movement to the xylem. J Exp Bot 52(358):891–899
White PJ (2009) Depolarization-activated calcium channels shape the calcium signatures induced by low temperature stress. New Phytol 183(1):6–8
White PJ, Broadley MR (2003) Calcium in plants. Ann Bot 92(4):487–511
Xie T, Ren R, Zhang YY, Pang Y, Yan C, Gong X, He Y, Li W, Miao D, Hao Q, Deng H (2012) Molecular mechanism for inhibition of a critical component in the *Arabidopsis thaliana* abscisic acid signal transduction pathways, SnRK2. 6, by protein phosphatase ABI1. J Biol Chem 287(1):794–802

Xing T, Wang XJ, Malik K, Miki BL (2001) Ectopic expression of an Arabidopsis calmodulin-like domain protein kinase-enhanced NADPH oxidase activity and oxidative burst in tomato protoplasts. MPMI 14(10):1261–1264
Xiong L, Zhu JK (2001) Abiotic stress signal transduction in plants: molecular and genetic perspectives. Physiol Plant 112(2):152–166
Xiong L, Schumaker KS, Zhu JK (2002) Cell signaling during cold, drought, and salt stress. Plant Cell 14(Suppl 1):S165–S183
Xiong TC, Bourque S, Lecourieux D, Amelot N, Grat S, Briere C, Mazars C, Pugin A, Ranjeva R (2006) Calcium signaling in plant cell organelles delimited by a double membrane. BBA-Mol Cell Res 1763(11):1209–1215
Xu J, Tian YS, Peng RH, Xiong AS, Zhu B, Jin XF, Gao F, Fu XY, Hou XL, Yao QH (2010) AtCPK6, a functionally redundant and positive regulator involved in salt/drought stress tolerance in Arabidopsis. Planta 231(6):1251–1260
Xu GY, Roch PS, Wang ML, Xu ML, Cui YC, Li LY, Zhu YX, Xia X (2011) A novel rice calmodulin like gene, OsMSR2, enhances drought and salt tolerance and increases ABA sensitivity in Arabidopsis. Planta 234(1):47–59
Yang T, Poovaiah BW (2003) Calcium/calmodulin-mediate signal network in plants. Trends Plant Sci 8(10):505–512
Yang T, Poovaiah BW (2000) An early ethylene up-regulated gene encoding a calmodulin-binding protein involved in plant senescence and death. J Biol Chem 275(49):38467–38473
Yang T, Poovaiah BW (2002) A calmodulin-binding/CGCG box DNA-binding protein family involved in multiple signaling pathways in plants. J Biol Chem 277(47):45049–45058
Yang WQ, Kong ZS, Omo-Ikerodah E, Xu WY, Li Q, Xue YB (2008) Calcineurin B-like interacting protein kinase OsCIPK23 functions in pollination and drought stress responses in rice (*Oryza sativa* L.). J Genet Genom 35(9):531-S2
Ye C-Y, Xia X, Yin W (2013) Evolutionary analysis of CBL-interacting protein kinase gene family in plants. Plant Growth Regul 71(1):49–56
Yildizli A, Çevik S, Ünyayar S (2018) Effects of exogenous myo-inositol on leaf water status and oxidative stress of *Capsicum annuum* under drought stress. Acta Physiol Plant 40(6):122
Yoo CY, Pence HE, Jin JB, Miura K, Gosney MJ, Hasegawa PM, Mickelbart MV (2010) The Arabidopsis GTL1 transcription factor regulates water use efficiency and drought tolerance by modulating stomatal density via transrepression of SDD1. Plant Cell 110
Young JJ, Mehta S, Israelsson M, Godoski J, Grill E, Schroeder JI (2006) CO_2 signaling in guard cells: calcium sensitivity response modulation, a Ca^{2+}-independent phase, and CO_2 insensitivity of the gca2 mutant. PNAS 103(19):7506–7511
Yu Y, Xia X, Yin W, Zhang H (2007) Comparative genomic analysis of CIPK gene family in Arabidopsis and Populus. Plant Growth Regul 52(2):101–110
Zhang H, Yang B, Liu WZ, Li H, Wang L, Wang B, Deng M, Liang W, Deyholos M, Jiang YQ (2014) Identification and characterization of CBL and CIPK gene families in canola (*Brassica napus* L.). BMC Plant Biol 14(1):8
Zhou L, Lan W, Chen B, Fang W, Luan S (2015a) A calcium sensor-regulated protein kinase CIPK19 is required for pollen tube growth and polarity1. Plant Physiol 114
Zhou X, Hao H, Zhang Y, Bai Y, Zhu W, Qin Y, Yuan F, Zhao F, Wang M, Hu J, Xu H (2015b) PKS5/CIPK11, a SnRK3-type protein kinase, is important for ABA responses in Arabidopsis through phosphorylation of ABI5. Plant Physiol 114
Zhu K, Chen F, Liu J, Chen X, Hewezi T, Cheng ZMM (2016) Evolution of an intron-poor cluster of the CIPK gene family and expression in response to drought stress in soybean. Sci Rep 6:28225
Zhu SY, Yu XC, Wang XJ, Zhao R, Li Y, Fan RC, Shang Y, Du SY, Wang XF, Wu FQ, Xu YH (2007) Two calcium-dependent protein kinases, CPK4 and CPK11, regulate abscisic acid signal transduction in Arabidopsis. Plant Cell 19(10):3019–3036
Zielinski RE (1998) Calmodulin and calmodulin-binding proteins in plants. Annu Rev Plant Biol 49(1):697–725

Zou JJ, Wei FJ, Wang C, Wu J, Ratnasekera D, Liu WX, Wu WH (2010) Arabidopsis calcium dependent protein kinase AtCPK10 functions in ABA and Ca^{2+}-mediated stomatal regulation in response to drought stress. Plant Physiol 110
Zou JJ, Li XD, Ratnasekera D, Wang C, Liu WX, Song LF, Zhang WZ, Wu WH (2015) Arabidopsis CALCIUM-DEPENDENT PROTEIN KINASE8 and CATALASE3 function in abscisic acid mediated signaling and H_2O_2 homeostasis in stomatal guard cells under drought stress. Plant Cell 15

ROS Signalling in Modulating Salinity Stress Tolerance in Plants

Mohsin Tanveer and Hassan Ahmed Ibraheem Ahmed

Abstract Salinity stress is a major environmental hurdle that restricts agricultural crop production across the globe. Amongst other factors, salinity stress-induced reactive oxygen species (ROS) production is considered to be one of the major constraints limiting plant growth and productivity in saline soils. Well before ROS can cause oxidative damage to plants, elevation in ROS levels cause a major disturbance to plant ionic homeostasis, thus affecting cell metabolism. However recently, salinity stress tolerance phenomena is linked to the enhanced production of ROS and, specifically, hydrogen peroxide (H_2O_2), that operate upstream of signaling and the associated regulation of gene expression through the redox signaling hub. In this book chapter, role of ROS signaling has been discussed in relation to growth, ion channels regulation, and protein accumulation, which are involved in inducing salinity tolerance in plants. Studies showed that ROS interacts with different important hormones such as abscisic acid, brassinolide (BL), or gibberellic acid (GA) to improve salinity tolerance. Reactive oxygen species especially H_2O_2 regulate the activities of different TFs by regulating different processes such as increase in mRNA stability and translation, or by facilitating nucleo-cytoplasmic traffic or by mediating DNA binding. Future research issues will also be raised to further speculate the role of ROS signaling and salinity tolerance in plants.

1 Introduction

Global trends in the climate change have a major negative impact on agricultural productivity and every year, massive amounts of crop yield are lost due to soil salinity (Tanveer and Shah 2017; Liu et al. 2018). Salinity stress reduces plant growth and development by disrupting different physiological and biochemical processes (Tanveer et al. 2018a). Salinity induced growth and yield reduction has been reported in number of plant species such as wheat, rice, barley, sorghum, and horticultural crops as well (Kerepesi and Galiba 2000; Uchida et al. 2002; Netondo et al. 2004;

M. Tanveer (✉) · H. A. I. Ahmed
School of Land and Food, University of Tasmania, Hobart, Australia
e-mail: Mohsin.tanveer@utas.edu.au

M. Hasanuzzaman and M. Tanveer (eds.), *Salt and Drought Stress Tolerance in Plants*, Signaling and Communication in Plants, https://doi.org/10.1007/978-3-030-40277-8_11

Liu et al. 2018). However, plants have developed different adaptive mechanisms to cope and survive under these stress conditions. One of these mechanisms is using reactive oxygen species (ROS) as signaling moiety to activate salinity defense system and regulate numerous physiological and transcriptional processes (Miller et al. 2010). Under multiple stress conditions such as salinity, excessive production of ROS damages cellular membranes and organelles (Baier et al. 2005). Thus a tight balance should be maintained between ROS production and scavenging in order to improve tolerance in plants (Tanveer and Shabala 2018). Besides the oxidative role of ROS, it has also been considered as "good guys", playing an important role as signaling molecules in plant responses to stress conditions (Mittler 2017). Beside this, different studies also showed that under multiple stress conditions, levels of ROS production and expression of ROS scavenging enzymes showed a unique pattern that is clearly different than that observed to be induced by single stress. Changes in such pattern can be reflected in the levels of different ROS production, i.e., enhanced activity of antioxidant production, osmolytes accumulation flavonoid, and tocopherol accumulations (Keles and Oncel 2002; Giraud et al. 2008; Vile et al. 2012; Vuleta et al. 2015; Martinez et al. 2016). As stated above, ROS act as signaling element, thus it is more likely that ROS signature can activate or deactivate numerous physiological processes is unique. This chapter focuses on specific details of the ROS signaling and associated downstream signaling to various genes and membrane proteins involved in plant adaptation to hostile environment.

2 ROS Mediated Plant Growth Response

Reactive oxygen species occurred long time ago in concomitant to the emerging and evolution of aerobic life and it is not a surprise nowadays to know their beneficial and essential role in normal processes in most aerobic organisms (Mittler 2017). Plants although being sessile organisms, they were able to adapt at cellular and organ levels to diverse biotic and abiotic stresses and ROS were a part of such systemic defense responses triggered by these smart organisms (Perez and Brown 2014). Reactive oxygen species was found to be produced in different cellular compartments (peroxisomes, mitochondria, chloroplast, and cell wall peroxidases) in living cells and stressful condition (biotic and abiotic) will increase their production rates (Sewelam et al. 2016). These new levels of ROS under stress circumstances subsequently regulate gene expression leading to an alteration of plant growth at the tissue and organ levels that might open or help in an adaptation mechanism (Tsukagoshi 2016).

Plants not only are able to achieve well-adaptation strategies in order to minimize ROS-induced injury under natural growth conditions but also can make much more benefit of the inevitable ROS production by directing them to do signaling functions (Sewelam et al. 2016). It is now an old-fashioned notion to keep considering ROS as toxic by-products of plant metabolism and neglect recent experimental data that assured ROS as highly controlled signaling molecules capable of transferring the environmental signals, with other signaling intermediates, to the genetic machinery

(Sewelam et al. 2016). Plants possess almost all of the components that mediate ROS homeostasis and redox signaling including NOX-like proteins (RBOHs), superoxide dismutase (SOD), catalase (CAT), peroxidase (PRX), iron uptake/storage regulating mechanisms, and a network of thio- and glutaredoxins (Mittler 2017). Governing plant growth, ROS also acts as secondary messenger that signals key cellular functions like cell proliferation, apoptosis, and necrosis (Choudhury et al. 2013). In this regard, ROS seemed to be core signal molecule in crosstalk between other growth signaling phyto-compound and salinity tolerance has been suggested to result partly from overlap between ROS signaling mechanisms. Hence and from an applicable point of view, agronomists have pursued the possibility of inducing systemic resistance in crops through chemical means "defense-inducers" such as hydrogen peroxide (H_2O_2), Ozone, SA (salicylic Acid), herbicides and fungicides, aluminum (Al^{+3}) (Perez and Brown 2014).

One of the most important functions of ROS is their suggested role as a plant growth regulator (Tsukagoshi 2016). Therefore, ROS essentially support normal physiological and metabolic functions, as well as promote cellular proliferation and differentiation. These beneficial roles of ROS could be exploited in a bad way, for example by cancer cells that tilt the ROS balance in their favor and maintain a higher basal level of ROS that promotes cellular proliferation. On the other hand, and smartly, necrotrophic plant pathogens were able to use ROS to induce plant necrosis and feed on the dead plant tissue (Mittler 2017). In higher plants, ROS at redox biology level were found to regulate diverse functions such as: cellular proliferation, stress acclimation, stress signaling transduction, development, differentiation, metabolic regulation, interactions with other organisms and pathogen defense, systemic responses, and physiological cell death. ROS signaling is mediated by a highly regulated process of ROS accumulation at specific cellular compartments and maintaining normal physiological functions mainly via interacting with cysteine (Cys) residues of proteins (Mittler 2017). In addition to RBOHs plasma membrane-bound NADPH-oxidases (respiratory burst oxidase homolog, RBOH; termed NOX in animals), ROS signaling was found to be mediated by peroxidases at the apoplast, as well as by the accumulation of ROS in different cellular compartments including the chloroplast, mitochondria, peroxisomes, and nuclei (Mittler 2017).

Currently, ROS were suggested as key molecules in directing plant root and shoot development, and thus, performing a similar regulatory role as plant hormones (Tsukagoshi 2016). This arisen as a result of the significant evidences that have accumulated supporting the assumption that ROS can act as signaling molecules serving a functional role similar to hormones (Diaz-Vivancos et al. 2013; Tsukagoshi 2016). Reactive oxygen species have been found to be particularly important for maintaining the balance between cell proliferation and differentiation, such an important action to achieve a good spatial distribution of lateral roots and/or triggering the formation of any stress-adaptive important organ or sub-organ in plant shoot. They do accumulate in root cells even under normal growth conditions although their potential harmful action (Perez and Brown 2014), and this assure their need for root growth and development (Tsukagoshi 2016). For achieving the best root development, one of the important roles of ROS in is their function in altering cell walls triggering lignin formation. This takes place by facilitating the conversion of H_2O_2 into H_2O and O_2 through Peroxidases. Hence, due to this process an electron is also produced which is utilized in whole plant cells to modify both primary and secondary cell walls. This modification process involves the transfer of the electron to monolignol molecules in cells within the maturation zone, where monolignol can be used to form the polymer, lignin (Tsukagoshi 2016).

In addition, ROS, play a crucial role in the formation of the casparian strip; which is a lignin-based diffusion barrier present in the endodermis in plant root cells that control apoplastic movement of solutes and water. For specifying the casparian strip formation in the endodermis, casparian strip domain proteins (CASPs), which express specifically in the endodermis, guide NADPH oxidase, peroxidase, and other enzymes localized in the plasma membrane in a region of endodermal cell walls (Tsukagoshi 2016). Recently, ROS has already been established as a cell cycle regulator in animal cells. In plants, exogenous H_2O_2 treatment has been shown to restrain the expression of cell cycle-related genes and reduce the meristem size of the root. Hence ROS seemed to play a role in determining the size of the meristimatic zone by regulating the cell cycle (Tsukagoshi 2016).

Reactive oxygen species regulates also the transition from cell proliferation to cell differentiation by controlling the transition of cells from a zone of cell division to a zone of cell elongation and differentiation. In this regard, superoxide was found to accumulate in the meristematic zone, while H_2O_2 accumulate in the elongation zone (Tsukagoshi 2016; Liu et al. 2019). A such function that might be similar to cytokinin role in controlling and achieving the best spatial distribution of lateral roots by inhibiting cell cycle activity as well as primordium development and stimulates cell elongation in developing lateral roots (Julkowska and Testerink 2015). Therefore, ROS take part in lateral root (LR) development for example, AtrbohD and F (NADPH oxidase proteins in Arabidopsis) negatively regulate LR formation and in double mutants of Atr-bohD and F, peroxidase activity in roots was higher than in wild type roots, and LR density increased (Tsukagoshi 2016).

Root hair development also appeared to be managed through ROS activity. Evidence was observed when the Loss of Root Hair Defective 2 (RHD2), which encodes a RbohC or NADPH oxidase, fails to accumulate ROS at the tip of an initial root hair. As a result, the incipient root hair does not develop (Tsukagoshi 2016). Additionally, wild type Arabidopsis treated with diphenyleneiodonium (DPI), an inhibitor of NADPH oxidase activity, also failed to develop root hairs and does not accumulate ROS at the tip of the root hair. These results strongly indicate that ROS accumulation at the tip of root hair is required for normal root hair development (Zhao et al. 2001; Tamás et al. 2010). Reactive oxygen species signaling, however, also regulates plant shoot development. As a result, cross talk systems may exist which affects the growth of whole plants (Tsukagoshi 2016).

3 ROS-Mediated Ion Transport in Plants

Ion channels are hydrophilic pathways with pretentious nature occurred in the membranes of all living organisms and are able to govern and mediate electrogenic transport of ions and metabolites (Scholz-Starke et al. 2005). They have important roles in several physiological processes in plant cells (e.g., maintenance of the turgor pressure, stomatal movements, and nutrient absorption by the roots); due to their unique biophysical properties such as showing high selectivity for an ion permeating, high turnover rate, and modulation by physico-chemical parameters (e.g., membrane potential, calcium concentration) (Hasegawa et al. 2000; Scholz-starke et al. 2005; Shabala 2009). Evidence is emerging that ROS can function as cellular second messengers that are likely to modulate many different proteins (especially membranes-bond ones) leading to a variety of responses (Mori and Schroeder 2004). Recently, findings indicate the involvement of ion channels in redox-dependent processes and are capable of sensing ROS (Kurusu et al. 2015).

The generation of ROS and the increment in the cytoplasmic calcium concentration are typically common early events in many stress and developmental responses of plant cells (Liu et al. 2019). As a matter of fact, oxidative stress not only lead to a damage effect on plant cell but can also signal prime stresses to proper gene expression through activation of Ca^{2+} influx and K^+ efflux ion channels (Demidchik 2010). Hence, the regulation of different cellular functions obviously relies upon the complicated network arisen from the interaction between Ca^{2+} and ROS at multiple levels (Swanson and Gilroy 2010). In this regard Calcium (Ca^{2+}) has both positive and negative regulatory roles on cellular ROS levels. For instance, a Ca^{2+}-binding EF-hand motif was found in ROS-producing NADPH oxidases, and NAD kinase, providing the NADPH substrate, is determined by calmodulin. Furthermore, calmodulin was shown to induce the activation of catalase, the major H_2O_2 scavenging enzyme, of *Arabidopsis thaliana* (Scholz-Starke et al. 2005). Moreover, ROS can influence cytosolic Ca^{2+} signaling in several ways. Hydrogen peroxide was denoted to bring about an oxidative methionine modification on calmodulin, thus inhibiting calmodulin-dependent activation of the plasma membrane Ca-ATPase (Hardin et al. 2009). Bovine calcineurin, a Ca^{2+}- and calmodulin-dependent protein phosphatase, also was suppressed by H_2O_2 by oxidative formation of a disulfide bridge (Scholz-Starke et al. 2005).

The modification in ion transport mechanisms results in changes in plant signaling system primarily Ca^{2+} homeostasis, which further augments the abnormal electrical activity and distortion of signal transduction, causing cell dysfunction (Kourie 1998). Ion channels activated by (ROS have been found in the plasma membrane of charophyte *Nitella flixilis*, dicotyledon *Arabidopsis thaliana*, *Pyrus pyrifolia* and *Pisum sativum*, and the monocotyledon *Lilium longiflorum* and hydrogen peroxide and hydroxyl radicals were recorded as major activating species for these channels (Demidchik 2018). Reactive oxygen species are important for cellular transport via ion channels because of their interaction with various ion transport proteins underlying the transmembrane signal transduction, namely, (1) ion channels, such as Ca^{2+}

channels (including voltage sensitive L-type Ca^{2+} currents, dihydropyridine receptor voltage sensors, ryanodine receptor Ca^{2+}-release channels, and D-*myo*-inositol 1,4,5-trisphosphate receptor Ca^{2+}-release channels), K^+ channels (such as Ca^{2+}-activated K^+ channels, inward and outward K^+ currents, and ATP-sensitive K^+ channels), Na1 channels, and Ca^{2+} channels; (2) ion pumps, such as sarcoplasmic reticulum and sarcolemmal Ca^{2+} pumps, Na^+-K^+-ATPase (Na1 pump), and H^+-ATPase (H1 pump); (3) ion exchangers such as the Na^+/Ca^{2+} exchanger and Na^+/H^+ exchanger; and (4) ion co-transporters (Kourie 1998). The first evidence for stimulation of ion fluxes in plant tissues treated with ROS was obtained from measurements of so-called electrolyte leakage caused by a transition metal copper ions $Cu^{+}/^{2+}$, which is a free radical capable of donating and accepting electrons from hydrogen peroxide, superoxide anion radicals, nitric oxide and other ROS and reactive nitrogen species (RNS) (Rodrigo-Moreno et al. 2013; Demidchik 2018). As a consequent, when *Chlorella vulgaris* cells were treated with copper ions and found a dramatic loss of K^+ from these cells, accounting for 94% of all cell K^+ at approximately 0.3 mM $CuSO_4$ (Demidchik 2018). The first record for ROS-induced ion fluxes in higher plants was probably obtained in the study of electrolyte leakage from Ca^{2+}-treated *Agrostis tenuis* plants (Demidchik 2018).

Under salinity condition, the mechanism of ROS-induced alterations in ion transport pathways involves (1) oxidation of sulfhydryl groups located on the ion transport proteins, (2) peroxidation of membrane phospholipids, and (3) inhibition of membrane-bound regulatory enzymes and modification of the oxidative phosphorylation and ATP levels (Kourie 1998). Three groups were categorized for plant ROS-activated ion channels including inwardly-rectifying, outwardly-rectifying, and voltage-independent groups. The inwardly rectifying ROS-activated ion channels mediate Ca^{2+}-influx for growth and development in roots and pollen tubes. The outwardly rectifying group K^+ facilitates efflux for the regulation of osmotic pressure in guard cells, induction of programmed cell death, and autophagy in roots (Gilroy et al. 2014; Demidchik 2018). The voltage-independent group mediates both Ca^{2+} influx and K^+ efflux (Demidchik 2018). The ROS-activated K^+ channels have recently been identified as products of Stellar K^+ Outward Rectifier (SKOR) and Guard cell Outwardly Rectifying K^+ channel (GORK) genes (Shabala and Pottosin 2014). Another target of ROS signal transduction is the activation of Ca^{2+}-permeable channels in plant membranes. Elevation in the cytosolic free calcium is crucial for plant growth, development, and adaptation. These channels are encoded by the following gene families: (1) cyclic nucleotide-gated channels (CNGCs), (2) ionotropic glutamate receptors (GLRs), (3) annexins, (4) 'mechanosensitive channels of small (MscS) conductance'-like channels (MSLs), (5) 'mid1-complementing activity' channels (MCAs), Piezo channels, and hyperosmolality-induced $[Ca^{2+}]$cyt channel 1 (OSCA1). Moreover a 'tandem-pore channel1' (TPC1) catalyzes Ca^{2+} efflux from the vacuole in response to the plasma membrane-mediated Ca^{2+} elevation (Demidchik and Shabala 2018). Calcium influx into plant cells is mediated by Ca^{2+} depolarisation-activated, hyperpolarisation-activated and voltage-independent Ca^{2+}-permeable channels (DACCs, HACCs and VICCs respectively) while extrusion of Ca^{2+} from the cytosol is mediated by Ca^{2+}-ATPases and Ca^{2+}/H^+ exchangers (Demidchik et al. 2002).

In plant cells, the calcium ion is a ubiquitous intracellular second messenger involved in numerous signalling pathways. Variations in the cytosolic concentration of Ca^{2+} ([Ca^{2+}]cyt) couple a large array of signals and responses (Lecourieux et al. 2006). ROS activation of Ca^{2+} channels may be a central step in many ROS-mediated processes, such as stress and hormone signaling, polar growth, development, and possibly during mechanotransduction (Mori and Schroeder 2004). The molecular nature of ROS-activated Ca^{2+} influx channels remains poorly understood, although annexins and cyclic nucleotide-gated channels have been proposed for this role (Swarbreck et al. 2013). BCC1, a calcium channel from tendrils of *Bryonia dioica*, is inhibited by H_2O_2 applied on the cytoplasmic side. However, since H_2O_2 treatment triggered transient [Ca^{2+}]cyt elevations in *Commelina* guard cells and tobacco, it is reasonable to expect other calcium channels to be activated by ROS (Scholz-Starke et al. 2005). Calcium-activated NADPH oxidases and ROS-activated Ca^{2+} conductances form a self-amplifying 'ROS-Ca^{2+}hub', enhancing and transducing Ca^{2+} and redox signals. The ROS-Ca^{2+} hub contributes to physiological reactions controlled by ROS and Ca^{2+}, demonstrating synergism and unity of Ca^{2+} and ROS signalling mechanisms (Demidchik and Shabala 2018). ABA-induced H_2O_2 production and the H_2O_2-activated Ca^{2+} channels are important mechanisms for ABA-induced stomatal closing (Pei et al. 2000). The H_2O_2-activated Ca^{2+} channels mediate both influx of Ca^{2+} in protoplasts and increases in [Ca^{2+}]cyt in intact guard cells. Abscisic acid induces the production of H_2O_2 in guard cells. If H_2O_2 production is blocked, ABA-induced closure of stomata is inhibited (Pei et al. 2000).

4 ROS Signaling and Hormonal Response

Phyto-hormones are very crucial, especially under stress conditions, as they can induce tolerance in plants by shaping the acclimation response (Tanveer et al. 2018b). Recent studies showed an interesting intricate relationship between the role of ROS as signaling molecules and hormones induced tolerance in plants. Moreover, in addition to controlling growth, ionic homeostasis and activation of stress proteins, ROS can also induce cross tolerance in plants via interplaying with hormones. Studies also showed that exogenous application of hormones induces tolerance to multiple abiotic stresses in different plant species. In the following, we have discussed interaction of ROS with different hormones such as auxin, gibberellins (GA), and brassinolides (BL).

Auxin (also known as indole 3 acetic acid) is an important plant hormone that plays numerous roles in plant growth and development such as root formation, stem elongation and gravitropism (Kleine-Vehn et al. 2008). Similar to BL, auxin produces ROS as second messengers by NADPH oxidase activation under stress conditions (Joo et al. 2001; Xia et al. 2011). Studies also showed that balance between ROS and antioxidant production regulates auxin biosynthesis and transport under stress conditions (Cosio and Dunand 2009; Chen and Xiong 2009; Tognetti et al. 2010). Genes encoding antioxidant enzymes such as glutathione (GSH), glutaredoxins (Grx), thioredoxins (Trxs) also trigger the transportation of auxin to auxin targets. For instance, ROS function as downstream components in auxin-mediated signal transduction to control gravitropism responses in roots (Joo et al. 2001). According to Tognetti et al. (2012), ROS and auxin have reciprocal interaction under abiotic stress conditions. In a study, Duan et al. (2010) showed that auxin activates Rho-GTPase, which interacts with NADPH oxidases, thus results in ROS production. Contrarily, ROS triggers MAPK which inhibits auxin dependent signaling while activating ROS stress signaling (Kovtun et al. 2000). According to Schopfer et al. (2001) auxin induced superoxide production facilities cell wall structure during cell elongation.

Brassinolide is also an important hormone which plays an important role in different biochemical and physiological processes under stress conditions. Numerous studies showed that tolerance to abiotic stress conditions correlates with high BL synthesis. Nie et al. (2013) showed that BL induces RBOH transcription and increases NADPH oxidase activity and H_2O_2 production. Nonetheless, it has been reported that 24-epibrasinolide (an active BL) improves tolerance in plants by reducing ROS production and increasing antioxidant activity (Tanveer et al. 2018a). They also proposed that ROS at low concentration may activate numerous genes encoding for BL biosynthesis and following this, BL activates antioxidants to scavenge ROS (Tanveer et al. 2018b). This mechanism enables plants to tolerate multiple stresses. Xia et al. (2009) documented another mechanism in cucumber; BL treatment induced expression of some regulatory genes, such as *RBOH*, *MAPK1*, and *MAPK3*. These results strongly suggest that elevated H_2O_2 levels resulting from enhanced NADPH oxidase activity are involved in the BR-induced stress tolerance to cold stress, light stress and paraquat stress. In *Brassica juncea*, BL improves salt and Ni tolerance by reducing ROS production and electrolytic leakage (Ali et al. 2008).

Gibberllins play very important role in numbers of plant responses to abiotic stresses and their mode of action is linked with the control of growth via controlling cell division and cell elongation (Colebrook et al. 2014). Gibberllins perform their function by regulating DELLA proteins and binding of Gas to a nuclear receptor GID1 causes conformational modifications in protein, facilitating its interaction with DELLA (Achard et al. 2008). Consequently, DELLA is ubiquitinated and targeted for degradation via the 26S proteasome (Colebrook et al. 2014). Gibberllins induce stress tolerance in plant controlling cellular redox homeostasis. In maize leaves, drought stress reduced GA contents, leading to an increase in DELLA activity, which resulted in increased ROS scavenging and improved tolerance (Wang et al. 2008). Under salt stress, ROS accumulation led to higher DELLA activity, suggesting role of ROS as signaling molecules in improving tolerance in plants. Achard et al. (2008) reported that Arabidopsis DELLA mutants showed high level of salt stress tolerance due to an increase in the expression of genes encoding for antioxidants. Similarly, expression of SUB1A in rice improves oxidative stress and restricts high accumulation of ROS (Fukao et al. 2011). SUB1A improves submergence stress in rice by increasing the accumulation of negative regulators of GAs signaling, DELLA and SLR1 protein (Fukao et al. 2006).

5 ROS Induced Post-translational Regulation and Their Role in Salinity Stress Tolerance

Reactive oxygen species act as both bad guy and good guy, their role primarily depends on their source of production and amount of ROS production. Reactive oxygen species act as signaling elements at low concentration and trigger numerous metabolic mechanisms in order to induce tolerance in plants under abiotic stress conditions. Contrarily, ROS can cause reversible or irreversible alterations of proteins, inducing in turn alterations in the control and regulation of plant metabolism. Under abiotic stress conditions, ROS activates numerous stress proteins which play crucial role in transcriptional regulatory networks and induce cross tolerance in plants. Therefore, studies of ROS induced protein modifications or ROS induced stress protein activation, are important to further enhance our understanding of how ROS could change or activate different proteins in order to improve cross tolerance in plants. Reactive oxygen species induced modifications at port-transcriptional levels have been encountered in different studies and among all of those, most important include glutathionylation, sulfonylation, and S-nitrosylation. In the following, ROS induced modifications in these processes have been discussed.

Sulfonylation is one of important mechanisms, which regulates the activity of different TFs and enzymes in plants under stress conditions. This process involves the oxidation of sulfhydryl groups, mainly regulated by H_2O_2 generating sulfenic acid (R-SOH). Formation of R-SOH results in the formation of disulfide bond between cysteine residues, which concomitantly induces conformational changes in proteins. The recovery of proteins from this oxidation process is mainly controlled by GSH and thioredoxin system, which respond to stress and regulate redox homeostasis. Similarly, numerous enzymes of Benson-Calvin cycle are regulated in response to oxidation (Scheibe et al. 2005). Another type of sulfonylation occurs with methionine oxidation, which produces methionine sulfoxide. Enzymes influenced by such modification can be reactivated by a reduction process mediated by methionine sulfoxide reductase (Gustavsson et al. 2002). This is providing an explicit example of Trx pathway in controlling the activities of numerous enzymes involved in stress tolerance. As mentioned above, first step in ROS dependent redox signaling pathway starts from the oxidation of cysteine to sulfenic acid. Sulfenic acid is highly reactive and can bind with thiol to develop S-glutathionylation. This reaction acts as redox driven regulator of signal transduction cascades (Fratelli et al. 2004).

Glutahtionylation can be reversed using thioltransferase (Gallogly and Mieyal 2007). S-Gluathionylation is also very important as it helps in buffering GSSG: GSH pool under stress conditions (Schafer and Buettner 2001). In addition to this, oxidation of tryptophan (Trp) to tryptophan hydroperoxide yields kynurenine and

N-formylkynurenine (NFK), which play an important role in the regulation of photosynthesis (Ronsein et al. 2008). According to Kasson and Barry (2012) oxidation of Trp365 to NFK in PSII correlates with high light stress and increased photoinhibition. In *Arabidopsis*, Trp oxidation was noted in Mn-SOD, mitochondrial peptidase and glycine decarboxylase (Choudhury et al. 2017).

S-nitrosylation involves the covalent bonding of NO to thiol group of Cys and is very important in regulating the function of different stress proteins under stress conditions. Camejo et al. (2013) reported that different enzymes involved in photorespiration and ROS scavenging were S-nitrosylated under salt stress. Likewise, under low temperature stress, main S-nitrosylated proteins were ROS responsive and were related to carbon metabolism (Puyaubert et al. 2014). S-Nitrosylation of proteins is an essential mechanism in plants for metabolic reprogramming to keep redox homeostasis and improve tolerance in plants under multiple stress conditions. S-Nitrosylation also induces modification in TFs which may affect their binding to DNA (Tanou et al. 2012). Moreover, S-Nitrosylation can act as a negative regulator of MYB TFs, which are essential regulators of abiotic and stress responses (Tavares et al. 2014).

6 ROS Induced Transcription Factors and Salinity Stress Tolerance

Regulation of gene expression at the transcriptional level governs numerous metabolic processes in a cell or organism, such as cell cycle, redox homeostasis, ion homeostasis, hormone accumulation and others (Riechmann et al. 2000). Such plant responses under abiotic stress conditions are mediated by multiple signaling pathways, which interact with each other in a complex way to improve tolerance in plants. Reactive oxygen species have been considered as one of the trickiest players, which act as signaling molecules in the regulation of different transcription factors (TFs), involved in stress acclimation. In *Arabidopsis*, expression of 1500 TFs has been examined in response to different ROS (Gadjev 2006). They noted that different ROS repressed or induced of about 500 TFs in *Arabidopsis* and revealing the role of ROS in regulating TFs and plant growth. Though, transcriptional regulation mechanism mediating ROS signaling is not fully explored however it has been suggested that ROS especially H_2O_2 regulate the activities of different TFs by regulating different processes (Sewelam et al. 2016). These processes include ROS induced (1) increase in mRNA stability and translation, (2) increase in TFs stability by inhibiting ubiquitin E3 ligase complex, (3) facilitate nucleo-cytoplasmic traffic (4) mediate DNA binding by identifying TF affinity towards DNA and (5) provide signal to target specific regions of chromatin to activate individual gene (Marinho et al. 2014).

Numerous examples have been reported showing the role of ROS in the activation of different stress-related TFs. Even simple organisms such as bacteria or yeast, sense high ROS production under stress conditions and activate redox sensitive TFs (Mittler et al. 2004). In bacteria, peroxidase regulon repressor and oxygen related (OxyR) TFs are directly activated by H_2O_2. OxyR protein is an important regulatory protein which also activates numerous other H_2O_2 responsive proteins to enhance tolerance (Storz and Imlayt 1999). Likewise in yeast, four TFS such as Maf1, Hsf1, Msn2, and Yap1 have been found to be regulated by ROS (Delaunay et al. 2000). According to Marinho et al. (2014), Yap1 plays a very crucial role in regulating the transcription of genes encoding antioxidant enzymes and oxidative stress response. In multicellular organism, nine different TFs have been reported to be regulated by ROS and which are SP1, TP53, HSF1, HIF1, CREB, AP1, SCREB1, NOTCH, and NRF2 (Marinho et al. 2014).

Besides these, genes such as basic helix loop helix and WRKY also enhanced the understanding of cross talks between oxidative and osmotic stress responsive signaling pathways. In *Arabidopsis*, WRKY33 responds to osmotic stress and oxidative stress (Miller et al. 2008). Under oxidative stress, it has been showed that regulatory functions of bHLH92 and WRKY33 were triggered by targeting GST and POD, linking ROS scavenging with ROS induced signaling (Miller et al. 2008; Jing et al. 2009). Vanderauwera et al. (2012) found a close link between ROS induced WRKY15 and mitrocondrial retrograde signaling in context of salt and osmotic stress tolerance. Xie et al. (2011) discovered a coupled function of plastid heme oxygenases and ROS production in salt acclimation and corroborated the importance of ROS signaling in plant salinity stress responses. Additionally, work on cross-species expression of a SUMO conjugating enzyme has provided considerable insight into the links of ROS, ABA-dependent signaling and the sumoylation pathway in plant salt and drought tolerance (Karan and Subudhi 2012). Moreover, functional characterization of maize ankyrin-repeat protein ITN1, glutamate carboxypeptidase *AMP1*, and bZIP transcription factor *ABP9*, in relation to ROS scavenging and ROS production is further supporting the concept of role of ROS signaling in inducing cross tolerance to salt and drought stress (Sakamoto et al. 2008; Zhang et al. 2011). C_2H_2-zinc finger proteins that contain the EAR repressor domain are thought to play a key role in modulating the defense response of plants to abiotic stress (Kazan 2006). Constitutive expression of the C_2H_2-EAR zinc finger protein *Zat10* in *Arabidopsis* was found to elevate the expression of reactive oxygen-defense transcripts and to enhance the tolerance of plants to salinity, heat and osmotic stress (Mittler et al. 2006). Surprisingly, knockout and RNAi mutants of *Zat10* were also more tolerant to osmotic and salinity stress (Mittler et al. 2006). These results suggest that *Zat10* plays a key role as both a positive and a negative regulator of plant defenses (Mittler et al. 2006).

7 Conclusion

Salinity tolerance is a very complex phenomenon in plants. Different metabolic processes contribute in either inducing or improving salinity tolerance in plants. Under salinity conditions, ROS are produced, which have been showed as very important signaling molecules, which played diverse role in different metabolic processes.

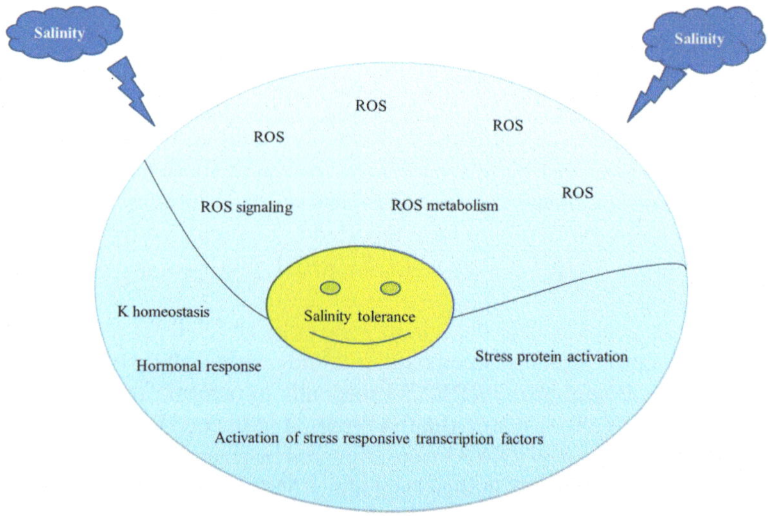

Fig. 1 Role of ROS in inducing salinity tolerance by playing signaling moiety in different processes

Under different abiotic stress conditions, ROS interact with different physiological processes such as growth regulation, ion homeostasis, accumulation of stress hormones or stress proteins and plays crucial role in stress acclimation (Fig. 1). It can be concluded from above discussion that ROS can induce cross tolerance in plants by activating different transcription factors and genes encoding for enhanced protein accumulation, hormone production, and plant growth regulation.

References

Achard P, Gong F, Cheminant S, Alioua M, Hedden P, Genschik P (2008) The cold-inducible CBF1 factor—dependent signaling pathway modulates the accumulation of the growth-repressing DELLA proteins via its effect on gibberellin metabolism. The Plant Cell 20(8):2117–2129

Ali B, Hayat S, Fariduddin Q, Ahmad A (2008) 24-Epibrassinolide protects against the stress generated by salinity and nickel in *Brassica juncea*. Chemosphere 72:1387–1392

Baier M, Kandlbinder A, Golldack D, Dietz KJ (2005) Oxidative stress and ozone: perception, signalling and response. Plant Cell Environ 28:1012–1020

Camejo D, Romero-Puertas Mdel C, Rodriguez-Serrano M, Sandalio LM, Lazaro JJ, Jimenez A, Sevilla F (2013) Salinity-induced changes in S-nitrosylation of pea mitochondrial proteins. J Proteomics 79:87–99

Chen H, Xiong L (2009) The short-rooted vitamin B6-deficient mutant pdx 1 has impaired local auxin biosynthesis. Planta 229:1301–1310

Choudhury S, Panda P, Sahoo L, Panda SK (2013) Reactive oxygen species signaling in plants under abiotic stress. Plant Signal Behav 8(4):e23681

Choudhury FK, Rivero RM, Blumwald E, Mittler R (2017) Reactive oxygen species, abiotic stress and stress combination. Plant J 90:856–867

Colebrook EH, Thomas SG, Phillips AL, Hedden P (2014) The role of gibberellin signalling in plant responses to abiotic stress. J Exp Biol 217:67–75

Cosio C, Dunand C (2009) Specific functions of individual class III peroxidase genes. J Exp Bot 60:391–440

Delaunay A, Isnard AD, Toledano MB (2000) H_2O_2 sensing through oxidation of the Yap1 transcription factor. EMBO J 19:5157–5166

Demidchik V (2010) Reactive oxygen species, oxidative stress and plant ion channels. In: Ion channels and plant stress responses. Springer, pp 207–232

Demidchik V (2018) ROS-activated ion channels in plants: biophysical characteristics, physiological functions and molecular nature. Int J Mol Sci 19:1263

Demidchik V, Shabala S (2018) Mechanisms of cytosolic calcium elevation in plants: The role of ion channels, calcium extrusion systems and NADPH oxidase-mediated 'ROS-Ca2+ Hub'. Funct Plant Biol 45:9–27

Demidchik V, Bowen HC, Maathuis FJ, Shabala SN, Tester MA, White PJ, Davies JM (2002) *Arabidopsis thaliana* root non-selective cation channels mediate calcium uptake and are involved in growth. Plant J 32:799–808

Diaz-Vivancos P, Barba-Espín G, Hernández JA (2013) Elucidating hormonal/ROS networks during seed germination: insights and perspectives. Plant Cell Rep 32:1491–1502

Duan Q, Kita D, Li C, Cheung AY, Wu HM (2010) FERONIA receptor-like kinase regulates RHO GTPase signaling of root hair development. Proc Natl Acad Sci USA 107:17821–17826

Fratelli M, Gianazza E, Ghezzi P (2004) Redox proteomics: identification and functional role of glutathionylated proteins. Expert Rev Proteomics 1:365–376

Fukao T, Xu K, Ronald PC, Bailey-Serres J (2006) A variable cluster of ethylene response factor-like genes regulates metabolic and developmental acclimation responses to submergence in rice. Plant Cell 18:2021–2034

Fukao T, Yeung E, Bailey-Serres J (2011) The submergence tolerance regulator SUB1A mediates crosstalk between submergence and drought tolerance in rice. Plant Cell 23:412–427

Gadjev I (2006) Transcriptomic footprints disclose specificity of reactive oxygen species signaling in *Arabidopsis*. Plant Physiol 141:436–445

Gallogly MM, Mieyal JJ (2007) Mechanisms of reversible protein glutathionylation in redox signaling and oxidative stress. Curr Opin Pharmacol 7:381–391

Gilroy S, Suzuki N, Miller G, Choi WG, Toyota M, Devireddy AR, Mittler R (2014) A tidal wave of signals: calcium and ROS at the forefront of rapid systemic signaling. Trends Plant Sci 19:623–630

Giraud E, Ho LH, Clifton R et al (2008) The absence of ALTERNATIVE OXIDASE1a in Arabidopsis results in acute sensitivity to combined light and drought stress. Plant Physiol 147:595–610

Gustavsson N, Kokke BP, Harndahl U, Silow M, Bechtold U, Poghosyan Z, Murphy D, Boelens WC, Sundby C (2002) A peptide methionine sulfoxide reductase highly expressed in photosynthetic tissue in Arabidopsis thaliana can protect the chaperone-like activity of a chloroplast-localized small heat shock protein. Plant J 29:545–553

Hardin SC, Larue CT, Oh MH, Jain V, Huber SC (2009) Coupling oxidative signals to protein phosphorylation via methionine oxidation in Arabidopsis. Biochem J 422:305–312

Hasegawa PM, Bressan RA, Zhu JK, Bohnert HJ (2000) Plant cellular and molecular responses to high salinity. Annu Rev Plant Biol 51:463–499

Jing S, Zhou X, Song Y, Yu D (2009) Heterologous expression of OsWRKY23 gene enhances pathogen defense and dark-induced leaf senescence in Arabidopsis. Plant Growth Regul 58:181–190

Joo JH, Bae YS, Lee JS (2001) Role of auxin-induced reactive oxygen species in root gravitropism. Plant Physiol 126:1055–1060

Julkowska MM, Testerink C (2015) Tuning plant signaling and growth to survive salt. Trends Plant Sci 20:586–594

Karan R, Subudhi PK (2012) A stress inducible SUMO conjugating enzyme gene (SaSce9) from a grass halophyte Spartina alterniflora enhances salinity and drought stress tolerance in Arabidopsis. BMC Plant Biol 12:187

Kasson TM, Barry BA (2012) Reactive oxygen and oxidative stress: N-formyl kynurenine in photosystem II and non-photosynthetic proteins. Photosynth Res 114:97–110

Kazan K (2006) Negative regulation of defence and stress genes by EAR-motif-containing repressors. Trends Plant Sci 11:109–112

Keles Y, Oncel I (2002) Response of antioxidative defense system to temperature and water stress combinations in wheat seedlings. Plant Sci 163:783–790

Kerepesi I, Galiba G (2000) Osmotic and salt stress-induced alteration in soluble carbohydrate content in wheat seedlings. Crop Sci 40:482–487

Kleine-Vehn J, Leitner J, Zwiewka M, Sauer M, Abas L, Luschnig C, Friml J (2008) Differential degradation of PIN2 auxin efflux carrier by retromer-dependent vacuolar targeting. Proc Natl Acad Sci USA 105:17812–17817

Kourie JI (1998) Interaction of reactive oxygen species with ion transport mechanisms. Am J Physiol-Cell Physiol 275:1–24

Kovtun Y, Chiu WL, Tena G, Sheen J (2000) Functional analysis of oxidative stress-activated mitogen-activated protein kinase cascade in plants. Proc Natl Acad Sci USA 97:2940–2945

Kurusu T, Kuchitsu K, Tada Y (2015) Plant signaling networks involving Ca2+ and Rboh/Nox-mediated ROS production under salinity stress. Front Plant Sci 6:427

Lecourieux D, Ranjeva R, Pugin A (2006) Calcium in plant defence-signalling pathways. New Phytol 171:249–269

Liu R, Wang L, Tanveer M, Song J (2018) Seed heteromorphism: an important adaptation of halophytes for habitat heterogeneity. Front Plant Sci 9:1515. https://doi.org/10.3389/fpls.2018.01515

Liu Y, Yu Y, Sun J, Cao Q, Tang Z, Liu M, Sun J (2019) Root-zone-specific sensitivity of K+-and Ca2+-permeable channels to H_2O_2 determines ion homeostasis in salinized diploid and hexaploid *Ipomoea trifida*. J Exp Bot 70:1389–1405

Marinho HS, Real C, Cyrne L, Soares H, Antunes F (2014) Hydrogen peroxide sensing, signaling and regulation of transcription factors. Redox Biol 2:35–62

Martinez V, Mestre TC, Rubio F, Girones-Vilaplana A, Moreno DA, Mittler R, Rivero RM (2016) Accumulation of flavonols over hydroxycinnamic acids favors oxidative damage protection under abiotic stress. Front Plant Sci 7:838

Miller G, Shulaev V, Mittler R (2008) Reactive oxygen signaling and abiotic stress. Physiol Plant 133:481–489

Miller GAD, Suzuki N, Ciftci-Yilmaz S, Mittler R (2010) Reactive oxygen species homeostasis and signalling during drought and salinity stresses. Plant Cell Environ 33:453–467

Mittler R (2017) ROS are good. Trends Plant Sci 22:11–19

Mittler R, Vanderauwera S, Gollery M, Van Breusegem F (2004) Reactive oxygen gene network of plants. Trends Plant Sci 9:490–498

Mittler R, Kim Y, Song L, Coutu J, Coutu A, Ciftci-Yilmaz S, Zhu JK (2006) Gain-and loss-of-function mutations in Zat10 enhance the tolerance of plants to abiotic stress. FEBS Lett 580:6537–6542

Mori IC, Schroeder JI (2004) Reactive oxygen species activation of plant Ca2+ channels. A signaling mechanism in polar growth, hormone transduction, stress signaling, and hypothetically mechanotransduction. Plant Physiol 135:702–708

Netondo GW, Onyango JC, Beck E (2004) Sorghum and salinity: II. Gas exchange and chlorophyll fluorescence of sorghum under salt stress. Crop Sci 44:806

Nie WF, Wang MM, Xia XJ, Zhou YH, Shi K, Chen Z, Yu JQ (2013) Silencing of tomato RBOH1 and MPK2 abolishes brassinosteroid induced H_2O_2 generation and stress tolerance. Plant Cell Environ 36:789–803

Pei ZM, Murata Y, Benning G, Thomine S, Klüsener B, Allen GJ, Grill E, Schroeder JI (2000) Calcium channels activated by hydrogen peroxide mediate abscisic acid signalling in guard cells. Nature 406:731

Perez IB, Brown PJ (2014) The role of ROS signaling in cross-tolerance: from model to crop. Front Plant Sci 5:754

Puyaubert J, Fares A, Reze N, Peltier JB, Baudouin E (2014) Identification of endogenously S-nitrosylated proteins in Arabidopsis plantlets: effect of cold stress on cysteine nitrosylation level. Plant Sci 215:150–156

Riechmann JL, Heard J, Martin G, Reuber L, Jiang C, Keddie J et al (2000) *Arabidopsis* transcription factors: genome-wide comparative analysis among eukaryotes. Science 290:2105–2110

Rodrigo-Moreno A, Andrés-Colás N, Poschenrieder C, Gunse B, Penarrubia L, Shabala S (2013) Calcium-and potassium-permeable plasma membrane transporters are activated by copper in Arabidopsis root tips: linking copper transport with cytosolic hydroxyl radical production. Plant Cell Environ 36:844–855

Ronsein GE, Oliveira MC, Miyamoto S, Medeiros MH, Di Mascio P (2008) Tryptophan oxidation by singlet molecular oxygen [O2(1Deltag)]: mechanistic studies using 18O-labeled hydroperoxides, mass spectrometry, and light emission measurements. Chem Res Toxicol 21:1271–1283

Sakamoto M, Munemura I, Tomita R, Kobayashi K (2008) Involvement of hydrogen peroxide in leaf abscission signaling, revealed by analysis with an in vitro abscission system in Capsicum plants. Plant J 56:13–27

Schafer FQ, Buettner GR (2001) Redox environment of the cell as viewed through the redox state of the glutathione disulfide/glutathione couple. Free Radical Biol Med 30:1191–1212

Scheibe R, Backhausen JE, Emmerlich V, Holtgrefe S (2005) Strategies to maintain redox homeostasis during photosynthesis under changing conditions. J Exp Bot 56:1481–1489

Scholz-Starke J, Gambale F, Carpaneto A (2005) Modulation of plant ion channels by oxidizing and reducing agents. Arch Biochem Biophys 434:43–50

Schopfer P, Plachy C, Frahry G (2001) Release of reactive oxygen intermediates (superoxide radicals, hydrogen peroxide, and hydroxyl radicals) and peroxidase in germinating radish seeds controlled by light, gibberellin, and abscisic acid. Plant Physiol 125:1591–1602

Sewelam N, Kazan K, Schenk PM (2016) Global plant stress signaling: reactive oxygen species at the cross-road. Front Plant Sci 7:187

Shabala S (2009) Salinity and programmed cell death: unravelling mechanisms for ion specific signalling. J Exp Bot 60:709–712

Shabala S, Pottosin I (2014) Regulation of potassium transport in plants under hostile conditions: implications for abiotic and biotic stress tolerance. Physiol Plant 151:257–279

Storz G, Imlayt JA (1999) Oxidative stress. Curr Opin Microbiol 2:188–194

Swanson S, Gilroy S (2010) ROS in plant development. Physiol Plant 138:384–392

Swarbreck SM, Colaço R, Davies JM (2013) Plant calcium-permeable channels. Plant Physiol 163:514–522

Tamás L, Mistrík I, Huttová J, Halušková LU, Valentovičová K, Zelinová V (2010) Role of reactive oxygen species-generating enzymes and hydrogen peroxide during cadmium, mercury and osmotic stresses in barley root tip. Planta 231:221

Tanou G, Filippou P, Belghazi M, Job D, Diamantidis G, Fotopoulos V, Molassiotis A (2012) Oxidative and nitrosative-based signaling and associated post-translational modifications orchestrate the acclimation of citrus plants to salinity stress. Plant J 72:585–599

Tanveer M, Shabala S (2018) Targeting redox regulatory mechanisms for salinity stress tolerance in crops. In: Salinity responses and tolerance in plants, vol 1. Springer, Cham, pp 213–234

Tanveer M, Shah AN (2017) An insight into salt stress tolerance mechanisms of *Chenopodium album*. Environ Sci Pollut Res 24:16531–16535

Tanveer M, Shahzad B, Sharma A, Biju S, Bhardwaj R (2018a) 24-Epibrassinolide; an active brassinolide and its role in salt stress tolerance in plants: a review. Plant Physiol Biochem 130:69–79

Tanveer M, Shahzad B, Sharma A, Khan EA (2018b) 24-Epibrassinolide application in plants: an implication for improving drought stress tolerance in plants. Plant Physiol Biochem 135:295–303
Tavares CP, Vernal J, Delena RA, Lamattina L, Cassia R, Terenzi H (2014) S-nitrosylation influences the structure and DNA binding activity of AtMYB30 transcription factor from *Arabidopsis thaliana*. Biochim Biophys Acta (BBA)-Proteins Proteomics 1844:810–817
Tognetti VB, Van Aken O, Morreel K, Vandenbroucke K, van de Cotte B, De Clercq I, Chiwocha S, Fenske R, Prinsen E, Boerjan W, Genty B, Stubbs KA, Inzé D, Van Breusegem F (2010) Perturbation of indole-3-butyric acid homeostasis by the UDP-glucosyltransferase UGT74E2 modulates Arabidopsis architecture and water stress tolerance. Plant Cell 22:2660–2679
Tognetti VB, Mühlenbock P, Van Breusegem F (2012) Stress homeostasis–the redox and auxin perspective. Plant Cell Environ 35:321–333
Tsukagoshi H (2016) Control of root growth and development by reactive oxygen species. Curr Opin Plant Biol 29:57–63
Uchida A, Jagendorf AT, Hibino T, Takabe T, Takabe T (2002) Effects of hydrogen peroxide and nitric oxide on both salt and heat stress tolerance in rice. Plant Sci 163:515–523
Vanderauwera S, Vandenbroucke K, Inzé A, van de Cotte B, Mühlenbock P, De Rycke R, Naouar N, Van Gaever T, Van Montagu MCE, Van Breusegem F (2012) AtWRKY15 perturbation abolishes the mitochondrial stress response that steers osmotic stress tolerance in Arabidopsis. Proc Natl Acad Sci USA 109:20113–20118
Vile D, Pervent M, Belluau M, Vasseur F, Bresson J, Muller B, Granier C, Simonneau T (2012) Arabidopsis growth under prolonged high temperature and water deficit: independent or interactive effects? Plant Cell Environ 35:702–718
Vuleta A, Manitasevic Jovanovic S, Tucic B (2015) How do plants cope with oxidative stress in nature? A study on the dwarf bearded iris (*Iris pumila*). Acta Physiol Plant 37:1711–1719
Wang C, Yang A, Yin H, Zhang J (2008) Influence of water stress on endogenous hormone contents and cell damage of maize seedlings. J Integr Plant Biol 50:427–434
Xia XJ, Wang YJ, Zhou YH, Tao Y, Mao WH, Shi K, Asami T, Chen Z, Yu JQ (2009) Reactive oxygen species are involved in brassinosteroid-induced stress tolerance in cucumber. Plant Physiol 150:801–814
Xia XJ, Zhou YH, Ding J, Shi K, Asami T, Chen Z, Yu JQ (2011) Induction of systemic stress tolerance by brassinosteroid in *Cucumis sativus*. New Phytol 191:706–720
Xie YJ, Xu S, Han B, Wu MZ, Yuan XX, Han Y, Shen WB (2011) Evidence of Arabidopsis salt acclimation induced by up-regulation of HY1 and the regulatory role of RbohD-derived reactive oxygen species synthesis. Plant J 66:280–292
Zhang W, Jeon BW, Assmann SM (2011) Heterotrimeric G-protein regulation of ROS signaling and calcium currents in *Arabidopsis* guard cells. J Exp Bot 62:2371–2379
Zhao Z, Chen G, Zhang C (2001) Interaction between reactive oxygen species and nitric oxide in drought-induced abscisic acid synthesis in root tips of wheat seedlings. Funct Plant Biol 28:1055–1061

Phytohormone Signaling in Response to Drought

Geetha Govind, Vokkaliga T. Harshavardhan and Chwan-Yang Hong

Abstract Under natural conditions, plants are often exposed to both biotic (pathogens, wounding, insects) and abiotic stresses (drought, salinity, heat, high or low temperature, UV light, heavy metals, ozone) simultaneously. Among the various abiotic stresses, drought along with salinity and alkalinity confers osmotic stress and causes major yield losses, with water scarcity (drought) contributing to maximum losses. On perceiving stress, plants respond by activating signal transduction cascades that interact with other pathways (phytohormones) for an optimal response for survival. Hormones also control the response to environmental factors, but do so by influencing each other at a biosynthesis or signaling level rather than by discrete pathways. This coordination leads to a network of signal transduction that integrates various inputs leading to a comprehensive output resulting in a physiological adaptation to stress. The points of hormones interacting with one another are where crosstalk occurs, which helps in switching the response from one pathway to another. Published literature provide comprehensive information on the response of plants to drought and the role of hormones in regulating the response to drought from experiments directly imposing drought stress (dehydration stress-withholding water) or osmotic stress (turgor loss due to extracellular solutes) and experiments with mutants and others involving the functions of hormones and other signals. Changes in gene expression induced by drought overlap with gene expression regulated by hormones. This chapter presents the signal transduction pathways of hormones under drought stress and how they cross talk with each other in regulating the drought stress response in plants.

G. Govind
Department of Crop Physiology, College of Agriculture, Karekere, Hassan, University of Agricultural Sciences, Bangalore, Bengaluru 65, India
e-mail: geetha.govind@uasbangalore.edu.in

V. T. Harshavardhan · C.-Y. Hong (✉)
Department of Agricultural Chemistry, College of Bioresources and Agriculture, National Taiwan University, Taipei 10617, Taiwan
e-mail: cyhong@ntu.edu.tw

M. Hasanuzzaman and M. Tanveer (eds.), *Salt and Drought Stress Tolerance in Plants*, Signaling and Communication in Plants, https://doi.org/10.1007/978-3-030-40277-8_12

1 Introduction

A number of environmental stresses (abiotic—drought, salinity, high and low temperature—and biotic) affect plant growth and productivity. Under stress, both hormonal and redox processes are imbalanced. Plants alter their gene and protein expression in response to signal cascades that are transduced when plants perceive stress (Munns and Tester 2008; Mittler et al. 2011; Munné-Bosch et al. 2013), which leads to a change in metabolic and physiological responses that help plants survive under stressful situations. Phytohormones, that under optimal growth conditions play a critical role in plant growth and development, also play a central role in integrating signal transduction occurring under stressful conditions. Phytohormones help in linking and reprogramming the stress-adaptive signaling cascades (Ma et al. 2006; Golldack et al. 2014).

The growth regulators abscisic acid (ABA) and ethylene (ET) play a vital role in signaling drought stress and conferring drought tolerance. Jasmonic acid (JA) and salicylic acid (SA), known for their role in biotic stress signaling and tolerance, also play a role in abiotic stress signaling and tolerance. In addition, the growth regulators cytokinins (CKs), auxin (AUX), gibberellic acid (GA), brassinosteroids (BRs), and strigolactones (SLs) play a role in abiotic stress tolerance (Zhao and Schaller 2004; Cheng et al. 2013; Wu et al. 2007; Cela et al. 2011; Jayakannan et al. 2015; Magome et al. 2008; Wu et al. 2014a, b; He et al. 2005; Divi et al. 2010). Most of the time, these phytohormones are involved in crosstalk with other phytohormones and other signaling cascades.

Among the abiotic stress, drought stress negatively affects crop growth and productivity (Anjum et al. 2017). Plants exposed to drought stress adapt a drought-escape, avoidance or tolerance mechanism to overcome the stress (Hussain et al. 2018). Under drought escape, plants maintain a short life cycle or adapt developmental plasticity. Under drought avoidance, plants increase their water uptake and decrease water loss such that they avoid physiological drought. However, under drought tolerance, cells tolerate drought by regulating mechanisms such as osmoprotection, antioxidative capacity, and tolerance. Plants maintain a balance among the different mechanisms.

Drought tolerance is a complex trait that operates at morphological, physiological and molecular levels (Tanveer et al. 2018). Traits such as earliness; stomata shape, size and structure; leaf number, orientation, and rolling; the presence of a cuticle; wax on leaf and stem; growth habit; and rooting pattern contribute to morphological characters associated with drought resistance and tolerance. Traits such as photosynthetic rate, transpiration rate, osmotic adjustment, and stomatal closure contribute to physiological characters. Regulation of these traits at the molecular level by regulating gene expression and protein synthesis contribute to molecular characteristics.

Functional genomic approaches (gene expression analysis) have aided in understanding the molecular mechanism of stress tolerance. A number of genes are regulated under stress conditions and can be grouped broadly in two categories based on function. One group involves the regulation of genes encoding proteins involved in cellular homeostasis and stress protection such as osmoprotectants, water channels, antioxidative enzymes, metabolic enzymes, and lipid-transfer proteins. The other group involves genes that code for kinases, transcription factors (TFs) and proteins or products that mediate stress signal transduction and result in the expression of stress-responsive genes. Among them are also genes that are associated with signaling pathway(s) of various phytohormones. Auxins, GA, CKs, ABA, ET, SA, JA, BR, and SLs are the major hormones produced by plants. Although all play a role in the stress response, the hormones ABA, SA, JA, and ET play a major role in the plant response to biotic and abiotic stresses (Bari and Jones 2009; Nakashima and Yamaguchi-Shinozaki 2013). On sensing stress (abiotic or biotic stress), several signaling pathways are initiated in plant cells. Some lead to the induction and alteration of intercellular levels of Ca^{2+}, reactive oxygen species (ROS), and inositol phosphate and activation of the kinase cascade.

2 Abscisic Acid (ABA)

Abscisic acid plays major role in the plant response to osmotic stress induced by cold, high salinity and drought. It is an endogenous messenger of the plant's water status. Plant hormone, ABA is typically synthesized in response to stress due to enhanced expression of its biosynthetic genes (Swamy and Smith 1999; Zhu 2002). Treating plants with ABA mimics exposure to stress. Abscisic acid plays a critical role in response to various stress signals. It triggers reprogramming of the transcriptome and stomatal closure to control transpirational water loss (Christmann et al. 2007; Cutler et al. 2010; Raghavendra et al. 2010).

Signaling under stress involves various proteins leading to regulation of stress-responsive gene expression. Under stress, signaling pathways occurring under stress (osmotic or drought stress) can be broadly classified as ABA-dependent- or independent. TFs that bind to ABA-responsive element (ABRE) *cis*-elements play a major role in ABA-dependent gene expression and those that bind to the dehydration-responsive element/C-repeat (DRE/CRT) *cis* element are involved in ABA-independent gene expression in response to drought stress (Fujita et al. 2011; Qin et al. 2011).

Abscisic acid signaling has been extensively studied, and various components have been identified. The proteins PYRABACTIN RESISTANCE/PYR1-LIKE/REGULATORY COMPONENTS OF ABA RECEPTORS (PYR/PYL/RCARs) function as soluble ABA receptors, as demonstrated by chemical genetic and biochemical approaches. Besides their identification in *Arabidopsis*, they have been identified in rice (*Oryza sativa*), maize (*Zea mays*), sorghum (*Sorghum bicolor*), soybean (*Glycine max*), grapevine (*Vitis vinifera*), citrus (*Citrus sinensis*), and tomato (*Solanum lycopersicum*) (Klingler et al. 2010; Sun et al. 2011; Boneh et al. 2012; Kim et al. 2011; Romero et al. 2012). GPCR-type G protein 1 (GTG1)/GTG2 are membrane proteins with homology to G-protein coupled receptors (GPCRs). ChlH (or GENOMES UNCOUPLED 5 [GUN5]) and GTG1/GTG2 may be ABA receptors. However, neither ChlH/GUN5 nor GTG1/GTG2 can bind the biologically active ABA isomer (Shen et al. 2006; Pandey et al. 2009).

Forward genetic screening using mutants resistant to the ABA agonist pyrabactin and a search for biochemical interacting partners of ABA insensitive 1 (ABI1) and ABI2 led to the identification of PYR1 and RCAR1 among the 14 genes that code for PYR/PYL/RCARs (Ma et al. 2009; Nishimura et al. 2009). On activation, the receptor interacts with partners such as type 2C protein phosphatase (PP2C) and sucrose nonfermenting 1-related protein kinase 2 (SnRK2) complex (Umezawa et al. 2009, 2010). In the absence of ABA, PP2C binds to SnRK2 and mediates its dephosphorylation. Binding of ABA to its receptor (PYR/PYL/RCARs) fits well into the active site of PP2C enzyme, thereby inhibiting the phosphatase activity of PP2C. This situation results in phosphorylated SnRK2s, which can activate relevant TFs, and in turn leads to activation of an ABA-induced transcriptional cascade (Kline et al. 2010). Therefore, an increase in endogenous ABA due to environmental and/or developmental cues leads to sequestration of PP2C and ABA-induced release or activation of SnRK2.6. Autophosphorylation of SnRK2s then phosphorylates downstream proteins such as ABA-responsive element binding proteins/ABA-responsive element binding factor/basic-leucine zipper (AREB/ABF/bZIP) proteins and anion channels (Cutler et al. 2010; Klingler et al. 2010; Umezawa et al. 2010; Seiler et al. 2014). SnRK2s can be further regulated by a positive regulator, ARK (B3-MAPKKKs), and a negative regulator, casein kinase 2 (CK2) (Saruhashi et al. 2015; Meggio and Pinna 2003; Vilela et al. 2015).

The *Arabidopsis* genome encodes 112 phosphatases, of which 76 code for PP2Cs and at least 6 are involved in ABA signaling with well-characterized ABI1/2 and homolog of ABI1 (HAB1). The mutants *abi1* and *abi2* display ABA hypersensitivity, insensitive phenotypes and reduced seedling growth, seed dormancy, drought tolerance, and stomatal regulation. Other members of PP2Cs include ABA-hypersensitive germination 1 (AHG1), AHG3/AtPP2CA from *Arabidopsis*, and OsPP108 from rice (Leung et al. 1994, 1997; Meyer et al. 1994; Saez et al. 2004; Singh et al. 2015).

The protein kinase SnRK2 family is a group of plant-specific serine/threonine kinases that participate in cellular responses to drought and dehydration (Hrabak et al. 2003; Saruhashi et al. 2015). Depending on the affinity toward ABA, the SnRK2 family is divided into three subgroups. Subgroup I does not respond to ABA, subgroup II includes weak respondents, and subgroup III includes active respondents of ABA that are key players of ABA-dependent gene expression (Hrabak et al. 2003; Boudsocq et al. 2007). *Arabidopsis* has 10 SnRK2 members (SnRK2.1-2.10); ABA activates five SnRK2s (SnRK-2.2, -2.3, -2.6, -2.7, -2.8). Mutation in three SnRKs (*snrk2.2/2.3/2.6*) blocks all main ABA responses (Fujii and Zhu 2009; Fujita et al. 2009; Waadt et al. 2015). In addition, all SnRK2s are involved in ABA and stress signaling except SnRK2.9, which is activated by osmotic stress (Boudsocq et al. 2007; Furihata et al. 2006; Yoshida et al. 2006). ABA-activated protein kinase (AAPK), the ortholog of Arabidopsis open stomata 1 (OST1)/SnRK2.6, was the first protein kinase identified (from *Vicia faba*).

SnRK2 kinases phosphorylate the TFs ABFs/AREBs and ABI5, both in vitro and in vivo, involved in ABA responsive gene expression (Kobayashi et al. 2005; Furihata et al. 2006; Sirichandra et al. 2010; Nakashima et al. 2009). In addition, SnRK2 (OST1/SNRK2.6) phosphorylates plasma membrane proteins anion channel slow anion channel-associated 1 (SLAC1), potassium channel protein KAT1 and NADPH oxidase AtrbohF involved in regulating stomatal aperture (Geiger et al. 2009; Sato et al. 2009; Sirichandra et al. 2009). Phosphoproteomic and mutation study approach has identified ascorbate peroxidase, Ca^{2+}/H^{+} antiporter regulator protein, G-protein beta subunit-like protein, glyoxysomal malate dehydrogenase, manganese superoxide dismutase, triose phosphate isomerase and others as potential SnRK2 targets (He and Li 2008; Umezawa et al. 2013).

TFs that play a significant role in ABA-regulated gene expression include ABFs/AREBs, ABI5, myeloblastosis (MYB), myelocytomatosis (MYC), NAM: no apical meristem (NAC), *Arabidopsis* transcription activation factor (ATAF), cup-shaped cotyledon (CUC), and ethylene response factor (ERF). ABFs/AREBs are regulated by SnRK2-mediated phosphorylation (Fujita et al. 2011; Nakashima et al. 2012; Rushton et al. 2012). ABFs/AREBs and ABI5 encode proteins belonging to the bZIP family regulate ABA and stress responses by binding to ABRE in promoters of target genes (Choi et al. 2000; Uno et al. 2000; Lopez-Molina et al. 2001). AREB1/ABF2, AREB2/ABF4, ABF3 and ABRE binding protein 9 (ABP9) are involved in signaling drought stress (Riechmann et al. 2000; Yoshida et al. 2010). MYC and MYB, belonging to the basic helix loop helix (bHLH) subfamily, also regulate ABA signaling by regulating stress-inducible genes. AtMYB60, AtMYB44, AtMYB15, and AtMYC2 mediate drought stress signaling (Abe et al. 1997, 2003, Jaradat et al. 2013). In addition, NAC (AtNAC) and ERF (AtERF7)-type TFs are involved in regulating ABA signal transduction under drought stress (Song et al. 2005; Zhang and Gan 2012).

3 Ethylene (ET)

This simple two-carbon molecule plays a major role in regulating biotic and abiotic stress signaling (Morgan and Drew 1997; van Loon et al. 2006). Biosynthesis of ET is simple, consisting of three steps. Methionine is used as precursor and is converted to S-adenosyl methionine (SAM), which is converted to ET by the action of ACC synthase (ACS) and ACC oxidase (ACO). ACS is regulated by drought stress, thereby regulating ET levels under drought stress.

Endoplasmic reticulum membrane-localized proteins perceive ET, are negative regulators of ET signaling, and play a role as ET receptors. Arabidopsis has five ET receptors: ET response 1 (ETR1) and ET response sensor 1 (ERS1), belonging to subfamily I, and ETR2, ERS2, and ET insensitive 4 (EIN4), belonging to subfamily II (Chen et al. 2005; Lacey and Binder 2014). Similarly, rice has five receptors: OsERS1 and OsERS2, belonging to subfamily I, and OsETR2, OsETR3, and OsETR4, belonging to subfamily II (Watanabe et al. 2004; Yau et al. 2004). Constitutive triple response 1 (CTR1), also a membrane protein and central transducer, is a negative regulator of ET signaling and acts downstream of the receptor, which is a Raf-like Ser/Thr protein kinase. Downstream of this is ETI2, also localized to the endoplasmic reticulum membrane that activates EIN3 and EIN3-like 1 (EIL1), which function as TFs inducing the expression of ERF1 and inducing or repressing that of many ET response genes (Kendrick and Chang 2008; Stepanova and Alonso 2009).

Therefore, in the absence of ET, CTR1 kinase activity is induced, which phosphorylates EIN2 and prevents its localization to the nucleus (Kieber et al. 1993; Bisson and Groth 2010; Ju et al. 2012). In addition, two F-box proteins—ETI2-targeting protein 1 (ETP1) and ETP2—reduce levels of EIN2 protein (Qiao et al. 2009). EIN2 is localized to the nucleus on its dephosphorylation and cleavage of its C-terminus, in the presence of ET, and it regulates activation of EIN3/EIL1 leading to the ET response (Ju et al. 2012). Similar to EIN2, two F-box proteins—ETI3 binding F-box protein 1 (EBF1) and EBF2—that are downregulated by ET, reduce EIN3/EIL1 protein levels in the absence of ET. Among the ET signal transduction pathway, ERFs are key players of abiotic-stress tolerance induced by drought stress. ERFs are also induced by various other abiotic stresses. ERFs belong to the APETALA2/ERF TF family. The expression of ERFs increases under drought stress, and many ERFs bind to DREs and regulate tolerance to abiotic stress.

ERF induces the expression of P5CS, germin-like protein 9, osmotin 34, similar to RCD one 5 (SRO5), responsive to dessication 29B (RD29B), early response to dehydration (ERD7) and RD20, leading to drought tolerance (Cheng et al. 2013). In addition, OsDERF1 directly binds promoters of the ERF repressors *OsERF3* and *OsAP2-39* that suppress ET synthesis and regulate drought tolerance (Wan et al. 2011; Zhang et al. 2013). Although most ERFs are transcriptional activators, two (ERF8 and ERF9) are transcriptional repressors. ERF8 is an inhibitor of cell division and leaf growth and thereby regulates the drought stress response (Thao et al. 2015; Dubois et al. 2013, 2018).

4 Jasmonic Acid (JA)

The JAs collectively include JA and its conjugates methyl jasmonate (MeJA) and jasmonyl-isoleucine (JA-Ile) (Ghasemi et al. 2014). They are lipid-derived oxylipins, synthesized from linolenic acid via the octadecanoid pathway as a result of oxidation by lipoxygenases (Feussner and Wasternack 2002).

Jasmonic acids are well known for their role in developmental processes, secondary metabolite synthesis and the plant defense response against pathogens and wounding (McConn et al. 1997; Reymond and Farmer 1998; Wasternack and Hause 2002; Lorenzo and Solano 2005; Schommer et al. 2008). Despite reports of the role of JA in the abiotic stress response, its role in drought stress is little known (Maksymiec et al. 2005; Brossa et al. 2011; Dong et al. 2013; Qiu et al. 2014; Zhao et al. 2013; Clarke et al. 2009). Jasmonic acid basically imparts drought tolerance by modulating stomatal closure and minimizing transpiration water loss. In addition, MeJA increases root hydraulic conductivity and helps plants uptake more water under drought stress (Tanaka et al. 2005; Munemasa et al. 2007; Daszkowska-Golec and Szarejko 2013; Sánchez-Romera et al. 2014). In addition, JAs play an important role in signaling the drought-induced antioxidant response by enhancing the expression and activity of the antioxidant enzymes superoxide dismutase, peroxidase, and catalase (Nafie et al. 2011; Anjum et al. 2011). However, JA biosynthesis, crosstalk and signal transduction is more or less similar between biotic and abiotic stress, with unique aspects being identified for abiotic stress response.

An F-box protein coronatine insensitive 1 (COI1), jasmonate ZIM domain (JAZ) proteins, multi-component E3 ubiquitin ligase-Skp/Cullin/F-box complex (SCF-COI1) and 26S proteasome are key players in JA signaling (Thines et al. 2007; Deshaies 1999; Turner et al. 2002). JA-Ile functions as a ligand and promotes the assembly of COI1 and JAZ proteins (Thines et al. 2007). JAZ proteins are repressors or negative regulators of JA signaling that are targeted for degradation by the SCF^{COI1}-dependent 26S proteasome pathway. In addition, JAZ proteins interact with co-repressor Topless (TPL) and TPL-related proteins (TPRs) via an adaptor protein, novel interactor of JAZ (NINJA), in JA signaling (Pauwels et al. 2010). Hence, JA-mediated destruction of JAZ transcription repressors leads to the activation of downstream genes (Staswick 2008; Thines et al. 2007; Pauwels and Goossens 2011). JAZ proteins target and regulate TFs belonging to bHLH factors (AtMYC2, OsBHLH1, RERJ1, MYC3, MYC4, Glabra3-GL3, enhancer of glabra 3 1 [EGL1] and transparent testa [TT8]). A few are regulated under drought stress (RERJ1, OsbHLH148), and their overexpression results in enhanced drought tolerance (Kiribuchi et al. 2005; Seo et al. 2011). JAZ proteins interact with the Apetala 2 group of TFs—target of eat 1 (TOE1) and TOE2—and repress transcription (Zhai et al. 2015). An intermediate of JA biosynthesis, 12-OPDA, also plays a role in JA signaling and has a major role in drought response (Savchenko et al. 2014).

5 Salicylic Acid (SA)

Similar to JA, the phytohormone SA plays an important role in response to biotic stresses, pathogenesis, plant growth and development, although antagonizing gene induction by JA. In addition, it participates in signaling abiotic stress responses (drought, salinity, high and low temperature, ozone, UV light and heavy metals) (Rivas-San Vicente and Plasencia 2011; Hara et al. 2012). However, its utility depends on its concentration: low concentration enhancing antioxidant capacity and high concentration inducing oxidative stress and leading to cell death (Hara et al. 2012).

The synthesis of endogenous SA levels is increased in plants exposed to drought stress, and plants susceptible to drought show reduced SA level. SA probably protects plants under stress by regulating the induction of SA-inducible PR genes (Munné-Bosch and Penuelas 2003; Sawada et al. 2006; Miura et al. 2013).

In plants, SA is synthesized in plastids via two distinct pathways: the isochorismate (IC) pathway, the major pathway, and the phenylalanine ammonia-lyase (PAL) pathway. Both pathways begin with chorismic acid, an end product of the shikimate pathway (Wildermuth et al. 2001; Catinot et al. 2008). IC synthase (ICS), responsible for converting chorismic acid to IC, is regulated by biotic stress and also drought stress (Nawrath and Métraux 1999; Hunter et al. 2013). Mutants (*adr1, myb96-1d, siz1, acd6*, and *cpr5*) that accumulated endogenous SA showed both SA-dependent disease resistance and drought tolerance (Seo and Park 2010; Miura et al. 2013).

Well-known systemic acquired resistance is induced by SA and leads to an immune response against pathogens via transcriptional reprogramming (Durrant and Dong 2004). Under drought stress, plants are protected against oxidative stress induced by drought stress by the induction of stress proteins such as glutathione S-transferases, ascorbate peroxidase (APX), and 2-cysteine peroxiredoxin by enhanced SA levels (Kang et al. 2012). Exogenous application of SA and high endogenous SA level promote stomatal closure during water deficit to conserve water, which leads to reduced photosynthesis likely because of ROS induced by SA (Senaratna et al. 2000; Korkmaz et al. 2007).

Jasmonic acid binds to effector proteins such as catalase, APX, methyl SA esterase, and carbonic anhydrase (Sanchez-Casas and Klessig 1994; Du and Klessig 1997; Slaymaker et al. 2002; Forouhar et al. 2005). However, mutant studies predicted that non-expressor of PR genes 1 (NPR1) was an SA receptor candidate (Delaney et al. 1995; Cao et al. 1997). NPR1 functions as a positive regulator of the SA-mediated defense response. In addition, NPR3 and NPR4, paralogs of NPR1, are SA receptors (Fu et al. 2012). They have a direct role in defense gene expression and affect disease resistance (Fu et al. 2012; Wu et al. 2012). They interact with cullin 3 (CUL3) ubiquitin E3 ligase and recruit NPR1 for proteasome-mediated degradation dependent on SA. Studies of mutants of the mitochondrial succinate dehydrogenase (SDH) pathway, disrupted in stress response 1 (*dsr1*) and SDH assembly factor 2 (*sdhaf2*), demonstrated that SA directly contributes to stress signaling by increasing mitochondrial H_2O_2 production.

6 Brassinosteroids (BRs)

Brassinosteroids are a plant-specific steroid group of plant hormones known for their role in growth, development, reproduction, and stress response (Sharma et al. 2015; Vardhini and Anjum 2015; Tang et al. 2016; Wei and Li 2016; Xia et al. 2009, 2014). Brassinolide (BL) is synthesized from campesterol via a series of reductions, hydroxylations, epimerizations, and oxidations, whereas the mevalonic acid-dependent triterpenoid pathway is used for synthesis of other major BRs (Fujioka et al. 1998; Fujioka and Yokota 2003; Chung and Choe 2013). BRs actively participate in drought stress response and tolerance, as demonstrated by enhanced drought of plants overexpressing the BR biosynthetic gene encoding cytochrome p450 (Tiwari et al. 2017). Brassinosteroids also help reduce the photoinhibition that commonly occurs under abiotic stress conditions (Ahammed et al. 2015)

The BRs are recognized by the BRASSINOSTEROID INSENSITIVE 1 (BRI1) cell-surface leucine-rich repeat kinase group of receptor proteins. BRI further interacts with co-receptor BRI-associated receptor kinase 1 (BAK1) and initiates a series of phosphorylation and dephosphorylation events resulting in the regulation of gene expression (Belkhadir and Jaillais 2015; Nakamura et al. 2017). BR-mediated signaling is transduced by the bHLH group of TFs such as CESTA and TEOSINTE BRANCHED 1, CYCLOIDEA, PCF1 (TCP) (Poppenberger et al. 2011).

Similarly, the TF BRASSINAZOLE RESISTANT 1 (BZR1), induced by BRs, plays a role in maintaining cell cycle progression even under stress (Hacham et al. 2011). In contrast, BRASSINOSTEROIDS AT VASCULAR AND ORGANIZING CENTER (BRAVO), an R2R3 MYB transcription factor, inhibits stem cell proliferation. In the root quiescent center, interaction with BRI1-EMS SUPRESSOR 1 (BES1) represses the activity of BRAVO (Vilarrasa-Blasi et al. 2014). In addition, BRs remodel the cell wall under abiotic stress by inducing genes responsible for cell wall remodeling: pectin lyase-like proteins, expansins, and xyloglucan endotransglucosylase/hydrolases (Rao and Dixon 2017). In addition, BRI1 receptor induces the expression of pectin methylesterases under stress, thereby resisting the loss of cell wall and maintaining the integrity of the cell wall via interaction of BAK1 and RECEPTOR-LIKE PROTEIN (RPL44) (Wolf et al. 2014). These mechanisms help plants in stress adaptation and increase longevity.

7 Hormonal Crosstalk

7.1 ABA Crosstalk

In response to environmental stress (drought), ABA regulates growth and development via crosstalk with other plant hormones. Abscisic acid interacts antagonistically with most plant hormones. ABA and auxin work in opposing directions in regulating growth and development. Both of these hormones control cell division, seed dormancy and germination, primary root growth, seedling growth, etc. Crosstalk between auxin and ABA in regulating growth occurs via regulation of auxin or ABA response pathways. Auxin activates the ABA response and or ABA biosynthesis. Under osmotic stress, ABA increases basipetal transport of auxin via AUX1 (an auxin influx transporter) and PIN2 (an auxin efflux transporter) (Sun and Li 2014; Wang et al. 2011; Belin et al. 2009; Liu et al. 2013; Xu et al. 2013). Similarly, ABA and CK crosstalk with each other. ABA regulates CK by regulating CK oxidase that degrades CK. Under stress, CK via cytokinin receptor kinases (AHK2 and AHK3) negatively regulates ABA, ABA signaling and stress gene expression (Tran et al. 2007; Jeon et al. 2010). ABA also acts antagonistically with ET in regulating stomatal closure. ET inhibits ABA signaling and thereby delays stomatal closure.

Similarly, ABA inhibits ET biosynthesis by regulating ERF1 and long hypocotyl 5 (HY5), thereby repressing ACS5 (ethylene biosynthesis gene). The TF HY5 is a key player in the crosstalk between ABA and ET (Harrison 2012; Tanaka et al. 2005; Li et al. 2011). However, the crosstalk between ABA and JA is complex. Reduction of ABA level upregulates JA. Multiple components of ABA and JA–ET signaling pathways antagonistically interact in regulating stress-responsive gene expression, with MYC2 playing a central role or being a major hub in the crosstalk (Anderson et al. 2004; Chen et al. 2011; Fernández-Calvo et al. 2011; Sun and Li 2014).

7.2 ET Crosstalk

The ERF key regulatory proteins of ET signal transduction are also the hub of the crosstalk between ET and ABA, JA, SA, BR, and redox signaling. Under abiotic stress, both ABA and JA regulate ERFs. However, we have few reports on the crosstalk under drought stress. JA positively regulates the induction of ERFs [ERF6, JERF1, JERF3, Tsi1, OPBP1, and GmERF3 (Sewelam et al. 2013; Wu et al. 2007, 2008; Wang et al. 2004; Park et al. 2001; Zhang et al. 2009a, b)], but ABA negatively regulates the induction of few ERFs (ERF1/6) (Cheng et al. 2013; Sewelam et al. 2013). In contrast, ABA induces other ERFs (CsERF, GmERF3, LchERF, JERF3, TSRF1, TaERF1 and JERF1) (Ma et al. 2014; Zhang et al. 2009a, b; Wu et al. 2007, 2008, 2014a, b; Quan et al. 2010; Xu et al. 2007). Overexpression of TSRF1 results in enhanced drought tolerance (Quan et al. 2010; Cheng et al. 2013). The negative regulator AtERF7 inhibits ABA-induced gene expression, thereby decreasing drought stress tolerance (Song et al. 2005).

7.3 JA Crosstalk

The JA crosstalk with ABA at MYC2 is involved in drought stress signaling and activates genes such as RD22 and RD26, which are also upregulated under drought stress (Abe et al. 2003; Fujita et al. 2004, 2006; Harb et al. 2010). JA and GA also crosstalk, which is mediated by direct interaction between DELLA and JAZ proteins, thereby competing for the JAZ–MYC2 interaction that activates JA-responsive genes (Boter et al. 2004; Hou et al. 2010). Expression of the MYC2 target RGL3, which encodes DELLA, is induced by JA signaling (Wild et al. 2012). MYC2 is released when JA degrades JAZ. DELLA is also the hub of crosstalk between GA and ABA under abiotic stress. In addition, there is crosstalk between JA and ET: both activate Arabidopsis AtERF1, which imparts drought tolerance (Cheng et al. 2013).

7.4 SA Crosstalk

SA interacts with other phytohormones, and depending on the hormone, has a synergistic or antagonistic relation under optimal and stressful conditions. SA-induced ABA improves osmotic adaptation and growth under normal, salt and other abiotic stress (Szepesi et al. 2009; Szalai et al. 2011). Mitogen-activated protein kinases are key players of signaling components where antagonistic interaction between SA and JA occurs (Khan et al. 2012a, b). In addition, NPR1, the master regulator of SA-mediated defense, is a critical component of the epibassinolide-mediated increase in abiotic stress tolerance (heat and salt) (Divi et al. 2010).

7.5 BR Crosstalk

BR and ABA crosstalk via ABI3/viviparous1 (ABI3/VP1) and related to ABI3/VP1 (RAV1) and induce BRI expression (Saini et al. 2015). Maintenance of cell wall integrity is via interaction of BR with other hormones (Wolf et al. 2014). The long-distance effect of BR is via crosstalk with multiple hormones (Vriet et al. 2013; Gudesblat and Russinova 2011). BRs antagonistically cross-interact with ABA, ET, SA, auxin, GA and polyamines (Vardhini and Anjum 2015). Analysis of *det-2* and *bri-1* BR mutants demonstrated that BIN2 negatively regulates BR signaling in the presence of ABA signaling, which is via ABI1/2 (PP2Cs) that function downstream of BRI1 and upstream of BIN2 and regulate BR signaling (Zhang et al. 2009a, b). ABI1/2 dephosphorylate BIN2, and ABA inhibits BIN2 phosphorylation by BES1, thereby reducing BR signaling (Wang et al. 2018). BIN2 also interacts with ABI5 and phosphorylates it and mediates ABA signaling (Hu and Yu 2014). Both ABA and BR induce the expression of BRASSINOSTEROID-SIGNALING KINASE 5 (BSK5), a key player in ABA signaling. Similar crosstalk between ABA and BR occurs with a TF, BZR1 (Yang et al. 2016). BR and ET crosstalk via H_2O_2 and regulate stress tolerance (Zhu et al. 2016). BR signaling interacts with SA signaling via the TF WRKY70, which negatively regulates SA biosynthesis. NPR1, regulating SA signaling, interacts with BIN2 and BZR1 and regulates stress tolerance (Divi et al. 2010; Li et al. 2013).

References

Abe H, Yamaguchi-Shinozaki K, Urao T, Iwasaki T, Hosokawa D, Shinozaki K (1997) Role of *Arabidopsis* MYC and MYB homologs in drought-and abscisic acid-regulated gene expression. Plant Cell 9:1859–1868

Abe H, Urao T, Ito T, Seki M, Shinozaki K, Yamaguchi-Shinozaki K (2003) *Arabidopsis* AtMYC2 (bHLH) and AtMYB2 (MYB) function as transcriptional activators in abscisic acid signaling. Plant Cell 15:63–78

Ahammed GJ, Li X, Xia X-J, Shi K, Zhou Y-H, Yu J-Q (2015) Enhanced photosynthetic capacity and antioxidant potential mediate brassinosteriod-induced phenanthrene stress tolerance in tomato. Environ Pollut 201:58–66

Anderson JP, Badruzsaufari E, Schenk PM, Manners JM, Desmond OJ, Ehlert C, Maclean DJ, Ebert PR, Kazan K (2004) Antagonistic interaction between abscisic acid and jasmonate-ethylene signaling pathways modulates defense gene expression and disease resistance in Arabidopsis. Plant Cell 16:3460–3479

Anjum SA, Xie X, Wang L, Saleem MF, Man C, Lei W (2011) Morphological, physiological and biochemical responses of plants to drought stress. Afr J Agric Res 6:2026–2032

Anjum SA, Ashraf U, Zohaib A, Tanveer M, Naeem M, Ali I, Nazir U (2017) Growth and development responses of crop plants under drought stress: a review. Zemdirbyste 104:267–276

Bari R, Jones JD (2009) Role of plant hormones in plant defence responses. Plant Mol Biol 69:473–488

Belin C, Megies C, Hauserová E, Lopez-Molina L (2009) Abscisic acid represses growth of the Arabidopsis embryonic axis after germination by enhancing auxin signaling. Plant Cell 21:2253–2268

Belkhadir Y, Jaillais Y (2015) The molecular circuitry of brassinosteroid signaling. New Phytol 206:522–540

Bisson MM, Groth G (2010) New insight in ethylene signaling: autokinase activity of ETR1 modulates the interaction of receptors and EIN2. Mol Plant 3:882–889

Boneh U, Biton I, Zheng C, Schwartz A, Ben-Ari G (2012) Characterization of potential ABA receptors in *Vitis vinifera*. Plant Cell Rep 31:311–321

Boter M, Ruíz-Rivero O, Abdeen A, Prat S (2004) Conserved MYC transcription factors play a key role in jasmonate signaling both in tomato and Arabidopsis. Genes Dev 18:1577–1591

Boudsocq M, Droillard M-J, Barbier-Brygoo H, Laurière C (2007) Different phosphorylation mechanisms are involved in the activation of sucrose non-fermenting 1 related protein kinases 2 by osmotic stresses and abscisic acid. Plant Mol Biol 63:491–503

Brossa R, López-Carbonell M, Jubany-Marí T, Alegre L (2011) Interplay between abscisic acid and jasmonic acid and its role in water-oxidative stress in wild-type, ABA-deficient, JA-deficient, and ascorbate-deficient Arabidopsis plants. J Plant Growth Regul 30:322–333

Cao H, Glazebrook J, Clarke JD, Volko S, Dong X (1997) The Arabidopsis NPR1 gene that controls systemic acquired resistance encodes a novel protein containing ankyrin repeats. Cell 88:57–63

Catinot J, Buchala A, Abou-Mansour E, Métraux J-P (2008) Salicylic acid production in response to biotic and abiotic stress depends on isochorismate in *Nicotiana benthamiana*. FEBS Lett 582:473–478

Cela J, Chang C, Munné-Bosch S (2011) Accumulation of γ-rather than α-tocopherol alters ethylene signaling gene expression in the vte4 mutant of *Arabidopsis thaliana*. Plant Cell Physiol 52:1389–1400

Chen YF, Etheridge N, Schaller GE (2005) Ethylene signal transduction. Ann Bot (Lond) 95:901–915

Chen Q, Sun J, Zhai Q, Zhou W, Qi L, Xu L, Wang B, Chen R, Jiang H, Qi J (2011) The basic helix-loop-helix transcription factor MYC2 directly represses PLETHORA expression during jasmonate-mediated modulation of the root stem cell niche in Arabidopsis. Plant Cell 23:3335–3352

Cheng M-C, Liao P-M, Kuo W-W, Lin T-P (2013) The Arabidopsis ETHYLENE-RESPONSE-FACTOR1 regulates abiotic-stress-responsive gene expression by binding to different cis-acting elements in response to different stress signals. Plant Physiol 162:1566–1582

Choi H-i, Hong J-h, Ha J-o, Kang J-y, Kim SY (2000) ABFs, a family of ABA-responsive element binding factors. J Biol Chem 275:1723–1730

Christmann A, Weiler EW, Steudle E, Grill E (2007) A hydraulic signal in root-to-shoot signaling of water shortage. Plant J 52:167–174

Chung Y, Choe S (2013) The regulation of brassinosteroid biosynthesis in Arabidopsis. Crit Rev Plant Sci 32:396–410

Clarke SM, Cristescu SM, Miersch O, Harren FJ, Wasternack C, Mur LA (2009) Jasmonates act with salicylic acid to confer basal thermotolerance in *Arabidopsis thaliana*. New Phytol 182:175–187
Cutler SR, Rodriguez PL, Finkelstein RR, Abrams SR (2010) Abscisic acid: emergence of a core signaling network. Annu Rev Plant Biol 61:651–679
Daszkowska-Golec A, Szarejko I (2013) Open or close the gate–stomata action under the control of phytohormones in drought stress conditions. Front Plant Sci 4:138
Delaney T, Friedrich L, Ryals J (1995) Arabidopsis signal transduction mutant defective in chemically and biologically induced disease resistance. Proc Natl Acad Sci USA 92:6602–6606
Deshaies R (1999) SCF and Cullin/Ring H2-based ubiquitin ligases. Annu Rev Cell Dev Biol 15
Divi UK, Rahman T, Krishna P (2010) Brassinosteroid-mediated stress tolerance in Arabidopsis shows interactions with abscisic acid, ethylene and salicylic acid pathways. BMC Plant Biol 10:151
Dong W, Wang M, Xu F, Quan T, Peng K, Xiao L, Xia G (2013) Wheat oxophytodienoate reductase gene TaOPR1 confers salinity tolerance via enhancement of ABA signalling and ROS scavenging. Plant Physiol 161:1217–1218
Du H, Klessig DF (1997) Identification of a soluble, high-affinity salicylic acid-binding protein in tobacco. Plant Physiol 113:1319–1327
Dubois M, Skirycz A, Claeys H, Maleux K, Dhondt S, De Bodt S, Bossche RV, De Milde L, Yoshizumi T, Matsui M (2013) The ETHYLENE RESPONSE FACTOR 6 acts as central regulator of leaf growth under water limiting conditions in *Arabidopsis thaliana*. Plant Physiol 5:e70289
Dubois M, Van den Broeck L, Inzé D (2018) The pivotal role of ethylene in plant growth. Trends Plant Sci 23:311–323
Durrant WE, Dong X (2004) Systemic acquired resistance. Annu Rev Phytopathol 42:185–209
Fernández-Calvo P, Chini A, Fernández-Barbero G, Chico JM, Gimenez-Ibanez S, Geerinck J, Eeckhout D, Schweizer F, Godoy M, Franco-Zorrilla JM (2011) The Arabidopsis bHLH transcription factors MYC3 and MYC4 are targets of JAZ repressors and act additively with MYC2 in the activation of jasmonate responses. Plant Cell 23:701–715
Feussner I, Wasternack C (2002) The lipoxygenase pathway. Annu Rev Plant Biol 53:275–297
Forouhar F, Yang Y, Kumar D, Chen Y, Fridman E, Park SW, Chiang Y, Acton TB, Montelione GT, Pichersky E (2005) Structural and biochemical studies identify tobacco SABP2 as a methyl salicylate esterase and implicate it in plant innate immunity. Proc Natl Acad Sci USA 102:1773–1778
Fu ZQ, Yan S, Saleh A, Wang W, Ruble J, Oka N, Mohan R, Spoel SH, Tada Y, Zheng N (2012) NPR3 and NPR4 are receptors for the immune signal salicylic acid in plants. Nature 486(7402):228
Fujii H, Zhu J-K (2009) Arabidopsis mutant deficient in 3 abscisic acid-activated protein kinases reveals critical roles in growth, reproduction, and stress. Proc Natl Acad Sci USA 106:8380–8385
Fujioka S, Yokota T (2003) Biosynthesis and metabolism of brassinosteroids. Annu Rev Plant Biol 54:137–164
Fujioka S, Noguchi T, Yokota T, Takatsuto S, Yoshida S (1998) Brassinosteroids in *Arabidopsis thaliana*. Phytochemistry 48:595–599
Fujita M, Fujita Y, Maruyama K, Seki M, Hiratsu K, Ohme-Takagi M, Tran LSP, Yamaguchi-Shinozaki K, Shinozaki K (2004) A dehydration-induced NAC protein, RD26, is involved in a novel ABA-dependent stress-signaling pathway. Plant J 39:863–876
Fujita Y, Nakashima K, Yoshida T, Katagiri T, Kidokoro S, Kanamori N, et al (2009) Three SnRK2 protein kinases are the main positive regulators of abscisic acid signaling in response to water stress in Arabidopsis. Plant Cell Physiol 50:2123–2132. https://doi.org/10.1093/pcp/pcp147
Fujita M, Fujita Y, Noutoshi Y, Takahashi F, Narusaka Y, Yamaguchi-Shinozaki K, Shinozaki K (2006) Crosstalk between abiotic and biotic stress responses: a current view from the points of convergence in the stress signaling networks. Curr Opin Plant Biol 9:436–442
Fujita Y, Fujita M, Shinozaki K, Yamaguchi-Shinozaki K (2011) ABA-mediated transcriptional regulation in response to osmotic stress in plants. J Plant Res 124:509–525

Furihata T, Maruyama K, Fujita Y, Umezawa T, Yoshida R, Shinozaki K, Yamaguchi-Shinozaki K (2006) Abscisic acid-dependent multisite phosphorylation regulates the activity of a transcription activator AREB1. Proc Natl Acad Sci USA 103:1988–1993
Geiger D, Scherzer S, Mumm P, Stange A, Marten I, Bauer H, Ache P, Matschi S, Liese A, Al-Rasheid KA (2009) Activity of guard cell anion channel SLAC1 is controlled by drought-stress signaling kinase-phosphatase pair. Proc Natl Acad Sci USA 106:21425–21430
Ghasemi PA, Sajjadi SE, Parang K (2014) A review (research and patents) on jasmonic acid and its derivatives. Arch Pharm 347:229–239
Golldack D, Li C, Mohan H, Probst N (2014) Tolerance to drought and salt stress in plants: unraveling the signaling networks. Front Plant Sci 5:151
Gudesblat GE, Russinova E (2011) Plants grow on brassinosteroids. Curr Opin Plant Biol 14:530–537
Hacham Y, Holland N, Butterfield C, Ubeda-Tomas S, Bennett MJ, Chory J, Savaldi-Goldstein S (2011) Brassinosteroid perception in the epidermis controls root meristem size. Development 138:839–848
Hara M, Furukawa J, Sato A, Mizoguchi T, Miura K (2012) Abiotic stress and role of salicylic acid in plants. In: Abiotic stress responses in plants. Springer, pp 235–251
Harb A, Krishnan A, Ambavaram MM, Pereira A (2010) Molecular and physiological analysis of drought stress in Arabidopsis reveals early responses leading to acclimation in plant growth. Plant Physiol 154:1254–1271
Harrison MA (2012) Cross-talk between phytohormone signaling pathways under both optimal and stressful environmental conditions. In: Phytohormones and abiotic stress tolerance in plants. Springer, pp 49–76
He H, Li J (2008) Proteomic analysis of phosphoproteins regulated by abscisic acid in rice leaves. Biochem Biophys Res Commun 371:883–888
He XJ, Mu RL, Cao WH, Zhang ZG, Zhang JS, Chen SY (2005) AtNAC2, a transcription factor downstream of ethylene and auxin signaling pathways, is involved in salt stress response and lateral root development. Plant J 44:903–916
Hou X, Lee LYC, Xia K, Yan Y, Yu H (2010) DELLAs modulate jasmonate signaling via competitive binding to JAZs. Dev Cell 19:884–894
Hrabak EM, Chan CW, Gribskov M, Harper JF, Choi JH, Halford N, Kudla J, Luan S, Nimmo HG, Sussman MR (2003) The Arabidopsis CDPK-SnRK superfamily of protein kinases. Plant Physiol 132:666–680
Hunter LJ, Westwood JH, Heath G, Macaulay K, Smith AG, MacFarlane SA, Palukaitis P, Carr JP (2013) Regulation of RNA-dependent RNA polymerase 1 and isochorismate synthase gene expression in Arabidopsis. PLoS ONE 8:e66530
Hussain M, Farooq S, Hasan W, Ul-Allah S, Tanveer M, Farooq M, Nawaz A (2018) Drought stress in sunflower: Physiological effects and its management through breeding and agronomic alternatives. Agric Water Manag 201:152–166
Jaradat MR, Feurtado JA, Huang D, Lu Y, Cutler AJ (2013) Multiple roles of the transcription factor AtMYBR1/AtMYB44 in ABA signaling, stress responses, and leaf senescence. BMC Plant Biol 13:192
Jayakannan M, Bose J, Babourina O, Shabala S, Massart A, Poschenrieder C, Rengel Z (2015) The NPR1-dependent salicylic acid signalling pathway is pivotal for enhanced salt and oxidative stress tolerance in Arabidopsis. J Exp Bot 66:1865–1875
Jeon J, Kim NY, Kim S, Kang NY, Novák O, Ku SJ, Cho C, Lee DJ, Lee EJ, Strnad M (2010) A subset of cytokinin two-component signaling system plays a role in cold temperature stress response in Arabidopsis. J Biol Chem 285:23371–23386
Ju C, Yoon GM, Shemansky JM, Lin DY, Ying ZI, Chang J, Garrett WM, Kessenbrock M, Groth G, Tucker ML (2012) CTR1 phosphorylates the central regulator EIN2 to control ethylene hormone signaling from the ER membrane to the nucleus in Arabidopsis. Proc Natl Acad Sci USA 109:19486–19491

Kang G, Li G, Xu W, Peng X, Han Q, Zhu Y, Guo T (2012) Proteomics reveals the effects of salicylic acid on growth and tolerance to subsequent drought stress in wheat. J Proteome Res 11:6066–6079

Kendrick MD, Chang C (2008) Ethylene signaling: new levels of complexity and regulation. Curr Opin Plant Biol 11:479–485

Khan MIR, Syeed S, Nazar R, and Anjum NA (2012a) An insight into the roleofsalicylicacidandjasmonicacidinsaltstresstolerance. In: Khan NA, Nazar R, Iqbal N, Anjum NA (eds) Phytohormones and abiotic stress tolerance in plants. Springer, Berlin, pp 277–300. https://doi.org/10.1007/978-3-642-25829-9_12

Khan NA, Nazar R, Iqbal N, Anjum NA (2012b) Phytohormones and abiotic stress tolerance in plants. Springer, Berlin. https://doi.org/10.1007/978-3-642-25829-9

Kieber JJ, Rothenberg M, Roman G, Feldmann KA, Ecker JR (1993) CTR1, a negative regulator of the ethylene response pathway in Arabidopsis, encodes a member of the raf family of protein kinases. Cell 72:427–441

Kim H, Hwang H, Hong J-W, Lee Y-N, Ahn IP, Yoon IS, Yoo S-D, Lee S, Lee SC, Kim B-G (2011) A rice orthologue of the ABA receptor, OsPYL/RCAR5, is a positive regulator of the ABA signal transduction pathway in seed germination and early seedling growth. J Exp Bot 63:1013–1024

Kiribuchi K, Jikumaru Y, Kaku H, Minami E, Hasegawa M, Kodama O, Seto H, Okada K, Nojiri H, Yamane H (2005) Involvement of the basic helix-loop-helix transcription factor RERJ1 in wounding and drought stress responses in rice plants. Biosci Biotechnol Biochem 69:1042–1044

Kline KG, Barrett-Wilt GA, Sussman MR (2010) In planta changes in protein phosphorylation induced by the plant hormone abscisic acid. Proc Natl Acad Sci USA 107:15986–15991

Klingler JP, Batelli G, Zhu JK (2010) ABA receptors: the START of a new paradigm in phytohormone signalling. J Exp Bot 61:3199–3210

Kobayashi Y, Murata M, Minami H, Yamamoto S, Kagaya Y, Hobo T, Yamamoto A, Hattori T (2005) Abscisic acid-activated SNRK2 protein kinases function in the gene-regulation pathway of ABA signal transduction by phosphorylating ABA response element-binding factors. Plant J 44:939–949

Korkmaz A, Uzunlu M, Demirkiran AR (2007) Treatment with acetyl salicylic acid protects muskmelon seedlings against drought stress. Acta Physiol Plant 29:503–508

Lacey RF, Binder BM (2014) How plants sense ethylene gas—the ethylene receptors. J Inorg Biochem 133:58–62

Leung J, Bouvier-Durand M, Morris P-C, Guerrier D, Chefdor F, Giraudat J (1994) Arabidopsis ABA response gene ABI1: features of a calcium-modulated protein phosphatase. Science 264(5164):1448–1452

Leung J, Merlot S, Giraudat J (1997) The Arabidopsis ABSCISIC ACID-INSENSITIVE2 (ABI2) and ABI1 genes encode homologous protein phosphatases 2C involved in abscisic acid signal transduction. Plant Cell 9:759–771

Li Z, Zhang L, Yu Y, Quan R, Zhang Z, Zhang H, Huang R (2011) The ethylene response factor AtERF11 that is transcriptionally modulated by the bZIP transcription factor HY5 is a crucial repressor for ethylene biosynthesis in Arabidopsis. Plant J 68:88–99

Li J, Besseau S, Törönen P, Sipari N, Kollist H, Holm L, Palva ET (2013) Defense-related transcription factors WRKY70 and WRKY54 modulate osmotic stress tolerance by regulating stomatal aperture in Arabidopsis. New Phytol 200:457–472

Liu X, Zhang H, Zhao Y, Feng Z, Li Q, Yang H-Q, Luan S, Li J, He Z-H (2013) Auxin controls seed dormancy through stimulation of abscisic acid signaling by inducing ARF-mediated ABI3 activation in Arabidopsis. Proc Natl Acad Sci USA 110:15485–15490

Lopez-Molina L, Mongrand S, Chua N-H (2001) A postgermination developmental arrest checkpoint is mediated by abscisic acid and requires the ABI5 transcription factor in Arabidopsis. Proc Natl Acad Sci USA 98:4782–4787

Lorenzo O, Solano R (2005) Molecular players regulating the jasmonate signalling network. Curr Opin Plant Biol 8:532–540

Ma S, Gong Q, Bohnert HJ (2006) Dissecting salt stress pathways. J Exp Bot 57:1097–1107

Ma Y, Szostkiewicz I, Korte A, Moes D, Yang Y, Christmann A, Grill E (2009) Regulators of PP2C phosphatase activity function as abscisic acid sensors. Science 324:1064–1068

Ma Y, Zhang L, Zhang J, Chen J, Wu T, Zhu S, Yan S, Zhao X, Zhong G (2014) Expressing a Citrus ortholog of Arabidopsis ERF1 enhanced cold-tolerance in tobacco. Sci Hortic 174:65–76

Magome H, Yamaguchi S, Hanada A, Kamiya Y, Oda K (2008) The DDF1 transcriptional activator upregulates expression of a gibberellin-deactivating gene, GA2ox7, under high-salinity stress in Arabidopsis. Plant J 56:613–626

Maksymiec W, Wianowska D, Dawidowicz AL, Radkiewicz S, Mardarowicz M, Krupa Z (2005) The level of jasmonic acid in *Arabidopsis thaliana* and *Phaseolus coccineus* plants under heavy metal stress. J Plant Physiol 162:1338–1346

McConn M, Creelman RA, Bell E, Mullet JE (1997) Jasmonate is essential for insect defense in Arabidopsis. Proc Natl Acad Sci USA 94:5473–5477

Meggio F, Pinna LA (2003) One-thousand-and-one substrates of protein kinase CK2? FASEB J 17:349–368

Meyer K, Leube MP, Grill E (1994) A protein phosphatase 2C involved in ABA signal transduction in *Arabidopsis thaliana*. Science 264(5164):1452–1455

Mittler R, Vanderauwera S, Suzuki N, Miller G, Tognetti VB, Vandepoele K, Gollery M, Shulaev V, Van Breusegem F (2011) ROS signaling: the new wave? Trends Plant Sci 16:300–309

Miura K, Okamoto H, Okuma E, Shiba H, Kamada H, Hasegawa PM, Murata Y (2013) SIZ1 deficiency causes reduced stomatal aperture and enhanced drought tolerance via controlling salicylic acid-induced accumulation of reactive oxygen species in Arabidopsis. Plant J 73:91–104

Morgan PW, Drew MC (1997) Ethylene and plant responses to stress. Physiol Plant 100:620–630

Munemasa S, Oda K, Watanabe-Sugimoto M, Nakamura Y, Shimoishi Y, Murata Y (2007) The coronatine-insensitive 1 mutation reveals the hormonal signaling interaction between abscisic acid and methyl jasmonate in Arabidopsis guard cells. Specific impairment of ion channel activation and second messenger production. Plant Physiol 143:1398–1407

Munné-Bosch S, Penuelas J (2003) Photo-and antioxidative protection, and a role for salicylic acid during drought and recovery in field-grown *Phillyrea angustifolia* plants. Planta 217:758–766

Munné-Bosch S, Queval G, Foyer CH (2013) The impact of global change factors on redox signaling underpinning stress tolerance. Plant Physiol 161:5–19

Munns R, Tester M (2008) Mechanisms of salinity tolerance. Annu Rev Plant Biol 59:651–681

Nafie E, Hathout T, Mokadem A, Shyma A (2011) Jasmonic acid elicits oxidative defense and detoxification systems in *Cucumis melo* L. cells. Braz J Plant Physiol 23:161–174

Nakamura A, Tochio N, Fujioka S, Ito S, Kigawa T, Shimada Y, Matsuoka M, Yoshida S, Kinoshita T, Asami T (2017) Molecular actions of two synthetic brassinosteroids, iso-carbaBL and 6-deoxoBL, which cause altered physiological activities between Arabidopsis and rice. PLoS ONE 12:e0174015

Nakashima K, Yamaguchi-Shinozaki K (2013) ABA signaling in stress-response and seed development. Plant Cell Rep 32:959–970

Nakashima K, Fujita Y, Kanamori N, Katagiri T, Umezawa T, Kidokoro S, Maruyama K, Yoshida T, Ishiyama K, Kobayashi M (2009) Three Arabidopsis SnRK2 protein kinases, SRK2D/SnRK2. 2, SRK2E/SnRK2. 6/OST1 and SRK2I/SnRK2. 3, involved in ABA signaling are essential for the control of seed development and dormancy. Plant Cell Physiol 50:1345–1363

Nakashima K, Takasaki H, Mizoi J, Shinozaki K, Yamaguchi-Shinozaki K (2012) NAC transcription factors in plant abiotic stress responses. Biochim Biophys Acta (BBA)-Gene Regul Mech 1819:97–103

Nawrath C, Métraux J-P (1999) Salicylic acid induction–deficient mutants of Arabidopsis express PR-2 and PR-5 and accumulate high levels of camalexin after pathogen inoculation. Plant Cell 11:1393–1404

Nishimura N, Hitomi K, Arvai AS, Rambo RP, Hitomi C, Cutler SR, Schroeder JI, Getzoff ED (2009) Structural mechanism of abscisic acid binding and signaling by dimeric PYR1. Science 326(5958):1373–1379

Pandey S, Nelson DC, Assmann SM (2009) Two novel GPCR-type G proteins are abscisic acid receptors in Arabidopsis. Cell 136:136–148
Park JM, Park C-J, Lee S-B, Ham B-K, Shin R, Paek K-H (2001) Overexpression of the tobacco Tsi1 gene encoding an EREBP/AP2–type transcription factor enhances resistance against pathogen attack and osmotic stress in tobacco. Plant Cell 13:1035–1046
Pauwels L, Barbero GF, Geerinck J, Tilleman S, Gurnewald W, Perez AC, Chico JM, Bossche RV, Sewell J, Gil E, Garcia-Casado G, Wltters E, Inze D, Long JA, De Jaeger G, Solano R, Goossens A (2010) NINJA connects the co-repressor TOPLESS to jasmonate signaling. Nature 464:788–791
Pauwels L, Goossens A (2011) The JAZ proteins: a crucial interface in the jasmonate signaling cascade. Plant Cell 23:3089–3100
Poppenberger B, Rozhon W, Khan M, Husar S, Adam G, Luschnig C, Fujioka S, Sieberer T (2011) CESTA, a positive regulator of brassinosteroid biosynthesis. EMBO J 30:1149–1161
Qiao H, Chang KN, Yazaki J, Ecker JR (2009) Interplay between ethylene, ETP1/ETP2 F-box proteins, and degradation of EIN2 triggers ethylene responses in Arabidopsis. Genes Dev 23:512–521
Qin F, Shinozaki K, Yamaguchi-Shinozaki K (2011) Achievements and challenges in understanding plant abiotic stress responses and tolerance. Plant Cell Physiol 52:1569–1582
Qiu Z, Guo J, Zhu A, Zhang L, Zhang M (2014) Exogenous jasmonic acid can enhance tolerance of wheat seedlings to salt stress. Ecotoxicol Environ Saf 104:202–208
Quan R, Hu S, Zhang Z, Zhang H, Zhang Z, Huang R (2010) Overexpression of an ERF transcription factor TSRF1 improves rice drought tolerance. Plant Biotechnol J 8:476–488
Raghavendra AS, Gonugunta VK, Christmann A, Grill E (2010) ABA perception and signalling. Trends Plant Sci 15:395–401
Rao X, Dixon RA (2017) Brassinosteroid mediated cell wall remodeling in grasses under abiotic stress. Front Plant Sci 8:806
Reymond P, Farmer EE (1998) Jasmonate and salicylate as global signals for defense gene expression. Curr Opin Plant Biol 1:404–411
Riechmann JL, Heard J, Martin G, Reuber L, Jiang C-Z, Keddie J, Adam L, Pineda O, Ratcliffe O, Samaha R (2000) Arabidopsis transcription factors: genome-wide comparative analysis among eukaryotes. Science 290(5499):2105–2110
Rivas-San Vicente M, Plasencia J (2011) Salicylic acid beyond defence: its role in plant growth and development. J Exp Bot 62:3321–3338
Romero P, Lafuente MT, Rodrigo MJ (2012) The Citrus ABA signalosome: identification and transcriptional regulation during sweet orange fruit ripening and leaf dehydration. J Exp Bot 63:4931–4945
Rushton DL, Tripathi P, Rabara RC, Lin J, Ringler P, Boken AK, Langum TJ, Smidt L, Boomsma DD, Emme NJ (2012) WRKY transcription factors: key components in abscisic acid signalling. Plant Biotechnol J 10:2–11
Saez A, Apostolova N, Gonzalez-Guzman M, Gonzalez-Garcia MP, Nicolas C, Lorenzo O, Rodriguez PL (2004) Gain-of-function and loss-of-function phenotypes of the protein phosphatase 2C HAB1 reveal its role as a negative regulator of abscisic acid signaling. Plant J 37:354–369
Saini S, Sharma I, Pati PK (2015) Versatile roles of brassinosteroid in plants in the context of its homoeostasis, signaling and crosstalks. Front Plant Sci 6:950
Sanchez-Casas P, Klessig DF (1994) A salicylic acid-binding activity and a salicylic acid-inhibitable catalase activity are present in a variety of plant species. Plant Physiol 106:1675–1679
Sánchez-Romera B, Ruiz-Lozano JM, Li G, Luu DT, Martínez-Ballesta Mdel C, Carvajal M, Zamarreño AM, García-Mina JM, Maurel C, Aroca R (2014) Enhancement of root hydraulic conductivity by methyl jasmonate and the role of calcium and abscisic acid in this process. Plant Cell Environ 37:995–1008
Saruhashi M, Ghosh TK, Arai K, Ishizaki Y, Hagiwara K, Komatsu K, Shiwa Y, Izumikawa K, Yoshikawa H, Umezawa T (2015) Plant Raf-like kinase integrates abscisic acid and hyperosmotic

stress signaling upstream of SNF1-related protein kinase 2. Proc Natl Acad Sci USA 112:E6388–E6396

Sato A, Sato Y, Fukao Y, Fujiwara M, Umezawa T, Shinozaki K, Hibi T, Taniguchi M, Miyake H, Goto DB (2009) Threonine at position 306 of the KAT1 potassium channel is essential for channel activity and is a target site for ABA-activated SnRK2/OST1/SnRK2. 6 protein kinase. Biochem J 424:439–448

Savchenko T, Kolla V, Wang C-Q, Nasafi Z, Hicks D, Phadungchob B, Chehab W, Brandizzi F, Froehlich J, Dehesh K (2014) Functional convergence of oxylipin and ABA pathways controls stomatal closure in response to drought. Plant Physiol 164:1151–1160

Sawada H, Shim I-S, Usui K (2006) Induction of benzoic acid 2-hydroxylase and salicylic acid biosynthesis—modulation by salt stress in rice seedlings. Plant Sci 171:263–270

Schommer C, Palatnik JF, Aggarwal P, Chételat A, Cubas P, Farmer EE, Nath U, Weigel D (2008) Control of jasmonate biosynthesis and senescence by miR319 targets. PLoS Biol 6(9):e230

Seiler C, Harshavardhan VT, Reddy PS, Hensel G, Kumlehn J, Eschen-Lippold L, Rajesh K, Korzun V, Wobus U, Lee J (2014) Abscisic acid flux alterations result in differential ABA signalling responses and impact assimilation efficiency in barley under terminal drought stress. Plant Physiol 164:1677–1696

Senaratna T, Touchell D, Bunn E, Dixon K (2000) Acetyl salicylic acid (Aspirin) and salicylic acid induce multiple stress tolerance in bean and tomato plants. Plant Growth Regul 30:157–161

Seo PJ, Park CM (2010) MYB96-mediated abscisic acid signals induce pathogen resistance response by promoting salicylic acid biosynthesis in Arabidopsis. New Phytol 186:471–483

Seo J, Joo J, Kim M, Kim Y, Nahm BH, Song SI, Cheong JJ, Lee JS, Kim JK, Choi YD (2011) OsbHlH148, a basic helix-loop-helix protein, interacts with OsJaZ proteins in a jasmonate signaling pathway leading to drought tolerance in rice. Plant J 65:907–921

Sewelam N, Kazan K, Thomas-Hall SR, Kidd BN, Manners JM, Schenk PM (2013) Ethylene response factor 6 is a regulator of reactive oxygen species signaling in Arabidopsis. PLoS ONE 8:e70289

Sharma I, Bhardwaj R, Pati PK (2015) Exogenous application of 28-homobrassinolide modulates the dynamics of salt and pesticides induced stress responses in an elite rice variety Pusa Basmati-1. J Plant Growth Regul 34:509–518

Shen Y-Y, Wang X-F, Wu F-Q, Du S-Y, Cao Z, Shang Y, Wang X-L, Peng C-C, Yu X-C, Zhu S-Y (2006) The Mg-chelatase H subunit is an abscisic acid receptor. Nature 443(7113):823

Singh A, Jha SK, Bagri J, Pandey GK (2015) ABA inducible rice protein phosphatase 2C confers ABA insensitivity and abiotic stress tolerance in Arabidopsis. PLoS ONE 10:e0125168

Sirichandra C, Davanture M, Turk BE, Zivy M, Valot B, Leung J, Merlot S (2010) The Arabidopsis ABA-activated kinase OST1 phosphorylates the bZIP transcription factor ABF3 and creates a 14-3-3 binding site involved in its turnover. PLoS ONE 5:e13935

Sirichandra C, Gu D, Hu H, Davanture M, Lee S, Djaoui M, et al (2009) Phosphorylation of the Arabidopsis AtrbohF NADPH oxidase by OST1 protein kinase. FEBS Lett 583:2982–2986. https://doi.org/10.1016/j.febslet.2009.08.033

Slaymaker DH, Navarre DA, Clark D, del Pozo O, Martin GB, Klessig DF (2002) The tobacco salicylic acid-binding protein 3 (SABP3) is the chloroplast carbonic anhydrase, which exhibits antioxidant activity and plays a role in the hypersensitive defense response. Proc Natl Acad Sci USA 99:11640–11645

Song C-P, Agarwal M, Ohta M, Guo Y, Halfter U, Wang P, Zhu J-K (2005) Role of an Arabidopsis AP2/EREBP-type transcriptional repressor in abscisic acid and drought stress responses. Plant Cell 17:2384–2396

Staswick PE (2008) JAZing up jasmonate signaling. Trends Plant Sci 13:66–71

Stepanova AN, Alonso JM (2009) Ethylene signaling and response: where different regulatory modules meet. Curr Opin Plant Biol 12:548–555

Sun J, Li C (2014) Cross talk of signaling pathways between ABA and other phytohormones. In: Abscisic acid: metabolism, transport and signalling. Springer, pp 243–253

Sun L, Wang Y-P, Chen P, Ren J, Ji K, Li Q, Li P, Dai S-J, Leng P (2011) Transcriptional regulation of SlPYL, SlPP2C, and SlSnRK2 gene families encoding ABA signal core components during tomato fruit development and drought stress. J Exp Bot 62:5659–5669

Swamy PM, Smith BN (1999) Role of abscisic acid in plant stress tolerance. Curr Sci 76:1220–1227

Szalai G, Pál M, Janda T (2011) Abscisic acid may alter the salicylic acid-related abiotic stress response in maize. J Agron Crop Sci 197:368–377

Szepesi Á, Csiszár J, Gémes K, Horváth E, Horváth F, Simon ML, Tari I (2009) Salicylic acid improves acclimation to salt stress by stimulating abscisic aldehyde oxidase activity and abscisic acid accumulation, and increases Na^+ content in leaves without toxicity symptoms in *Solanum lycopersicum L.* J Plant Physiol 166:914–925

Tanaka Y, Sano T, Tamaoki M, Nakajima N, Kondo N, Hasezawa S (2005) Ethylene inhibits abscisic acid-induced stomatal closure in Arabidopsis. Plant Physiol 138:2337–2343

Tang J, Han Z, Chai J (2016) Q&A: what are brassinosteroids and how do they act in plants? BMC Biol 14(1):113

Tanveer M, Shahzad B, Sharma A, Khan EA (2018) 24-Epibrassinolide application in plants: An implication for improving drought stress tolerance in plants. Plant Physiol Biochem 135:295–303

Thao NP, Khan MIR, Thu NBA, Hoang XLT, Asgher M, Khan NA, Tran L-SP (2015) Role of ethylene and its cross talk with other signaling molecules in plant responses to heavy metal stress. Plant Physiol 169:73–84

Thines B, Katsir L, Melotto M, Niu Y, Mandaokar A, Liu G, Nomura K, He SY, Howe GA, Browse J (2007) JAZ repressor proteins are targets of the SCF COI1 complex during jasmonate signaling. Nature 448(7154):661

Tiwari S, Lata C, Singh Chauhan P, Prasad V, Prasad M (2017) A functional genomic perspective on drought signaling and its crosstalk with phytohormone-mediated signaling pathways in plants. Curr Genomics 18:469–482

Tran LSP, Urao T, Qin F, Maruyama K, Kakimoto T, Shinozaki K, Yamaguchi-Shinozaki K (2007) Functional analysis of AHK1/ATHK1 and cytokinin receptor histidine kinases in response to abscisic acid, drought, and salt stress in Arabidopsis. Proc Natl Acad Sci USA 104:20623–20628

Turner JG, Ellis C, Devoto A (2002) The jasmonate signal pathway. Plant Cell 14(suppl 1):S153–S164

Umezawa T, Sugiyama N, Mizoguchi M, Hayashi S, Myouga F, Yamaguchi-Shinozaki K, Ishihama Y, Hirayama T, Shinozaki K (2009) Type 2C protein phosphatases directly regulate abscisic acid-activated protein kinases in Arabidopsis. Proc Natl Acad Sci USA 106:17588–17593

Umezawa T, Nakashima K, Miyakawa T, Kuromori T, Tanokura M, Shinozaki K, Yamaguchi-Shinozaki K (2010) Molecular basis of the core regulatory network in ABA responses: sensing, signaling and transport. Plant Cell Physiol 51:1821–1839

Umezawa T, Sugiyama N, Takahashi F, Anderson JC, Ishihama Y, Peck SC, Shinozaki K (2013) Genetics and phosphoproteomics reveal a protein phosphorylation network in the abscisic acid signaling pathway in *Arabidopsis thaliana*. Sci Signal 6(270):rs8

Uno Y, Furihata T, Abe H, Yoshida R, Shinozaki K, Yamaguchi-Shinozaki K (2000) Arabidopsis basic leucine zipper transcription factors involved in an abscisic acid-dependent signal transduction pathway under drought and high-salinity conditions. Proc Natl Acad Sci USA 97:11632–11637

van Loon LC, Geraats BP, Linthorst HJ (2006) Ethylene as a modulator of disease resistance in plants. Trends Plant Sci 11:184–191

Vardhini BV, Anjum NA (2015) Brassinosteroids make plant life easier under abiotic stresses mainly by modulating major components of antioxidant defense system. Front Environ Sci 2:67

Vilarrasa-Blasi J, González-García M-P, Frigola D, Fàbregas N, Alexiou KG, López-Bigas N, Rivas S, Jauneau A, Lohmann JU, Benfey PN (2014) Regulation of plant stem cell quiescence by a brassinosteroid signaling module. Dev Cell 30:36–47

Vilela B, Nájar E, Lumbreras V, Leung J, Pagès M (2015) Casein kinase 2 negatively regulates abscisic acid-activated SnRK2s in the core abscisic acid-signaling module. Mol Plant 8:709–721

Vriet C, Russinova E, Reuzeau C (2013) From squalene to brassinolide: the steroid metabolic and signaling pathways across the plant kingdom. Mol Plant 6:1738–1757

Waadt R, Manalansan B, Rauniyar N, Munemasa S, Booker MA, Brandt B, Waadt C, Nusinow DA, Kay SA, Kunz H-H (2015) Identification of open stomatal-interacting proteins reveals interactions with sucrose non-fermenting1-related protein Kinases2 and with Type 2A protein phosphatases that function in abscisic acid responses. Plant Physiol 169:760–779

Wan L, Zhang J, Zhang H, Zhang Z, Quan R, Zhou S, Huang R (2011) Transcriptional activation of OsDERF1 in OsERF3 and OsAP2-39 negatively modulates ethylene synthesis and drought tolerance in rice. PLoS ONE 6(9):e25216

Wang H, Huang Z, Chen Q, Zhang Z, Zhang H, Wu Y, Huang D, Huang R (2004) Ectopic overexpression of tomato JERF3 in tobacco activates downstream gene expression and enhances salt tolerance. Plant Mol Biol 55:183–192

Wang L, Hua D, He J, Duan Y, Chen Z, Hong X, Gong Z (2011) Auxin Response Factor2 (ARF2) and its regulated homeodomain gene HB33 mediate abscisic acid response in Arabidopsis. PLoS Genet 7(7):e1002172

Wang H, Tang J, Liu J, Hu J, Liu J, Chen Y, Cai Z, Wang X (2018) Abscisic acid signaling inhibits brassinosteroid signaling through dampening the dephosphorylation of BIN2 by ABI1 and ABI2. Mol Plant 11:315–325

Wasternack C, Hause B (2002) Jasmonates and octadecanoids: signals in plant stress responses and development. Prog Nucleic Acid Res Mol Biol 72:165–221

Watanabe H, Saigusa M, Hase S, Hayakawa T, Satoh S (2004) Cloning of a cDNA encoding an ETR2-like protein (Os-ERL1) from deep water rice (Oryza sativa L.) and increase in its mRNA level by submergence, ethylene, and gibberellin treatments. J Exp Bot 55:1145–1148

Wei Z, Li J (2016) Brassinosteroids regulate root growth, development, and symbiosis. Mol Plant 9:86–100

Wild M, Davière J-M, Cheminant S, Regnault T, Baumberger N, Heintz D, Baltz R, Genschik P, Achard P (2012) The Arabidopsis DELLA RGA-LIKE3 is a direct target of MYC2 and modulates jasmonate signaling responses. Plant Cell 24:3307–3319

Wildermuth MC, Dewdney J, Wu G, Ausubel FM (2001) Isochorismate synthase is required to synthesize salicylic acid for plant defence. Nature 414(6863):562

Wolf S, Van Der Does D, Ladwig F, Sticht C, Kolbeck A, Schürholz A-K, Augustin S, Keinath N, Rausch T, Greiner S (2014) A receptor-like protein mediates the response to pectin modification by activating brassinosteroid signaling. Proc Natl Acad Sci USA 111:15261–15266

Wu L, Chen X, Ren H, Zhang Z, Zhang H, Wang J, Wang X-C, Huang R (2007) ERF protein JERF1 that transcriptionally modulates the expression of abscisic acid biosynthesis-related gene enhances the tolerance under salinity and cold in tobacco. Planta 226:815–825

Wu L, Zhang Z, Zhang H, Wang X-C, Huang R (2008) Transcriptional modulation of ethylene response factor protein JERF3 in the oxidative stress response enhances tolerance of tobacco seedlings to salt, drought, and freezing. Plant Physiol 148:1953–1963

Wu Y, Zhang D, Chu JY, Boyle P, Wang Y, Brindle ID, De Luca V, Després C (2012) The Arabidopsis NPR1 protein is a receptor for the plant defense hormone salicylic acid. Cell Rep 1:639–647

Wu D, Ji J, Wang G, Guan C, Jin C (2014a) LchERF, a novel ethylene-responsive transcription factor from *Lycium chinense*, confers salt tolerance in transgenic tobacco. Plant Cell Rep 33:2033–2045

Wu X, He J, Chen J, Yang S, Zha D (2014b) Alleviation of exogenous 6-benzyladenine on two genotypes of eggplant (*Solanum melongena* Mill.) growth under salt stress. Protoplasma 251:169–176

Xia XJ, Wang YJ, Zhou YH, Tao Y, Mao WH, Shi K, Asami T, Chen Z, Yu JQ (2009) Reactive oxygen species are involved in brassinosteroid-induced stress tolerance in cucumber. Plant Physiol 150:801–814

Xia X, Gao CJ, Song LX, Zhou Y, Shi K, Yu JQ (2014) Role of H2O2 dynamics in brassinosteroid-induced stomatal closure and opening in Solanum lycopersicum. Plant Cell Environ 37:2036–2050

Xu Z-S, Xia L-Q, Chen M, Cheng X-G, Zhang R-Y, Li L-C, Zhao Y-X, Lu Y, Ni Z-Y, Liu L (2007) Isolation and molecular characterization of the *Triticum aestivum* L. ethylene-responsive factor 1 (TaERF1) that increases multiple stress tolerance. Plant Mol Biol 65:719–732
Xu W, Jia L, Shi W, Liang J, Zhou F, Li Q, Zhang J (2013) Abscisic acid accumulation modulates auxin transport in the root tip to enhance proton secretion for maintaining root growth under moderate water stress. New Phytol 197:139–150
Yang X, Bai Y, Shang J, Xin R, Tang W (2016) The antagonistic regulation of abscisic acid-inhibited root growth by brassinosteroids is partially mediated via direct suppression of ABSCISIC ACID INSENSITIVE 5 expressions by BRASSINAZOLE RESISTANT 1. Plant Cell Environ 39:1994–2003
Yau CP, Wang L, Yu M, Zee SY, Yip WK (2004) Differential expression of three genes encoding an ethylene receptor in rice during development, and in response to indole-3-acetic acid and silver ions. J Exp Bot 55:547–556
Yoshida T, Nishimura N, Kitahata N, Kuromori T, Ito T, Asami T, Shinozaki K, Hirayama T (2006) ABA-hypersensitive germination3 encodes a protein phosphatase 2C (AtPP2CA) that strongly regulates abscisic acid signaling during germination among Arabidopsis protein phosphatase 2Cs. Plant Physiol 140:115–126
Yoshida T, Fujita Y, Sayama H, Kidokoro S, Maruyama K, Mizoi J, Shinozaki K, Yamaguchi-Shinozaki K (2010) AREB1, AREB2, and ABF3 are master transcription factors that cooperatively regulate ABRE-dependent ABA signaling involved in drought stress tolerance and require ABA for full activation. Plant J 61:672–685
Zhai Q, Zhang X, Wu F, Feng H, Deng L, Xu L, Zhang M, Wang Q, Li C (2015) Transcriptional mechanism of jasmonate receptor COI1-mediated delay of flowering time in Arabidopsis. Plant Cell 27:2814–2828
Zhang K, Gan SS (2012) An abscisic acid-AtNAP transcription factor-SAG113 protein phosphatase 2C regulatory chain for controlling dehydration in senescing Arabidopsis leaves. Plant Physiol 158:961–969
Zhang G, Chen M, Li L, Xu Z, Chen X, Guo J, Ma Y (2009a) Overexpression of the soybean GmERF3 gene, an AP2/ERF type transcription factor for increased tolerances to salt, drought, and diseases in transgenic tobacco. J Exp Bot 60:3781–3796
Zhang S, Cai Z, Wang X (2009b) The primary signaling outputs of brassinosteroids are regulated by abscisic acid signaling. Proc Natl Acad Sci USA 106:4543–4548
Zhang H, Zhang J, Quan R, Pan X, Wan L, Huang R (2013) EAR motif mutation of rice OsERF3 alters the regulation of ethylene biosynthesis and drought tolerance. Planta 237:1443–1451
Zhao XC, Schaller GE (2004) Effect of salt and osmotic stress upon expression of the ethylene receptor ETR1 in Arabidopsis thaliana. FEBS Lett 562:189–192
Zhao ML, Wang JN, Shan W, Fan JG, Kuang JF, Wu KQ, Li XP, Chen WX, He FY, Chen JY (2013) Induction of jasmonate signalling regulators MaMYC2s and their physical interactions with MaICE1 in methyl jasmonate-induced chilling tolerance in banana fruit. Plant Cell Environ 36:30–51
Zhu JK (2002) Salt and drought stress signal transduction in plants. Annu Rev Plant Biol 53:247–273
Zhu T, Deng X, Zhou X, Zhu L, Zou L, Li P, Zhang D, Lin H (2016) Ethylene and hydrogen peroxide are involved in brassinosteroid-induced salt tolerance in tomato. Sci Rep 6:35392

Physiological Role of Gamma-Aminobutyric Acid in Salt Stress Tolerance

Syed Uzma Jalil and Mohammad Israil Ansari

Abstract Extreme salt environments are destructive to several plant and limiting the crop productivity. The effect of salt stress involves modifications in different metabolic and physiological processes. Salinity stress inhibits plant growth, to inhibit photosynthesis, to alter metabolism of reactive oxygen species, and to cause cell death that finally obstructs the production of crops. Plants synthesized various compounds that involve in signaling and play important role in stress tolerance. Gamma-aminobutyric acid (GABA) is a four-carbon, ubiquitous, non-protein amino acid. The synthesis of GABA occurs in the cytosol by an irreversible alpha-decarboxylation of L-glutamate by L-glutamate decarboxylase. During salt stress condition, the accumulation of GABA increases that helps in to stabilizing the intracellular pH of cell and carbon/nitrogen metabolism for the Krebs cycle. Exogenic GABA treatment enhanced plant growth, improved stress tolerance via scavenges free radicals, regulating enzyme activities and stabilization in plant against salinity stress. In this chapter, we highlight on the physiological and metabolic role of GABA in relation to salt stress tolerance.

1 Introduction

Salinity has come to be a foremost problem restraining the developmental mechanism of several plants. It has been accounted that 7% of the total global land is adversely influenced by salinity stress (Munns 2002). Impacts of salinity stress harm on plants are associated with different morphological, physiological, and biochemical process (Ashraf 1993). For instance, plant roots can adversely influence by increased amount of sodium and chlorine ions. The over accumulation of sodium and chlorine ions in plants leads to ion toxicity and furthermore diminishes the

S. U. Jalil
Amity Institute of Biotechnology, Amity University Uttar Pradesh, Lucknow Campus, Lucknow 226028, India

M. I. Ansari (✉)
Department of Botany, University of Lucknow, Lucknow 226007, India
e-mail: ansari_mi@hotmail.com; ansari_mi@lkouniv.ac.in

M. Hasanuzzaman and M. Tanveer (eds.), *Salt and Drought Stress Tolerance in Plants*, Signaling and Communication in Plants, https://doi.org/10.1007/978-3-030-40277-8_13

retention of other ions prompting development restraint and metabolic disruption during salinity stress. Furthermore, salt stress earnestly decreases the capacity of plants taking up water bringing about drought condition (Al-Karaki 1997; Bybordi et al. 2010a, b; Woodrow et al. 2016). Those negative impacts specifically or in a roundabout way causes generation of reactive oxygen species (ROS) that results to oxidative harm in plants during high salinity condition (Liu et al. 2007). It is notable that salt stress contrarily influences every single vegetative phase of plants, particularly for seed germination (Cuartero et al. 2006). Earlier studies have demonstrated that excessive salinity fundamentally decreased the germination rate, shoot, and root length of various plants (Zhang et al. 2007; Duan et al. 2007; Barbagallo et al. 2010). Recently, nano-biotechnology is also receiving attraction of researchers for abiotic stresses to achieve an improved agricultural production throughout the world (Jalil and Ansari 2019).

The γ-aminobutyric acid (GABA) is non-protein, ubiquitous, and one of the special interest of molecule that found in all organisms (Shelp et al. 1999). In plants, GABA involved in C/N metabolism and essential for plant development (Ansari et al. 2014). The accumulation of GABA in plants is rapidly increased due to different biotic and abiotic stress conditions such as low temperature, γ radiation, low pH, hypoxia, and darkness including leaf senescence process (Lane and Stiller 1970; Wallace et al. 1984; Ansari and Chen 2009; Shelp et al. 2012). GABA in plants acts as an active osmolyte devoid of harmful effects under the salinity-induced desiccation and had the free radicals scavenging activity in stress conditions (Carillo 2018). It has been considered that during environmental stress conditions, GABA shunt pathway can potentially involve in the regulation of GABA metabolism as shown in Fig. 1 (Al-Quraan and Al-Share 2016; Jalil et al. 2016, 2017).

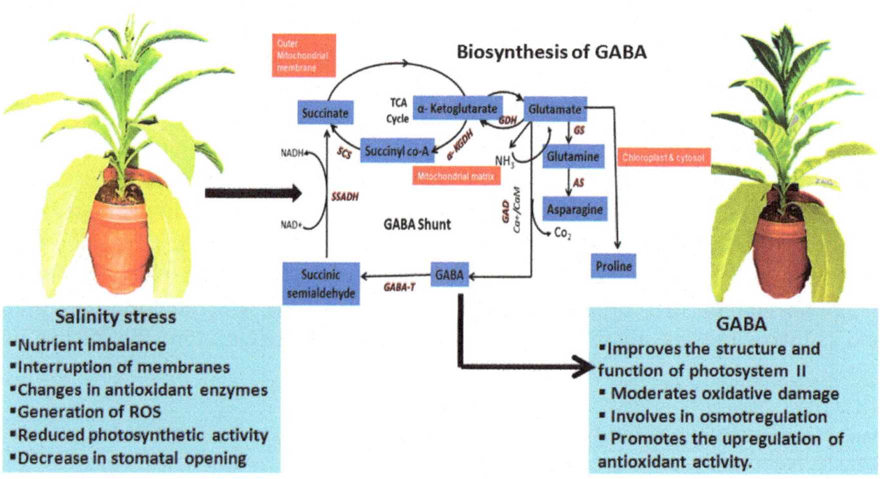

Fig. 1 The accumulation of GABA in plants is rapidly increased in salt stress. GABA in plants acts as an active osmolyte reduced harmful effects under the salinity-induced damage and had the free radicals scavenging activity in stress conditions

Moreover, the application of GABA encourages plant development, antioxidant metabolism, and expression of antioxidant genes in dose-dependent manner, thus reducing stress induced oxidative harm in plants (Shi et al. 2010; Vijayakumari et al. 2016; Li et al. 2017). Application of GABA likewise directs the osmoregulation in plant adding to the upgrade of stress resistance (Yu et al. 2014; Vijayakumari and Puthur 2015). More prominently, exogenous GABA successfully hindered the generation of free radicals and decreased oxidative harm through controlling the expression of important genes of free radical generation and genes encoding antioxidant enzyme in salinity stress (Shi et al. 2010). The treatment of GABA on seed also improved seed germination rate and diminished the salinity-induced damage during seeds germination of *Triticum aestivum* and *Zea mays* (Luo et al. 2011).

The exogenous application of GABA improves the tolerance in several plants against abiotic stresses, but diminutive data is accessible in relation to GABA-regulated salinity tolerance. This chapter exposes the GABA-regulated salinity tolerance related with antioxidative defense system, osmoregulation, and relevant genes expression during salinity stress.

2 Salt Stress Damage to Plants

The prolong exposure of salinity stress shows common symptoms of damage such as growth inhibition, senescence, and death. Growth inhibition is the key damage that prompts different side effects in spite of the fact that programmed cell death may likewise happen under extreme salt condition. Salinity stress instigates the production of abscisic acid (ABA) that close stomata while transferred to guard cells, which decreases the photosynthesis and photo inhibition and causes oxidative stress (Assmann 2003; Christmann et al. 2007). A quick influence of osmotic stress on plant development is its hindrance of cell structure directly or indirectly by ABA. Intemperate Na^+ ions on the root disturb K^+ sustenance of plant. Due to the comparative synthetic characteristic of Na^+ and K^+ ions, Na^+ ions have suppresses the absorption K^+ by the root (Christmann et al. 2007). Plant utilizes low and high-affinity approach for K^+ absorption. Na^+ ions have an even more negative impact on the low-affinity approach that has a low K^+/Na^+ selectivity (Cheng et al. 2018). Under high NaCl condition, it is fundamental for plants to activate the particular high-affinity K^+ absorption mechanism to sustain sufficient K^+ nutrition. Potassium is the most plentiful cell cation involves in retaining cell turgidity, membrane stability, and enzyme activities, and its deficiency certainly prompts to growth inhibition. Under salt stress, Na^+ enters into the cytoplasm that hinders various enzymes activities. This hindrance is also dependent on how much K^+ is available: a high Na^+/K^+ proportion is the most injurious (Jouyban 2012).

3 Salinity Stress Effects on Nutrient Absorption

Salinity stress causes nutrient deficiencies in the cell that decrease plant development by influencing the availability, transportation, and distributing of nutrients. Conversely, salt condition differentially effect the mineral nourishment of plants and causes nutrient imbalances, due to the rivalry of Na^+ and Cl^- with supplements, for example, K^+, Ca^{2+}, and NO_3^-. In salinity environments, plant development decreases due ion toxicities and ionic uneven characters following up on biophysical as well as metabolic segments of plant development (Grattan and Grieves 1999). Elevated NaCl level accounted for to prompt increments in Na^+ and Cl^- and in addition diminishes in nitrogen, phosphorus, calcium, potassium and magnesium content in *Foeniculum vulgare* (Abd El-Wahab 2006); Ajwain (Ashraf and Orooj 2006); *Mentha piperata, Aloysia citrodora* (Tabatabaie and Nazari 2007), Chamomile (Baghalian et al. 2008), and *Achillea* (Abd EL-Azim and Ahmed 2009).

4 Metabolic Changes Under Salinity Stress

Salinity stress inspired a transformation from C3 into the crassulacean acid metabolism (CAM) system in *Mesembryanthemum crystallinum* (Ishimaru 1999). A portion of the enzyme mechanism for CAM metabolism is actuated by a very short period of salinity stress condition. The primary preferred standpoint of the CAM metabolism is elevated moisture usage efficiency due to the stomata possibly open around at dark hours while minimum water is evaporated. A metabolic change in plants is an aggregation of low molecular weight organic solutes during extreme salinity conditions. These solutes incorporate direct polyols (glycerol, mannitol, or sorbitol), cyclic polyols (inositol or pinitol and other mono-and dimethylated inositol subsidiaries), amino acids (glutamate or proline), and betaines (glycine betaine or alanine betaine) (Ishimaru 1999). Plants that regularly encounter nitrogen confinement may aggregate dimethyl sulfonium propionate that proportionate to the nitrogen-containing betaines. The organic solutes at increased concentration are not destructive cell assemblies. Consequently, these are regularly suggested as good osmolytes (Jouyban 2012).

Compatible solutes at large amount positively involve in osmoregulation. This is particularly valid in halophytes, which frequently accumulate osmolytes in the range of 0.5–4.0 mol L^{-1} in the cells (Al Hassan et al. 2017). The increase concentrations of osmolytes exist in cytoplasm, to equilibrium the increase conc. of salt exterior to the cell also hindering the high concentrations of Na^+ and Cl^- ions in the vacuole. Not all the organic solutes destructive but some have a defensive impact against harm by toxic ions or lack of hydration. Genes for proline, polyol biosynthesis was overexpressed in transgenic *Nicotiana tabacum* and *Arabidopsis thaliana* (Lilius et al. 1996; Hayashi et al. 1997). The genetically engineered plants constitutively creating osmolytes execute superior in comparison to control plants in salinity stress. The defensive impact cannot be completely clarified by the osmotic change hypothesis because the transgenic plants just deliver a few mmol L^{-1} a greater amount of the designed osmolytes, focuses too low to even think about contributing fundamentally to osmoregulation (Jouyban 2012). It recommends that the low concentration of osmolytes can secure plants by scavenges ROS induced by salinity stress. Significance of antioxidants in salinity stress resistance is additionally bolstered by trials appearing progressive outcome of genetically engineered development of ROS scavenging enzymes on plant execution during high salt condition.

5 Biosynthesis and Metabolism of GABA in Plants

GABA is one of the special interest of molecule among stress responsive metabolites and synthesize in prokaryotes, yeast, animals, and plants (Shelp et al. 1999). GABA is involved as neurotransmitter in nervous system of mammals and its metabolization was done in cytosol (Breitbreuz and Shelp 1995; Kinnersley and Turano 2000; Reimer et al. 2001; Bouche and Fromm 2004). GABA is highly water-soluble and presents in both bound and unbound forms presents in almost all plant tissues (Satyanarayan and Nair 1990). Inactive concentration of unbound GABA normally present in the concentration of 0.03–2.00 μmol g^{-1} FW in plant cells (Shelp et al. 1999), whereas bound forms present in 6.6% of the total dry weight of root nodules in legumes (Satyanarayan and Nair 1990). GABA involved in many mechanisms of plants such as metabolic activities and signaling. The accumulation of GABA in plants is rapidly increased due to different stress conditions (Kinnersley and Turano 2000; Lane and Stiller 1970; Wallace et al. 1984; Bown and Shelp 1997). GABA metabolism was generally upregulated under NaCl induced stress in *A. thaliana* (Renault et al. 2010). These stresses initiate signal transduction pathway in which increased cytosolic calmodulin-dependent activity of the GAD. However, it appears that GABA is important for vegetative development.

GABA metabolism occur in two compartments of cell; synthesis of GABA synthesis takes place in the cytosol whereas degradation of GABA in the mitochondria. GABA is an intermediate product in the synthesis of succinate from L-glutamate via a set of reactions collectively called GABA shunt pathway (Fig. 1). The three reactions in the pathway are: in the first reaction, glutamate is irreversibly decarboxylated by enzyme glutamate decarboxylase to form GABA. In second step, reversible transamination of GABA takes place by enzyme GABA-transaminase that utilized α-ketoglutarate for transferring of amino group in mammalian cells, while GABA-transaminase used pyruvate to form succinic semialdehyde in plant. The third reaction starts with the irreversible oxidation of succinic semialdehyde by succinic semialdehyde dehydrogenase to form succinate that enter into Krebs cycle (Bown and Shelp 1997). Glutamate decarboxylase enzyme existed in cytosol (Breitbreuz and Shelp 1995) and precise for L-glutamate, retains an auto inhibitory calmodulin-binding domain, and has an optimum pH of 5.8. GABA-transaminase and succinic semialdehyde are present in mitochondria (Breitbreuz and Shelp 1995) with optimum pH of 8–10 and 9, respectively. Synthesis of GABA occurs in the cytosol by glutamate decarboxylase and transferred to the mitochondrion for consequent conversion into succinate by GABA-transaminase and succinic semialdehyde dehydrogenase. GABA shunt components and GABA has performs various functions, depending upon species to species on the basis of the compartment in the cell where they active.

6 Mechanism and Function of GABA in Plants in Salinity Stress

Studies in plants stressed principally on the mechanism of GABA metabolism, generally in the perspective of stress effect, till some finding were reported to indicate a signaling function of GABA in plants in salt stress condition.

6.1 GABA Promotes Photosynthesis and Chlorophyll Fluorescence in Salt Stress

Photosynthesis is an integral mechanism used by plants to transform light energy into chemical energy that can be released as organic complexes via CO_2 fixation (Omoto et al. 2012). Salinity stress obstructs absorption ability of plants, caused deficiency of nutrients, and declines chlorophyll content, therefore hindering photosynthesis activity (Buttery and Buzzell 1977). NaCl hinders photosynthetic activity by decreasing CO_2 fixation, stomatal conductance, and transpiration. A sensible clarification is the photosynthetic activity experienced harm from salinity stress, bringing about the decrease of CO_2 assimilation. PSII is deliberated as the essential location of the photosynthetic system damage in stress condition (Sharkey and Zhang 2010). Damage to PSII can prompt an adjustment in chlorophyll fluorescence. In this manner, chlorophyll fluorescence utilized as an influential process to evaluate modifications in the purpose of PSII and to study the essential photosynthetic methods in stressful environment (Hajiboland et al. 2010). Maximal fluorescence (Fm), most extreme quantum proficiency of PSII photochemistry (Fv/Fm) and electron transport rate diminished and minimal fluorescence (F0) expanded in maize seedlings and coastal plant species during salt stress (Wang et al. 2017; Naumann et al. 2007).

Extreme salt condition harms the oxygen-advancing system and disturbs the electron transport and photochemical efficiency (Steponkus and Lanphear 1967). Declined PSII activity in salt stress is viewed as the aftereffect of reduced excitation energy getting PSII reaction centres and modifications in protein-pigment complexes of thylakoids membrane (Misra et al. 2001). Few studies revealed that plants that exposed with GABA + NaCl showed higher net photosynthetic rate and stomatal conductance than those that treated only with NaCl (Jiang et al. 2006) and that exogenous GABA improved transpiration rate and intercellular CO_2 concentration in under different level of salinity stress (Li et al. 2015). The GABA + NaCl plants had higher Fv/Fm and Fm during extreme salt condition in comparison to NaCl-treated plants (Karaba et al. 2007). Moreover, exogenous GABA recovers the assembly and role of photosystem II in muskmelon seedlings under salt stress (Xiang et al. 2016). In addition, chlorophyll fluorescence improved in GABA + NaCl treated plants as compared to only NaCl-treated plants. Exogenous GABA can decrease the accumulation of toxic constituents (Cao et al. 2012), sustain cell morphology (Almeselmani et al. 2009), and enhance the cell function under salinity stress. These defensive impacts can defend photosystem II from salinity stress induced damage and increase chlorophyll fluorescence parameters. Exogenous GABA can facilitate the salinity stress induced damage and escalated leaf relative water content. Furthermore, due to the enhancement of cell function, net photosynthetic rate was increased. This impact enhanced the use of CO_2 and diminished intercellular CO_2 concentration in maize seedling leaves (Wang et al. 2017).

6.2 GABA Moderates Oxidative Damage in Plants Under Salt Stress

Salinity stress may transport calcium ions from plasma membrane binding sites, prompting membrane leakiness that elevates the transudation of ions, commencing an expansion in leaf relative conductivity (Pitann et al. 2013). Toxicity of NaCl in plant damages cellular organelles, decreasing cellular reduction ability and degrades chlorophyll prior to plant demise (Davenport et al. 2005; Quintero et al. 2007). The halotolerant cyanobacterium *Aphanothece halophytica* revealed high amount of GABA under salt stress proposing the prospective of genetic manipulation of the genes involved in GABA synthesis to increase GABA content (Boonburapong et al. 2016). Exogenic treatment of GABA can control the osmotic equilibrium in plant for improving the resistance against adverse environmental conditions (Vijayakumari and Puthur 2015; Yu et al. 2014; Malekzadeh et al. 2014). Seed treated with GABA improved seed germination rate and decrease the salinity-induced damage of *Triticum aestivum* and *Zea mays* (Luo et al. 2011). It has been observed that the exogenous GABA enhanced the shoot length, root length, and fresh weight of *Z. mays* L. seedling in salinity stress (Wang et al. 2017). Some studies shows that salinity stress escalates the membrane damage and declines the relative leaf water in NaCl treated *Z. mays* seedlings (Wang et al. 2017).

During high temperature or hypoxia stress, membrane permeability, leaf pigment, and role of mitochondria were refurbished of rice seedlings by the treatment of GABA (Nayyar et al. 2014). Succinic semialdehyde dehydrogenase (SSADH) mutants of *A. thaliana* are oversensitive to stress conditions because of defective scavenging ability for hydrogen peroxide. In the final reaction, the GABA shunt pathway provided succinate and NADH to electron transport chain. Thus, it was proposed that the degradation of GABA could restrict the generation of reactive oxygen species during oxidative stress that limits some of the enzymes of the TCA cycle. It has been reported that GABA shunt genes knocked out mutants of yeast and show more sensitivity to hydrogen peroxide (Coleman et al. 2001; Bouché et al. 2003). The mechanism of these refurbished functions under salinity-stressed cells by the application of GABA still needs to be investigated and might be related to enhancement of water eminence followed by decrease in oxidative damage to the cell structure.

6.3 GABA Involve in Osmotic Adjustment via Accumulation of the Osmolytes Under Salt Stress

Osmotic adjustment is a strategic acclimatization of plants cells to limit the impacts of salt-actuated drought stress. GABA can work as an osmolyte or can encourage the production of osmolytes during drought condition (Shelp et al. 1999). Proline synthesis escalates in *Z. mays* under salinity stress (Kaya et al. 2010). Additionally, it has been observed that there is an increase of osmolytes (proline and glycine betaine) in *Z. mays* seedlings during salinity stress (Mansour et al. 2005). Furthermore, it has been reported that the proline and soluble sugar considerably escalates in 150 mM NaCl-treated *Z. mays* seedlings but significantly diminishes in 300 mM NaCl-treated seedlings (Wang et al. 2017). The impact of salinity stress causes osmotic stress at the primary level and ion toxicity at the secondary level (Fortmeier and Schubert 1995). It has been suggested that the GABA play key role as an effective osmolyte at cellular level in plants without harmful impact during the dehydration stress induced by high salt condition (Carillo 2018).

Exogenous treatment of GABA augmented the proline and soluble sugar accumulation in 150 and 300 mM NaCl treated *Z. mays* plants. 150 mM NaCl-treated plants regulates GABA; accumulates high proline and soluble sugar as compared to 300 mM NaCl-treated plants (Wang et al. 2017). These findings show that application of GABA-enhanced osmolytes level under adverse condition and developed the capability of plants to mitigate with abiotic stresses (Chen et al. 2007). The Krebs cycle is the fundamental method for aerobic organisms to get energy and a distinctive metabolic pathway of carbohydrates, fatty acids, and proteins oxidation. Succinate and α-ketoglutarate are essential transitional products of the Krebs cycle and synthesized via oxidation of GABA. GABA also transported via proline transporters from AtProT2 of *A. thaliana* and LeProT1 of tomato. During water or salinity stress, AtProT2 is highly triggered. Proline-GABA transporters also play protective role during osmotic stress by transporting organic osmolytes (Breitkreuz et al. 1999; Schwacke et al. 1999). Enhancement of proline and soluble sugar level by exogenous GABA may be indirectly related with its involvement in the Krebs cycle.

6.4 GABA Promotes the Upregulation of Antioxidant Activity

The ROS generated by salinity stress causes membrane damage and lipid peroxidation by extreme escalation of malonaldehyde (MDA) content (Mittler et al. 2004). MDA accumulation is viewed as a marker of plant oxidative oxidative stress (Taulavuori et al. 2001). The antioxidant system, including superoxide dismutase (SOD), ascorbate peroxidase (APX), catalase (CAT) and peroxidase (POD), are involved in ROS regulation (Qiu et al. 2013). The activity of antioxidant enzymes such as SOD, POD, CAT, and APX increased in maize seedlings in salinity stress (Wang et al. 2017). Few studies also observed that CAT activity reduced whereas SOD activity enhanced under salinity (Zhu et al. 2004; Li et al. 2010). Furthermore, MDA and superoxide anion ($O_2^{\bullet -}$) enhanced directly proportional to salinity stress. The activity of antioxidant enzymes increased to decrease damage from oxidative stress in NaCl-treated plants (Wang et al. 2017). Maize plants confronting salinity stress utilize an assortment of adaptive systems at the molecular level to oppose the negative impacts of salt stress. It shows that the upregulation and downregulation of various gene transcriptions are essential (Zörb et al. 2004). Transcriptomic analysis revealed that the antioxidant defense system plays significant roles in *Medicago sativa* L. in response to salt stress by analysing the expression pattern of ROS-detoxification-related genes and their physiological changes (Luo et al. 2019).

The antioxidant genes transcripts are prompted to defend cells from oxidative damage during salt stress. CAT activity escalates because of the stimulation of mRNA due to increased free radicals levels during salinity stress in *Z. mays* (Menezes-Benavente et al. 2004). The gene expressions of superoxide dismutase, ascorbate peroxidase, and peroxidase elevated but that of catalase declined during salinity stress in *Solanum lycopersicum* plants. The gene expressions of ascorbate peroxidase and peroxidases escalated in 200 mM NaCl but were lesser than in 100 mM NaCl treated plants (Zhang et al. 2015). High NaCl concentrations can inhibit gene expression, ensuing in a decrease in antioxidant enzyme activities. Exogenous GABA reduced MDA and $O_2^{\bullet -}$ level in 150 and 300 mM NaCl-treated maize seedlings (Wang et al. 2017). The seedlings treated with GABA + NaCl showed increased antioxidant enzyme activity as compared to NaCl-treated plants (Wang et al. 2017). Moreover, some investigations shows that MDA content was lesser in the GABA + NaCl-treated plants in comparison to the NaCl-treated plants that constant with high antioxidant enzymes activities (Li et al. 2015). Further, it has been observed that an appropriate amount of GABA application can stimulate plant development, antioxidant regulation, and transcription levels of antioxidant genes, thus lessening oxidative destruction by salt stress in plant (Li et al. 2017).

GABA reveals the key function of ROS scavenges activity in plants (Carillo 2018). It has been showed that exogenic GABA can considerably increases the various antioxidant enzyme activities in *Oryza sativa* (Nayyar et al. 2014), *Piper nigrum* seedlings (Vijayakumari and Puthur 2015), and *Lolium perenne* (Krishnan et al. 2013). Application of 1–2 mmol/L GABA on *Z. mays* seeds reduced the impact of high NaCl level on seed germination by increasing the antioxidant enzyme activities

(Tian et al. 2014). Exogenous GABA also decreases the low light stress damage by regulating the antioxidant activity (Li et al. 2017). Furthermore, GABA-treated enhanced antioxidant enzymes activities in *S. lycopersicum* seedlings in relation with decreases in the ROS generation and oxidative damage during NaCl stress (Luo et al. 2011). Additionally, application of GABA efficiently repressed the accumulation of hydrogen peroxide free radicals and decrease the oxidative damage by regulation of the genes expression of hydrogen peroxide synthesis and antioxidant enzyme in *Caragana intermedia* roots, proposing that GABA acts as a signalling ligand to regulate the gene expression under salinity stress (Shi et al. 2010). Further, exogenous GABA application induced the accretion of endogenous GABA, proline and other compounds, thus increasing the antioxidant activity in germinated hulless barley in salt stress (Maa et al. 2018). The mechanism that allows GABA to prompt the upregulation of antioxidant enzymes need to be studied. Exogenous GABA application alleviated oxidative stress in roots of *Hordeum vulgare* seedlings under Aluminium and proton stress (Song et al. 2010) and their study revealed that antioxidant enzymes (SOD, CAT, and POD) activities unregulated in response to declined MDA and ROS level after the application of GABA (Beuve et al. 2004).

7 Conclusions

Salinity stress restricts the growth and production of crops, and comes to be severe problem for the agronomists. GABA involved in many mechanisms of plants such as metabolic activities and signaling. It has been considered that during environmental stress conditions, GABA shunt pathway can potentially involve in the regulation of GABA metabolism. The accumulation of GABA in plants is rapidly increased due to different biotic and abiotic stress conditions. The application of GABA encourages plant development, antioxidant metabolism, and expression of antioxidant genes in dose-dependent manner and can alleviate the growth restriction and oxidative destruction induced by salinity stress and prompts the accretion of endogenous GABA and other osmolytes. GABA alleviates the effect of salinity stress on plants that improves crop productivity.

References

Abd EL-Azim WM, Ahmed STh (2009) Effect of salinity and cutting date on growth and chemical constituents of *Achillea fragratissima* Forssk, under Ras Sudr conditions. Res J Agr Biol Sci 5:1121–1129

Abd El-Wahab MA (2006) The efficiency of using saline and fresh water irrigation as alternating methods of irrigation on the productivity of *Foeniculum vulgare* Mill subsp. vulgare var. vulgare under North Sinai conditions. Res J Agr Biol Sci 2:571–577

Al Hassan M, Estrelles E, Soriano P, López-Gresa MP, Bellés JM, Boscaiu M, Vicente O (2017) Unraveling salt tolerance mechanisms in halophytes: a comparative study on four Mediterranean Limonium species with different geographic distribution patterns. Front Plant Sci 8:1438

Al-Karaki G (1997) Barley response to salt stress at varied levels of phosphorus. J Plant Nutr 20:1635–1643
Almeselmani M, Deshmukh PS, Sairam RK (2009) High temperature stress tolerance in wheat genotypes: role of antioxidant defence enzymes. Acta Agron Hung 57:1–14
Al-Quraan NA, Al-Share AT (2016) Characterization of the γ-aminobutyric acid shunt pathway and oxidative damage in *Arabidopsis thaliana* pop 2 mutants under various abiotic stresses. Biol Plantarum 60:132–138
Ansari MI, Chen SCG (2009) Biochemical characterization of gamma-aminobutyric acid (GABA): pyruvate transaminase during rice leaf senescence. Int J Integr Biol 16:27–32
Ansari MI, Hasan S, Jalil SU (2014) Leaf senescence and GABA shunt. Bioinformation 10:730–732
Ashraf M (1993) Effect of sodium chloride on water relations and some organic osmotica in arid zone plant species *Melilotus indica* (L.). All Tropenlandwirt 94:95–102
Ashraf M, Orooj A (2006) Salt stress effects on growth, ion accumulation and seed oil concentration in an arid zone traditional medicinal plant ajwain (*Trachyspermum ammi* [L.] Sprague). J Arid Environ 64:209–220
Assmann SM (2003) OPEN STOMATA1 opens the door to ABA signaling in *Arabidopsis* guard cells. Trends Plant Sci 8:151–153
Baghalian K, Haghiry A, Naghavi MR, Mohammadi A (2008) Effect of saline irrigation water on agronomical and phytochemical characters of chamomile (*Matricaria recutita* L.). Scientia Hort 116:437–441
Barbagallo A, Nicola AD, Missikoff M (2010) The influence of salt stress on seed germination, growth and yield of canola cultivars. Notulae Bot Horti Agrobotanici Cluj-Napoca 38:128–133
Beuve N et al (2004) Putative role of γ-amino butyric acid (GABA) as a long distance signal in upregulation of nitrate uptake in *Brassica napus* L. Plant Cell Environ 27:1035–1046
Boonburapong B, Laloknam S, Incharoensakdi A (2016) Accumulation of gamma-aminobutyric acid in the halotolerant cyanobacterium *Aphanothece halophytica* under salt and acid stress. J Appl Phycol 28:141–148
Bouche N, Fromm H (2004) GABA in plants: just a metabolite? Trends Plant Sci 9:110–115
Bouché N, Fait A, Bouché D, Moller SG, Fromm H (2003) Mitochondrial succinate-semialdehyde dehydrogenase of the gamma-aminobutyrate shunt in required to restricts levels of reactive oxygen intermediates in plants. Proc Natl Acad Sci USA 100:6843–6848
Bown AW, Shelp BJ (1997) The metabolism and function of gamma aminobutyric acid. Plant Physiol 115:1–5
Breitbreuz KE, Shelp BJ (1995) Subcellular compartmentation of the 4-aminobutyric shunt in protoplast from developing soybean cotelydons. Plant Physiol 108:99–103
Breitkreuz KE, Shelp BJ, Fischer WN, Schwacke R, Rentsch D (1999) Identification and characterization of GABA, proline and quaternary ammonium compound transporters from Arabidopsis thaliana. FEBS Lett 450 (3):280–284
Buttery BR, Buzzell RI (1997) The relationship between chlorophyll content and rate of photosynthesis in soybeans. Can J Plant Sci 57:1–5
Bybordi A, Tabatabaei SJ, Ahmedov A (2010a) Effect of salinity on the growth and peroxidase and iaa oxidase activities in canola. J Food Agric Environ 8:109–112a
Bybordi A, Tabatabaei SJ, Ahmedov A (2010b) Effects of salinity on fatty acid composition of canola (*Brassica napus* L.). J Food Agric Environ 8:113–115b
Cao SF, Cai YT, Yang ZF, Zheng YH (2012) MeJA induces chilling tolerance in loquat fruit by regulating proline and γ-aminobutyric acid contents. Food Chem 133:1466–1470
Carillo P (2018) GABA shunt in durum wheat. Front Plant Sci 9:100
Chen Z et al (2007) Compatible solute accumulation and stress-mitigating effects in barley genotypes contrasting in their salt tolerance. J Exp Bot 58:4245–4255
Cheng B, Li Z, Liang L, Cao Y, Zeng W, Zhang X, Ma X, Huang L, Nie G, Liu W, Peng Y (2018) The γ-Aminobutyric Acid (GABA) alleviates salt stress damage during seeds germination of white clover associated with Na+/K+ transportation, dehydrins accumulation, and stress-related genes expression in white clover. Int J Mol Sci 19(9):2520

Christmann A, Weiler EW, Steudle E, Grill E (2007) A hydraulic signal in root-to-shoot signaling of water shortage. Plant J 52:167–174

Coleman ST, Fang TK, Rovinsky SA et al (2001) Expression of a glutamate decarboxylase homologue is required for normal oxidative stress tolerance in *Saccharomyces cerevisiae*. J Biol Chem 276:244–250

Cuartero J, Bolarín MC, Asíns MJ, Moreno V (2006) Increasing salt tolerance in the tomato. J Exp Bot 57:1045–1058

Davenport R, James RA, Zakrisson-Plogander A, Tester M, Munns R (2005) Control of sodium transport in durum wheat. Plant Physiol 137:807–818

Duan DY, Li WQ, Liu XJ, Ouyang H, An P, Duan DY, Liu WQ (2007) Seed germination and seedling growth of Suaeda salsa under salt stress. Ann Bot Fenn 44:161–169

Fortmeier R, Schubert S (1995) Salt tolerance of maize (Zea mays L.): the role of sodium exclusion. Plant Cell Environ 18 (9):1041–1047

Grattan SR, Grieves CM (1999) Salinity-mineral nutrient relations in horticultural crops. Scientia Hort 78:127–157

Hajiboland R, Aliasgharzadeh N, Laiegh SF, Poschenrieder C (2010) Colonization with arbuscular mycorrhizal fungi improves salinity tolerance of tomato (*Solanum lycopersicum* L.) plants. Plant Soil 331:313–327

Hayashi H, Alia Mustardy L, Deshnium P, Ida M, Murata N (1997) Transformation of *Arabidopsis thaliana* with the codA gene for choline oxidase; accumulation of glycinebetaine and enhanced tolerance to salt and cold stress. Plant J 12:133–142

Ishimaru K (1999) Transformation of a CAM plant, the facultative halophyte *Mesembryanthemum crystallinum* by *Agrobacterium tumefaciens*. Plant Cell Tissue Organ Cult 57:61–63

Jalil SU, Ansari MI (2019) Nanoparticles and abiotic stress tolerance in plants: synthesis, action and signalling mechanisms. In: Khan MIR, Sudhakar P, Ferrante RA, Khan N (eds) Plant signaling molecules-role and regulation under stressful environments, 1st edn. Elsevier, pp 549–561

Jalil SU, Ahmad I, Ansari MI (2016) Screening of *Arabidopsis thaliana* GABA—transaminase mutant and its root pattern analysis. J Biol Chem Res 33:909–917

Jalil SU, Ahmad I, Ansari MI (2017) Functional loss of GABA transaminase (GABA-T) expressed early leaf senescence under various stress conditions in *Arabidopsis thaliana*. Curr Plant Biol 9–10:11–22

Jiang Q, Roche D, Monaco TA, Hole D (2006) Stomatal conductance is a key parameter to assess limitations to photosynthesis and growth potential in barley genotypes. Plant Biol 8:515–521

Jouyban Z (2012) The effects of salt stress on plant growth. Tech J Eng Appl Sci 2:7–10

Karaba A et al (2007) Improvement of water use efficiency in rice by expression of HARDY, an *Arabidopsis* drought and salt tolerance gene. Proc Natl Acad Sci USA 104:15270–15275

Kaya C, Tuna AL, Okant AM (2010) Effect of foliar applied kinetin and indole acetic acid on maize plants grown under saline conditions. Turk J Agric For 34:529–538

Kinnersley AM, Turano FJ (2000) g-Aminobutyric acid (GABA) and plant responses to stress. Crit Rev Plant Sci 19:479–509

Krishnan S, Laskowski K, Shukla V, Merewitz EB (2013) Mitigation of drought stress damage by exogenous application of a non-protein amino acid gamma aminobutyric acid on perennial ryegrass. J. Am Soc Hort Sci 138:358–366

Lane TR, Stiller M (1970) Glutamic acid decarboxylation in Chlorella. Plant Physiol 45:558–562

Li Y, Bai Q, Jin X, Wen H, Gu Z (2010) Effects of cultivar and culture conditions on γ-aminobutyric acid accumulation in germinated fava beans (*Vicia faba* L.). J Sci Food Agric 90:52–57

Li Z et al (2015) Exogenously applied spermidine improves drought tolerance in creeping bentgrass associated with changes in antioxidant defense, endogenous polyamines and phytohormones. Plant Growth Regul 76:71–82

Li Y, Fan Y, Ma Y, Zhang Z, Yue H, Wang L, Li J, Jiao Y (2017) Effects of exogenous-aminobutyric acid (GABA) on photosynthesis and antioxidant system in pepper (*Capsicum annuum* L.) seedlings under low light stress. J Plant Growth Regul 36:1–14

Lilius G, Holmberg N, Bülow L (1996) Enhanced NaCl stress tolerance in transgenic tobacco expressing bacterial choline dehydrogenase. Biotechnology (NY) 14:177–180

Liu K, Xu S, Xuan W, Ling T, Cao Z, Huang B, Sun Y, Fang L, Liu Z, Zhao N (2007) Carbon monoxide counteracts the inhibition of seed germination and alleviates oxidative damage caused by salt stress in *Oryza sativa*. Plant Sci 172:544–555

Luo HY, Yang LW, Gao HB, Xiao-Lei WU, Liu HH (2011) Physiological mechanism of GABA soaking to tomato seed germination and seedling development under NaCl stress. Acta Bot Boreal-Occident Sin 31:2235–2242

Luo D, Zhou Q, Wu Y, Chai X et al (2019) Full-length transcript sequencing and comparative transcriptomic analysis to evaluate the contribution of osmotic and ionic stress components towards salinity tolerance in the roots of cultivated alfalfa (*Medicago sativa* L.). BMC Plant Biol 19:32

Maa Y, Wanga P, Chena Z, Gua Z, Yanga R (2018) GABA enhances physio-biochemical metabolism and antioxidant capacity of germinated hulless barley under NaCl stress. J Plant Physiol 231:192–201

Malekzadeh P, Khara J, Heydari R (2014) Alleviating effects of exogenous gamma-aminobutiric acid on tomato seedling under chilling stress. Physiol Mol Biol Plants 20:133–137

Mansour MMF, Salama KHA, Ali FZM, Abou HAF (2005) Cell and plant responses to NaCl in *Zea mays* cultivars differing in salt tolerance. Gen Appl Plant Physiol 31:29–41

Menezes-Benavente KSP, Margis-Pinheiro M, Scandalios JG (2004) Salt-induced antioxidant metabolism defenses in maize (*Zea mays* L.) seedlings. Redox Rep 9:29–36

Misra AN, Srivastava A, Strasser RJ (2001) Utilization of fast chlorophyll a fluorescence technique in assessing the salt/ion sensitivity of mung bean and brassica seedlings. J Plant Physiol 158:1173–1181

Mittler R, Vanderauwera S, Gollery M, Van Breusegem F (2004) Reactive oxygen gene network of plants. Trends Plant Sci 9:490–498

Munns R (2002) Comparative physiology of salt and water stress. Plant Cell Environ 25:239–250

Naumann JC, Young DR, Anderson JE (2007) Linking leaf chlorophyll fluorescence properties to physiological responses for detection of salt and drought stress in coastal plant species. Physiol Plant 131:422–433

Nayyar H, Kaur R, Kaur S, Singh R (2014) g-Aminobutyric acid (GABA) imparts partial protection from heat stress injury to rice seedlings by improving leaf turgor and upregulating osmoprotectants and antioxidants. J Plant Growth Regul 33:408–419

Omoto E, Taniguchi M, Miyake H (2012) Adaptation responses in C4 photosynthesis of maize under salinity. J Plant Physiol 169:469–477

Pitann B, Mohamed AK, Neubert AB, Schubert S (2013) Tonoplast Na+/H+ antiporters of newly developed maize (*Zea mays*) hybrids contribute to salt resistance during the second phase of salt stress. J Plant Nutr Soil Sci 176:148–156

Qiu ZB, Li JT, Zhang MM, Bi ZZ, Li ZL (2013) He-Ne laser pretreatment protects wheat seedlings against cadmium-induced oxidative stress. Ecotoxicol Environ Saf 88:135–141

Quintero JM, Fournier JM, Benlloch M (2007) Na+ accumulation in shoot is related to water transport in K+-starved sunflower plants but not in plants with a normal K+ status. J Plant Physiol 164:60–67

Reimer RJ, Fremeau RT, Bellocchio EE, Edwards RH (2001) The senesence of excitation. Curr Opin Cell Biol 13:417–420

Renault H, Roussel V, El-Amrani A, Arzel M, Renault D, Bouchéreau A, Deleu C (2010) The Arabidopsis pop2-1 mutant reveals the involvement of GABA transaminase in salt stress tolerance. BMC Plant Biol 10:20

Satyanarayan V, Nair PM (1990) Metabolism, enzymology and possible roles of 4-aminobutyrate in higher plants. Phytochemistry 29:367–375

Schwacke R et al (1999) LeProT1, a transporter for proline, glycine betaine, and gamma-amino butyric acid in tomato pollen. Plant Cell 11:377–392

Sharkey TD, Zhang R (2010) High temperature effects on electron and proton circuits of photosynthesis. J Integr Plant Biol 52:712–722

Shelp BJ, Bown AW et al (1999) Metabolism and function of gamma aminobutyric acid. Trends Plant Sci 4:446–452

Shelp BJ, Bozzo GG, Trobacher CP, Zarei A, Deyman KL, Brikis CJ (2012) Hypothesis/review: contribution of putrescine to 4-aminobutyrate (GABA) production in response to abiotic stress. Plant Sci 193–194:130–135

Shi SQ, Zheng S, Jiang ZP, Qi LW, Sun XM, Li CX, Liu JF, Xiao WF, Zhang SG (2010) Effects of exogenous GABA on gene expression of Caragana intermedia roots under NaCl stress: regulatory roles for H2O2 and ethylene production. Plant Cell Environ 33:149–162

Song H, Xu X, Wang H, Wang H, Tao Y (2010) Exogenous gamma-aminobutyric acid alleviates oxidative damage caused by aluminium and proton stresses on barley seedlings. J Sci Food Agric 90:1410–1416

Steponkus PL, Lanphear FO (1967) Refinement of the triphenyl tetrazolium chloride method of determining cold injury. Plant Physiol 42:1423–1426

Tabatabaie SJ, Nazari J (2007) Influence of nutrient concentration and NaCl salinity on growth, photosynthesis and essential oil content of peppermint and lemon verbena. Turk J Agric 31:245–253

Taulavuori E, Hellström EK, Taulavuori K, Laine K (2001) Comparison of two methods used to analyse lipid peroxidation from *Vaccinium myrtillus* (L.) during snow removal, reacclimation and cold acclimation. J Exp Bot 52:2375–2380

Tian Y, Guan B, Zhou D, Yu J, Li G, Lou Y (2014) Responses of seed germination, seedling growth, and seed yield traits to seed pretreatment in maize (*Zea mays* L.). Sci World J 2014:8

Vijayakumari K, Puthur JT (2015) g-Aminobutyric acid (GABA) priming enhances the osmotic stress tolerance in *Piper nigrum* linn. plants subjected to PEG-induced stress. Plant Growth Regul 78:1–11

Vijayakumari K, Jisha C, Puthur JT (2016) GABA/BABA priming: a means for enhancing abiotic stress tolerance potential of plants with less energy investments on defence cache. Acta Physiol Plant 38:230

Wallace W, Secor J, Schrader LE (1984) Rapid accumulation of gamma-aminobutyric acid and alanine in soybean leaves in response to an abrupt transfer to lower temperature, darkness, or mechanical manipulation. Plant Physiol 75:170–175

Wang Y, Gu W, Yao M, Xie T, Li L, Jing L, Shi W (2017) g-Aminobutyric acid imparts partial protection from salt stress injury to maize seedlings by improving photosynthesis and upregulating osmoprotectants and antioxidants. Sci Rep 7:43609

Woodrow P, Ciarmiello LF, Annunziata MG, Pacifico S et al (2016) Durum wheat seedling responses to simultaneous high light and salinity involve a fine reconfiguration of amino acids and carbohydrate metabolism. Physiol Plant 159:290–312

Xiang L, Hu L, Xu W, Zhen A, Zhang L, Hu X (2016) Exogenous γ-aminobutyric acid improves the structure and function of photosystem II in muskmelon seedlings exposed to salinity-alkalinity stress. PLoS ONE 11:e0164847

Yu C, Zeng L, Sheng K, Chen F, Zhou T, Zheng X, Yu T (2014) g-Aminobutyric acid induces resistance against *Penicillium expansum* by priming of defence responses in pear fruit. Food Chem 159:29–37

Zhang S, Hu J, Zhang Y, Xie XJ, Knapp A (2007) Seed priming with brassinolide improves lucerne (*Medicago sativa* L.) seed germination and seedling growth in relation to physiological changes under salinity stress. Aust J Agric Res 58:811–815

Zhang ZP, Miao MM, Wang CL (2015) Effects of ALA on photosynthesis, antioxidant enzyme activity, and gene expression, and regulation of proline accumulation in tomato seedlings under NaCl stress. J Plant Growth Regul 34:637–650

Zhu ZJ, Wei GQ, Li J, Qian QQ, Yu JQ (2004) Silicon alleviates salt stress and increases antioxidant enzymes activity in leaves of salt-stressed cucumber (*Cucumis sativus* L.). Plant Sci 167:527–533

Zörb C et al (2004) The biochemical reaction of maize (*Zea mays* L.) to salt stress is characterized by a mitigation of symptoms and not by a specific adaptation. Plant Sci 167:91–100

NAC Transcription Factors in Drought and Salinity Tolerance

Xuan Lan Thi Hoang, Yen-Nhi Hoang Nguyen, Nguyen Phuong Thao and Lam-Son Phan Tran

Abstract Water deficit and salinity are known as the top serious abiotic stresses affecting quality and productivity of many crop plants. Expanding knowledge in plant science and advances in plant biotechnology have brought more practical opportunities to improve plant tolerance capacity toward drought and/or salinity conditions in a number of major crops. Compared with manipulating a functional gene, manipulation of a regulatory gene like transcription factor-encoding gene could alter more plant characteristics, and these changes together can protect plants against either single or multiple stresses. The purpose of this chapter is to update the readers with the latest reports on the involvement of NAC (NAM, ATAF1/2, and CUC2) transcription factors in drought- and salinity-responsive regulation. Although most studies have focused on laboratory-scale experiments, the present findings are expected to provide an overview about the progress on our understanding of NAC transcription factors-associated mechanisms underlying plant responses to drought and salinity, as well as to evaluate the potential applications of the recently characterized NAC transcription factors in development of transgenic plants with improved tolerance to water scarcity and/or salinity.

Abbreviations

ABA	Abscisic acid
ARF	Auxin response factor
AP2/ERF	APETALA2/ethylene-responsive element binding factor

X. L. T. Hoang (✉) · Y.-N. H. Nguyen · N. P. Thao
Applied Biotechnology for Crop Development, School of Biotechnology, International University, Ho Chi Minh City 700000, Vietnam
e-mail: htlxuan@hcmiu.edu.vn

Vietnam National University, Ho Chi Minh City 700000, Vietnam

L.-S. P. Tran
Stress Adaptation Research Unit, RIKEN Center for Sustainable Resource Science, 1-7-22, Suehiro-cho, Tsurumi, Yokohama 230-0045, Japan
e-mail: son.tran@riken.jp

M. Hasanuzzaman and M. Tanveer (eds.), *Salt and Drought Stress Tolerance in Plants*, Signaling and Communication in Plants, https://doi.org/10.1007/978-3-030-40277-8_14

bZIP	Basic leucine zipper
LEA	Late embryogenesis abundant proteins
MDA	Malondialdehyde
MeJA	Methyl jasmonate
MYB	Myeloblastosis
NAC	NAM, ATAF1/2, and CUC2
NACBS	NAC binding site
NACRS	NAC recognition sequence
ROS	Reactive oxygen species
RT-qPCR	Reverse transcription-quantitative polymerase chain reaction
TF	Transcription factor

1 Introduction

A major challenge for current agricultural production is the inability to keep in pace with the increasing food demand to satisfy the growing world population (Tester and Langridge 2010). Dramatic losses of cultivable land and severe soil destruction are a few reasons among the causes of this inability (Golldack et al. 2011). What is more, abiotic and biotic stresses can also decrease crop yields and quality with various extents (Zhu 2016). Among the abiotic stressors, drought and salinity are the two common ones that have negative effects on most aspects of plant physiology and metabolism (de Oliveira et al. 2013). These stressors prevent plants to get sufficient water accommodating their needs while triggering the over-production of intracellular reactive oxygen species (ROS), which in turn results in oxidative stress to plants. The non-manageable high levels of ROS production cause cell damage (Miller et al. 2010), cellular homeostasis disruption (Fleury et al. 2010; Shabala and Munns 2012), and cellular activity impediment (Anjum et al. 2011). Therefore, the consequences of either drought or salinity are abnormal plant growth and development, lower biomass accumulation, lower plant productivity, and even plant death in severe cases (de Oliveira et al. 2013).

In general, plant stress response is the outcome of a three-stage process, beginning with perception of the external stimulus (e.g., the stressor), followed by the transduction of the signal from membrane receptors toward the nucleus through different mediators, and finally the regulation of gene expression activities to improve stress adaptation/acclimatization (Hoang et al. 2014). Under drought or salinity conditions, plant defense activities include minimizing homeostasis fluctuation, preventing cellular injuries, and redirecting plant growth and developmental processes (Roychoudhury et al. 2013). Among the phytohormones, abscisic acid (ABA) has been identified as a key signaling molecule playing important role in regulating expression of many osmotic stress-responsive genes, which can be functional or regulatory genes (Finkelstein 2013; Yoshida et al. 2014). Examples for several stress-responsive components that are under ABA regulation are LEA (late embryogenesis abundant) proteins, kinases, phosphatases, water transporters like aquaporins, detoxifying enzymes like peroxidase, catalase and superoxide dismutase, and transcription factors (TFs) like those belonging to basic leucine zipper (bZIP), APETALA2/ethylene-responsive element binding factor (AP2/ERF), myeloblastosis (MYB) and NAM/ATAF1/2/CUC2 (NAC) TF families (Fujita et al. 2011; Nakashima et al. 2012; Hoang et al. 2014; Vishwakarma et al. 2017). In the signal transduction cascades, TFs are usually the terminal signal transducer. Their capability to interact with promoter regions to synchronously regulate the expression of downstream genes makes them the key components determining the successful responses or adaptations of plants to the ever-changing environmental conditions (Liang et al. 2016; Hoang et al. 2017). Members of various TF families have been identified and characterized for their mode of action in mediating plant responses to environmental stresses. In this chapter, we focus on the TFs belonging to the NAC family, which have been identified to play key regulatory functions in plant responses to drought and/or salinity. Some of the latest findings on stresses-related functions of NAC TFs are summarized in Table 1.

2 NAC TFs—An Overview

NACs are unique plant TFs. In addition to analyses of NAC TFs in common model and crop plant species (Thao et al. 2013), whole genome search for genes encoding NAC TFs including membrane-bound NACs has been recently performed in *Camellia sinensis* (tea, 45 genes found) (Wang et al. 2016b), *Raphanus sativus* (radish, at least 98 found) (Karanja et al. 2017), *Triticum turgidum* (durum wheat, 168 found) (Saidi et al. 2017); *Cucumis sativus* (cucumber, 82 found) (Zhang et al. 2017), *Sesamum indicum* (sesame, 87 found) (Zhang et al. 2018), *Pyrus pyrifolia* (Asian pear, 185 found) (Ahmad et al. 2018), groundnut (*Arachis* species, 76 found) (Suchithra et al. 2018) and *Fragaria vesca* (wild strawberry, 112 found) (Moyano et al. 2018). Moreover, specific *NAC* expression patterns in various plant organs in responses to various stresses have also been comprehensively examined by using RNA-sequencing and reverse transcription-quantitative polymerase chain reaction (RT-qPCR) in a number of plants species, such as soybean (*Glycine max*) (Le et al. 2011), chickpea (*Cicer arietinum*) (Ha et al. 2014), Asian pear (Ahmad et al. 2018), and rice (*Oryza sativa*) (Chung et al. 2018).

Structurally, the NAC proteins consist of a highly conserved N-terminal DNA-binding domain and a variable C-terminal regulatory domain (Tran et al. 2010). The binding specificity of different NAC TFs might be defined by the sequences flanking the core site in the promoters of target genes (Shao et al. 2015). The NAC TFs bind to a consensus sequence called NAC recognition sequence (NACRS) or NACRS-like sequence containing the "CACG" core-DNA binding motif in the promoters of downstream genes (Shao et al. 2015). Additionally, other *cis*-elements containing the core sequence "CGT(G/A)", "GCTT" or "(T/A)NN(C/T)(T/C/G)TNNNNNNNA(A/C)GN(A/C/T)(A/T)" are also reported as NAC-binding sites (NACBS) (Zhong et al. 2010; Puranik et al. 2012; Olins et al.

Table 1 The NAC (NAM/ATAF1/2/CUC2) transcription factors that were reported to be involved in plant responses to drought and/or salinity

TF origin	Gene name	Responses to stresses (in wild-type plants)	Genetically modified plant system	Characterized TF functions using transgenic/mutant plants	References
In planta **studies using homologous plant systems**					
Arabidopsis	*ANAC032*	Salinity, mannitol	*Arabidopsis*	Promotes senescence and accumulation of H_2O_2, regulates expression of senescence-related genes like *AtNYE1, SAG113* and *SAUR36/SAG201*	Mahmood et al. (2016)
Rice	*OsNAC6*	Drought, salinity	Rice	Positive regulator of drought response: increases root system development and yield, regulates genes related to nicotianamine production, glycosylation, membrane and transportation activities, and carbohydrate metabolism	Lee et al. (2017), Chung et al. (2018)
	ONAC066	Polyethylene glycol, salinity, H_2O_2	Rice	Positive regulator of drought response: decreases water loss rate, increases proline and soluble sugar contents	Yuan et al. (2019)
	ONAC022	Drought, salinity	Rice	Positive regulator of responses to drought and salinity: lowers Na^+ accumulation, reduces transpiration, increases proline and soluble sugar contents	Hong et al. (2016)
	OsNAC2	Drought, salinity	Rice	Negative regulator of responses to drought and salinity: decreases yield, down-regulates *LEA3*, GTPase-related genes (e.g. *RAB21*, *RAB16C* and *RAB16D*) and kinase-related genes (e.g. *SAPK1* and *SAPK10*)	Shen et al. (2017)

(continued)

Table 1 (continued)

TF origin	Gene name	Responses to stresses (in wild-type plants)	Genetically modified plant system	Characterized TF functions using transgenic/mutant plants	References
Tomato	*SlNAC11*	Dehydration, salinity, cold, heat	Tomato	Positive regulator of responses to drought and salinity: *SlNAC11*-RNAi tomato plants became less tolerant, had lower chlorophyll content and germination rate, higher malondialdehyde (MDA)	Wang et al. (2017)
	SlJUB1	Dehydration, polyethylene glycol, salinity, H_2O_2	Tomato	Positive regulator of drought response: reduces cellular damage, regulates expression of *DREB*s and *DELLA*	Thirumalaikumar et al. (2018)
Banana	*MusaNAC042*	Drought, salinity	Banana	Positive regulator of responses to drought and salinity: increases chlorophyll content, lowers MDA content	Tak et al. (2017)
Poplar	*PeNAC045*	Drought, salinity	Poplar	Negative regulator of responses to drought and salinity: inhibits root growth, lowers photosynthesis	Lu et al. (2018)
***In planta* studies using heterologous plant systems**					
Tomato	*SlNAC35*	Dehydration, salinity	Tobacco	Positive regulator of responses to drought and salinity: promotes root development, reduces reactive oxygen species (ROS) levels, increases anti-oxidative enzyme activities	Wang et al. (2016a)
Trifoliate orange	*PtrNAC72*	Drought, cold	Tobacco	Negative regulator of drought response: increases water loss rate, levels of MDA and electrolyte leakage (EL)	Wu et al. (2016)

(continued)

Table 1 (continued)

TF origin	Gene name	Responses to stresses (in wild-type plants)	Genetically modified plant system	Characterized TF functions using transgenic/mutant plants	References
Suaeda liaotungensis	*SlNAC8*	Drought, salinity	*Arabidopsis*	Positive regulator of responses to drought and salinity: increases proline content, chlorophyll fluorescence and activities of superoxide dismutase (SOD), catalase (CAT) and peroxidase (POD), lowers MDA and EL; up-regulates *RD20*, *GSTF6*, *COR47*, *RD29A*, *RD29B* and *NYC1*	Wu et al. (2018)
Pumpkin	*CmNAC1*	Salinity, dehydration, cold, H_2O_2	*Arabidopsis*	Positive regulator of responses to multiple stresses: promotes root growth; reduces Na^+ accumulation in leaves under salinity, increases survival rate under drought	Cao et al. (2017)
Chickpea	*CarNAC4*	Drought, salinity, H_2O_2	*Arabidopsis*	Positive regulator of responses to drought and salinity: increases proline content, lowers water loss rate and MDA content	Yu et al. (2016)
Maize	*ZmNAC55*	Drought, salinity	*Arabidopsis*	Positive regulator of drought response: lowers water loss rate, up-regulates ABA-responsive genes (e.g. *RD17*, *PP2CA*, *RAB18* and *NCED3*)	Mao et al. (2016)
Foxtail millet	*SiNAC110*	Dehydration, salinity	*Arabidopsis*	Positive regulator of responses to dehydration and salinity: increases germination rate, root growth, plant fresh and dry weights	Xie et al. (2017)

(continued)

Table 1 (continued)

TF origin	Gene name	Responses to stresses (in wild-type plants)	Genetically modified plant system	Characterized TF functions using transgenic/mutant plants	References
Pearl millet	*PgNAC21*	Salinity	*Arabidopsis*	Positive regulator of response to salinity: increases germination rate, fresh weight and root length, up-regulates *GSTF6*, *RD20* and *COR47*	Shinde et al. (2019)
Finger millet	*EcNAC67*	Salinity	Rice	Positive regulator of responses to drought and salinity: increases root length, shoot biomass, relative water content, post-stress recovery rate and yield	Rahman et al. (2016)
Miscanthus lutarioriparius	*MlNAC10*	Drought, salinity	*Arabidopsis*	Positive regulator of responses to drought and salinity: increases proline content and anti-oxidative enzyme activities, reduces ROS and MDA accumulation	He et al. (2019)
	MlNAC9	Drought, salinity, cold	*Arabidopsis*	Positive regulator of responses to drought and salinity: increases activities of SOD, CAT and POD, lowers MDA accumulation	Zhao et al. (2016)
Cotton	*GhSNAC3*	Drought, salinity	Tobacco	Positive regulator of responses to drought and salinity: increases primary root length and fresh weight	Liu et al. (2018)
Soybean	*GmNAC085*	Dehydration	*Arabidopsis*	Positive regulator of drought response: increases activities of SOD, CAT and peroxidase ascorbate, enhances glutathione-dependent detoxifications of ROS and methylglyoxal, reduces water loss rate and biomass reduction rate	Nguyen et al. (2018a, b)

(continued)

Table 1 (continued)

TF origin	Gene name	Responses to stresses (in wild-type plants)	Genetically modified plant system	Characterized TF functions using transgenic/mutant plants	References
Poplar	*PeNAC034*	Drought, salinity	*Arabidopsis*	Negative regulator of responses to drought and salinity: inhibits root development, down-regulates *COR47, RD29B, ERD11, RD22* and *DREB2A*	Lu et al. (2018)
	PeNAC036	Drought, salinity	*Arabidopsis*	Positive regulator of responses to drought and salinity: promotes root development, up-regulates *COR47, RD29B, ERD11, RD22* and *DREB2A*	

The table listed only those reported from 2016

2018). Therefore, NAC TFs can potentially regulate a large number of downstream genes. Many studies have documented divergent functions of NACs associated with various biological processes in plant growth and development (Zhong et al. 2010; Bollhöner et al. 2012; Nakashima et al. 2012; Hussey et al. 2013) as well as plant responses to a wide range of biotic and abiotic stresses (Puranik et al. 2012; Nuruzzaman et al. 2013; Thu et al. 2014). These features make NAC TF family a highly potential resource of valuable candidates for genetic engineering.

3 Drought- and Salinity-Related NAC TF Characterization Using Homologous Plant Systems

Early attempts in identification and functional analyses of stress-related *NAC* genes, such as *ANAC019*, *ANAC055* and *ANAC072* (Tran et al. 2004), and *AtNAC2* (*ANAC092/ORE1*) (He et al. 2005), have been conducted in the model plant *Arabidopsis thaliana*. Following the identification of the regulatory role of AtNAC2 in regulating ethylene-/auxin-dependent pathways, root development (He et al. 2005) and leaf senescence (Balazadeh et al. 2010) in response to salinity in *Arabidopsis*, ectopic expression of *AtNAC2* gene in groundnut (*A. hypogaea*) could provide the transgenic plants with an enhanced drought and salinity resistance as evidenced by significantly higher chlorophyll stability index, lower rate of water loss, better growth and yield characteristics (Patil et al. 2014). ANAC032 was a more recently characterized *Arabidopsis* NAC member. The expression analysis of *ANAC032* showed that this gene was induced by various abiotic stresses, including salinity (Mahmood et al. 2016). Also from this study, overexpression of *ANAC032* was reported to promote senescence process in responses to osmotic stress and salinity conditions and accumulate higher level of hydrogen peroxide (H_2O_2) in the transgenic lines. Meanwhile, the chimeric-repression *Arabidopsis* lines of *ANAC032* displayed a more tolerant phenotype to salinity and lower accumulation of H_2O_2 compared with the wild-type plants.

In rice, the model monocot crop plant, many NAC members have been shown to play a part in plant responses to drought and salinity, such as SNAC1 (Stress-responsive NAC1) (Hu et al. 2006), OsNAC10 (Jeong et al. 2010), OsNAC9 (Redillas et al. 2012), ENAC1 (early NAC-domain protein induced by abiotic stress 1) (Sun et al. 2012) and SNAC3 (Fang et al. 2015). Among these, *SNAC1*, a drought-inducible *NAC* gene found to express mainly in rice guard cells, has been well documented to be an excellent candidate for genetic manipulation (Hu et al. 2006). The employment of *SNAC1* overexpression was reported to enhance the drought and salinity adaptations in various crop plants, including rice (Hu et al. 2006), wheat (*T. aestivum*) (Saad et al. 2013), cotton (*Gossypium hirsutum*) (Liu et al. 2014) and ramie (*Boehmeria nivea*) (An et al. 2015). Recently, Hong et al. (2016) found that the presence of overexpressed *ONAC022* in the transformed rice could equip the plants with a better drought and salinity resistance abilities, at least due to lowered Na^+ accumulation in the root and shoot tissues compared with that in the non-transgenic plants. Likewise, molecular analysis of *ONAC066* illustrated that this gene was induced significantly by polyethylene glycol and NaCl, and *ONAC066*-overexpressing transgenic rice showed higher tolerance to drought as indicated by decreased ROS accumulation and water loss rate, and increased proline and soluble sugar contents (Yuan et al. 2019). Examination on OsNAC6 function also revealed that this rice NAC TF is a positive regulator of plant response to drought. Overexpression of *OsNAC6* specifically in the root tissues could enlarge the stele and aerenchyma cells in the root tissues of the transgenic rice, and thus enable the plants to be more resistant to water deficit under both laboratory and field testing conditions (Lee et al. 2017). Analyzing the gene expression to identify its downstream genes revealed that OsNAC6 is likely the direct regulator of certain genes associating with membrane-transporter proteins, vesicle, phytohormones and carbohydrate metabolism (Lee et al. 2017; Chung et al. 2018). In contrast, the rice OsNAC2 was demonstrated to function as a negative regulator. *OsNAC2* expression was up-regulated by osmotic stress conditions; however, its overexpression led to reduced resistance and yields of the transgenic rice plants under drought and salinity (Shen et al. 2017).

In tomato (*Solanum lycopersicum*), SlNAC11 was recently found to function as a stress-responsive TF. Molecular analyses revealed that *SlNAC11* had significantly increased expression patterns by various abiotic stresses, including dehydration (Wang et al. 2017). *SlNAC11*-silencing tomato plants became more vulnerable under drought as well as salinity conditions, as they displayed lower photosynthetic performance and higher stress-induced levels of cellular injuries than their non-transgenic counterparts (Wang et al. 2017). Similarly silencing of another drought-inducible tomato *NAC* gene (the so-called *Jungbrunnen1* (*JUB1*)) made the tomato plants more susceptible to drought with escalation of ion leakage and H_2O_2 levels, and the reduction in transcriptional levels of drought-responsive *DREB* and *DELLA* genes (Thirumalaikumar et al. 2018). From a study in banana (*Musa acuminata*), Tak et al. (2017) reported that the expression of *MusaNAC042* gene was positively associated with banana resistance to salinity and drought. According to their results, the *MusaNAC042*-overexpressing transgenic banana displayed higher contents of proline and chlorophyll, and lower level of malondialdehyde (MDA) compared with the control plants under the stress conditions.

4 Drought- and Salinity-Related NAC TF Characterization Using Heterologous Plant Systems

It is well noted that the employment of NACs from other plants species could produce similar levels of enhanced tolerance as observed in the tested transgenic plant systems. Tomato SlNAC35 appeared to be a positive regulator of plant responses to drought and salinity as ectopic expression of *SlNAC35* in the transgenic tobacco stimulated the root development under these stress conditions, suggestively due to modulating *auxin response factor* (*ARF*) expression, and thus the auxin signaling pathway (Wang et al. 2016a). In chickpea, ectopic expression of *CarNAC4* in *Arabidopsis* could increase the tolerance of transgenic plants to both drought and salinity, indicating that CarNAC4 functions as a positive mediator (Yu et al. 2016). Meanwhile, in maize (*Zea mays*), the expression analysis of *ZmNAC55* showed that this gene was induced by drought and high salinity (Mao et al. 2016). Ectopic expression of *ZmNAC55* in *Arabidopsis* demonstrated its involvement in the ABA-mediated pathway as the seed germination was more sensitive to ABA treatment and a number of ABA-responsive genes were shown to be up-regulated. The positive regulatory role of ZmNAC55 in lowering transpiration rate under drought conditions was also elucidated (Mao et al. 2016). In a recent study, functional characterization of the dehydration-inducible *GmNAC085* isolated from soybean revealed the positive role of this TF in drought response by controlling the ROS removal-related activities. Specifically, ectopic expression of this gene in *Arabidopsis* led to improved drought tolerance of the transgenic lines, which was associated with mitigation of drought-induced oxidative stress through enhanced activities of antioxidant enzymes like superoxide dismutase, catalase and ascorbate peroxidase (Nguyen et al. 2018a), and glutathione-dependent detoxifications of ROS and methylglyoxal (Nguyen et al. 2018b).

In *Miscanthus lutarioriparius*, *MlNAC10* was indicated to have important role in enhancing drought and salinity tolerance of the transgenic *Arabidopsis* plants ectopically expressing *MlNAC10* by increasing proline content and antioxidant enzyme activities, and reducing ROS and MDA accumulations (He et al. 2019). Searching for stress-related *NAC* genes in upland cotton revealed that *GhNAC18* gene was up-regulated by drought (Evans et al. 2016). Additionally, phylogenetic analysis revealed that GhNAC18 was a close relative of the wheat TaNAC67, which was known as a positive regulator of *Arabidopsis* plant responses to drought and salinity (Mao et al.

2014; Evans et al. 2016). A more recent report indicated that transgenic tobacco plants ectopically expressing *GhSNAC3* showed enhanced tolerance by promoting primary root elongation under drought and salinity conditions (Liu et al. 2018). Another stress-related NAC confirmed by *in planta* study was the finger millet (*Eleusine coracana*) EcNAC67 (Rahman et al. 2016). Genetic analysis of *EcNAC67* showed that this gene is involved in regulating plant responses to drought and high salinity. Transgenic rice ectopically expressing *EcNAC67* was reported to have longer root, higher relative water content, better shoot biomass and post-stress recovery as well as higher yields under these adverse conditions (Rahman et al. 2016). In trifoliate orange (*Poncirus trifoliata*), PtrNAC72 was defined as a repressor for drought tolerance as ectopic expression of *PtrNAC72* in tobacco induced higher levels of water loss, electrolyte leakage and MDA in the transgenic lines under drought or dehydration treatment (Wu et al. 2016). In pumpkin (*Cucurbita moschata*), *CmNAC1* was indicated as a potential candidate with positive role in root growth and salinity stress tolerance in transgenic *Arabidopsis* plants (Cao et al. 2017). In foxtail millet (*Setaria italica*), *SiNAC110* was illustrated to improve the tolerance of *SiNAC110*-expressing transgenic *Arabidopsis* to drought and high salinity as evidenced by their better seed germination rate, root growth, fresh and dry weights of the whole plant by controlling various cell regulation pathways (Xie et al. 2017). In pearl millet (*Pennisetum glaucum*), *PgNAC21* was revealed as a potential candidate in improving salinity stress tolerance. The transgenic *Arabidopsis* plants ectopically expressing *PgNAC21* exhibited several better physiological and phenotypic traits, including enhanced seed germination, plant fresh weight and root length (Shinde et al. 2019). It is also worth noting that a number of studies on stress-responsive *NAC* genes have been conducted using perennial trees in order to learn how plants deal with a long duration of drought or salinity. For example, an investigation in a poplar species living in desert (e.g., *Populus euphratica* Olivier) showed that the expression of the three genes *PeNAC034*, *36*, and *45* was differentially altered in responses to drought and salinity (Lu et al. 2018). Detailed studies of these three genes using overexpression and mutant systems (*Arabidopsis* for *PeNAC034* and *36*, poplar for *PeNAC045*) demonstrated that PeNAC036 acts as a positive regulator, whereas PeNAC034 and PeNAC045 function as negative regulators of plant responses to drought and salinity (Lu et al. 2018).

5 Conclusions and Perspectives

Within recent decades, a large number of NAC TFs from various plant species have been reported to be involved in mediating plant responses to drought and salinity. Their important roles and diverse mode of actions under abiotic stress have been demonstrated. Functional characterization studies revealed many of them acting as positive regulators, which can be used for genetic engineering. However, it has to be admitted that the majority of *in planta* studies only have attempted to resolve the acting mechanisms of the TFs in regulating plant responses to environmental stresses, while values of agronomic traits in genetically modified plants, especially testing the plant performance and yield production under field conditions, have not been much achieved. Therefore, to more practically employ NAC TFs in generating elite crop cultivars with improved stress tolerance and minimized yield losses, more application-oriented research should be designed and conducted.

Acknowledgements This research is funded by Vietnam National University HoChiMinh City (VNU-HCM) under grant number C2018-28-04 and Vietnam National Foundation for Science and Technology Development (NAFOSTED) under grant number 106-NN.02-2015.85.

References

Ahmad M, Yan X, Li J, Yang Q, Jamil W, Teng Y, Bai S (2018) Genome wide identification and predicted functional analyses of NAC transcription factors in Asian pears. BMC Plant Biol 18:214

An X, Liao Y, Zhang J, Dai L, Zhang N, Wang B, Liu L, Peng D (2015) Overexpression of rice NAC gene *SNAC1* in ramie improves drought and salt tolerance. Plant Growth Regul 76:211–223

Anjum SA, Xie XY, Wang LC, Saleem MF, Man C, Lei W (2011) Morphological, physiological and biochemical responses of plants to drought stress. Afr J Agri Res 6:2026–2032

Balazadeh S, Siddiqui H, Allu AD, Matallana-Ramirez LP, Caldana C, Mehrnia M, Zanor MI, Köhler B, Mueller-Roeber B (2010) A gene regulatory network controlled by the NAC transcription factor ANAC092/AtNAC2/ORE1 during salt-promoted senescence. Plant J 62:250–264

Bollhöner B, Prestele J, Tuominen H (2012) Xylem cell death: emerging understanding of regulation and function. J Exp Bot 63:1081–1094

Cao H, Wang L, Nawaz MA, Niu M, Sun J, Xie J, Kong Q, Huang Y, Cheng F, Bie Z (2017) Ectopic expression of pumpkin NAC transcription factor *CmNAC1* improves multiple abiotic stress tolerance in *Arabidopsis*. Front Plant Sci 8:2052

Chung PJ, Jung H, Choi YD, Kim JK (2018) Genome-wide analyses of direct target genes of four rice NAC-domain transcription factors involved in drought tolerance. BMC Genomics 19:40

de Oliveira AB, Alencar NLM, Gomes-Filho E (2013) Comparison between the water and salt stress effects on plant growth and development. In: Akinci S (ed) Responses of organisms to water stress. InTech, pp 67–94

Evans O, Dou L, Guo Y, Pang C, Wei H, Song M, Fan S, Yu S (2016) *GhNAC18*, a novel cotton (*Gossypium hirsutum* L.) NAC gene, is involved in leaf senescence and diverse stress responses. Afr J Biotechnol 15:1233–1245

Fang Y, Liao K, Du H, Xu Y, Song H, Li X, Xiong L (2015) A stress-responsive NAC transcription factor SNAC3 confers heat and drought tolerance through modulation of reactive oxygen species in rice. J Exp Bot 66:6803–6817

Finkelstein R (2013) Abscisic acid synthesis and response. Arabidopsis Book 11:e0166

Fleury D, Jefferies S, Kuchel H, Langridge P (2010) Genetic and genomic tools to improve drought tolerance in wheat. J Exp Bot 61:3211–3222

Fujita Y, Fujita M, Shinozaki K, Yamaguchi-Shinozaki K (2011) ABA-mediated transcriptional regulation in response to osmotic stress in plants. Plant Res 124:509–525

Golldack D, Lüking I, Yang O (2011) Plant tolerance to drought and salinity: stress regulating transcription factors and their functional significance in the cellular transcriptional network. Plant Cell Rep 30:1383–1391

Ha CV, Esfahani MN, Watanabe Y, Tran UT, Sulieman S, Mochida K, Nguyen DV, Tran L-SP (2014) Genome-wide identification and expression analysis of the *CaNAC* family members in chickpea during development, dehydration and ABA treatments. PLoS ONE 9:e114107

He XJ, Mu RL, Cao WH, Zhang ZG, Zhang JS, Chen SY (2005) AtNAC2, a transcription factor downstream of ethylene and auxin signaling pathways, is involved in salt stress response and lateral root development. Plant J 44:903–916

He K, Zhao X, Chi X, Wang Y, Jia C, Zhang H, Zhou G, Hu R (2019) A novel *Miscanthus* NAC transcription factor MlNAC10 enhances drought and salinity tolerance in transgenic *Arabidopsis*. Plant Physiol 233:84–93

Hoang XLT, Thu NBA, Thao NP, Tran L-SP (2014) Transcription factors in abiotic stress responses: their potentials in crop improvement. In: Ahmad P, Wani MR, Azooz MM, Tran L-SP (eds) Improvement of crops in the era of climatic changes. Springer, New York, pp 337–366

Hoang XLT, Nhi DNH, Thu NBA, Thao NP, Tran L-SP (2017) Transcription factors and their roles in signal transduction in plants under abiotic stresses. Curr Genomics 18:483–497

Hong Y, Zhang H, Huang L, Li D, Song F (2016) Overexpression of a stress-responsive NAC transcription factor gene *ONAC022* improves drought and salt tolerance in rice. Front Plant Sci 7:4

Hu H, Dai M, Yao J, Xiao B, Li X, Zhang Q, Xiong L (2006) Overexpressing a NAM, ATAF, CUC (NAC) transcription factor enhances drought resistance and salt tolerance in rice. Proc Natl Acad Sci USA 103:12987–12992

Hussey SG, Mizrachi E, Creux NM, Myburg AA (2013) Navigating the transcriptional roadmap regulating plant secondary cell wall deposition. Front Plant Sci 4:325

Jeong JS, Kim YS, Baek KH, Jung H, Ha SH, Do CY, Kim M, Reuzeau C, Kim JK (2010) Root-specific expression of *OsNAC10* improves drought tolerance and grain yield in rice under field drought conditions. Plant Physiol 153:185–197

Karanja BK, Xu L, Wang Y, Muleke EMM, Jabir BM, Xie Y, Zhu X, Cheng W, Liu L (2017) Genome-wide characterization and expression profiling of *NAC* transcription factor genes under abiotic stresses in radish (*Raphanus sativus* L.). Peer J 5:e4172

Le DT, Nishiyama R, Watanabe Y, Mochida K, Yamaguchi-Shinozaki K, Shinozaki K, Tran LS (2011) Genome-wide survey and expression analysis of the plant-specific NAC transcription factor family in soybean during development and dehydration stress. DNA Res 18:263–276

Lee DK, Chung PJ, Jeong JS, Jang G, Bang SW, Jung H, Kim YS, Ha SH, Choi YD, Kim JK (2017) The rice OsNAC6 transcription factor orchestrates multiple molecular mechanisms involving root structural adaptions and nicotianamine biosynthesis for drought tolerance. Plant Biotechnol J 15:754–764

Liang C, Meng Z, Meng Z, Malik W, Yan R, Lwin KM, Lin F, Wang Y, Sun G, Zhou T, Zhu T, Li J, Jin S, Guo S, Zhang R (2016) GhABF2, a bZIP transcription factor, confers drought and salinity tolerance in cotton (*Gossypium hirsutum* L.). Sci Rep 6:35040

Liu G, Li X, Jin S, Liu X, Zhu L, Nie Y, Zhang X (2014) Overexpression of rice NAC gene *SNAC1* improves drought and salt tolerance by enhancing root development and reducing transpiration rate in transgenic cotton. PLoS ONE 9:e86895

Liu Z-J, Li F, Wang L-G, Liu R-Z, Ma J-J, Fu M-C (2018) Molecular characterization of a stress-induced NAC gene, *GhSNAC3*, from *Gossypium hirsutum*. J Genet 97:539–548

Lu X, Zhang X, Duan H, Lian C, Liu C, Yin W, Xia X (2018) The stress-responsive NAC transcription factors from *Populus euphratica* differentially regulate salt and drought tolerance in transgenic plants. Physiol Plant 162:73–97

Mahmood K, El-Kereamy A, Kim SH, Nambara E, Rothstein SJ (2016) ANAC032 positively regulates age-dependent and stress-induced senescence in *Arabidopsis thaliana*. Plant Cell Physiol 57:2029–2046

Mao X, Chen S, Li A, Zhai C, Jing R (2014) Novel NAC transcription factor TaNAC67 confers enhanced multi-abiotic stress tolerances in *Arabidopsis*. PLoS ONE 9:e84359

Mao H, Yu L, Han R, Li Z, Liu H (2016) ZmNAC55, a maize stress-responsive NAC transcription factor, confers drought resistance in transgenic *Arabidopsis*. Plant Physiol Biochem 105:55–66

Miller G, Suzuki N, Ciftci-Yilmaz S, Mittler R (2010) Reactive oxygen species homeostasis and signalling during drought and salinity stresses. Plant Cell Environ 33:453–467

Moyano E, Martínez-Rivas FJ, Blanco-Portales R, Molina-Hidalgo FJ, Ric-Varas P, Matas-Arroyo AJ, Caballero JL, Muñoz-Blanco J, Rodríguez-Franco A (2018) Genome-wide analysis of the NAC transcription factor family and their expression during the development and ripening of the *Fragaria× ananassa* fruits. PLoS ONE 13:e0196953

Nakashima K, Takasaki H, Mizoi J, Shinozaki K, Yamaguchi-Shinozaki K (2012) NAC transcription factors in plant abiotic stress responses. Biochim Biophys Acta 1819:97–103

Nguyen KH, Mostofa MG, Li W, Van Ha C, Watanabe Y, Le DT, Nguyen PT, Tran LSP (2018a) The soybean transcription factor GmNAC085 enhances drought tolerance in *Arabidopsis*. Environ Exp Bot 151:12–20

Nguyen KH, Mostofa MG, Watanabe Y, Tran CD, Rahman MM, Tran LSP (2018b) Overexpression of *GmNAC085* enhances drought tolerance in *Arabidopsis* by regulating glutathione biosynthesis, redox balance and glutathione-dependent detoxification of reactive oxygen species and methylglyoxal. Environ Exp Bot 161:242–254

Nuruzzaman M, Sharoni AM, Kikuchi S (2013) Roles of NAC transcription factors in the regulation of biotic and abiotic stress responses in plants. Front Microbiol 4:248

Olins JR, Lin L, Lee SJ, Trabucco GM, MacKinnon KJ-M, Hazen SP (2018) Secondary wall regulating NACs differentially bind at the promoter at a *cellulose synthase A4 cis*-eQTL. Front Plant Sci 9:1895

Patil M, Ramu SV, Jathish P, Sreevathsa R, Reddy PC, Prasad TG, Udayakumar M (2014) Overexpression of *AtNAC2* (*ANAC092*) in groundnut (*Arachis hypogaea* L.) improves abiotic stress tolerance. Plant Biotechnol Rep 8:161–169

Puranik S, Sahu PP, Srivastava PS, Prasad M (2012) NAC proteins: regulation and role in stress tolerance. Trends Plant Sci 17:369–381

Rahman H, Ramanathan V, Nallathambi J, Duraialagaraja S, Muthurajan R (2016) Overexpression of a NAC 67 transcription factor from finger millet (*Eleusine coracana* L.) confers tolerance against salinity and drought stress in rice. BMC Biotechnol 16:35

Redillas MC, Jeong JS, Kim YS, Jung H, Bang SW, Choi YD, Ha S-W, Reuzeau C, Kim JK (2012) The overexpression of *OsNAC9* alters the root architecture of rice plants enhancing drought resistance and grain yield under field conditions. Plant Biotechnol J 10:792–805

Roychoudhury A, Paul S, Basu S (2013) Cross-talk between abscisic acid-dependent and abscisic acid-independent pathways during abiotic stress. Plant Cell Rep 32:985–1006

Saad ASI, Li X, Li HP, Huang T, Gao CS, Guo MW, Cheng W, Zhao GY, Liao YC (2013) A rice stress-responsive *NAC* gene enhances tolerance of transgenic wheat to drought and salt stresses. Plant Sci 203:33–40

Saidi MN, Mergby D, Brini F (2017) Identification and expression analysis of the NAC transcription factor family in durum wheat (*Triticum turgidum* L. ssp. durum). Plant Physiol Biochem 112:117–128

Shabala S, Munns R (2012) Salinity stress: physiological constraints and adaptive mechanisms. In: Shabala S (ed) Plant stress physiology. CABI, pp 59–93

Shao H, Wang H, Tang X (2015) NAC transcription factors in plant multiple abiotic stress responses: progress and prospects. Front Plant Sci 6:902

Shen J, Lv B, Luo L, He J, Mao C, Xi D, Ming F (2017) The NAC-type transcription factor *OsNAC2* regulates ABA-dependent genes and abiotic stress tolerance in rice. Sci Rep 7:40641

Shinde H, Dudhate A, Tsugama D, Gupta SK, Liu S, Takano T (2019) Pearl millet stress-responsive NAC transcription factor *PgNAC21* enhances salinity stress tolerance in *Arabidopsis*. Plant Physiol Biochem 135:546–553

Suchithra B, Devaraj VR, Babu RN (2018) Genome wide analysis of NAC transcription factors and their expression pattern during high temperature and drought stress in groundnut. Afr Crop Sci J 26:327–348

Sun H, Huang X, Xu X, Lan H, Huang J, Zhang H-S (2012) ENAC1, a NAC transcription factor, is an early and transient response regulator induced by abiotic stress in rice. Mol Biotechnol 52:101–110

Tak H, Negi S, Ganapathi TR (2017) Banana NAC transcription factor MusaNAC042 is positively associated with drought and salinity tolerance. Protoplasma 254:803–816

Tester M, Langridge P (2010) Breeding technologies to increase crop production in a changing world. Science 327:818–822

Thao NP, Thu NBA, Hoang XLT, Ha CV, Tran L-SP (2013) Differential expression analysis of a subset of drought-responsive *GmNAC* genes in two soybean cultivars differing in drought tolerance. Int J Mol Sci 14:23828–23841

Thirumalaikumar VP, Devkar V, Mehterov N, Ali S, Ozgur R, Turkan I, Mueller-Roeber B, Balazadeh S (2018) NAC transcription factor JUNGBRUNNEN1 enhances drought tolerance in tomato. Plant Biotechnol J 16:354–366

Thu NBA, Hoang XLT, Doan H, Nguyen TH, Bui D, Thao NP, Tran L-SP (2014) Differential expression analysis of a subset of *GmNAC* genes in shoots of two contrasting drought-responsive soybean cultivars DT51 and MTD720 under normal and drought conditions. Mol Biol Rep 41:5563–5569

Tran L-SP, Nakashima K, Sakuma Y, Simpson SD, Fujita Y, Maruyama K, Fujita M, Seki M, Shinozaki K, Yamaguchi-Shinozaki K (2004) Isolation and functional analysis of *Arabidopsis* stress-inducible NAC transcription factors that bind to a drought-responsive *cis*-element in the *early responsive to dehydration stress 1* promoter. Plant Cell 16:2481–2498

Tran LS, Nishiyama R, Yamaguchi-Shinozaki K, Shinozaki K (2010) Potential utilization of NAC transcription factors to enhance abiotic stress tolerance in plants by biotechnological approach. GM Crops 1:32–39

Vishwakarma K, Upadhyay N, Kumar N, Yadav G, Singh J, Mishra RK, Kumar V, Verma R, Upadhyay RG, Pandey M, Sharma S (2017) Abscisic acid signaling and abiotic stress tolerance in plants: a review on current knowledge and future prospects. Front Plant Sci 20:161

Wang G, Zhang S, Ma X, Wang Y, Kong F, Meng Q (2016a) A stress-associated NAC transcription factor (SlNAC35) from tomato plays a positive role in biotic and abiotic stresses. Physiol Plant 158:45–64

Wang Y-X, Liu Z-W, Wu Z-J, Li H, Zhuang J (2016b) Transcriptome-wide identification and expression analysis of the NAC gene family in tea plant [*Camellia sinensis* (L.) O. Kuntze]. PLoS ONE 11:e0166727

Wang L, Hu Z, Zhu M, Zhu Z, Hu J, Qanmber G, Chen G (2017) The abiotic stress-responsive NAC transcription factor SlNAC11 is involved in drought and salt response in tomato (*Solanum lycopersicum* L.). Plant Cell Tiss Organ Cult 129:161–174

Wu H, Fu B, Sun P, Xiao C, Liu J-H (2016) A NAC transcription factor represses putrescine biosynthesis and affects drought tolerance. Plant Physiol 172:1532–1547

Wu D, Sun Y, Wang H, Shi H, Su M, Shan H, Li T, Li Q (2018) The *SlNAC8* gene of the halophyte *Suaeda liaotungensis* enhances drought and salt stress tolerance in transgenic *Arabidopsis thaliana*. Gene 662:10–20

Xie L-n, Chen M, Min D-h, Feng L, Xu Z-s, Zhou Y-b, Xu D-b, Li L-c, Ma Y-z, Zhang X-h (2017) The NAC-like transcription factor *SiNAC110* in foxtail millet (*Setaria italica* L.) confers tolerance to drought and high salt stress through an ABA independent signaling pathway. J Integr Agr 16:559–571

Yoshida T, Mogami J, Yamaguchi-Shinozaki K (2014) ABA-dependent and ABA-independent signaling in response to osmotic stress in plants. Curr Opin Plant Biol 21:133–139

Yu X, Liu Y, Wang S, Tao Y, Wang Z, Shu Y, Peng H, Mijiti A, Wang Z, Zhang H, Ma H (2016) CarNAC4, a NAC-type chickpea transcription factor conferring enhanced drought and salt stress tolerances in *Arabidopsis*. Plant Cell Rep 35:613–627

Yuan X, Wang H, Cai J, Bi Y, Li D, Song F (2019) Rice NAC transcription factor ONAC066 functions as a positive regulator of drought and oxidative stress response. BMC Plant Biol 19:278

Zhang XM, Yu HJ, Sun C, Deng J, Zhang X, Liu P, Li YY, Li Q, Jiang WJ (2017) Genome-wide characterization and expression profiling of the *NAC* genes under abiotic stresses in *Cucumis sativus*. Plant Physiol Biochem 113:98–109

Zhang Y, Li D, Wang Y, Zhou R, Wang L, Zhang Y, Yu J, Gong H, You J, Zhang X (2018) Genome-wide identification and comprehensive analysis of the NAC transcription factor family in *Sesamum indicum*. PLoS ONE 13:e0199262

Zhao X, Yang X, Pei S, He G, Wang X, Tang Q, Jia C, Lu Y, Hu R, Zhou G (2016) The *Miscanthus* NAC transcription factor MlNAC9 enhances abiotic stress tolerance in transgenic *Arabidopsis*. Gene 586:158–169

Zhong R, Lee C, Ye ZH (2010) Evolutionary conservation of the transcriptional network regulating secondary cell wall biosynthesis. Trends Plant Sci 15:625–632

Zhu JK (2016) Abiotic stress signaling and responses in plants. Cell 167:313–324

Genetic Manipulation of Drought Stress Signaling Pathways in Plants

Sadia Latif, Tariq Shah, Fazal Munsif and Roberto D'Amato

Abstract Stress signaling networks in drought responses are composed of intracellular signaling systems, transcriptional regulatory complexes, and intercellular communication systems. The signaling mechanisms underlying changes in gene expression enable plant responses to drought stress. Signaling factors and transcription factors are themselves regulated transcriptionally and/or post-translationally (e.g., phosphorylation or proteolysis) in response to drought stress. Abscisic acid (ABA) is a key phytohormone, and ABA signaling is a major part of the drought response regulatory networks. However, ABA-independent pathways are also involved. The complexity and the cross talk between ABA-dependent and ABA-independent pathways in drought stress signaling networks have been extensively analyzed at the cellular level, but not at the intercellular level. Intercellular signaling in response to water deficit has to be elucidated for a comprehensive understanding of plant responses and adaptation to drought stress.

1 Introduction

Our society is at present facing various challenges, such as climate change, competition for land use, rising demands for food, etc. Addressing these problems will depend in part upon expanding our knowledge of plant biology, because plants provide us

S. Latif
Department of Biology and Environmental Sciences, Allama Iqbal Open University, Islamabad, Pakistan

T. Shah (✉)
Department of Agronomy, Faculty of Crop Production Sciences, University of Agriculture Peshawar, Peshawar, Pakistan
e-mail: tariqshah@aup.edu.pk

F. Munsif
Department of Agronomy, Faculty of Crop Production Sciences, University of Agriculture Peshawar, Amir Muhammad Khan Campus, Peshawar, Pakistan

R. D'Amato
Department of Environmental and Agricultural Sciences, University of Perugia, Perugia, Italy

M. Hasanuzzaman and M. Tanveer (eds.), *Salt and Drought Stress Tolerance in Plants*, Signaling and Communication in Plants, https://doi.org/10.1007/978-3-030-40277-8_15

with food and other sustainable resources. Drought is an abiotic stress responsible for some of the greatest crop losses (Anjum et al. 2016). Water deficiency limits plant yields; therefore, improvement in drought stress tolerance may markedly increase crop yields (Hussain et al. 2018).

Understanding drought stress responses in plants have been an active area of plant research. Historically, drought stress responses have been observed through numerous physiological experiments in various plants, including crops, vegetables, trees, and horticulture plants. In addition to classical studies, there has been marked progress in molecular biological analyses over the last decade. Particularly, the availability of the *Arabidopsis* genome sequence has had a major impact on all fields of plant research and has resulted in rapid developments in plant molecular biology. The availability of the complete genome sequence has provided much information regarding all genes, including predicted genes, cis elements, gene structures, and gene families. In addition to *Arabidopsis*, the complete genome sequences of other plant species have been reported. Comparative analysis between species has resulted in significant progress in research regarding the mechanisms of plant drought response and tolerance in nature. Plant genome resources have been elaborated upon through comprehensive transcriptome, proteome, and metabolome analyses. Integration of data obtained from each omics hierarchy has provided a more comprehensive picture of drought stress responses (Hirayama and Shinozaki 2010).

When plants respond to drought, stress signals are issued immediately after perception of the change in water status, transcription of stress response genes is induced according to these signals, and further signals are spread systemically throughout the plant (Fig. 1). During drought stress, the phytohormone, abscisic acid (ABA), accumulates and plays important roles in various responses to drought stress, such as stomata closure, induction of stress proteins, and accumulation of various metabolites for the protection of cells from water-deficit stress (Umezawa et al. 2010). In addition, molecular analyses have revealed ABA-independent processes in drought stress responses. This entry describes the drought stress response signaling networks, which consist of intracellular signal transduction and transcriptional regulatory and intercellular communication networks.

2 Signal Transduction Networks in Drought Stress

In general, signal transduction pathways consist of a number of well-ordered processes, i.e., perception, signal transduction at the protein level or by second messengers, and changes in target gene expression. The response to drought stress involves highly complex and diverse signaling pathways as this type of stress involves various stimuli, e.g., osmotic shock, oxidative burst, and sometimes strong light, heat, wounding responses, etc., in association with dynamic changes in cellular metabolism, phytohormones, or other signaling molecules (Tanveer et al. 2018). Therefore, it is necessary to dissect the various components of such complex pathways.

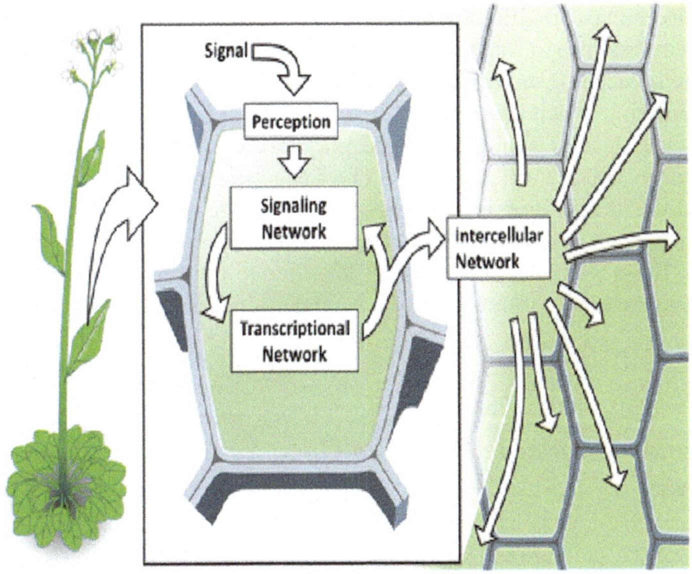

Fig. 1 Outline of signal transduction in plants drought response. The signaling network is activated after perception of stress, which enhances transcriptional networks to produce molecules required to cope with the stress, and induces signaling to distant tissues to make adjustments at the whole-plant level (modified from Kuromori et al. 2014)

First, drought stress signals can be propagated through one or both of two pathways, an ABA-dependent pathway and an ABA-independent pathway. The major ABA-signaling pathway was clarified recently. Although a number of factors involved in other signaling pathways have been identified, not all such factors have yet been determined.

3 Transcriptional Factors Involved in Drought Stress Signaling

The genome controls the regulation of the response to water deficit as well as the effectiveness of the response. Microarrays, largely performed using Arabidopsis thaliana as model plant, have been used to catalogue the many genes that are induced or repressed in response to conditions that may lead to cellular water-deficit stress (Seki et al. 2002). These genes can be placed in at least four different functional groups: signal transduction, transcriptional regulation, cellular metabolism and transport, and protection of cellular structures.

There are at least six different classes of TFs that participate in gene induction or repression in response to water deficit. Homeobox domain and NAC domain containing TFs are induced by multiple treatments that mimic water-deficit stress. Accumulation of proteins which have metabolic or structural functions promote adaptation to stress. One class of genes that could play a role in protection is called the late embryogenesis abundant (Lea) genes. The Leagenes are also developmentally programmed for expression in desiccating seeds. These genes encode small hydrophilic proteins that are predicted to protect proteins and membranes through chaperone-like functions. These proteins were thought to improve the performance of rice plants by protecting cell membranes from injury under abiotic stress (Chandra Babu et al. 2004).

3.1 Transcriptional Factors and Gene Regulation in Drought Stress

A review (Shinozaki and Yamaguchi-Shinozaki 2007) on analysis of gene expression during drought stress response in plants show and summarize the functions of some genes in both stress response and tolerance. Microarray analysis performed on wheat genome, showed that among 300 unique single expressed sequences tag (ESTs), the 30% of genes were significantly up-regulated and the 18% were down-regulated under drought stress (Way et al. 2005).

Potential functions of approximatively 130 genes of *A. thaliana* up-regulated in water-deficit was reported by Bray (2002). These genes are involved in cellular response to drought stress by signalling events, detoxification and other functions. cDNA microarray analysis on 7000 Arabidopsis full-length cDNAs clarify relationship between rehydratation-, proline- and water-treatment inducible genes. Among the 152 rehydratation-inducible genes, 58 genes contained in their promoter regions the ACTCAT sequence involved in proline- and hypoosmolarity-inducible gene expression, suggesting that this motif is a major cis-activing element involved in rehydratation-inducible gene expression (Oono et al. 2003).

Moreover, microarray analysis performed on two moderately drought-tolerant native Andeon potato clones revealed that there was 1713 differentially expressed genes with 186 up-regulated involved in drought tolerance by inducing of osmotic adjustment, changes in carbohydrate metabolism, membrane modifications and cell rescue mechanisms, such as detoxification of oxygen radicals and protein stabilization (Schafleitner et al. 2007).

These recent studies underline how the expression of genes in response to water deficit is complex and can be regulated at the transcriptional, post-transcriptional and translational levels. Two major transcriptional regulatory pathways of gene expression play an important role in response to water-deficit stress: the ABA-independent pathway and ABA-dependent pathway. The first is controlled largely by a family

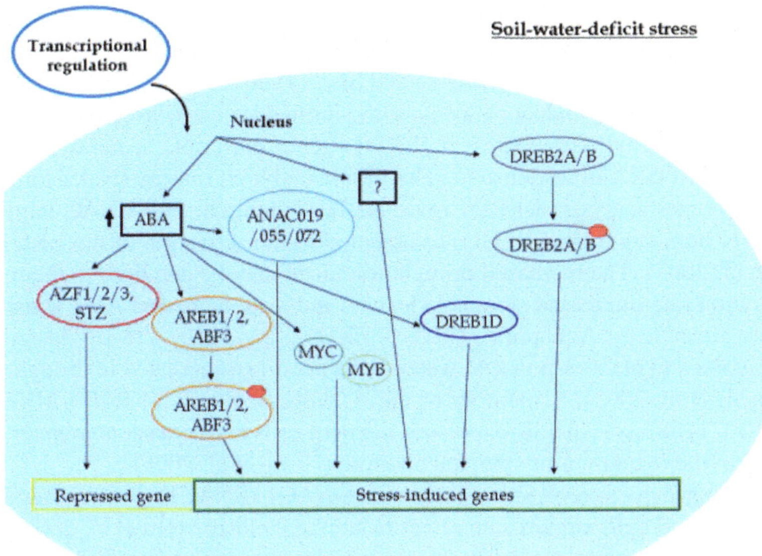

Fig. 2 Different types of transcription factors involved in induction/repression of regulons. Classes of genes that are induced by water-deficit stress

of TFs called dehydration response element binding protein (DREB), which contains a DNA binding motif originally identified in a flower patterning protein called APETALA2 (AP2) (Fig. 2), while transcription factor families known to be as the most responsive to ABA signalling under drought are NAC, AREB/ABF, and MYB.

3.1.1 ABA-Independent Pathway

DREB are important TFs, which induce a set of abiotic stress-related genes and confer stress resistance to plants. The DREB TFs could be divided into two groups: DREB1, involved signal transduction pathways under low temperature; DREB2, involved in signal transduction pathways under dehydration. They belong to the ethylene responsive element binding factors (ERF) family of TFs. ERF proteins are a sub-family of the AP2/ethylene responsive element binding protein (EREBP) TFs that is distinctive to plants. ERF proteins share a conserved 58–59 amino acid domain (the ERF domain) that binds to cis-elements, the GCC box, found in many pathogens-related (PR) gene promoters conferring ethylene responsiveness (Gu et al. 2000), and to the C-repeat CRT/dehydration responsive element (DRE) motif involved in the expression of cold and dehydration responsive genes (Agarwal et al. 2006).

The DREB proteins contain an ERF/AP2DNA-binding domain quite conserved: amino acid alignment shows high sequence similarity in the nuclear localization signal at the N-terminal region and some similarity in the C-terminal acidic domain (Agarwal et al. 2006). Indeed, TFs containing ERF/AP2DNA-binding domain are widely found in many plants such as *Arabidopsis* (Okamuro et al. 1997), tomato (Zhou et al. 1997), tobacco (Ohme-Takagi and Shinshi 1995), rice (Sasaki et al. 1994; Weigel 1995) and maize (Moose and Sisco 1996).

Another ABA-independent pathway was identified after the observation that Early Responsive to Dehydration Stress 1 (ERD1) gene transcripts accumulated before any increase of ABA in response to dehydration and high salinity (Nakashima et al. 1997). Promoter analysis of ERD1 revealed TFs belonging to the NAC family and zinc finger homeodomain (ZF-HD) as essential to the activation of the ERD1 gene (Tran et al. 2007). The increased drought tolerance may be due both to the reduced transpiration rate (increased stomatal closure) and to an increased ABA sensitivity.

Many genes (e.g., Aquaporin, ERD10, ERD13, and ERF) already described as being involved in plant response to water stress are down-regulated in drought stress (Cominelli et al. 2005). A member of the *A. thaliana* family of R2R3-MYB TFs, AtMYB61, is also specifically expressed in guard cells in a consistent manner, being involved in the regulation of stomatal aperture (Liang et al. 2005).

The strong induction of Stress Responsive NAC1 (SNAC1) gene expression by drought in guard cells suggests an effect in stomatal closure (Hu et al. 2006). It has been reported that modulation of transcription plays an important role in controlling guard cell activity. Recently two MYB-type TFs were identified as regulators of stomatal movements.

3.1.2 ABA-Dependent Pathway

ABA-dependent gene induction during water deficit is controlled by at least five different classes of TFs. The ABA response element (ABRE) with the consensus ACGTGG/TC is bound by basic Leucine Zipper Domain (bZIP-type) TFs (Fig. 2). Three Arabidopsis bZIP TFs (AREB1/ABF2, AREB2/ABF4, and ABF3) are expressed in response to water-deficit stress and ABA treatment. Activation of the TFs requires ABA accumulation and the induction of an ABA-responsive protein kinase which activates the TF through phosphorylation.

Other TFs are also involved in ABA regulation of gene expression during cellular water deficit. Three genes encoding a class of TFs that is unique to plants, the NAC domain proteins ANAC019, ANAC055, and ANAC072 are induced by water deficit and ABA treatment. The NAC domain is a 60 bp DNA binding domain that is predicted to form a helix-turn-helix motif.

MYB, MYC and homeodomain TFs, and a family of transcriptional repressors (Cys2/His2-type zinc-finger proteins) are also involved in the ABA response to water deficit. Expression of the drought-inducible gene Responsive to Dehydration 22 (RD22) from *Arabidopsis* was found to be induced by ABA. The promoter region of RD22 contains MYC (CANNTG) and MYB (C/TAACNA/G) cis-element recognition sites. MYC and MYB TFs only accumulate after an increase of ABA concentration. Over-expression of these TFs result in enhanced sensitivity to ABA and drought tolerance (Abe et al. 2003).

4 Other Important Signaling Pathways in Drought Stress

4.1 Early Osmotic Stress Signaling

In the 1990s, osmosensors and osmotic-stress signaling were characterized in yeast cells in which it was shown that Sln1 encodes a membrane-spanning histidine kinase that regulates a MAPK cascade, the Hog1 pathway that directly controls osmotic-stress-responsive gene expression. In *Arabidopsis*, AHK1/AtHK1 were identified by their ability to complement Sln1 in yeast and are therefore candidate plant osmosensors. Reverse genetic studies revealed that AHK1/AtHK1 are involved in osmotic-stress signaling. Although the signal transduction mechanism is unknown, response-regulator-like proteins (ARRs) may function downstream of AHK1/AtHK1.

Several protein kinases have been suggested to be involved in osmotic-stress signaling. For example, SnRK2s are also activated in response to osmotic stress. In contrast to ABA signaling, all subclasses of SnRK2 show osmotic-stress-responsive activation. Among them, SRK2C/SnRK2.8, a subclass II SnRK2, is activated by osmotic stress and positively regulates drought tolerance in *Arabidopsis*. In SRK2C-overexpressing plants, a number of drought stress-responsive genes, including the DREB1A/CBF3 transcription factor, were hyperinduced by drought stress. In addition, a reverse genetic study suggested that the subclass II SnRK2s, SRK2C/SnRK2.8, and SRK2F/SnRK2.7, may be the major factors involved in osmotic-stress signaling, although also involved in ABA signaling. Finally, a pivotal role of SnRK2s in osmotic-stress signaling was reported by a recent study in which an *Arabidopsis*-decuple mutant in all ten SnRK2s was developed. The decuple mutant was hypersensitive to osmotic stress, suggesting that SnR2s positively regulates plant tolerance to osmotic stress.

As described above, SnRK2s are activated by both ABA and osmotic stress. Although the mechanism of ABA-dependent activation of SnRK2 has been established, it remains unclear how osmotic-stress signals activate SnRK2s (Kulik et al. 2011). It is likely that osmotic-stress signaling involves a number of mechanisms of SnRK2 activation, because osmotic-stress-dependent activation is more rapid than ABA-dependent activation. In addition, subclass III SnRK2s are negatively regulated by group A PP2Cs, as described above, while subclasses I and II SnRK2s show no or only weak interactions, respectively, with PP2C. Therefore, some as-yet-unidentified negative regulators may function upstream of SnRK2 in the osmotic-stress response. A candidate regulator is the calcium-binding protein designated SnRK2-interacting calcium sensor (SCS), which interacts with a tobacco SnRK2, NtOSAK. An in vitro kinase assay demonstrated that SCS inhibits SnRK2 activity in a Ca^{2+}-dependent manner, suggesting that SCS acts as a negative regulator of SnRK2 activation.

4.2 Calcium-Dependent Pathway

Among the several second messengers involved in drought stress signaling, Ca^{2+} plays an important role in both ABA-dependent and ABA-independent pathways by affecting different types of protein kinases, including the calcium-dependent protein kinase (CDPK) family. CDPK contains internal EF-hand motif(s), and its activity is directly regulated by Ca^{2+} binding. ABA or osmotic stress induces an intracellular Ca^{2+} spike or oscillation, which likely affects CDPK activity. Several studies have indicated that CDPKs are also involved in ABA or osmotic-stress signaling in plants.

Several CDPK members are involved in typical ABA responses, e.g., stomatal closure or gene expression control. In addition to the SnRK2 family, recent studies highlighted the possible involvement of CDPKs in the core ABA pathway. Interestingly, SnRK2s and CDPKs may share substrates in ABA signaling, e.g., the *Arabidopsis* CDPKs, AtCPK4, and AtCPK11, positively regulate ABA responses and can phosphorylate AREB/ABF transcription factors. As described above, AREB/ABFs are major substrates of SnRK2 in the current ABA-signaling model. Another well-known SnRK2 substrate is SLAC1, a slow anion channel, and its homologs, which play an essential role in the regulation of ion concentration in guard cells for stomatal closure. Recent studies demonstrated that SLAC1 is also phosphorylated by some CDPKs, i.e., AtCPK6, AtCPK21, and AtCPK23. In addition, recent studies revealed that some CDPKs functionally and physically interact with group A PP2Cs, as well as SnRK2s, suggesting that CDPKs may function as an alternative pathway in core ABA signaling.

Another Ca^{2+}-regulated protein kinase family, i.e., the CBL-interacting protein kinase (CIPK)/SnRK3 family, plays significant roles in ABA and osmotic-stress signaling. The protein kinase activity of CIPK is regulated in a Ca^{2+}-dependent manner. In the absence of Ca^{2+}, CIPK activity is inhibited by autoregulation. When intracellular Ca^{2+} is elevated, calcineurin B-like (CBL) proteins bind to Ca^{2+} and then associate with CIPKs to mediate their autoregulation. The CBL-CIPK system participates in various signal transduction pathways, e.g., salt, ABA, or osmotic stress, and has a crucial role in salt tolerance in plants. In the early 2000s, a genetic screen identified *Arabidopsis* mutants, salt-overly-sensitive 2 (sos2) and sos3, and SOS2 and SOS3 encode CIPK24 and CBL4, respectively. In this case, SOS2 activity is upregulated by the Ca^{2+}-bound form of SOS3, and then SOS2 phosphorylates membrane-bound Na^+/H^+ antiporters to maintain cellular ion homeostasis by excluding Na^+. Another example is CBL1/9-CIPK23, which regulates the AKT1 potassium channel. In addition, the CBL-CIPK system has been implicated in ABA and osmotic-stress signaling.

4.3 Mitogen-Activated Protein Kinase (MAPK)-Mediated Signaling

The MAPK family is a group of protein kinases that is well conserved in eukaryotes. In general, MAPK activity is regulated by MAPK cascade(s), consisting of MAPK, MAPK kinase (MAPKK), and MAPKK kinase (MAPKKK). MAPKKs activate MAPKs by phosphorylation of the TxY motif, while MAPK phosphatases dephosphorylate and inactivate MAPKs. Plant MAPKs play key roles in various signaling pathways in responses to pathogens, phytohormones (ethylene, jasmonate, etc.), cell differentiation, etc. Several MAPK cascades are involved in ABA or osmotic stress, e.g., AtMPK1, AtMPK2, AtMPK3, and AtMPK6 are activated by ABA or osmotic stress. However, our knowledge of the molecular mechanism of MAPK signaling in ABA or osmotic-stress responses is limited. MAPKs can be activated by ABA or osmotic stress in several ways. For example, plant MAPKs are activated by reactive oxygen species (ROS), and previous studies demonstrated that ABA and osmotic stress cause an oxidative burst in plant cells. In addition, a recent study indicated that MAPK(s) can be regulated directly by calmodulin(s). The in vivo targets of MAPKs should be identified to understand MAPK-mediated ABA or osmotic-stress signaling. SnRK2 has been shown to activate MAPK1 and MAPK2 in response to osmotic stress.

4.4 Proteolysis

In addition to posttranslational modifications and second messengers, proteolysis is a major signal transduction regulatory mechanism. Proteolysis affects signal transduction systems by specifically controlling the amounts of signaling factors at the protein level. Such selective proteolysis depends on ubiquitination of target proteins by E3 ubiquitin ligases. Several types of E3 ligases exist in plants, a number of which are implicated in ABA signaling, e.g., RING E3 ligases, such as KEEP ON GOING (KEG), RHA2a/b, SDIR1, AIP2, etc.; or components of CUL4-based E3 ligases, DWA1 and DWA2. It will be important to identify such proteolysis pathways to gain insight into feedback regulation in ABA signaling. In addition to ubiquitination, some other modifications, such as SUMOylation, have also been implicated in ABA and osmotic-stress signaling. The SUMO ligase AtSIZ1 plays multiple roles not only in ABA signaling but also in phosphate deficiency and flowering. For both ubiquitination and SUMOylation, target proteins should be identified to gain further understanding of these systems.

4.5 Phospholipids

In addition to Ca^{2+}, phospholipid signaling had been well studied as a second messenger system in the context of ABA or osmotic-stress signaling. The several classes of phospholipid-derived signal messengers, such as phosphatidic acid (PA), diacylglycerol, inositol polyphosphates, etc., are generated by modification of membrane lipids by several types of enzymes, including phospholipases, lipid kinases, and/or phosphatases. One such enzyme, phospholipase D (PLD), is a major family of phospholipases in plants, which hydrolyses phospholipids to generate PA. Recent studies indicated that PA production by PLDs plays a significant role in plant drought responses. Different PLDs and their PAs likely have different functions, e.g., AtPLDα1 and AtPLDδ are involved in the ABA and ROS responses, respectively. PA binds to several target proteins, such as ABI1 (PP2C), phosphoinositide-dependent protein kinase 1 (PDK), and NADPH oxidases. In addition, PA can activate H^+-ATPase or MAPKs. Further studies are required to determine the functional roles of PA or PLDs in drought stress signaling in plants.

4.6 Reactive Oxygen Species (ROS)

As described above, ROS may function as signal mediators in ABA or osmoticstress signaling. There are several sources of ROS generation in plant cells. NADPH oxidase is involved in ROS generation. In Arabidopsis, two NADPH oxidases, AtrbohD and AtrbohF, have been implicated in ABA responses in guard cells. In vitro analysis showed that ABA-activated SnRK2 phosphorylates AtrbohF, suggesting that ROS generation may be affected by ABA signaling.

In general, ROS attack proteins, thus inducing functional changes and affecting signal transduction systems. Therefore, it is important to identify which proteins are attacked by ROS; however, our knowledge regarding those attacked in ABA and osmotic-stress signaling is limited. Biochemical studies have shown that ABI1 and ABI2, group A PP2Cs, are sensitive to H_2O_2 at the protein level. However, the functional relationships between such biochemical evidence and in vivo ABA signaling are unclear.

Nitric oxide (NO) is another oxidant that functions as a signaling molecule. The importance of NO was originally identified in the blood vessels of animals, and it was then confirmed to be an endogenous signal in plant cells. For example, nitratereductase-mediated NO generation is required for ABA-induced stomatal closure, by affecting ion channels in guard cells. Furthermore, NO stimulates MAPK activation as well as ROS, suggesting its involvement in the responses to abiotic stress, phytohormones, or developmental cues. However, our knowledge of the mechanism of NO-mediated signaling in plants is inadequate. As in the case of ROS, identifying the targets of NO in plant cells is vital.

5 Enhancement in Drought Signaling: From Genetics Perspective

The selection of an appropriate breeding strategy to develop drought tolerant cultivars is the key step for a successful breeding program. Therefore, strategies which can transfer specific genes, exploit wild relatives of crops, identify and transfer genes with ease, require less time, and labor to develop cultivars are of great value. The advances in genetics, genomics, and functional genomics have enabled researchers to combine one or more advantages of different strategies to develop drought tolerant wheat. Here, we have described the advances in these methods.

5.1 Improving Drought Tolerance Through Classical Breeding

Classical breeding dates back to 5,000–10,000 years when man domesticated selective plant species based on their better taste. Domestication was followed by selection of high yielding genotypes and cross-hybridization to recombine tolerance genes from different sources. The presence of genetic variation in wheat is the key to identify the contrasting parents for classical breeding (Shelden and Roessner 2013). Significant genetic variation in wheat for drought tolerance has been identified for selection of diverse parents. Cross-hybridization of wheat diploid progenitors produce drought-tolerant synthetic hexaploid (SHs) wheat (see the section Introgression of Drought Tolerance Genes from Wild Species) with introgression of several novel drought tolerance genes (Zhang et al. 2005). Most of plant breeders selected the drought and other stress tolerant wheat varieties on the basis of higher yields and ignored the physiological mechanism behind it. Therefore, few cultivars having drought and abiotic stress tolerance have been developed in comparison to the ones improved for high yield (Hussain et al. 2015). Under drought, plant machinery shifts its focus to ABA production for downstream activation of signaling and tolerance mechanisms which lowers the grain filling and yield. Therefore, the balance between yield and drought tolerance needs to be investigated (Alvarez et al. 2014). Abscisic acid content has been used as selection index for screening wheat under drought and contrasting parents for its production have been crossed. Several QTLs and genes involved in signaling pathways have been identified in subsequent segregating populations (Iehisa et al. 2014). Longer time, intensive labor requirements, transfer of non-desirable genes, limited resistance resources, genetic barriers and limited understanding of tolerance mechanisms are problems associated with classical breeding. Since breeding to date has not been focused on signaling pathways, there is a need to combine the available information collected by QTLs, MAS, and Omics tools with traditional traits for improved wheat drought tolerance.

5.2 Molecular Markers for the Identification of Drought Signaling Genes

Selection in classical breeding is performed on the basis of morphological and economical traits which are highly influenced by the environment. Environmental influence on phenotypic expression creates confusion in selection of desirable traits. Discovery of DNA markers for economic and stress-related crop traits has helped to select the desirable traits and parents with ease, efficacy, and reliability in remarkably shorter time. Therefore, markers, especially the SNPs have added more power to identify the genes linked to drought and other stresses. DNA markers for various genes involved in drought signaling have been reported, e.g., RAPD markers by using P21F/P21R and P25F/PR primers in A genome; and P18F/P18R primer in B genome mapped DREB1 on 3A chromosome (Huseynova and Rustamova 2010). In wheat, DREBs were tagged with five SNPs in A (P21F/P21R and P25F/PR primers), B (P18F/P18R primer), and D genome (P20F/P20R and P22F/PR primers). DREB1 gene was tagged on chromosome 3A, 3B, and 3D. S646 and S770 SNPs were used and SNP S770 mapped DREB-B1 between markers Xfbb117 and Xmwg818 on chromosome 3BL (Wei et al. 2009). SNPs identified the involvement of five signaling genes in yield and drought-tolerance pathways. The DREB1Acorrelated with heading date, vegetation index and biomass, while flag leaf width, harvest index and leaf senescence were associated with ERA1-B and ERA1-D (enhanced response to ABA) genes. Other signaling genes, 1-FEH-A and 1-FEH-B (fructan-1-exohydrolase) were linked to yield and thousand kernel weight (Edae et al. 2013). Significant relationships between morpho-physiological traits and SNPs suggested key role of detected SNPs in drought tolerance.

High-resolution melting (HRM) technology is the most powerful tool to identify the allelic variations. Use of HRM found that allelic variation in DREB TFs identified by SNPs led to variation in peptide sequences as well. The variation in peptide sequences was linked with differences in protein geometry and recognition of cis-elements involved in ABA signaling (Mondini et al. 2015). Two important TFs, i.e., DREB1, WRKY1, and a Na^+ transporter, HKT-1 conferring drought and salt tolerance were also mapped by SNPs (Mondini et al. 2012). SNPs were used to map TaSnRK2.8 gene, which plays important role in carbohydrate metabolism, protein–protein interaction, and ABA signaling (Zhang et al. 2013). Chromosome locations and primers for these markers are given in Table 1. It can be concluded that MAS has a lot of promise to identify the signaling genes. Strong association between signaling genes and drought-related physiological traits suggest that MAS should be focused to identify the signaling genes (Edae et al. 2013). This can help to improve the drought tolerance with less effort, time, and resources and can speed up the breeding programs in future.

Table 1 Identified molecular markers for plants drought signaling genes

Marker type	Primer	Species	Chromosome location	Target drought signaling gene	Citations
SNP	DREB3a and DREB3b	Wheat		DREB3	Mondini et al. (2015)
SNP	DREB2a and DREB2b	Maize	5A	DREB2	Mondini et al. (2015)
SNP	M13	Rice		TaSnRK2.8	Zhang et al. (2013)
SNP	W32	Rice	6A	1-FEH-B	Edae et al. (2013)
SNP	W12	Wheat		1-FEH-A	Edae et al. (2013)
SNP	ERA1D	Wheat	3A, 3B, 3D	ERA1-D	Edae et al. (2013)
SNP	ERA1B	Wheat	3A	ERA1-B	Edae et al. (2013)
SNP	P21	Wheat	AF303376.1[a]	DREB1A	Mondini et al. (2012)
SNP	HKT-1	Wheat	DQ323885.1[a]	HKT-1	Mondini et al. (2012)
SNP	WRKY1	Wheat	U16709.1[a]	WRKY1	Mondini et al. (2012)
SNP	DREB1a and DREB1b	Wheat	3A	DREB1	Huseynova and Rustamova (2010)
RAPD	P18F/P18R	Wheat	3A	DREB1	Huseynova and Rustamova (2010)
RAPD	P25F/PR	Wheat	3D	DREB1	Wei et al. (2009)
SNP	P20F/P20R and P22F/PR	Maize	Xfbb117-Xmwg818	DREB1	
SNP S770		Wheat	3BL	DREB1	Wei et al. (2009)
SNP	P21F/P21R and P25F/PR P18F/P18R	Wheat	3A	DREB1	Wei et al. (2009)

[a]Reference gene

6 Conclusion and Future Directions

This article focused on stress signaling networks in drought responses in terms of signal transduction, transcriptional regulation, and intercellular transmission. Drought perception is one of the major outstanding questions in drought signaling (Fig. 1). For example, which types of cells first sense or are sensitive to changes in the drought status, and how do these cells recognize and send drought signals? Some signal transduction molecules and the related transcriptional factors have been discovered, but their upstream regulators are not known.

Drought stress signaling comprises several complex networks, in which protein modification, proteolysis, and transcriptional and posttranscriptional regulations are involved. Although our knowledge of each signaling factor is increasing, it is still difficult to grasp the whole picture of drought stress signaling. Further investigations are required to interconnect known signaling factors, as well as efforts to discover novel factors. Various approaches, including either mature or emerging technologies, should be utilized in future research. For example, recent in-depth next-generation sequencing analyses identified a number of unknown noncoding RNAs and short ORFs expressed in response to ABA or osmotic stress. With regard to intracellular signal transduction at the protein level, recent advances in proteomics, such as modification-specific proteomics or interactomics, will facilitate further elucidation of drought stress signaling in plants.

Many of the links in intercellular signaling networks are not yet known. Several systemic compounds are candidates for cell-to-cell signaling, and specific transporters for these compounds are under investigation in roots, between roots and shoots, and within shoots. Some ideas for the operation of intercellular signaling networks have been proposed based on model plants, but significant variations in drought tolerance mechanisms among different plant species may exist.

The trade-off between growth ability and stress tolerance is a difficult issue, which will become more important in future studies. Improving stress tolerance through genetic engineering and modifications may have negative effects on plant growth and development. In particular, trade-offs between carbon assimilation and water loss related to drought responses, especially the closing of stomata, should be taken into account. Such studies using transgene expression systems and suitable promoters are at present underway. A simple and easy method of determining water-use efficiency (WUE) is required. Technologies for matching water input with plant requirements at the whole-plant level are essential to improve drought tolerance.

References

Abe H, Urao T, Ito T, Seki M, Shinozaki K, Yamaguchi-Shinozaki K (2003) *Arabidopsis* AtMYC2 (bHLH) and AtMYB2 (MYB) function as transcriptional activators in abscisic acid signaling. Plant Cell 15:63–78

Agarwal PK, Agarwal P, Reddy MK, Sopory SK (2006) Role of DREB transcription factors in abiotic and biotic stress tolerance in plants. Plant Cell Rep 25:1263–1274

Alvarez S, Roy Choudhury S, Pandey S (2014) Comparative quantitative proteomics analysis of the ABA response of roots of drought-sensitive and drought-tolerant wheat varieties identifies proteomic signatures of drought adaptability. J Proteome Res 13:1688–1701

Anjum SA, Tanveer M, Ashraf U, Hussain S, Shahzad B, Khan I, Wang L (2016) Effect of progressive drought stress on growth, leaf gas exchange, and antioxidant production in two maize cultivars. Environ Sci Pollut Res 23(17):17132–17141

Bray EA (2002) Classification of genes differentially expressed during water-deficit stress in *Arabidopsis thaliana*: an analysis using microarray and differential expression data. Ann Bot 89:803–811

Chandra Babu R, Jingxian Z, Blumc L, David-HHod T, Wue R, Nguyenf HT (2004) HVA1, a LEA gene from barley confers dehydration tolerance in transgenic rice (*Oryza sativa* L.) via cell membrane protection. Plant Sci 166:855–862

Cominelli E, Galbiati M, Vavasseur A, Conti L, Sala T, Vuylsteke M, Leonhardt N, Dellaporta SL, Tonelli C (2005) A guard-cell-specific MYB transcription factor regulates stomatal movements and plant drought tolerance. Curr Biol 151:196–1200

Edae EA, Byrne PF, Manmathan H, Haley SD, Moragues M, Lopes MS et al (2013) Association mapping and nucleotide sequence variation in five drought tolerance candidate genes in spring wheat. Plant Genome 6:13

Gu YQ, Yang C, Thara VK, Zhou J, Martin GB (2000) Pti4 is induced by ethylene and salicylic acid, and its product is phosphorylated by Pto kinase. Plant Cell 12:771–786

Hirayama T, Shinozaki K (2010) Research on plant abiotic stress responses in the post-genome era: past, present and future. Plant J 61:1041–1052

Hu H, Dai M, Yao J, Xaio B, Li X, Zhang Q, Xiong L (2006) Overexpressing a NAM, ATAF, and CUC (NAC) transcription factor enhances drought resistance and salt tolerance in rice. Proc Natl Acad Sci USA 103:12987–12992

Huseynova I, Rustamova S (2010) Screening for drought stress tolerance in wheat genotypes using molecular markers. Biol Sci 65:132–139

Hussain B, Khan AS, Ali Z (2015) Genetic variation in wheat germplasm for salinity tolerance at seedling stage: improved statistical inference. Turk J Agric For 39:182–192

Hussain M, Farooq S, Hasan W, Ul-Allah S, Tanveer M, Farooq M, Nawaz A (2018) Drought stress in sunflower: physiological effects and its management through breeding and agronomic alternatives. Agric Water Manag 201:152–166

Iehisa JCM, Matsuura T, Mori IC, Yokota H, Kobayashi F, Takumi S (2014) Identification of quantitative trait loci for abscisic acid responsiveness in the D-genome of hexaploid wheat. J Plant Physiol 171:830–841

Kulik A, Wawer I, Krzywin´ska E, Bucholc M, Dobrowolska G (2011) SnRK2 protein kinases–key regulators of plant response to abiotic stresses. OMICS 15:859–872

Kuromori T, Mizoi J, Umezawa T, Yamaguchi-Shinozaki K, Shinozaki K (2014) Drought stress signaling network. Mol Biol 383–409

Liang YK, Dubos C, Dodd IC, Holroyd GH, Hetherington AM, Campbell MM (2005) AtMYB61, an R2R3-MYB transcription factor controlling stomatal aperture in *Arabidopsis thaliana*. Curr Biol 15:1201–1206

Mondini L, Nachit M, Porceddu E, Pagnotta MA (2012) Identification of SNP mutations in DREB1, HKT1, and WRKY1 genes involved in drought and salt stress tolerance in durum wheat (*Triticum turgidum* L. var durum). OMICS 16:178–187

Mondini L, Nachit MM, Pagnotta MA (2015) Allelic variants in durum wheat (*Triticum turgidum* L. var. durum) DREB genes conferring tolerance to abiotic stresses. Mol Genet Genomics 290:531–544

Moose SP, Sisco PH (1996) Glossy15, an APETAL2-like gene from maize that regulates leaf epidermal cell identity. Genes Dev 10:3018–3027

Nakashima K, Kiyosue T, Yamaguchi-Shinozaki K, Shinozaki K (1997) A nuclear gene, erd1, encoding a chloroplast-targeted Clp protease regulatory subunit homolog is not only induced by water stress but also developmentally up-regulated during senescence in *Arabidopsis thaliana*. Plant J 12:851–861

Ohme-Takagi M, Shinshi H (1995) Ethylene-inducible DNA binding proteins that interact with an ethylene-responsive element. Plant Cell 7:173–182

Okamuro JK, Caster B, Villarroel R, Van Mantagu M, Jofuku KD (1997) The AP2 domain of APETELA2 defines a large new family of DNA binding proteins in *Arabidopsis*. Proc Natl Acad Sci USA 94:7076–7081

Oono Y, Seki M, Nanjo T, Narusaka M, Fujita M, Satoh R, Satou M, Sakurai T, Ishida J, Akiyama K, Iida K, Maruyama K, Satoh S, Yamaguchi- Shinozaki K, Shinozaki K (2003) Monitoring expression profiles of *Arabidopsis* gene expression during rehydratation process after dehydration using ca. 7000 full-length cDNA microarray. Plant J 34:868–887

Sasaki T, Song J, Koga-Ban Y et al (1994) Toward cataloguing all rice genes: large scale sequencing of randomly chosen rice cDNAs from a callus cDNA library. Plant J 6:615–624

Schafleitner R, Rosales ROG, Gaudin A, AliagaCAA MGN, Marca LRT, Bolivar LA, Delgado FM, Simon R, Bonierbale M (2007) Capturing candidate drought tolerance traits in two native Andean potato clones by transcription profiling of field grown plants under water stress. Plant Physiol Biochem 45:673–690

Seki M, Narusaka M, Ishida J, Nanjo T, Fujita M, Oono Y, Kamiya A, Nakajima M, Enju A, SakuraiT SM, Akiyama K, Taji T, YamaguchiShinozaki K, Carninci P, Kawai J, Hayashizaki Y, Shinozaki K (2002) Monitoring the expression profiles of 7000 *Arabidopsis* genes under drought, cold and high-salinity stresses using a full-length cDNA microarray. Plant J 31:279–292

Shelden MC, Roessner U (2013) Advances in functional genomics for investigating salinity stress tolerance mechanisms in cereals. Front Plant Sci 4:123

Shinozaki K, Yamaguchi-Shinozaki K (2007) Gene networks involved in drought stress response tolerance. J Exp Bot 58:221–227

Tanveer M, Shahzad B, Sharma A, Khan EA (2018) 24-Epibrassinolide application in plants: an implication for improving drought stress tolerance in plants. Plant Physiol Biochem 135:295–303

Tran LS, Nakashima K, Sakuma Y et al (2007) Co-expression of the stress-inducible zinc finger homeodomain ZFHD1 and NAC transcription factors enhances expression of the ERD1 gene in *Arabidopsis*. Plant J 49:46–63

Umezawa T, Nakashima K, Miyakawa T, Kuromori T, Tanokura M, Shinozaki K, YamaguchiShinozaki K (2010) Molecular basis of the core regulatory network in abscisic acid responses: sensing, signaling, and transport. Plant Cell Physiol 51:1821–1839

Way H, Chapman S, McIntyre L, Casu R, Xue GP, Manners J, Shorter R (2005) Identification of differentially expressed genes in wheat undergoing gradual water deficit stress using a subtractive hybridisation approach. Plant Sci 168:661–670

Wei B, Jing R, Wang C, Chen J, Mao X, Chang X et al (2009) Dreb1 genes in wheat (*Triticum aestivum* L.): development of functional markers and gene mapping based on SNPs. Mol Breed 23:13–22

Weigel D (1995) The APETELA2 domain is related to a novel type of DNA binding domain. Plant Cell 7:388–389

Zhang P, Dreisigacker S, Melchinger AE, Reif JC, Mujeeb Kazi A, Van Ginkel M et al (2005) Quantifying novel sequence variation and selective advantage in synthetic hexaploid wheats and their backcross-derived lines using SSR markers. Mol Breed 15:1–10

Zhang H, Mao X, Zhang J, Chang X, Jing R (2013) Single-nucleotide polymorphisms and association analysis of drought-resistance gene TaSnRK2.8 in common wheat. Plant Physiol Biochem 70:174–181

Zhou JM, Tang X, Martin GB (1997) The Pto kinase conferring resistance to tomato bacterial speck disease interacts with proteins that bind a cis-element of pathogenesis-related genes. EMBO J 16:3207–3218

QTL Mapping for Drought Stress Tolerance in Plants

Tayyaba Andleeb, Tariq Shah, Rizwana Nawaz, Iqra Munir, Fazal Munsif and Arshad Jalal

Abstract Drought tolerance is a significant quantitative trait with many phenotypes and is often complicated by plant phenology. Different types of environmental stresses, such as high irradiance, high temperature, nutrition deficiencies and toxicity, may challenge crops at the same time; therefore, the cultivation of drought resistance is very complicated. Interdisciplinary researchers have been trying to use various methods to analyze and understand the mechanisms of plant tolerance to drought stress. However, the limited success of molecular breeding and physiological approaches suggests that we need to rethink our strategies. Recent genetic techniques and genomics tools, coupled with advances in breeding methods and precise phenotypes, are likely to reveal candidate genes and metabolic pathways for drought tolerance in crops. In order to update the analysis methods of drought resistance, this chapter reviewed the latest research progress on precise phenotypes of plant physiology in response to drought. Finally, according to the physiological/morphological and molecular mechanism of drought-resistant parent lines, a strategy of selecting specific environment and making suitable germplasm adapt to the environment was proposed. This chapter mainly involves use of quantitative trait locus (QTL) analysis for traits that are related to drought stress. In this article, we briefly review the available literature on QTL analyses in wheat for traits, which respond to drought/water

T. Andleeb · R. Nawaz
Department of Plant Sciences, Faculty of Biological Sciences, Quaid-e-Azam University, Islamabad, Islamabad, Pakistan

T. Shah (✉)
Department of Agronomy, University of Agriculture Peshawar, Peshawar, Pakistan
e-mail: tariqshah@aup.edu.pk

I. Munir
The University of Lahore, Sargodha Campus, Sargodha, Pakistan

F. Munsif
Department of Agronomy, University of Agriculture Peshawar, Amir Muhammad Khan Campus, Peshawar, Pakistan

A. Jalal
Department of Plant Health, Rural Engineering and Soils, Sao Paulo State University, Ilha Solteira, Brazil

M. Hasanuzzaman and M. Tanveer (eds.), *Salt and Drought Stress Tolerance in Plants*, Signaling and Communication in Plants, https://doi.org/10.1007/978-3-030-40277-8_16

stress. The outlook for future research in this area and the possible approaches for utilizing the available information on genetics of drought tolerance for wheat breeding are also discussed.

1 Introduction

Major yield loss in agricultural crops is primarily due to biotic and abiotic stress across the globe (Kang et al. 2009). Genetic basis of abiotic and biotic resistance in almost all crops are quite elaborative. Sufficient relevant information is available on basic mechanisms involved to cope up with both biotic and abiotic stresses in plants specifically in crops (Collins et al. 2008). Most drastic abiotic stress which causes maximum grain yield reduction in important crops is drought stress (Collins et al. 2008; Anjum et al. 2017a). Drought in general term can be defined as deficient precipitation in extended period which may be several years or one year or one season long ultimately resulting in water shortage due to climatic variations or human activity. According to agricultural criteria, drought can be defined as insufficient soil moisture which could not meet the water requirement of a specific crop/plant in specific time period (Hussain et al. 2018). Drastic climatic variations with unpredictable rainfall ultimately resulted in long episodes of water scarcity in form of drought in almost all crops including wheat, rice, maize, sorghum, etc. The intensity and timing of drought stress have range of impact on crop plant development along with other environmental conditions (Calderini et al. 2001a, b; Araus et al. 2002; Tanveer et al. 2018). The erratic environmental conditions resulted in unreliable heritability estimates, so considered as one of major hindrance for analyses of drought associated traits at genetic level (Ribaut et al. 1997; Sellammal et al. 2014). In recent years, severity and frequency of drought stress have been reported to increase due to drastic change in climatic conditions which is one of the major concerns worldwide. According to one estimate, 40–60% yield losses in major crops worldwide are mainly due to drought stress (Shao et al. 2008). In the present scenario, crop plants which are most efficient in available water use to maintain desired yields will be considered as superior (https://www.liverpool.ac.uk). A huge amount of information about adaptation of plants under water-stressed conditions has seldom benefit for crop improvement under drought stress due to complex genetic basis of yield and other related traits. Thus in past, improvement and stabilization in crop performance under drought stress remains slow and laborious (Wu et al. 2017).

Selection of drought-resistant genotypes that can be better adapted even in rainfed areas as well as in partially irrigated areas could be a good breeding strategy to overcome water stress in crop plants (Fahad et al. 2017). Agronomic traits associated with water stress are mostly polygenic in nature (Gupta et al. 2017). In crop plants, genetic basis of traits which are linked with drought stress are available in already-reported literature (Joshi et al. 2007). Under drought conditions conventional breeding approaches are not that much beneficial due to poor heritability and instability of yield and related agronomic traits. Therefore, molecular breeding approaches like marker-assisted selection approach could be deployed to develop various crop populations which will be more efficient in terms of drought tolerance (Bankole et al. 2017). In addition to molecular breeding, other strategies including phenomics, genomics, transcriptomics, and proteomics can be exploited to develop drought-tolerant cultivars. Basically drought-tolerant cultivars maintain sufficient plant water status in order to identify and map QTLs-related drought tolerance, water-stressed conditions are necessary to understand genetic variations expressed in plant water use and water retentions at cellular level (de Miguel et al. 2014; Kapanigowda et al. 2014; Iglesias-García et al. 2015).

Numerous studies comprising genome wide association studies and QTL mapping have been conducted in plants under drought stress (Gupta et al. 2017). Due to which large number of QTLs have been identified for some particular traits associated with water stress tolerance specifically in crop plants. Some of these traits are coleoptile length, root architecture, plant biomass, water soluble carbohydrates membrane stability index and grain yield. Some physiological parameters including estimation of superoxide dismutase (SOD), peroxide dismutase (POD), and proline can also be used for precise selection of water stress tolerant lines/genotypes as well as in QTL and GWAS analysis (Gupta et al. 2012). Some of these QTLs either linked to agronomic or physiological traits associated with drought stress can contribute up to 20% phenotypic variation (Budak et al. 2015). This chapter covers review of literature on QTL mapping in plant specifically in crops. It also gives brief review of future promises of genetic research from already available information to develop drought-tolerant cultivars in different crop plants. So the present chapter provides an updated and detailed account, with an emphasis on QTL analysis of drought stress in plants.

2 Morphological, Agronomic, and Growth Responses Under Drought Stress

Under drought stress, different physiological, morphological, defense and metabolic systems have responded to assist plant not only to survive but also sustain growth and development (Valliyodan and Nguyen 2006; Anjum et al. 2016). Genetic basis of drought tolerance or susceptibility are very complex in nature and associated with traits which are controlled by more than one gene (polygenic nature) thus selection of drought tolerant genotypes is a difficult task (Obidiegwu et al. 2015). Plants being sessile in nature responded to drought stress by production of different stress signals, then these signals are perceived and transduced through different transcription factors by plant defense system ultimately leading to expression of genes in form of protein that provide drought tolerance (Gahlaut et al. 2016; Anjum et al. 2017b). Plants generally adapted two mechanisms upon encountering drought stress, i.e., either drought tolerance or drought avoidance (Price et al. 2002). Some morphological features which are adapted by majority of cereals as drought avoidance behavior are leaf rolling, leaf glaucousness, shoot vigor, transpirational cooling, stomatal conductance, early maturation, membrane stability, green flag leaf area, root vigor, and architecture (Chourasia 2017). In addition to above-mentioned traits, an important adaptation by plants under water shortage is efficient use of available water, i.e., is termed water use efficiency (WUE) (Passioura 1977). While physiological mechanism adapted by crop plant as drought tolerance strategy includes photosynthetic rate, chlorophyll content, ABA accumulation, osmoprotectant accumulation, soluble sugar content, generation of reactive oxygen species (ROS) (Gill and Tuteja 2010; Fig. 1). Different morphological and physiological traits with their associated QTLs that are involved in mechanisms for drought adaptation and tolerance are listed in Table 1. Such information helps in understanding of genetic system involved in providing resistance against drought stress. Markers linked with these QTLs could be used for marker assisted selection in future breeding program to develop crop cultivars that would be drought tolerant.

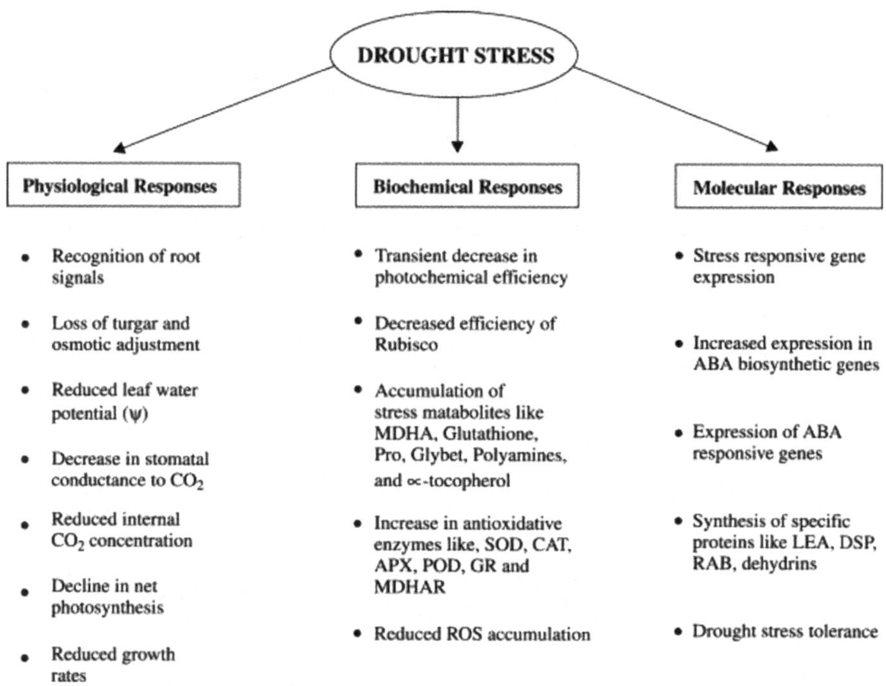

Fig. 1 Physiological, biochemical and molecular responses of drought stress in plants (adapted from Reddy et al. 2004)

Table 1 QTLs identified for drought tolerance in cereals

Traits	QTL number	Linkage group	Phenotypic variation (%)	Citations
Grain yield	77	All except 12	7.5–55.7	Ciuca and Petcu (2009)
	1 (qtl12.1)	12	51	Al-Khatib and Paulsen (1999)
	2	2, 3	13.0–31.0	Innes et al. (1984)
	1 (qDTY12.1)	12	23.8	Abebe et al. (2003)
	1 (qDTY3.2)	3	23.4	Rosa et al. (2009)
	14	All except 9, 11	13.3	Gill and Tuteja (2010)
	1 (qDTY2.3)	2	9	Rivero et al. (2007)
	1 (qDTY2.2)	2	6	Rivero et al. (2009)
	4	3	18.8–31.8	Araus et al. (2002)
	7	1, 2, 3, 9, 12	31–77	Bennett et al. (2011)
Osmotic adjustment	1 (OA_{70})	8	Major	Richards (1996)
Drought avoidance	17	All except 9	4.4–25.6	EL-Hafid et al. (1998)
Root traits	18	All chromosomes	1.2–18.5	Pinto et al. (2010)
	42	All chromosomes	6.0–24.4	Hurd (1974)
Coleoptile length and drought resistance index	15	All except 3, 8, 11	4.9–22.7	Al-Khatib and Paulsen (1990)
Relative growth rate and specific water use	7	2, 4, 5, 6, 7, 8	10.0–22.0	Kohli et al. (1991)
Basal root thickness and 100-grain weight	2	4, 6	20.6–33.4	Calderini et al. (2001a, b)
Plant production	24	1, 6	14–20.9	Calderini et al. (2001a, b)

(continued)

Table 1 (continued)

Traits	QTL number	Linkage group	Phenotypic variation (%)	Citations
Physio-morphological	9	4	36.8	Calderini et al. (2001a, b)
1000-grain weight	21	All except 12	50.3	Gill and Tuteja (2010)
Panicle number per plant	14	All except 1, 3, 8		Gill and Tuteja (2010)
Filled grain number per panicle	23	All except 12	33.3	Gill and Tuteja (2010)

3 Nature of Genes Expressed Under Drought Stress

Genes involved in drought tolerance mechanism includes genes which are expressed on signal perception of drought are membrane transporters and ion channels followed by expression of receptor-like protein kinases, and calcium-binding proteins such as calcineurin and calmodulin for transduction of perceived signal along with regulation of transcriptional factors respectively. A group of genes belonging to different classes like AP2, bHLH, bZIP, C2H2, ABI3VP1, MYB, zinc fingers, MADS, NAC, and WRKY are involved in encoding transcription factors and involved in downstream signal transduction ultimately binds with cis-regulatory sequences of specific structural genes (Table 2). Some genes which are main targets to develop drought-resistant crop cultivars are involved in encoding osmoprotectants/antioxidants and ROS as they are vital for maintenance and protection of cellular structure under drought stress. During varying ranges of drought episodes; most commonly involved are genes which reduced cell growth, synthesize hormones like ABA, metabolize proline content, scavenge ROS, involved in root cell expansion in the form of proteins such as expansins, extensins, xyloglucans, and cellulose synthase, protective protein production such as late embryogenesis abundant (LEA) and chaperons, sopentyltransferase production to delay senescence and maintain photosynthetic activity comparatively for long duration (Rivero et al. 2007, 2009; Table 2).

4 Mapping QTLs in Plants

One of the important focuses of biotechnology research in these days is to figure out the genetic basis of quantitative traits in crop plant by using various approaches. Various QTL mapping studies have been carried out in this regard in order to improve our understanding about the nature of inheritance of quantitative traits. It was reported that most of important agronomic traits including drought stress are quantitative

Table 2 Drought stress and their related candidate genes of model plant

Gene	Accession number	Plant species	Function/Plant trait	Citations
AIRP1	AT4G23450	*Arabidopsis thaliana*	Regulatory proteins, post-translational modification, ubiquitin ligases	Ryu et al. (2010)
DREB2A	AT5G05410	*A. thaliana*	Regulatory proteins, transcription factors, DREB	Liu et al. (1998)
HYL1	AT1G09700	*A. thaliana*	Stomatal movement, ion and osmotic homeostasis	Lu and Fedoroff (2000)
LOS5	AT1G16540	*A. thaliana*	Stomatal movement, ion and osmotic homeostasis	Xiong et al. (2001)
AVP1	AT1G15690	*A. thaliana*	Regulatory proteins, acid anhydride hydrolases	Gaxiola et al. (2001)
NCED	AT3G14440	*A. thaliana*	Stomatal movement, ion and osmotic homeostasis, hormone signaling, ABA biosynthesis	Iuchi et al. (2001)
GolS1	AT2G47180	*A. thaliana*	Detoxification, osmolyte production, enzymes for osmolyte biosynthesis	Taji et al. (2002)
GolS2	AT1G56600	*A. thaliana*	Detoxification, osmolyte production, enzymes for osmolyte biosynthesis	Taji et al. (2002)
AnnAt1	AT1G35720	*A. thaliana*	Growth control	Konopka-Postupolska et al. (2009)

(continued)

Table 2 (continued)

Gene	Accession number	Plant species	Function/Plant trait	Citations
APX2	AT1G07890	*Arabidopsis A. thaliana*	Detoxification, removal of ROS, detoxification signaling	Rossel et al. (2006)
AREB1	AT1G45249	*A. thaliana*	Regulatory proteins, transcription factors, bZIP	Furihata et al. (2006)
AtATL78	AT1G49230	*A. thaliana*	Post-translational modification, ubiquitin ligases	Kim and Kim (2013)
AtBG1	AT1G52400	*A. thaliana*	Stomatal movement, ion and osmotic homeostasis, regulatory proteins, hormone and ABA signaling	Lee et al. (2006)
ATHB6	AT2G22430	*A. thaliana*	Stomatal movement, ion and osmotic homeostasis, regulatory proteins, transcription factors	Himmelbach et al. (2002)
AtNF-YB1	AT2G38880	*A. thaliana*	Regulatory proteins, transcription factors, NF-Y	Zhu et al. (2002)
AtrbohD	AT5G47910	*A. thaliana*	Stomatal movement, ion and osmotic homeostasis	Mustilli et al. (2002)
AtrbohF	AT1G64060	*A. thaliana*	Stomatal movement, ion and osmotic homeostasis	Mustilli et al. (2002)
OSM1/SYP61	AT1G28490	*A. thaliana*	Stomatal movement, ion and osmotic homeostasis	Zhu et al. (2002)

(continued)

Table 2 (continued)

Gene	Accession number	Plant species	Function/Plant trait	Citations
OST1/SRK2E	AT4G33950	*A. thaliana*	Stomatal movement, ion and osmotic homeostasis, regulatory proteins, signal transduction, protein kinases	Mustilli et al. (2002)
CBF4	AT5G51990	*A.thaliana*	Regulatory proteins, transcription factors, DREB	Haake et al. (2002)
MYC2	AT1G32640	*A. thaliana*	Regulatory proteins, transcription factors, MYC	Abe et al. (2003)
GORK	AT5G37500	*A. thaliana*	Ion and osmotic homeostasis, transporters, ion channels, cation channel	Hosy et al. (2003)
GCR1	AT1G48270	*A. thaliana*	Stomatal movement, ion and osmotic homeostasis	Pandey and Assmann (2004)
DREB1A/CBF3	AT4G25480	*A. thaliana*	Regulatory proteins, transcription factors, DREB	Kasuga et al. (2004)
CBP20	AT5G44200	*A. thaliana*	Stomatal movement, ion and osmotic homeostasis, regulatory proteins, miRNA	Kim et al. (2008)
CLCc	AT5G49890	*A. thaliana*	Ion and osmotic homeostasis, channels and transporters, ion and anion channel	Jossier et al. (2010)
CML9	AT3G51920	*A. thaliana*	Regulatory proteins, signal transduction	Jossier et al. (2010)
CPK21	AT4G04720	*A. thaliana*	Regulatory proteins, signal transduction, protein kinases	Tran et al. (2007)

in nature as these traits are controlled by more than one gene thus termed as polygenes. By using specific molecular markers like SSR, SNP, DaRT, EST, AFLP, RFLP, etc., and statistical tools or software including MAPMAKER/QTL, QTL Network, QTL Cartographer, QGene, MQTL, MapQTL, QTLMapper and PLABQTL, chromosomal positions (loci) of genes controlling quantitative traits are identified. These identified loci are basically termed as "QTLs" which is a genomic region which may be a single gene or cluster of genes that effect trait of our interest. Many diverse traits like, abiotic and biotic stress tolerance including drought in crop plants have been reported through QTL mapping studies till to date. The information mined from QTL analysis can be utilized to develop stress tolerant crops in long run through the aid of marker assisted selection (MAS) and advanced molecular breeding approaches. Many important QTLs and genes associated with drought tolerance which were mapped till to date in all major crops are available at (https://www.plantstress.com). The QTLs linked with drought traits are representative of only small part of phenotypic variation (PV) thus cannot be deployed directly for MAS in crop breeding programmes. Thus, QTL meta-analysis approach can be used for estimation of QTL effects along with advanced phenomic and genomic studies conducted on drought in plants (Rivero et al. 2009).

5 QTL Mapping Studies in Cereal Crops

Numerous studies approximately more than 50 have reported the complex genetic background of drought tolerance and associated traits through QTL or association mapping approach in cereal crops during last 5–10 years (Table 1). These studies have reported ~800 QTLs/MTAs (marker-trait associations). These QTLs along with their associated molecular markers are distributed uniformly or randomly on almost all the chromosomes in cereal crops, while maximum QTLs/MTAs numbers were reported for physiological traits (~429) followed by agronomic and root architecture associated traits, i.e., (318) and the (23) respectively (for a summary, see Table 1). Among these few were major QTLs (with $\geq$20% phenotypic variation) and termed as stable QTLs as these were detected in >50% environments.

6 QTL Mapping for Physiological and Agronomic Traits Under Drought Stress

Stable QTLs for agronomic traits associated with drought tolerance in major crops showed approximately 20–45% PV values which is significant value due to variable nature of drought stress phenomenon. Most of the major QTLs were identified for wide range of traits such as (i) yield and yield related traits, i.e., (days to heading, days to maturity, biomass, grain filling duration, plant height, grain number, spike density, 1000-kernal weight and grain yield), (ii) physiological responses, viz. relative water content, leaf osmotic potential, chlorophyll contents, flag leaf rolling index, osmotic potential, carbon isotope ratio, osmotic adjustment, grain carbon isotope distinction, canopy temperature, water-soluble carbohydrates, and (iii) root architecture related traits like root length and number of roots, root angle and network area, etc., QTLs for three traits viz. stem reserve mobilization, water soluble carbohydrates and chlorophyll content explaining approximately 20–60% PV can be exploited for MAS resulting yield improvement under water-stressed conditions in near-future for cereal crops. Besides already mentioned three physiological traits, QTLs for the abscisic acid (ABA) accumulation can be major breeding target, as ABA is involved in physiological processes and gene expression regulation in plants when exposed to drought conditions. Although 17 QTLs for ABA accumulation under drought stress have been reported in *Triticum aestivum* L. till now (Rivero et al. 2009).

7 QTL Mapping for Drought Tolerance in Wheat

Wheat is a major crop occupying largest cropped area in the world. Throughout the world, it is widely cultivated in arid, semi-arid and irrigated areas of the world. The most important objective of wheat breeders is the development of high yield cultivars with desirable traits combination. In wheat, a lot of work for drought tolerance has been done for the improvement of root architecture. In hexaploid bread wheat, a QTL has been reported for deep root ratio, a QTL for root dry weight (QRdw.ccsu-2A.1, QRdw.ccsu-2A.2) and root length (QRl.ccsu-2B.1) on chromosome No 2A and 2B respectively. Similarly, Cristopher et al. (2013) found the QTLs (QRA.qgw-2A, QRA.qgw-3D, qRA.qgw-5D) for root angles and a QTL for root numbers (qRN.qgw-1B). Moreover, Sharma et al. (2014) reported QTLS for root anatomical characteristics, i.e., characteristics of xylem vessels. Fine mapping for a major chromosome on 3B is in progress for durum wheat, which will affect the grain yield across a wide range of soil moisture regimes. Five major markers, i.e., Xwmc11, Xgwm314, Xwmc296, and Xgwm400 which were reported to be linked with important agro-physiological traits can deployed in MAS of wheat crop to find major controlling genes for drought mechanism (Rivero et al. 2009; Pinto et al. 2010).

8 QTL Mapping for Drought Tolerance in Maize

Drought is a prime abiotic stress factor that has a significant impact on yield in maize grown throughout South-East Asia and South. In maize, drought tolerance QTL studies and the strategies for its application in MAS in breeding has been reported by Christopher et al. (2013). QTLs associated with drought-related traits were identified by a genetic linkage map developed using RFLP markers. 22 QTLs were detected for drought-related traits, with a minimum of 1 and a maximum of 9. For osmotic potential, 4 QTLs were detected on chromosomes 1, 3 and 9 together accounting for 50% of the variation in phenotype. 9 QTLs for leaf surface area on chromosomes 3 and 9 were detected, with various degrees of phenotypic variation, ranging from 25.8 to 42.2% in maize, stay green is a desirable character for crop production. Mano et al. (2005) mapped QTLs on chromosome 4 and 8 in maize for adventitious root formation in waterlogged conditions. Tolerance to abiotic soil stresses can be improved by root traits, by increasing the metabolic efficiency of soil exploration. Root respiration is reduced by root cortical parenchyma and improves the yield and performance of plant under reduced water conditions in maize. In maize near isogenic lines, two major QTLs have been derived on chromosome bins 2.04 and 1.06 that influence grain yield and root architecture under varying water regimes. The genetic dissection for drought tolerance was reported in maize, which accounts for successful utilization of identified QTLS in the improvement programs of maize.

9 Quantitative Trait Loci for Drought Tolerance in Rice

Rice is consumed as staple food by 60% population worldwide. Drought stress severely impacted rice production in sub-Saharan Africa and Asia. Mapping of QTLs associated with high grain yield under drought stress in rice crop and their validation have been considered as better breeding strategy to understand drought tolerance and adaptation mechanism. In rice almost 19 studies on QTL mapping have been conducted till to date which has identified many QTLs linked with drought tolerance (Table 2). While small number of QTL studies identified major QTLs related to drought stress with their ultimate impact on grain yield of rice crop for instance, Vikram et al. (2011), reported qDTY1.1 (major QTL) flanked by markers, i.e., RM11943 and RM431 with major additive effect of on grain yield due to water stress particularly at reproductive stage in three different mapping population viz. N22/MTU1010, N22/Swarna and N22/IR64.

10 QTL Mapping for Drought Tolerance in Sorghum and Pearl Millet

Sorghum and pearl millet are often known as dryland cereals, as these are cultivated in areas with minimal rainfall. In these two crops terminal drought stress significantly impacted grain yield thus emphasized on genetic improvement under water-stressed condition by using QTL mapping approach. In sorghum, drought adaptation mechanism at post flowering stage is found to be associated with stay green trait (Xu 2002). Additionally, in sorghum, 3 QTLs (PV = 17–21%) linked with CO_2 assimilation and transpiration have identified and reported to be co-localized with yield-related traits including biological yield and leaf area under drought conditions (Kapanigowda et al. 2014). Nodal root angle are identified to be linked with drought adaptation mechanism in sorghum under drought conditions; 4 QTLs linked with nodal root angle were mapped by Mace et al. (2012) in addition to 2 QTLs for shoot dry weight, 3 for root dry weight, 3 for leaf area.

Pearl millet is a climate-resilient crop, thus could be served as valuable sources of novel genes-associated stress tolerance mechanism that should be characterized and validated as need of present time. Significant efforts have been put forward to map QTLs associated with stover and grain yields in pearl millet under drought conditions in order to maintain yield under terminal water stress (Yadav et al. 2004; Bidinger et al. 2005). A QTL associated with grain yield with 32% PV have been identified in pearl millet under terminal water stress on LG-2 (Yadav et al. 2011). Additionally, a QTL for low transpiration rate under drought stress in pearl millet was reported by Kholová et al. (2012).

11 QTL Mapping for Drought Tolerance in Barley

Barely is ranked fourth in term of production and cultivation after rice, maize and wheat across the globe. Barely is used as food (5%), beverages (20%) and animal feed (75%) (Sreenivasulu et al. 2008). In poaceae family, barely is an ideal species for genetic studies due to its simple genetic background as compared to other species. Barely is more adapted to environmental stresses as compared to wheat being its close relative. During last two decades, several studies have been conducted to identify QTLs linked with yield- and yield-related traits as well as physiological traits under drought stress in barely crop (Teulat et al. 1998, 2001a, b, 2002; Baum et al. 2003; Talamé et al. 2004; von Korff et al. 2008; Cuesta-Marcos et al. 2009; Kalladan et al. 2013; Mansour et al. 2014; Tondelli et al. 2014; Diab et al. 2004; Szira et al. 2011; Wójcik-Jagła et al. 2013; Honsdorf et al. 2014; Fan et al. 2015; Mora et al. 2016). In contrast, only individual studies have detected QTLs for water-soluble carbohydrates (WSC) (Teulat et al. 2001a; Diab et al. 2004). By using comparative genomic approach, it has been identified that many QTLs associated with drought tolerance in wheat also co-localized with QTLs identified in barely under drought condition Peighambari et al. (2005). Drought tolerance associated QTLs such as QDT.TxFr.2H on chromosome 2H and QDT.TxFr.5H on chromosome 5H have been identified by Fan et al. (2015). While QTLs linked with proline content, i.e., QPC-D.TxFr.3H and QPC-S.TxFr.3H and QTl for moisture content have been reported in barely under drought stress in recent year (Sayed et al. 2012; Fan et al. 2015) which could be exploited for MAS in barely drought tolerance breeding programmes.

12 QTL Mapping for Drought Tolerance in Lentil

Lentil is small crop compared to other major cereal crops like wheat, maize, soybean and rice and it is cultivated on 4.08 million hectares annually worldwide (https://apps.fao.org/faostat). Being cool season cereal legume, it is ranked 3rd followed by pea and chickpea. Lentils are mostly cultivated in semiarid regions (Kahriman et al. 2015). Among various production constraints, drought is one of the main stresses that can result in approximately 70% yield losses in lentil crop (Rana et al. 2017). The marker assisted selection is not commonly implemented in lentils breeding program till now as compared to other crop plants. For efficient selection of drought resistant cultivars in lentils, cost-effective utilization of molecular marker through QTL mapping is key to research in Lentils (Muehlbauer et al. 2006). To date four QTL mapping studies on drought stress in lentils have been conducted so far. All these studies reported 129 QTLs linked to different agronomic and morpho-physiological traits in different mapping populations and environment under drought stress. Mukeshimana et al. (2014), reported 14 stable QTLs related to different agronomic traits such as flowering days, maturity days, under drought stress in common bean (*Phaseolus vulgaris* L.). Another study on drought QTL mapping in *Lens culinaris* Medikus reported 75 QTLs linked to 27-Agro-morphological traits. Among these QTLs, 13 were termed as stable QTLs with PV ranged from 5.4 to 45.9% (Rana et al. 2017). Two more respective studies conducted on *P. vulgaris* L. and *P. vulgaris* L. reported 22 and 18 QTLS respectively.

13 Meta-QTLs and Their Associated Candidate Genes

A meta-QTL analysis is robust approach that could be used for analysis of non-dependent datasets obtained from different populations in various environments for traits of interest. Different software like Biomercator for comparing QTLs positions to identify MQTLs for traits that have no direct relation. For accuracy of MQTLs, confidence interval values were carefully considered (Arcade et al. 2004). This approach of meta-analysis for authentication of drought responsive regions was conducted in wheat, which involved 502 QTLs for agronomic as well as physiological traits already reported in 30 studies conducted under conditions of drought. As many as 19 MQTLs for drought tolerance spread over 13 chromosomes were reported. Each MQTL corresponded to 2–8 individual QTLs and had narrow confidence interval of 5.8 cM. In rice, QTLs related to yield which have been identified in 15 independent mapping population were projected on consensus map with 531 genetic markers and map length is 1821 cM. Fourteen MQTLs were obtained on 7 chromosomes with genetic distance ranging from 1.8 to 5 cM and they can be exploited MAS in rice. As MQTLs have small physical and genetic intervals so these could be deployed for MAS. After MQTL identification and validation, their consistency can be confirmed between or within species by using comparative genomic approaches. The candidate genes associated with MQTLs can be cloned to study underlying molecular mechanism responsible to regulate yield under drought stress.

14 Future Perspectives of QTL Mapping in Various Crops

Significant progresses have been made in recent years to not only improve genotyping and phenotyping methods but also the statistical tools in order to increase speed and accuracy of conducting QTL analysis. Future possibilities of some of these improvements include;

(a) High throughput phenotyping by using non-invasive, highly automated and integrated imaging techniques for large plant population under drought stress could be done through fluorescence, thermal infrared, visible light, and multi-spectral imaging. Other advanced techniques of above ground trait phenotyping technologies include phenomobiles, pheno-fields, breedvision, phenocart, pheno-towers, blimps and infrared imagery (Rivero et al. 2009).
(b) High throughput genotyping can be done through GBS and SNP chips more frequently in recent years which facilitated molecular markers identification associated with or in close vicinity with targeted QTL which will lead to candidate gene identification underlying desired QTLs.
(c) Genes underlying the QTLs linked to drought tolerance associated traits have been cloned in sorghum, rice, wheat, and maize (Pinto et al. 2010). Cloning of these genes ultimately help in understanding drought mechanism as well as function markers development based on genes of interest so they can be directly used in breeding programmes for drought tolerance improvement.
(d) The variation in gene expression could be identified through expression QTL (eQTL) mapping while large-scale analyses of genetic variation of entire transcriptomes could be deployed to interpret the biochemical pathways on the basis of variants at transcript levels. Thus, more studies are required in the area of genetical genomics and eQTL analyses to help elucidate the regulation of individual gene expression and the biochemical pathways of the interacting genes in wheat under drought.
(e) The epigenetic control of genes expressed under water-stressed conditions due to histone modification or DNA methylation have been reported in some plant species including rice (Rivero et al. 2009; Rana et al. 2017). So accurate and precise identification of epi-QTLs associated with drought tolerance must be done in order to understand role of epigenetics on quantitative traits associated with drought stress in different plant species.

15 Conclusions

As discussed in this chapter, some major and stable QTLs and MQTLs along with candidate genes have been identified for drought tolerance in important crop plants which were found to be associated with yield and other important agronomic and physiological traits. Therefore, it is need of time to exploit this huge amount of information on genetic basis of drought tolerance mechanism as well as genomic selection and other strategies like epi-QTLs, e-QTLs, cloning, etc., to design programs for breeding crops tolerant to drought.

References

Abe H, Urao T, Ito T, Seki M, Shinozaki K, Yamaguchi-Shinozaki K (2003) *Arabidopsis* AtMYC2 (bHLH) and AtMYB2 (MYB) function as transcriptional activators in abscisic acid signaling. Plant Cell 15(1):63–78

Abebe T, Guenzi AC, Martin B, Cushman CJ (2003) Tolerance of mannitol-accumulating transgenic wheat to water stress and salinity. Plant Physiol 131:1748–1755

Al-Khatib K, Paulsen G-M (1990) Photosynthesis and productivity during high temperature stress of wheat genotypes from major world regions. Crop Sci 30:1127–1132

Al-Khatib K, Paulsen G-M (1999) High-temperature effects on photosynthetic processes in temperate and tropical cereals. Crop Sci 39:119–125

Anjum S-A, Tanveer M, Ashraf U, Hussain S, Shahzad B, Khan I, Wang L (2016) Effect of progressive drought stress on growth, leaf gas exchange, and antioxidant production in two maize cultivars. Environ Sci Pollut Res 23(17):17132–17141

Anjum S-A, Ashraf U, Tanveer M, Khan I, Hussain S, Zohaib A, Abbas F, Saleem M-F, Wang L (2017a) Drought tolerance in three maize cultivars is related to differential osmolyte accumulation, antioxidant defense system, and oxidative damage. Front Plant Sci 8:69–

Anjum S-A, Ashraf U, Zohaib A, Tanveer M, Naeem M, Ali I, Tabassum T, Nazir U (2017b) Growth and development responses of crop plants under drought stress: a review. Zemdirbyste 104(3):267–276

Araus JL, Slafer GA, Reynolds MP, Royo C (2002) Plant breeding and drought in C3 cereals: what should we breed for? Ann Bot 89:925–940

Arcade A, Labourdette A, Falque M, Mangin B, Chardon F, Charcosset A, Joets J (2004) BioMercator: integrating genetic maps and QTL towards discovery of candidate genes. Bioinformatics 20:2324–2326

Bankole F, Menkir A, Olaoye G, Crossa J, Hearne S, Unachukwu N, Gedil M (2017) Genetic gains in yield and yield related traits under drought stress and favorable environments in a maize population improved using marker assisted recurrent selection. Front Plant Sci 8:808

Baum M, Grando S, Backes G, Jahoor A, Sabbagh A, Ceccarelli S (2003) QTLs for agronomic traits in the Mediterranean environment identified in recombinant inbred lines of the cross' Arta' × *H. spontaneum* 41-1. Theor App Gen 107(7):1215–1225

Bennett D, Izanloo A, Edwards J, Kuchel H, Chalmers K, Tester M, Reynolds MP, Schnurbusch T, Langridge P (2011) Identification of novel quantitative trait loci for days to ear emergence and flag leaf glaucousness in a bread wheat (*Triticum aestivum* L.) population adapted to southern Australian conditions. Theor Appl Genet 124:1–15

Bidinger FR, Serraj R, Rizvi SMH, Howarth C, Yadav RS, Hash CT (2005) Field evaluation of drought tolerance QTL effects on phenotype and adaptation in pearl millet [*Pennisetum glaucum* (L.) R. Br.] topcross hybrids. Field Crops Res 94(1):14–32

Budak H, Hussain B, Khan Z, Ozturk N-Z, Ullah N (2015) From genetics to functional genomics: improvement in drought signaling and tolerance in wheat. Front Plant Sci 6:1012

Calderini D, Savin R, Abeledo L, Reynolds M, Slafer G (2001a) The importance of the period immediately preceding anthesis for grain weight determination in wheat. Euphytica 119:199–204

Calderini D, Savin R, Abeledo L, Reynolds M, Slafer G (2001b) The importance of the period immediately preceding anthesis for grain weight determination in wheat. In: Bedö Z, Láng L (eds) Wheat in a global environment. Springer, Dordrecht, The Netherlands, pp 503–509

Chourasia KN (2017) Resistance/Tolerance mechanism under water deficit (Drought) condition in plants. Int J Curr Microbiol App Sci 6:66–78

Christopher J, Christopher M, Jennings R, Jones S, Fletcher S, Borrell A, Hammer G (2013) QTL for root angle and number in a population developed from bread wheats (*Triticum aestivum*) with contrasting adaptation to water-limited environments. Theor App Gen 126(6):1563–1574

Ciuca M, Petcu E (2009) SSR markers associated with membrane stability in wheat (*Triticum aestivum* L.). Rom Agric Res 26:21–24

Collins NC, Tardieu F, Tuberosa R (2008) Quantitative trait loci and crop performance under abiotic stress: where do we stand? Plant physiol 147:469–486

Cuesta-Marcos A, Casas AM, Hayes PM, Gracia MP, Lasa JM, Ciudad F. Igartua E (2009). Yield QTL affected by heading date in Mediterranean grown barley. Plant Breed 128(1):46–53

de Miguel M, Cabezas J-A, de María N, Sánchez-Gómez D, Guevara M-Á, Vélez M-D, Sáez-Laguna E, Díaz LM, Mancha JA, Barbero MC (2014) Genetic control of functional traits related to photosynthesis and water use efficiency in Pinus pinaster Ait. drought response: integration of genome annotation, allele association and QTL detection for candidate gene identification. BMC Genomics 15:464

Diab AA, Teulat-Merah B, This D, Ozturk NZ, Benscher D, Sorrells ME (2004) Identification of drought-inducible genes and differentially expressed sequence tags in barley. Theor App Gen 109(7):1417–1425

El-Hafid R, Smith D-H, Karrou M, Sami K (1998) Morphological attributes associated with early-season drought tolerance in spring durum wheat in Mediterranean environment. Euphytica 101:273–282

Fahad S, Bajwa A-A, Nazir U, Anjum S-A, Farooq A, Zohaib A, Sadia S, Nasim W, Adkins S, Saud S (2017) Crop production under drought and heat stress: plant responses and management options. Front Plant Sci 8:1147

Fan Y, Shabala S, Ma Y, Xu R, Zhou M (2015) Using QTL mapping to investigate the relationships between abiotic stress tolerance (drought and salinity) and agronomic and physiological traits. BMC Genomics 16(1):43

Furihata T, Maruyama K, Fujita Y, Umezawa T, Yoshida R, Shinozaki K, Yamaguchi-Shinozaki K (2006) Abscisic acid-dependent multisite phosphorylation regulates the activity of a transcription activator AREB1. Proc Natl Acad Sci USA 103(6):1988–1993

Gahlaut V, Jaiswal V, Kumar A, Gupta PK (2016) Transcription factors involved in drought tolerance and their possible role in developing drought tolerant cultivars with emphasis on wheat (*Triticum aestivum* L.). Theor App Genetics 129:2019–2042

Gaxiola RA, Li J, Undurraga S, Dang LM, Allen GJ, Alper SL, Fink G-R (2001) Drought-and salt-tolerant plants result from overexpression of the AVP1 H^{+}-pump. Proc Natil Acad Sci USA 98(20):11444–11449

Gill SS, Tuteja N (2010) Reactive oxygen species and antioxidant machinery in abiotic stress tolerance in crop plants. Plant Physiol Biochem 48:909–930

Gupta P, Balyan H, Gahlaut V, Kulwal P (2012) Phenotyping, genetic dissection, and breeding for drought and heat tolerance in common wheat: status and prospects. Plant Breed Rev 36:85–168

Gupta P, Balyan H, Gahlaut V (2017) QTL analysis for drought tolerance in wheat: present status and future possibilities. Agronomy 7:5

Haake V, Cook D, Riechmann J, Pineda O, Thomashow M-F, Zhang J-Z (2002) Transcription factor CBF4 is a regulator of drought adaptation in *Arabidopsis*. Plant Physiol 130(2):639–648

Himmelbach A, Hoffmann T, Leube M, Höhener B, Grill E (2002) Homeodomain protein ATHB6 is a target of the protein phosphatase ABI1 and regulates hormone responses in *Arabidopsis*. EMBO J 21(12):3029–3038

Honsdorf N, March TJ, Berger B, Tester M, Pillen K (2014) High-throughput phenotyping to detect drought tolerance QTL in wild barley introgression lines. PLoS One 9(5)

Hosy E, Vavasseur A, Mouline K, Dreyer I, Gaymard F, Porée F, Boucherez J, Lebaudy A, Bouchez D, Véry A-A, Simonneau T (2003) The *Arabidopsis* outward K+ channel GORK is involved in regulation of stomatal movements and plant transpiration. Proc Natl Acad Sci USA 100(9): 5549-5554

Hurd EA (1974) Phenotype and drought tolerance in wheat. Agric Meteorol 14:39–55

Hussain M, Farooq S, Hasan W, Ul-Allah S, Tanveer M, Farooq M, Nawaz A (2018) Drought stress in sunflower: physiological effects and its management through breeding and agronomic alternatives. Agr Water Manag 201:152–166

Iglesias-García R, Prats E, Fondevilla S, Satovic Z, Rubiales D (2015) Quantitative trait loci associated to drought adaptation in pea (*Pisum sativum* L.). Plant Mol Biol Rep 33:1768–1778

Innes P, Blackwell RD, Quarrie S-A (1984) Some effects of genetic variation in drought-induced abscisic acid accumulation on the yield and water use of spring wheat. J Agric Sci 102:341–351

Iuchi S, Kobayashi M, Taji T, Naramoto M, Seki M, Kato T, Tabata S, Kakubari Y, Yamaguchi-Shinozaki K, Shinozaki K (2001) Regulation of drought tolerance by gene manipulation of 9-cis-epoxycarotenoid dioxygenase, a key enzyme in abscisic acid biosynthesis in *Arabidopsis*. Plant J 27(4):325–333

Joshi A, Mishra B, Chatrath R, Ferrara G-O, Singh R-P (2007) Wheat improvement in India: present status, emerging challenges and future prospects. Euphytica 157:431–446

Jossier M, Kroniewicz L, Dalmas F, Le Thiec D, Ephritikhine G, Thomine S, Barbier-Brygoo H, Vavasseur A, Filleur S, Leonhardt N (2010) The Arabidopsis vacuolar anion transporter, AtCLCc, is involved in the regulation of stomatal movements and contributes to salt tolerance. Plant J 64(4):563–576

Kahriman A, Temel HY, Aydoğan A, Tanyolac MB (2015) Major quantitative trait loci for flowering time in lentil. Turk J of Agric For 39:588–595

Kalladan R, Worch S, Rolletschek H, Harshavardhan VT, Kuntze L, Seiler C, Röder MS (2013) Identification of quantitative trait loci contributing to yield and seed quality parameters under terminal drought in barley advanced backcross lines. Mol Breed 32(1):71–90.

Kang Y, Khan S, Ma X (2009) Climate change impacts on crop yield, crop water productivity and food security—a review. Proc Natl Acad Sci USA 19:1665–1674

Kapanigowda M-H, Payne W-A, Rooney W-L, Mullet J-E, Balota M (2014) Quantitative trait locus mapping of the transpiration ratio related to preflowering drought tolerance in sorghum (*Sorghum bicolor*). Funct Plant Biol 41:1049–1065

Kasuga M, Miura S, Shinozaki K, Yamaguchi-Shinozaki K (2004) A combination of the *Arabidopsis* DREB1A gene and stress-inducible rd29A promoter improved drought-and low-temperature stress tolerance in tobacco by gene transfer. Plant Cell Physiol 45(3):346–350

Kholová J, Nepolean T, Hash C-T, Supriya A, Rajaram V, Senthilvel S, Kakkera A, Yadav R, Vadez V (2012) Water saving traits co-map with a major terminal drought tolerance quantitative trait locus in pearl millet [*Pennisetum glaucum* (L.) R. Br.]. Mol Breed 30(3): 1337–1353

Kim SJ, Kim WT (2013) Suppression of *Arabidopsis* RING E3 ubiquitin ligase AtATL78 increases tolerance to cold stress and decreases tolerance to drought stress. FEBS Lett 587(16):2584–2590

Kim S, Yang J-Y, Xu J, Jang I-C, Prigge M-J, Chua N-H (2008) Two cap-binding proteins CBP20 and CBP80 are involved in processing primary MicroRNAs. Plant Cell Physiol 49(11):1634–1644

Kohli MM, Mann CE, Rajaram S (1991) Global status and recent progress in breeding wheat for the warmer areas. In: Saunders DA (ed) Wheat for non-traditional, warm areas. CIMMYT, El Batan, Mexico, pp 96–112

Konopka-Postupolska D, Clark G, Goch G, Debski J, Floras K, Cantero A, Fijolek B, Roux S, Hennig J (2009) The role of annexin 1 in drought stress in Arabidopsis. Plant Physiol 150(3):1394–1410

Lee KH, Piao HL, Kim HY, Choi SM, Jiang F, Hartung W, Hwang I, Kwak JM, Lee IJ, Hwang I (2006) Activation of glucosidase via stress-induced polymerization rapidly increases active pools of abscisic acid. Cell 126(6):1109–1120

Liu Q, Kasuga M, Sakuma Y, Abe H, Miura S, Yamaguchi-Shinozaki K, Shinozaki K (1998) Two transcription factors, DREB1 and DREB2, with an EREBP/AP2 DNA binding domain separate two cellular signal transduction pathways in drought-and low-temperature-responsive gene expression, respectively. Arabidopsis. Plant Cell 10(8):1391–1406

Lu C, Fedoroff N (2000) A mutation in the Arabidopsis HYL1 gene encoding a dsRNA binding protein affects responses to abscisic acid, auxin, and cytokinin. Plant Cell 12(12):2351–2365

Mace ES, Singh V, Van Oosterom EJ, Hammer GL, Hunt CH, Jordan DR (2012) QTL for nodal root angle in sorghum (*Sorghum bicolor* L. Moench) co-locate with QTL for traits associated with drought adaptation. Theor App Gen 124(1):97–109

Mano Y, Omori F, Muraki M, Takamizo T (2005) QTL mapping of adventitious root formation under flooding conditions in tropical maize (*Zea mays* L.) seedlings. Breed Sci 55(3):343–347

Mansour E, Casas AM, Gracia MP, Molina-Cano JL, Moralejo M, Cattivelli L, Igartua E (2014) Quantitative trait loci for agronomic traits in an elite barley population for Mediterranean conditions. Mol Breed 33(2):249–265

Mora F, Quitral YA, Matus I, Russell J, Waugh R, Del Pozo A (2016) SNP-based QTL mapping of 15 complex traits in barley under rain-fed and well-watered conditions by a mixed modeling approach. Front Plant Sci 7:909

Muehlbauer FJ, Cho S, Sarker A, McPhee K-E, Coyne CJ, Rajesh P, Ford R (2006) Application of biotechnology in breeding lentil for resistance to biotic and abiotic stress. Euphytica 147:149–165

Mustilli AC, Merlot S, Vavasseur A, Fenzi F, Giraudat J (2002) Arabidopsis OST1 protein kinase mediates the regulation of stomatal aperture by abscisic acid and acts upstream of reactive oxygen species production. Plant Cell 14(12):3089–3099

Obidiegwu JE, Bryan GJ, Jones HG, Prashar A (2015) Coping with drought: stress and adaptive responses in potato and perspectives for improvement. Front Plant Sci 6:542

Pandey S, Assmann S-M (2004) The *Arabidopsis* putative G protein–coupled receptor GCR1 interacts with the G protein α subunit GPA1 and regulates abscisic acid signaling. Plant Cell 16(6):1616–1632

Passioura JB (1977) Grain yield, harvest index, and water use of wheat

Peighambari SA, Samadi BY, Nabipour AWCIT, Charmet G, Sarrafi A (2005) QTL analysis for agronomic traits in a barley doubled haploids population grown in Iran. Plant Sci 169(6):1008–1013

Pinto RS, Reynolds MP, Mathews KL, McIntyre CL, Olivares-Villegas JJ, Chapman SC (2010) Heat and drought adaptive QTL in a wheat population designed to minimize confounding agronomic effects. Theor Appl Genet 121:1001–1021

Price AH, Cairns JE, Horton P, Jones HG, Griffiths H (2002) Linking drought-resistance mechanisms to drought avoidance in upland rice using a QTL approach: progress and new opportunities to integrate stomatal and mesophyll responses. J Exp Bot 53:989–1004

Rana M, Verma P, Hussain W, Kaldate R, Shikha D, Kaachra A, Chahota R, Bhatia S, Sharma T (2017) Molecular mapping of QTLs for drought tolerance and yield traits in lentil

Reddy AR, Chaitanya KV, Vivekanandan M (2004) Drought-induced responses of photosynthesis and antioxidant metabolism in higher plants. J Plant Physiol 161(11):1189–1202

Ribaut JM, Jiang C, Gonzalez-de-Leon D, Edmeades G, Hoisington D (1997) Identification of quantitative trait loci under drought conditions in tropical maize. 2. Yield components and marker-assisted selection strategies. Theor Appl Genet 94:887–896

Richards RA (1996) Defining selection criteria to improve yield under drought. Plant Growth Reg 20:157–166

Rivero RM, Kojima M, Gepstein A, Sakakibara H, Mittler R, Gepstein S, Blumwald E (2007) Delayed leaf senescence induces extreme drought tolerance in a flowering plant. Proc Natl Acad Sci USA 104:19631–19636

Rivero RM, Shulaev V, Blumwald E (2009) Cytokinin-dependent photorespiration and the protection of photosynthesis during water deficit. Plant Physiol 150:1530–1540
Rosa M, Prado C, Podazza G, Interdonato R, González JA, Hilal M, Prado FE (2009) Soluble sugars. Plant Signal Behav 4:388–393
Rossel JB, Walter PB, Hendrickson L, Chow WS, Poole A, Mullineaux PM, Pogson BJ (2006) A mutation affecting ASCORBATE PEROXIDASE 2 gene expression reveals a link between responses to high light and drought tolerance. Plant Cell Environ 29(2):269–281
Ryu MY, Cho SK, Kim WT (2010) The Arabidopsis C3H2C3-type RING E3 ubiquitin ligase AtAIRP1 is a positive regulator of an abscisic acid-dependent response to drought stress. Plant Physiol 154:1983–1997
Sayed MA, Schumann H, Pillen K, Naz AA, Léon J (2012) AB-QTL analysis reveals new alleles associated to proline accumulation and leaf wilting under drought stress conditions in barley (*Hordeum vulgare* L.). BMC Genet 13(1):61
Sellammal R, Robin S, Raveendran M (2014) Association and heritability studies for drought resistance under varied moisture stress regimes in backcross inbred population of rice. Rice Sci 21:150–161
Shao HB, Chu LY, Jaleel CA, Zhao CX (2008) Water-deficit stress-induced anatomical changes in higher plants. C R Biol 331:215–225
Sharma PC, Singh D, Sehgal D, Singh G, Hash CT, Yadav RS (2014) Further evidence that a terminal drought tolerance QTL of pearl millet is associated with reduced salt uptake. Environ Exp Bot 102:48–57
Sreenivasulu N, Usadel B, Winter A, Radchuk V, Scholz U, Stein N, Graner A (2008) Barley grain maturation and germination: metabolic pathway and regulatory network commonalities and differences highlighted by new MapMan/PageMan profiling tools. Plant Physio 146(4):1738–1758
Szira F, Börner A, Neumann K, Nezhad KZ, Galiba G, Bálint AF (2011) Could EST-based markers be used for the marker-assisted selection of drought tolerant barley (*Hordeum vulgare*) lines? Euph 178(3):373–391
Taji T, Ohsumi C, Iuchi S, Seki M, Kasuga M, Kobayashi M, Yamaguchi-Shinozaki K, Shinozaki K (2002) Important roles of drought-and cold-inducible genes for galactinol synthase in stress tolerance in *Arabidopsis thaliana*. Plant J 29(4):417–426
Talamé V, Sanguineti MC, Chiapparino E, Bahri H, Ben Salem M, Forster BP, Ellis RP, Rhouma S, Zoumarou W, Waugh R, Tuberosa R (2004) Identification of *Hordeum spontaneum* QTL alleles improving field performance of barley grown under rainfed conditions. Ann Appl Biol 144(3):309–319
Tanveer M, Shahzad B, Sharma A, Khan E-A (2018) 24-Epibrassinolide application in plants: An implication for improving drought stress tolerance in plants. Plant Physiol Biochem
Teulat B, This D, Khairallah M, Borries C, Ragot C, Sourdille P, Leroy P, Monneveux P, Charrier A (1998) Several QTLs involved in osmotic adjustment trait variation in barley (*Hordeum vulgare* L.). Theor Appl Genet 96:688–698
Teulat B, Borries C, This D (2001a) New QTLs identified for plant water status, water-soluble carbohydrates, osmotic adjustment in a barley population grown in a growth-chamber under two water regimes. Theor Appl Genet 103:161–170
Teulat B, Merah O, Souyris I, This D (2001b) QTLs for agronomic traits from Mediterranean barley progeny grown in several environments. Theor Appl Genet 103:774–787
Teulat B, Merah O, Sirault X, Borries C, Waugh R, This D (2002) QTL for grain carbon isotope discrimination in field-grown barley. Theor Appl Genet 106:118–126
Tondelli A, Francia E, Visioni A, Comadran J, Mastrangelo AM, Akar T, van Eeuwijk FA (2014) QTLs for barley yield adaptation to Mediterranean environments in the 'Nure' × 'Tremois' biparental population. Euph 197(1):73–86
Tran LSP, Nakashima K, Sakuma Y, Osakabe Y, Qin F, Simpson S-D, Maruyama K, Fujita Y, Shinozaki K, Yamaguchi-Shinozaki K (2007) Co-expression of the stress-inducible zinc finger

homeodomain ZFHD1 and NAC transcription factors enhances expression of the ERD1 gene in *Arabidopsis*. Plant J 49(1):46–63

Valliyodan B, Nguyen H-T (2006) Understanding regulatory networks and engineering for enhanced drought tolerance in plants. Curr Opin Plant Biol 9:189–195

Vikram P, Swamy BM, Dixi S, Ahmed HU, Cruz MTS, Singh AK, Kumar A (2011) qDTY 1.1, a major QTL for rice grain yield under reproductive-stage drought stress with a consistent effect in multiple elite genetic backgrounds. BMC Genet 12(1):89

von Korff M, Wang H, Léon J, Pillen K (2008) AB-QTL analysis in spring barley: III. Identification of exotic alleles for the improvement of malting quality in spring barley (*H. vulgare* ssp. spontaneum). Mol Breed 21(1):81–93

Wójcik-Jagła M, Rapacz M, Tyrka M, Kościelniak J, Crissy K, Żmuda K (2013) Comparative QTL analysis of early short-time drought tolerance in Polish fodder and malting spring barleys. Theor App Gen 126(12):3021–3034

Wu S, Ning F, Zhang Q, Wu X, Wang W (2017) Enhancing omics research of crop responses to drought under field conditions. Front Plant Sci 8:174

Xiong L, Ishitani M, Lee H, Zhu JK (2001) The Arabidopsis LOS5/ABA3 locus encodes a molybdenum cofactor sulfurase and modulates cold stress–and osmotic stress–responsive gene expression. Plant Cell 13(9):2063–2083

Xu Y (2002) Global view of QTL: rice as a model. Quantitative genetics, genomics and plant breeding. CABI Publishing, Wallingford, pp 109–134

Yadav RS, Hash CT, Bidinger FR, Devos KM, Howarth CJ (2004) Genomic regions associated with grain yield and aspects of post-flowering drought tolerance in pearl millet across stress environments and tester background. Euph 136(3):265–277

Yadav RS, Sehgal D, Vadez V (2011) Using genetic mapping and genomics approaches in understanding and improving drought tolerance in pearl millet. J Exp Bot 62(2):397–408

Zhu J, Gong Z, Zhang C, Song CP, Damsz B, Inan G, Koiwa H, Zhu JK, Hasegawa P-M Bressan R-A (2002) OSM1/SYP61: a syntaxin protein in *Arabidopsis* controls abscisic acid–mediated and non-abscisic acid–mediated responses to abiotic stress. Plant Cell 14(12):3009–3028

Printed by Printforce, the Netherlands